Omics and Biotechnological Approaches for Product Profile-Driven Sorghum Improvement

Ephrem Habyarimana •
Muhammad Azhar Nadeem •
Faheem Shehzad Baloch • Nusret Zencirci
Editors

Omics and Biotechnological Approaches for Product Profile-Driven Sorghum Improvement

Springer

Editors
Ephrem Habyarimana
International Crops Research Institute for the Semi-Arid Tropics
Hyderabad, Telangana, India

Muhammad Azhar Nadeem
Faculty of Agri Sciences & Technologies
Sivas University of Science and Technology
Sivas, Türkiye

Faheem Shehzad Baloch
Faculty of Agri Sciences & Technologies
Sivas University of Science and Technology
Sivas, Türkiye

Nusret Zencirci
Department of Biology
Bolu Abant Izzet Baysal University
Bolu, Türkiye

ISBN 978-981-97-4346-9 ISBN 978-981-97-4347-6 (eBook)
https://doi.org/10.1007/978-981-97-4347-6

This Springer imprint is published by the registered company Springer Nature Singapore Pte Ltd.
The registered company address is: 152 Beach Road, #21-01/04 Gateway East, Singapore 189721, Singapore

Preface

The Food and Agriculture Organization reported that maize, rice, and wheat provide about 60% of the world's food energy intake, while soybean (an oil crop) is the most commonly used crop in animal feed since it has high protein content. The high demand for these four crops over the years explains the current several agricultural technologies and practices designed to increase their productivity and adaptation to different environmental conditions. In parallel to this, consumers developed food expectations tailored to these crop species that are hard to change overnight. However, the current and predicted climate change scenarios are placing much pressure on these major crops to the point of thinking they might become minor crops at some point in the future production environments. Most important climate change indicators include increased greenhouse gas concentrations in the atmosphere, global air temperature increase, drought events, sea surface temperature, and erratic precipitation patterns. These events harm crop production and setback the developed efforts to ensure a stable supply to the growing world's population.

Sorghum (*Sorghum bicolor* (L.) Moench) is one of the world's most important crops grown for food, feed, and biofuel. It was traditionally a staple food for hundreds of millions of people in Africa and Asia, but it is becoming more popular worldwide, including in developed countries, for its uses in food industry to satisfy a rise in demand for specialty grains, especially those that are gluten-free and rich in health-promoting compounds. Moreover, the importance of sorghum in agricultural ecosystems is expected to keep rising because this crop presents high resilience to the predicted food security threatening scenarios related to climate change such as drought and heat, it is adapted under lower and higher latitudes, and it outperforms several crop species in terms of productivity and energy balance. In recent years, the omics technologies have considerably contributed to studying biotic and abiotic stress responses in plants. Omics tools have played a significant role in crop quality improvement and protection, which has brought about increase in agricultural food production by enhancing the quality, taste, and nutritional composition of food crops. Through the use of genomics, transcriptomics, proteomics, and metabolomics, the consistency and predictability in plant breeding have been improved, reducing the time and expense of producing better quality food crops that are resistant to stress but still exhibit a high nutritional value. For example, omics technologies have been able to create information on possible plant–microbe and plant–pest interactions to transfer genes which can influence crop response to climatic conditions and tools now used in genetic engineering for biotechnological applications. Linking genes to traits provides more scientific certainty leading to improved cultivars and understanding the mechanisms of insect and weed resistance. Omics enables a systems biology approach toward understanding the complex interactions between genes, proteins, and metabolites within the resulting phenotype. This integrated approach relies heavily on chemical analytical methods, bioinformatics, computational analysis, and many other disciplines of biology, leading to crop protection and improvements.

According to IPCC, by the end of the twenty-first century, global surface temperatures are predicted to increase in the range of +1.4 to +5.8 °C, and this is expected to increase incidence of drought stress frequency worldwide. However, knowing that sorghum can withstand long drought condition including other abiotic and biotic stresses, understanding its morphophysiological and biochemical mechanisms will provide better insight toward development of robust crops that can tolerate and adapt to these adverse environmental conditions, alleviating the predicted and/or expected climate change in the future. Thus, in this book, sorghum is proposed as a model plant for improvement of other cereals crops such as maize, wheat, and rice that are often susceptible to these abiotic and biotic stresses. This book is offered to the readership in a set of 19 chapters that clearly present sorghum as a "camel" among crops owing to its very high water-use efficiency and its C4 photosynthesis. Further, sorghum is suggested as a plant resource to explore and identify genes and proteins involved in biotic and abiotic stress resistance. Chapters 1–6 describe the global importance, agronomy, adaptability, and genetic improvement in sorghum. Chapters 7–9 give up-to-date information on heterosis breeding, market segments, photoperiod, and temperature sensitivity in sorghum. Chapters 10 and 11 present the breeding status in Asia and Europe. Chapters 12–15 discuss molecular breeding, omics, and new breeding trends. Chapters 16 and 17 describe the sorghum breeding strategies under climate change and provide insights in harvest and post-harvest of sorghum. Chapters 18 and 19 provide guidelines in harnessing omics and G x M modeling toolbox across diverse environmental contexts. As the Food and Agriculture Organization and United Nations have recognized 2023 as International Year of Millets, this book is a contribution to scaling up the awareness about health and nutritional benefits of millets. This book was prepared by scientists and sorghum breeders and can be used as a research and teaching tool in academia and research and extension institutions. The wider community can use this book as a sorghum resource material.

Patancheru, Telangana, India Ephrem Habyarimana

Contents

Editors and Contributors

About the Editors

Ephrem Habyarimana is a Principal Scientist and Head of the Sorghum Breeding program at the International Crops Research Institute for the Semi-Arid Tropics, headquartered in Patancheru, Telangana, India. He is also a Professor of Agricultural Genetics at the University of Agricultural Sciences, Raichur, Karnataka, India. He has more than 25 years of experience in crop science, genetics, and plant breeding, with a particular focus on sorghum, integrating diverse technologies and skills, i.e., agroecological and genomic modeling, big data analytics, e.g., phenomics, genomics for sustainable, resilient, cost-effective, environment- and biodiversity-friendly crop improvement, and for optimizing the breeding program to deliver higher rates of genetic gains. He trained in North and Latin America, Africa, Asia, and Europe, and authored several specialized scientific articles and books, as documented under Scopus and other certified databases. He is recognized as an inventor by the European Commission in the fields of exploratory and market-ready DeepTech innovations, and he was a finalist for the 2019 European DatSci and Artificial Intelligence Awards.

Muhammad Azhar Nadeem, an Associate Professor at the Faculty of Agricultural Sciences and Technologies, Sivas University of Science and Technology, Türkiye, has documented great skills and expertise in crop science, plant genomics, and molecular plant breeding. Dr. Nadeem has been actively engaged in research activities involving the assessment of genetic diversity and population structure, as well as performing genome-wide association studies (GWAS) to identify genomic regions associated with traits of interest. Dr. Nadeem has numerous research publications and has served as the editor and co-editor of four published books.

Faheem Shehzad Baloch is working as a Professor of Plant Genetics and Breeding at Mersin University, Turkey. He received his Ph.D. with a dissertation on "QTL Mapping in Wheat" from University of Çukurova, Adana, Turkey. He has more than 11 years of experience in teaching and research in plant phenomics, genetics, genomics, and biotechnology. He has led and participated in many projects funded by national and international organizations. He has visited more than 30 countries across different continents for research and collaborative purposes and serves as a member of Management Committee and active member of various COST actions, playing important roles in different work packages within these initiatives. He has more than 120 SCI publications in WOS and has edited 9 books and written over 19 book chapters on important aspects of molecular genetics in relation to plant species. He has an extensive array of citations with over 4300 times as per Google Scholar and an h-index of 35. He serves as an Editorial Board Member for several high-impact factor journals.

Nusret Zencirci is a graduate of the Department of Field Crops, Faculty of Agriculture, Çukurova University. He worked as a wheat breeder at the Central Research Institute for Field Crops, Ankara, and served as the National and international Coordinator for wheat programs, including International Winter Wheat Improvement Program implemented by Türkiye, CIMMIYT, and ICARDA. He is now a Professor of Molecular Biology at the Department of Biology, Faculty of Science and Art, Bolu Abant İzzet Baysal University, Bolu, Turkey. He is a member of various editorial boards for many international journals. He is, along with his team, the breeder of 22 bread and durum wheat cultivars. His research interests are cereals, biotic and abiotic stresses, plant genetic resources, and wheat breeding.

Contributors

Victor Afari-Sefa International Crops Research Institute for the Semi-Arid Tropics, Patancheru, Telangana, India

Anisha Aghuru College of Agriculture, Jayashankar Telangana State Agricultural University, Rajendranagar, Telangana, India

Indian Institute of Millet Research, Hyderabad, Telangana, India

B. Amasiddha ICAR-Indian Institute of Millets Research (IIMR), Hyderabad, Telangana, India

Kanchanapally Anilkumar Genebank, International Crops Research Institute for the Semi-Arid Tropics (ICRISAT), Hyderabad, India

Ashok Kumar Are International Crops Research Institute for the Semi-Arid Tropics, Hyderabad, India

Yunus Emre Arvas Faculty of Sciences, Department of Biology, Karadeniz Technical University, Trabzon, Turkey

Faculty of Science, Department of Molecular Biology and Genetics, Van Yüzüncü Yıl University, Tusba, Van, Turkey

S. Baraskar Genebank, International Crops Research Institute for the Semi-Arid Tropics (ICRISAT), Hyderabad, India

Professor Jayashankar Telangana State Agricultural University (PJTSAU), Rajendranagar, Hyderabad, India

Rutwik Barmukh International Crops Research Institute for the Semi-Arid Tropics, Patancheru, Hyderabad, Telangana, India

Srikanth Bollam International Crops Research Institute for the Semi-Arid Tropics, Patancheru, India

Awio Bruno Professor Jayashankar Telangana State Agricultural University, College of Agriculture, Rajendranagar, Telangana, India

International Crops Research Institute for the Semi-Arid Tropics, Patancheru, Telangana, India

Matteo Busconi Department of Sustainable Crop Production, Università Cattolica del Sacro Cuore, Piacenza, Italy

Stanislaus Antony Ceasar Division of Plant Molecular Biology and Biotechnology, Department of Biosciences, Rajagiri College of Social Sciences, Cochin, Kerala, India

Visalakshi Chandra ICAR-Central Tuber Crops Research Institute, Trivandrum, Kerala, India

D. Kumara Charyulu International Crops Research Institute for the Semi-Arid Tropics, Patancheru, Telangana, India

Debamalya Chatterjee Department of Plant Science, Penn State University, University Park, PA, USA

Sonal Chavan Professor Jayashankar Telangana State Agricultural University, College of Agriculture, Rajendranagar, Telangana, India

International Crops Research Institute for the Semi-Arid Tropics, Patancheru, Telangana, India

Surinder Chopra Department of Plant Science, Penn State University, University Park, PA, USA

Sunita Choudhary International Crops Research Institute for the Semi-Arid Tropics, Patancheru, Telangana, India

Naresh Reddy Dagunapur Genebank, International Crops Research Institute for the Semi-Arid Tropics (ICRISAT), Hyderabad, India

Einstein Mariya David VIT School of Agricultural Innovations and Advanced Learning (VAIAL), Vellore Institute of Technology, Vellore, India

Ali Devlet Faculty of Agriculture and Natural Science, Department of Field Crops, Bilecik Seyh Edebali University, Bilecik, Turkey

Dilara Dorter Department of Molecular Biology and Genetics, Faculty of Arts and Sciences, Yildiz Technical University, Istanbul, Turkey

Dinakaran Elango Department of Agronomy, Iowa State University, Ames, IA, USA

Neethu Francis Department of Millets, Tamil Nadu Agricultural University, Coimbatore, India

Bonti Gogoi Assam Agricultural University, Jorhat, Assam, India

Ana Maria Pereia Gomes Universidade Católica Portuguesa, CBQF—Centro de Biotecnologia e Química Fina—Laboratório Associado, Escola Superior de Biotecnologia, Porto, Portugal

Sunita Gorthy International Crops Research Institute for the Semi-Arid Tropics (ICRISAT), Patancheru, India

Mahalingam Govindaraj Harvest Plus, Alliance CIAT-Bioversity International, Hyderabad, India

Kamaleeswari Govindarajan Faculty of Agriculture, SRM Institute of Science and Technology, Chennai, India

Murali Krishna Gumma International Crops Research Institute for the Semi-Arid Tropics, Patancheru, Telangana, India

Ephrem Habyarimana International Crops Research Institute for the Semi-Arid Tropics, Hyderabad, Telangana, India

Vishal Hivare International Crop Research Institute for the Semi-Arid Tropics, Patancheru, Telangana, India

Chaitra Hugar International Crops Research Institute for the Semi-Arid Tropics (ICRISAT), Patancheru, India

University of Agricultural Sciences, Raichur, Karnataka, India

Jagdish Jaba International Crops Research Institute for the Semi-Arid Tropics, Hyderabad, Telangana, India

K. Jagadeesh Genebank, International Crops Research Institute for the Semi-Arid Tropics (ICRISAT), Hyderabad, India

Professor Jayashankar Telangana State Agricultural University (PJTSAU), Rajendranagar, Hyderabad, India

Vanitha Jayaraman Faculty of Agriculture, SRM Institute of Science and Technology, Chennai, India

Yinping Jiao Department of Plant and Soil Science, Institute of Genomics for Crop Abiotic Stress Tolerance (IGCAST), Texas Tech University, Lubbock, TX, USA

Uttej Karla Maharana Pratap University of Agriculture and Technology, Udaipur, Rajasthan, India

Arunsaikumar Karrem International Crops Research Institute for the Semi-Arid Tropics, Hyderabad, Telangana, India

Department of Agricultural Entomology, University of Agricultural Sciences, Raichur, Karnataka, India

Krishna Kasanaboina Professor Jayashankar Telangana State Agricultural University, College of Agriculture, Rajendranagar, Telangana, India

International Crops Research Institute for the Semi-Arid Tropics, Patancheru, Telangana, India

Alessandra Lanubile Department of Sustainable Crop Production, Università Cattolica del Sacro Cuore, Piacenza, Italy

Jieqin Li College of Agriculture, Anhui Science and Technology University, Fengyang, China

Anhui Provincial Forage Biological Breeding International Cooperation Research Center, Fengyang, China

Sevgi Marakli Department of Molecular Biology and Genetics, Faculty of Arts and Sciences, Yildiz Technical University, Istanbul, Turkey

Murugan Marimuthu Department of Entomology, Tamil Nadu Agricultural University, Coimbatore, Tamil Nadu, India

Adriano Marocco Department of Sustainable Crop Production, Università Cattolica del Sacro Cuore, Piacenza, Italy

Ann Murithi Department of Agronomy, Iowa State University, Ames, IA, USA

Tharanya Murugesan International Crop Research Institute for the Semi-Arid Tropics, Patancheru, Telangana, India

Recep İrfan Nazlı Department of Field Crops, Faculty of Agriculture, Cukurova University, Adana, Turkey

Dhanyakumar Onkarappa International Crops Research Institute for the Semi-Arid Tropics, Hyderabad, Telangana, India

Department of Entomology, Tamil Nadu Agricultural University, Coimbatore, Tamil Nadu, India

Theivasigamani Parthasarathi VIT School of Agricultural Innovations and Advanced Learning (VAIAL), Vellore Institute of Technology, Vellore, India

Ana Sofia Pimenta-Martins Universidade Católica Portuguesa, CBQF—Centro de Biotecnologia e Química Fina—Laboratório Associado, Escola Superior de Biotecnologia, Porto, Portugal

Mihrab Polat Department of Field Crops, Faculty of Agriculture, Cukurova University, Adana, Turkey

Somashekhar Punnuri Agriculture Research Station, Fort Valley State University, Fort Valley, GA, USA

Y. Vinod Kumar Reddy International Crop Research Institute for the Semi-Arid Tropics, Patancheru, Telangana, India

Naini Agricultural Institute, Sam Higginbottom University of Agriculture Technology and Sciences (SHUATS), Prayagraj, Uttar Pradesh, India

Pradeep Ruperao International Crops Research Institute for the Semi-Arid Tropics, Patancheru, India

Pramod Sargar International Crops Research Institute for the Semi-Arid Tropics (ICRISAT), Patancheru, Hyderabad, Telangana, India

Vasantrao Naik Marathwada Krishi Vidyapeeth, Parbhani, Maharashtra, India

Kuldeep Singh Genebank, International Crops Research Institute for the Semi-Arid Tropics (ICRISAT), Hyderabad, India

Kaliamoorthy Sivasakthi International Crops Research Institute for the Semi-Arid Tropics, Patancheru, Telangana, India

Raja Srinivas Genebank, International Crops Research Institute for the Semi-Arid Tropics (ICRISAT), Hyderabad, India

Lorenzo Stagnati Department of Sustainable Crop Production, Università Cattolica del Sacro Cuore, Piacenza, Italy

Niranjan Thakur International Crops Research Institute for the Semi-Arid Tropics, Patancheru, Telangana, India

Vasantrao Naik Marathwada Agriculture University, Parbhani, Maharashtra, India

Mahendar Thudi Agriculture Research Station, Fort Valley State University, Fort Valley, GA, USA

A. V. Umakanth ICAR-Indian Institute of Millets Research (IIMR), Hyderabad, Telangana, India

Wilfred Vermerris Department of Microbiology and Cell Science and UF Genetics Institute, University of Florida, Gainesville, FL, USA

Mani Vetriventhan Genebank, International Crops Research Institute for the Semi-Arid Tropics (ICRISAT), Hyderabad, India

Catarina Vila-Real Universidade Católica Portuguesa, CBQF—Centro de Biotecnologia e Química Fina—Laboratório Associado, Escola Superior de Biotecnologia, Porto, Portugal

Lihua Wang College of Agriculture, Anhui Science and Technology University, Fengyang, China

Anhui Provincial Forage Biological Breeding International Cooperation Research Center, Fengyang, China

Wanyan Wang Ecosystem Science and Management, Penn State University, University Park, PA, USA

Trends in Global Sorghum Production: Perspectives and Limitations

1

D. Kumara Charyulu, Victor Afari-Sefa, and Murali Krishna Gumma

Abstract

Sorghum acts as a staple food for millions of people living in about 30 countries in the subtropical and semi-arid regions of Africa and Asia. Though sorghum cultivation is reported from more than 110 countries (FAOSTAT, https://www.fao.org/faostat/en/#home, 2023), only ten countries have more than one million ha area under sorghum, which together contribute to more than 68% of the world's total sorghum production. The global sorghum cropped area has declined marginally (39.9 m ha) during the last decade from 41.79 million ha in 2011–2013. However, global production was more stable during the past one-decade period. This might have been possible because of enhancement in sorghum productivity across different regions. Africa region is the largest producer of sorghum (46% of global sorghum) followed by Americas region (35%). Out of the global production, about 47.5% is used for food, while another 34.7% is used for feed. The non-food, seed, and processing use together consume about 10.5% of the total production. World trade in sorghum is strongly linked to demand for livestock products, which is dominated by feed requirements and prices etc. Although most of the sorghum continues to be consumed in the countries where it is produced, despite several opportunities for expansion of sorghum globally, there are an equal number of challenges threatening it.

Keywords

Sorghum · Sorghum production · Sorghum trade · Sorghum trend · Utilization

D. K. Charyulu (✉) · V. Afari-Sefa · M. K. Gumma
International Crops Research Institute for the Semi-Arid Tropics, Patancheru, Telangana, India
e-mail: Kumaracharyulu.deevi@icrisat.org

E. Habyarimana et al. (eds.), *Omics and Biotechnological Approaches for Product Profile-Driven Sorghum Improvement*,
https://doi.org/10.1007/978-981-97-4347-6_1

1.1 Importance of Sorghum

About 30 countries from the subtropical and semi-arid regions of Africa and Asia rely on sorghum as a basic dietary staple food. In the traditional smallholder farm and livestock sector, sorghum is the major source of food and fodder. Sorghum also finds a place in the high-input modern commercial farming sector as a feed crop and is fast emerging as a biofuel crop. Nearly 80% of global sorghum is cultivated in developing countries with low and marginal productivity levels contributing to slightly above half of the global grain output. About 20% of global sorghum area is from developed countries, which is categorized by high productivity levels and contributes to nearly 40% of global grain production (Hariprasanna and Rakshit 2016). According to FAOSTAT (2023), though sorghum cultivation is reported from more than 110 countries in the world, only ten countries out of them cultivated in more than one million ha. They together contribute to more than 68% of global sorghum production. Relatively, sorghum is widely grown and highly preferred major staple food grain in large parts of the African continent when compared with Asia region. However, it is not spread over uniformly in the region, and few countries exhibit a major share of the area under it. Despite its economic and nutritional importance, sorghum cropped areas around the world have declined over the past two decades. Countries such as Brazil, Ethiopia, Sudan, Australia, Mexico, Nigeria, and Burkina Faso showed an expansion of areas under sorghum mainly because of new land brought under cultivation or diversion of a portion of area planted to other crops such as maize and wheat.

Historically, global sorghum production peaked during the mid-1980s, and thereafter it declined by about 13–15%, but not steadily. In general, global sorghum productivity levels have increased with enhanced access to improved cultivars, better management practices, and higher input use except in Africa region. Most of the sorghum is consumed locally where it is produced. Hence, trade in sorghum grain across countries is highly limited. The global sorghum trade is mainly linked to demand for livestock products, which is governed by the feed requirements and prices in developed countries. Except in Africa region, consumption of sorghum for food is gradually declining because of a change in food preferences and tastes brought about by economic status and skewed policies. However, the alternate use of sorghum such as for animal feed and other industrial purposes is expanding. Sorghum is gaining importance and renewed interest under projected changing climate regime as well as for addressing malnutrition issues. Therefore concerted efforts are necessary on sorghum to ensure food and nutritional security as well as achieving the Sustainable Developmental Goals (SDGs) by 2030.

Sorghum (*Sorghum bicolor* (L.) Moench) is the fifth largest produced cereal crop in the world. It occupies second place after millets in the global coarse cereals production. Sorghum is a C4 carbon cycle plant with high photosynthetic efficiency and productivity (Tari et al. 2013). Among all cereals, sorghum is well-known for its strong resistance and wide adaptability to multiple biotic and abiotic stresses (Tari et al. 2013; Huang 2018; Zhang et al. 2019). The wider popularity of sorghum is due to its (1) diverse end-uses as human food, livestock feed, biofuel, and forage; (2)

high returns; (3) more resilience to adverse environmental conditions compared to many other cereal crops; and (4) good performance under conditions of water and temperature constraints, especially in marginal lands (Tari et al. 2013; Monk et al. 2014; Hao et al. 2021). All these characteristics have gained much attention toward sorghum as a potential "star" crop to tackle the anticipated challenges around global food security (Tari et al. 2013; Kadam et al. 2017; Hao et al. 2021; Ndlovu et al. 2021). Sorghum grain and stover are of economic value—the grain is used as food, cattle and poultry feed, and in alcohol manufacture, while its stover is an important dry fodder resource for large ruminants. In India, for example, the stover value accounts for nearly 30–40% of the total value of the crop (Parthasarathy Rao and Hall 2003).

Sorghum is grown for a variety of uses (bread, biscuits, starch, sugar, syrups, alcohol, beer, and malt products etc.) both in Africa and Asia regions. It is the staple food crop for millions of people in the developing countries. While it is an important source of animal feed in case of developed countries such as Thailand, Australia, America, Japan, and Europe, sorghum stalks are also used as fuel, fencing, and roofing material in Africa. Crop residues from the fields are used for livestock to graze upon. Sorghum is also grown for green forage purposes. Sweet sorghum, a variety of sorghum, is mainly used for ethanol extraction in the developed countries (China, Brazil, and United States). The industrial uses of sorghum is, however, limited to the Africa region (Nigeria, South Africa, and Zimbabwe).

1.2 Global Sorghum Area and Production Regions

Globally, sorghum is the fifth preferred cereal crop in the world, after wheat, maize, rice, and barley (Iqbal et al. 2010; Huang 2018; Hao et al. 2021; Ndlovu et al. 2021). It is cultivated in diverse geographic areas in the Americas, Africa, Asia, and the Pacific regions. Sorghum was grown in 110 countries of the world (Upadhyaya et al. 2019; Hao et al. 2021) during the year 2021, covering approximately a cropped area of 39.91 million ha during the triennium ending 2019–2021 (FAOSTAT 2023). About 70% of sorghum cropped area is concentrated in Africa region followed by Asia region (14.2%) and Americas region (14.1%). The sorghum cropped area share of European Union and Caribbean was almost negligible. Among the African regions, Western Africa has the lion's share with 35.0% followed by Northern Africa region (16.7%) (Table 1.1).

The global sorghum cropped area has declined marginally during the last decade from 41.79 million ha in 2011–2013. The sorghum cropped area under Asia region declined significantly (from 8.2 to 5.6 m ha) during the triennium between 2011–2013 and 2019–2021. The Americas region showed a similar declining trend from 6.3 to 5.6 m ha during the study period. The Caribbean region also lost reasonable sorghum cropped area during the past one-decade period. In contrast, the sorghum cropped area has marginally increased in case of Africa region (26.2–27.8 m ha). The decline in sorghum cropped area during the study period was more conspicuous in Caribbean region (−84.5%) followed by Asia region (−31.1%).

Table 1.1 Area planted with sorghum by region, 2011–2013 and 2019–2021

Region	Area ('000 ha)		% change
	2011–2013[a]	2019–2021[a]	
Asia	8238.8 (19.7)	5673.6 (14.2)	−31.1
Africa	26,273.2 (62.9)	27,887.9 (69.9)	6.1
Eastern Africa	5185.7 (12.4)	5046.0 (12.6)	−2.7
Western Africa	12,533.0 (30.0)	13,986.1 (35.0)	11.6
Northern Africa	6302.9 (15.0)	6674.0 (16.7)	5.8
Southern Africa	142.1 (0.3)	123.2 (0.3)	−15.3
Americas	6313.1 (15.1)	5614.0 (14.1)	−11.1
Caribbean	121.2 (0.3)	18.8 (0.0)	−84.5
European Union	318.9 (0.8)	310.6 (0.8)	−2.6
World	41,791.9 (100.0)	39,919.5 (100.0)	−4.5

Source: FAOSTAT (2023)
[a] Figures in parentheses indicate % share to column total

Table 1.2 Sorghum production by region, 2011–2013 and 2019–2021

Region	Production ('000 t)		% change
	2011–2013[a]	2019–2021[a]	
Asia	9718.4 (16.6)	8240.6 (14.0)	−15.2
Africa	24,286.8 (41.4)	27,482.8 (46.6)	13.2
Eastern Africa	6486.8 (11.1)	7729.4 (13.1)	19.2
Western Africa	10,866.6 (18.5)	13,093.7 (22.2)	20.5
Northern Africa	4586.8 (7.8)	4183.9 (7.0)	−8.7
Southern Africa	181.8 (0.3)	239.8 (0.4)	31.9
Americas	21,543.8 (36.7)	20,926.6 (35.5)	−2.9
Caribbean	109.1 (0.2)	18.8 (0.0)	−82.8
European Union	986.7 (1.7)	1276.8 (2.2)	29.4
World	58,674.6 (100.0)	58,997.2 (100.0)	0.5

Source: FAOSTAT (2023)
[a] Figures in parentheses indicate % share to column total

The global sorghum production across different regions between 2011–2013 and 2019–2021 is summarized in Table 1.2. The global sorghum production was more stable during the past one-decade period even though there was a decline in global sorghum cropped area. This might have been possible because of enhancement in sorghum productivity across different regions. Africa region alone contributes to closely 46% of global sorghum production followed by Americas region (35%). Asia region occupied the third position with 14% of total sorghum production during 2019–2021. Among Africa regions, Western Africa contributed about 22% of the total sorghum production followed by Eastern Africa (13%). The sorghum production has witnessed decline in the case of the Caribbean region followed by Asia and Americas regions. Both Africa and the European Union showed enhanced production during the study period.

Figure 1.1 summarizes the mean sorghum productivity levels during 2011–2021 in major sorghum producing regions of the world. The mean sorghum productivity levels are higher (2–3 times) in Europe and Americas regions when compared to Asia and Africa regions. The global average sorghum productivity is hovering in between these two categories. Except in Africa, the sorghum productivity levels in the rest of the regions are exhibiting increasing trend over time (2011–2021). The growth in sorghum productivity level during the study period is more stable in the Americas region when compared to Europe. The mean productivity level in the Asia region showed consistent improvement (1200–1500 kg/ha) over the decade period. This might be due to enhanced penetration of improved sorghum cultivars as well as adoption of scientific management practices in target ecologies. In contrast, the productivity per hectare hovered around 930 kg during the study period in Africa region. Lack of access to seed, poor seed systems, limited role of private sector, etc., are the major challenges noticed in Africa region. A marginal improvement (1350–1500 kg/ha) in global sorghum productivity was observed during the study period.

Table 1.3 furnishes the performance of sorghum during 2019–2021 in the top 15 countries of the world. Sudan stood on the top in terms of the highest sorghum cropped area (6.5 m ha) in the world. It alone accounts for nearly 16% of the global sorghum cropped area. But the mean sorghum productivity levels in case of Sudan was at 524 kg/ha. Nigeria occupied the second position with a sorghum cropped area of about 5.5 m ha (13.8%). The productivity performance (1197 kg/ha) of sorghum was reasonably good in Nigeria. The extent of sorghum cropped area in India was 4.3 m ha and ranked third globally (10.8%) in terms of area. The average sorghum productivity levels hover around 1000 kg/ha during the triennium ending 2019–2021.

Even though China secured 15th position in terms of global sorghum cropped area (0.5 m ha), its productivity level is the highest (5914 kg/ha) in the world. The

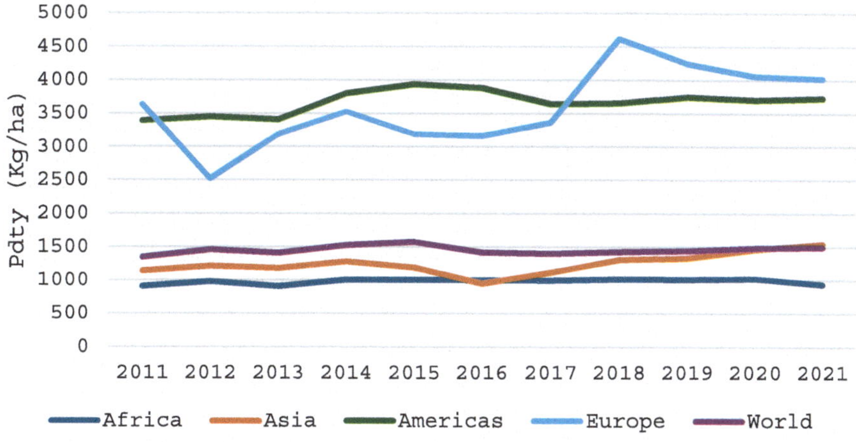

Fig. 1.1 Sorghum productivity trends across major regions

Table 1.3 Top 15 countries cultivating sorghum in the world, 2019–2021

Country	Area ('000 ha)	% share in global area	Production ('000 ton)	Yield (kg/ha)
Sudan	6513.7	16.1	3417.6	524.5
Nigeria	5565.7	13.8	6660.0	1197.2
India	4385.5	10.8	4352.4	990.9
Niger	3646.3	9.0	1745.4	476.6
United States	2193.4	5.4	9840.7	4503.3
Ethiopia	1756.0	4.3	4924.5	2801.1
Burkina Faso	1733.8	4.3	1776.1	1025.5
Mali	1626.2	4.0	1524.5	934.6
Mexico	1358.7	3.4	4475.6	3296.5
Chad	1134.9	2.8	946.2	833.8
United Republic of Tanzania	892.0	2.2	950.6	1073.7
Brazil	862.0	2.1	2649.4	3079.0
Cameroon	715.6	1.8	1216.2	1699.4
South Sudan	694.6	1.7	604.4	870.2
China	547.0	1.4	3037.8	5914.1

Source: FAOSTAT (2023)

United States (4503 kg/ha) and Mexico (3296 kg/ha) exhibited the second and third highest levels, respectively, of productivity globally. Among the top 15 countries cultivating sorghum globally, about ten fall within the Africa region while only two of them belong to the Asia region. This clearly highlights the importance of sorghum in the livelihoods of small and marginal farmers in the continent. Any crop genetic improvement in the case of sorghum is going to impact millions of families globally.

1.3 Global Sorghum Utilization Pattern

Sorghum is eaten in a variety of forms that vary from region to region. In general, it is consumed as whole grain or processed into flour, from which traditional meals are prepared (FAO and ICRISAT 1996). It is a key staple in many parts of the developing world, especially in the drier and more marginal areas of the semi-arid tropics. In general, the per capita consumption in rural areas will be higher than that in the urban areas. Among the rural areas, sorghum consumption tends to be the highest in the poorest, most food-insecure regions. It is interesting to understand the pattern of sorghum utilization among major growing regions globally. The diverse uses of sorghum and their demand determines the extent of cropped area spread in those regions.

Table 1.4 details the pattern of sorghum utilization among major producing regions during 2018–2020. Globally, approximately 58.3 million tons of sorghum are produced from nearly 110 sorghum producing countries belonging to different

Table 1.4 Utilization pattern of sorghum in different regions, 2018–2020 ('000 tons)

		% share to domestic supply			
Region	Domestic supply	Food	Feed	Other uses[a]	Losses
World	58,350	47.5	34.7	10.5	7.2
Africa	28,742	68.4	6.5	15.6	9.4
Americas	15,169	2.1	83.4	9.1	5.4
Asia	12,197	63.2	29.9	1.8	5.1
Europe	1465	0.1	97.4	2.1	2.3
Oceania	777	3.3	86.9	4.1	5.6

Source: FAOSTAT (2023)

[a] Includes non-food, seed, and processing

regions. Out of this, about 47.5% of the total production is used for food. Nearly 34.7% of global production is used as feed for livestock. The non-food, seed, and processing use together consume about 10.5% of total production. About 7.2% of the total production goes in the form of food losses.

Africa region alone contributes nearly half (28.7 million tons) of the global sorghum production. Nearly 68% of the total production is used for food consumption followed by non-food, seed, and processing (15.6%) purposes. The share of feed use is significantly lower (6.5%) in the region. The extent of food losses (9.4%) in the total production is higher in the Africa region. Partly, the pattern of sorghum utilization is also similar in the Asia region. A major chunk of total sorghum production in the Asia region is used for food (63.2%) purposes. About 30% of Asia's total production goes for feed use. The share of other uses is negligible (1.8%) and the extent of food losses is around 5% in the total production. The developing countries (Africa and Asia regions) exhibit higher importance of sorghum utilization for food use followed by feed purposes.

Specifically, in India, sorghum continues to be a crucial food security crop in few pockets of the country (e.g., Maharashtra, Karnataka, Rajasthan, Andhra Pradesh, and Telangana). Both the sorghum area and production has witnessed a fall in the country during the past two-decades period due to changes in consumer preferences toward fine cereals. The rise in household incomes accentuated consumer preferences to wheat and rice which taste better, and are easier and faster to cook. Furthermore, the rapid urbanization and growing availability of a range of convenience foods have shifted their consumption behavior. Government policies in several countries have also contributed to the decline in food utilization of sorghum. For instance, imports of relatively cheaper wheat and rice by many countries discouraged the consumption of locally produced cereals. In India and China, government purchases and sales of sorghum under public distribution systems (PDS) were discontinued. In several countries, consumer subsidies, overvalued currencies, or subsidized imports kept prices of wheat and/or rice relatively low, reducing the competitiveness of domestically produced coarse grains.

Source: Bhagavatula et al. (2013)

Developed regions such as Americas, Europe, and Oceania exhibit a diverse pattern of sorghum utilization when compared to developing regions. The share of total sorghum production used for food purposes in developed regions is less than 5%. The lion's share (>80%) of total production goes for feed use rather than food. The contribution toward non-food, seed, and processing use is negligible. This pattern of utilization clearly highlights the role of sorghum production in feed use in developed regions. Demand for animal feed is concentrated more in developed countries and middle-income countries in Latin America and Asia. The reason for high demand is the higher density of livestock population as well as increased consumption demand for meat. Another important outlet/use for grain sorghum, especially in Africa, is in the preparation of alcoholic beverages. However, the extent of utilization in the total production at country level is not available.

Sorghum vs maize: Competition between sorghum and maize is a key factor in feed utilization globally. In general, the feed characteristics of sorghum are quite similar to those of other cereals such as maize. Sorghum provides about as much metabolizable energy as maize, has a higher crude protein content (though of lower quality), and is relatively rich in niacin, an essential vitamin. However, large investments in maize crop improvement research have helped to increase yields and reduced growing cycles for this competing energy source. This has improved the competitiveness of maize prices in many countries globally. Feed industries in most countries apply the least-cost formulations to produce compound feeds, in which sorghum/maize is mixed with non-grain ingredients. The quantity of sorghum used in feed depends primarily on the relative prices of sorghum and maize, and on their relative feed value.

Another important factor is consumer preference for meat color. Maize contains higher carotene levels than sorghum, so meat from maize-fed animals tends to be more yellow than meat from sorghum -fed animals. In Japan, for example, consumers generally prefer white-colored meat. Therefore, sorghum is a valued ingredient in some compound feed rations (for poultry, pigs, and some breeds of beef cattle). In contrast, sorghum is discounted by producers in India because consumers there generally prefer poultry meat and egg yolks with a deeper yellow color. In addition, farmers in Asia have shown a growing interest in the sale and purchase of sorghum fodder. While the use of sorghum crop residues in Africa remains largely restricted to the farm, there is a large and growing market in Asia for traded sorghum residues to meet both rural feed shortfalls and urban agricultural demand, the latter largely for maintenance of dairy animals.

Source: FAO and ICRISAT (1996)

Table 1.5 furnishes the pattern of utilization of sorghum in major producing countries in the world during 2018–2020. Among all the study countries, the highest

Table 1.5 Utilization pattern of sorghum in major producing countries, 2018–2020 ('000 tons)

		% share to domestic supply			
Country	Domestic supply	Food	Feed	Other uses[a]	Losses
Sudan	4492	64.2	6.0	14.6	15.3
Nigeria	6629	85.7	5.4	4.3	4.6
India	4402	90.7	1.4	3.3	4.6
Niger	2086	72.9	8.4	4.6	14.1
United States	4649	4.2	59.5	27.4	8.9
Ethiopia	5136	60.8	1.9	32.6	4.5
Burkina Faso	1763	60.3	7.2	24.8	7.7
Mali	1564	61.1	7.2	17.8	16.1
Mexico	5025	0.0	95.1	0.4	4.5
Chad	1002	85.6	0.1	0.8	13.4
United Republic of Tanzania	719	42.5	1.5	43.5	12.4
Brazil	2549	0.0	97.1	0.5	2.4
Cameroon	1225	53.1	20.0	12.4	14.6
South Sudan	802	80.9	0.0	1.4	18.8
China	6416	49.9	43.4	0.8	5.9

[a] Includes non-food, seed, and processing
Source: FAOSTAT (2023)

share of sorghum production (90.7%) has been utilized for consumption was noticed in case of India. It was followed-up by Nigeria (85.7%) and Chad (85.6%) countries where closely 85% of total production goes for food consumption. Rest of all other African countries (Sudan, Niger, Ethiopia, Burkina Faso, and South Sudan) also exhibited higher use of sorghum for food consumption. The lowest share (among African countries) was observed in case of United Republic of Tanzania at 42.5%. As envisaged, the relative proportion of sorghum production has been utilized more for food purposes in developing countries while it was for feed purposes in the case of developed nations.

Brazil (97.1%), Mexico (95.1%), and United States (59.5%) have showed relatively higher consumption of sorghum for feed consumption rather than for food purpose. Nearly 43% of China's total sorghum production also diverted for feed purposes to livestock. The other uses (non-food, processing, and seed purposes) of sorghum were the highest (43.5%) noticed in case of Tanzania followed by Ethiopia (32.6%). The extent of post-harvest losses of sorghum was conspicuous in all developing countries while their losses were minimum in case of developed countries around the world.

1.4 Global Sorghum Trade

Historically, world trade in sorghum is strongly linked to demand for livestock products, dominated by feed requirements and prices etc. Although most sorghum continues to be consumed in the countries where it is produced. The details of global

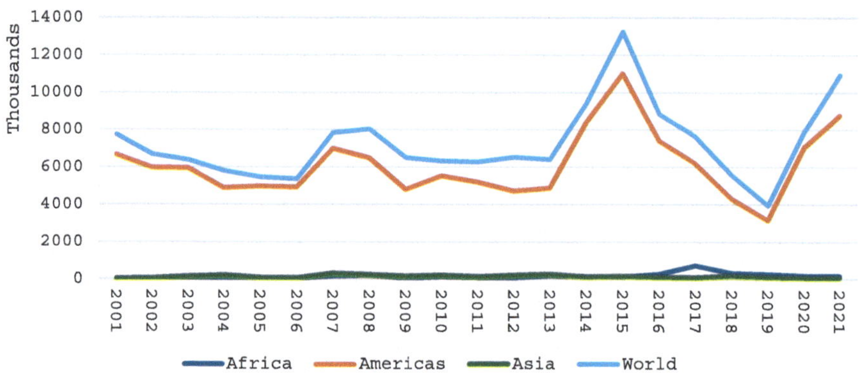

Fig. 1.2 Trends in global sorghum exports (tons)

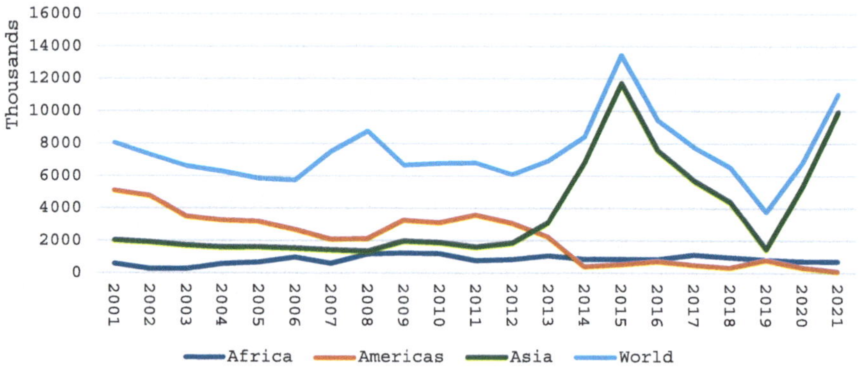

Fig. 1.3 Trends in global sorghum imports (tons)

sorghum exports during 2001–2021 across major producing regions are summarized in Fig. 1.2. The trend in global sorghum exports is marginally in increasing trend (from 8 to 11 m tons) with significant fluctuations over the study period. But the peak in sorghum exports were noticed during 2015 with value of more than 13 m tons. Exports from the Asia and Africa regions are almost negligible. Americas region has the lion's share (~80%) in the global sorghum exports from 2001 to 2021. Except the Americas region, none of the other sorghum producing regions in the world are participating in the global sorghum exports.

Similarly, the details about the global sorghum imports are summarized during 2001–2021 across major regions in Fig. 1.3. The global sorghum imports are highly concentrated toward the Asia (mainly to China) region when compared to other regions. Nearly 90% of total global imports have been directed toward the Asia region since 2015. A peak in global imports were observed during 2015 and touched close to 12 m tons. The Americas region played a dominant role in global sorghum imports from 2001 to 2013 and afterward its imports were gone down significantly. While the active role of Asia region in global imports picked up since 2013 and

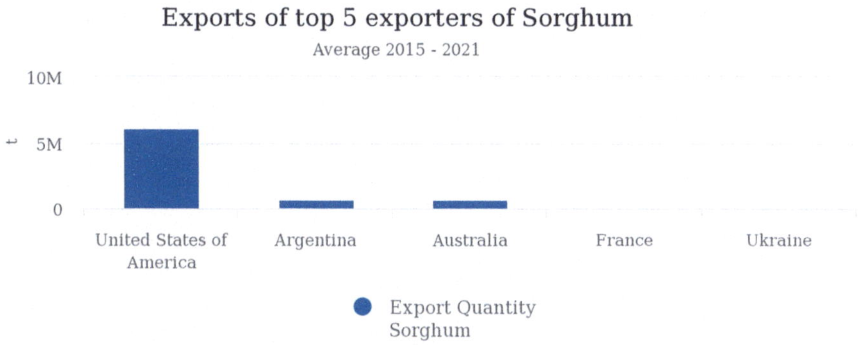

Fig. 1.4 Top five sorghum exporters in the world

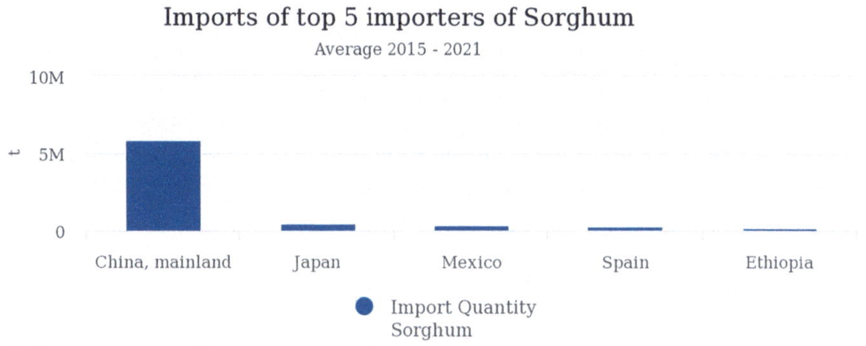

Fig. 1.5 Top five sorghum importers in the world

continued till 2021, Africa region played a negligible role in the total sorghum imports during the study period.

Figure 1.4 furnishes the information of the top five exporters of sorghum in the world during the period 2015 to 2021. The FAOSTAT trade data (average) have been used to prioritize the top countries in the world. The United States stood on the top with a mean export potential of sorghum grain to the tune of about 6 million tons. Argentina occupied the second place in sorghum exports with a quantity of nearly 0.7 million tons. Similarly, Australia secured the third place (closely 0.7 m tons) in the global exports. France and Ukraine placed at fourth and fifth positions, respectively, in terms of the total sorghum exports in the world.

Figure 1.5 furnishes the information of the top five importers of sorghum in the world during the period 2015 to 2021. China stood in the first place with almost close to 6 million tons of sorghum imports in the world. It was followed-up by Japan (0.52 m tons), Mexico (0.38 m tons), Spain (0.25 m tons), and Ethiopia (0.15 m tons). The sorghum import demand to China was high, as much as 10 million tons, during 2021. During COVID-19, sorghum imports to China were at the lowest because of import restrictions.

1.5 Perspectives and Limitations for Sorghum Expansion

The peculiar characteristics of sorghum, such as low water requirement crop, thrive on marginal lands, generate high biomass, and can withstand hardy climatic conditions, would enhance its potential for rapid expansion in the dryland regions of the world. There is ample scope for expansion of cropped area because of its diverse uses as food, feed, and fodder. The nutraceutical properties of sorghum also enhance its demand from urban and peri-urban sectors. Increased awareness as well as high susceptibility toward lifestyle diseases have slowly substituted the fine cereals from the common man food basket. However, the skewed policies across different sorghum growing countries favored irrigated ecosystems rather than the dryland ecosystem. The quantum of agricultural investments and the extent of input subsidies provided per hectare are remarkably higher in the case of fine cereals when compared to coarse cereals. Nagging research issues, such as resistance to gray mold, stem rot, and amenability to mechanization, in sorghum limit its expansion in major growing countries. The lack of research investments toward promotion of value addition opportunities also curtails its diversified uses in different fields. Sorghum is one of the potential crops for expanding as biofuel crop under mandatory (Euro standards set by different countries) blending policies. The absence of political will and bold policy initiatives across different regions limit its expansion further.

1.5.1 Perspectives for Sorghum Expansion

Sorghum mainly serves as a staple food for humans. But, it is primarily produced as a major feed crop in few developed countries such as the United States, Brazil, and Australia. Due to its great nutritional (gluten-free nature) and functional potential, numerous studies on sorghum for human consumption have been undertaken and proved it as the best (de Cardoso Morais et al. 2015; Rakshit et al. 2014; Habyarimana et al. 2019). Sorghum is a crucial dual-purpose crop and has the greatest drought tolerance among the world's top five carbohydrate-rich crops along with its reputation as a model crop. The cultivation height of sorghum ranges from the equator (about 50° latitude) to altitudes of 2500 m (Rao et al. 2015). It is remarkably tolerant to low input levels which gave it an indispensable feature for the areas getting little rainfall. Thus, sorghum can play a major role in nourishing the world's most helpless people under predicted climate change impacts and dwindling natural resources.

Sorghum has huge potential to expand to new environments, especially under water scarce drylands around the world. Sorghum is of great use in the regions where irregular rainfall distribution and high air temperature is a common phenomenon. This is the crop of excellence for semi-arid and arid regions because of its high tolerance to drought under present and future climatic situations. Sorghum farming can be expanded further as it has the remarkable potential to grow under hostile environments consuming minimum input and care. The yields of several foremost principal crops have been predicted to decline under the increasing desertification situations, especially under South Asia and sub-Saharan Africa regions. The International Food Policy Research Institute (IFPRI) (Rosegrant and Msangi 2014) research study has projected that sorghum is one among the few crops which have greater latent yield potential under extreme climatic situations.

Among cereals, sorghum can thrive well under harsh growing conditions. It can be considered an important line of defense system against food shortage under predicted climate change. Pushing growing conditions of the crops beyond their limits will jeopardize the food security of millions of people around the world. Sorghum yields have increased globally despite global warming due to improved cultivars and approaches, and offer hope and similar adaptation approaches that can be fruitful in sub-Saharan Africa. The lessons learnt from Asia can be easily transferred and scaled-up to Africa region under Global South–South collaboration. A combination of interventions including adequate use of fertilizers and technology adoption (e.g., tolerant cultivars), efficient management (e.g., improved irrigation), and better agronomic practices, are key to boost sorghum yields and ensuring the sustainability of this important crop under a warmer climate (Peñuelas et al. 2007).

The low productivity levels in many sub-Saharan Africa countries pose a serious challenge not only for food security but also for hunger and malnutrition. However, if addressed properly, it offers a huge opportunity to attain higher sorghum production from the same cultivated area (Khalifa et al. 2020). The projected calculations indicated that bridging the yield gap to the average maximum yield achieved in the 31 countries in sub-Saharan Africa (1.3 t/ha) has the potential to increase sorghum production in sub-Saharan Africa by 29.5% from the current production level, i.e., from 22.7 to 29.4 million t/ha. This is an incredible opportunity for enhancing sorghum production globally, especially it triggers the sorghum expansion in the Africa region.

1.5.2 Limitations for Sorghum Expansion

Great progress has been made in sorghum breeding in the past decades. However, the average grain yield of sorghum is still very low compared with other cereals. For instance, the global average yield of sorghum in 2018/19 was only ~1.49 MT/ha, which is much lower than that of rice (~4.58 MT/ha), wheat (~3.39 MT/ha), and maize (~5.86 MT/ha) (http://www.fao.org/faostat). There are serious regional imbalances in global sorghum production among different ecologies. Sorghum yield levels, which are highly influenced by genotype and environment, vary significantly among different growing countries (Pfeiffer et al. 2019; Müller et al. 2020; Hao et al. 2021; Ndlovu et al. 2021). Huge research investments and efforts are needed to address such challenges and harmonize sorghum production levels among regions.

Although USDA-NPGS, ICRISAT, and other institutes collected thousands of accessions, most of the accessions lack phenotypic and genotyping data. With the decrease in sequencing cost and the rapid development of phenomics, all collections can be genotyped by resequencing and phenotyped. Accession redundancy is another major problem in germplasm collections because it requires the maintenance and screening of accessions that do not contribute to collection diversity. The underutilization of wild sorghums, which has excellent properties and has great potential in sorghum breeding for various end uses (Abdelhalim et al. 2019).

There is an increasing trend in cultivating sorghum for biofuel production, generating competition with food production on limited water and land resources (Zhuang et al. 2011; Damerau et al. 2016). Some researchers argue that the rise in biofuel production is considered the main driver of the recent increase in food prices (Graham-Rowe 2011; Shrestha et al. 2019), in particular during the period 2001–2008 (Hochman et al. 2012). The increase in the prices of food commodities can be devastating for poor people, especially in developing countries (Stein 2007; Rosegrant and Msangi 2014). However, growing biofuel sorghum as a dual-purpose crop (grain for food and stalk for biofuel) was seen as a solution to reduce the competition with food production.

There are multiple biotic (e.g., pests, diseases, birds, and parasitic weeds) and abiotic (e.g., heat stress, soil moisture deficit, and waterlogging) that affect the health of the sorghum plant, and, in turn, might reduce its productivity (Perumal et al. 2018; Abreha et al. 2021; Fakrudin et al. 2021) or even result in crop failure in extreme cases.

Predicted impact of climate change on sorghum production is documented by several researchers. The various types of climate models and forecasting scenarios showed that the long-term impact on sorghum productivity is higher than short- and medium-term climate change. Based on the region and predicted model used, the predicted reductions of up to 41% of the current sorghum yield is expected in the future due to climate change (Ringler et al. 2010). These findings urge for the development and adoption of effective adaptation measures to sustain sorghum production in the warming world (see Table 1.6).

A large body of literature found important differences between sorghum varieties in terms of yield, nutritional content, and tolerance to climate risks (Amare et al. 2015; Ogbaga et al. 2016; Assefa et al. 2020; Tasie and Gebreyes 2020; Wendmu et al. 2022). Similarly, agronomic practices such as tillage, plant density, and sowing date, among others, are found to widely impact sorghum yield (Conley and Wiebold 2003; Ajaj et al. 2021; Carcedo et al. 2021; Zander et al. 2021; Gao et al. 2022; Naoura et al. 2023), indicating that adopting best agronomic practices is an indispensable factor for any program that aims to improve crop productivity and build climate resilience. This clearly indicates that sorghum is not an exceptional crop under predicted climate change impacts.

1.6 Conclusions

Sorghum is the hardiest dryland crop that can thrive well under marginal lands and with projected climate change conditions. It is a staple food for millions of small and marginal farmers living in nearly 30 countries of the subtropical and semi-arid regions under Africa and Asia. In developing countries with decline in food use of

Table 1.6 Studies investigated the impact of future climate change on sorghum yield

Authors	Country/region	Model/climate scenario	Main findings
Srivastava et al. (2010)	India	HadCM3 model A2a scenario	Climate change will decrease rainy-season sorghum yields by up to 14% and post-rainy-season sorghum yields by up to 7% by 2020
Butt et al. (2005)	Mali	HADCM and CGCM	A decrease of 11–17% in sorghum yield is expected in Mali by 2030
Tingem et al. (2008)	Cameroon	GISS 2080 scenarios	Sorghum yields are expected to decrease by 39.9% across the whole country
Chipanshi et al. (2003)	Botswana	Southern African core climate change scenario	Simulated sorghum yields declined by 36% under climate change
Sultan et al. (2013)	Sub-Saharan West Africa (Senegal, Mali, Burkina Faso, and Niger)	CMIP3, CMIP5	Millet and sorghum yields are likely to decrease by some 0–41% in the twenty-first century over West Africa because of the expected warming, irrespective of rainfall condition
Knox et al. (2012)	Africa South Asia	GCM	Across Africa, mean yield changes of −14.6% and −10.8% in sorghum were estimated for Africa and South Asia, respectively
Burke et al. (2009)	Africa	CCMA, CNRM, CSIRO, GFDL0, GFDL1, GISS. AOM, GISS.EH, GISS.ER, IAP, INMCM3, IPSL, MIROC.HIRES, MIROC. MEDRES, ECHAM, MRI, CCSM, PCM, HADCM3	Many of the African countries will have novel climates over at least half of their current crop area by 2050
Chadalavada et al. (2022)	India	MMM, CMIP5, CERES	On average, post-rainy season sorghum yields are projected to vary between −4% and +40% as well as +10% and +59% in the near future (2040–2069) for RCP 4.5 and RCP 8.5, respectively

Source: Khalifa and Eltahir (2023)

sorghum, its consumption is growing in alternative non-food uses like poultry feed, alcohol manufacture, etc. (In Africa, sorghum is still an important staple crop). Developing varieties with traits suitable for different uses should be an important priority of crop improvement programs to meet end user requirements (e.g., varieties for alcohol manufacture). The global sorghum production was more stable during the past one-decade period even though there was a decline in the global cropped area. This might have been possible because of enhancement in sorghum productivity across different regions. Nearly 48% of global production goes for food followed by 35% for feed, non-food and processing contribute 10%, and 7% as food losses. Although most sorghum continues to be consumed in the countries where it is produced, the global trade of sorghum is strongly linked to demand for livestock products, dominated by feed requirements and prices. The United States is the dominant exporter of sorghum in the world while China is identified as the largest importer.

Despite several opportunities for expansion of sorghum, there are an equal number of challenges threatening it. The wide yield gap in sub-Saharan Africa is a serious challenge for food security and malnutrition in the region. However, it highlights huge hidden opportunities available to boost global sorghum production in the region. Bridging the yield gap to the country-specific maximum achieved yield in 31 sub-Saharan African countries would increase sorghum production from 22.7 to 29.4 million tons. The potential production of 29.4 million tons from these countries represents more than 50.7% of the current global sorghum production. Despite the relative resilience of sorghum to different stresses, it can be affected by multiple biotic and abiotic stresses. Among abiotic stresses, heat stress, moisture deficit, and waterlogging might be the most pressing challenges for sorghum cultivation in the future under climate change. On the demand side, huge efforts are required to enhance the per capita consumption of sorghum globally by creating nutrition-sensitive education as well as correcting the policy bias against coarse cereals when compared with fine cereals. Similarly, massive strategic research investments are required to address the inherent challenges in sorghum grain consumption traits (such as hardness, cooking time, flour keeping quality, softness) when compared with other cereals. There is huge untapped potential in the crop, especially in the development of sorghum-based products and value creation along the supply chain. The enhanced demand from both domestic and export markets will create a huge impact on sorghum area expansion or stabilization and thereby generate remunerative prices to sorghum growers. Sorghum's higher price limits its substitution in East and South Asia (ESA) region for maize in the value chains for livestock feed, wheat flour, and maize meal. Breaking out of this trap requires a supply shift to increase yields and reduce wholesale prices. Country-specific policy incentives are required for promoting the alternative uses of sorghum for not only protecting the interest of the farmers but also enhancing food systems diversification, protecting the natural resources, and achieving the SDGs.

Future research should focus on knowledge regarding traits associated with gray mold, stem rot, heat, moisture deficit, and waterlogging tolerance. Heat, moisture and water logging stresses are three stresses. With enhanced labor costs as well as scarcity in different regions, amenability to mechanization is critical for sorghum rapid expansion. Traits related to grain softness, cooking time, and flour keeping quality etc. are going to be game-changing traits in the years to come.

References

Abdelhalim TS, Kamal NM, Hassan AB (2019) Nutritional potential of wild sorghum: grain quality of Sudanese wild sorghum genotypes (*Sorghum bicolor* L. Moench). Food Sci Nutr 7:1529–1539

Abreha KB, Enyew M, Carlsson AS, Vetukuri RR, Feyissa T, Motlhaodi T et al (2021) Sorghum in dryland: morphological, physiological, and molecular responses of sorghum under drought stress. Planta 255:20. https://doi.org/10.1007/s00425-021-03799-7

Ajaj HA, Mohammed YA, Alrubaya AA, Addaheri AM (2021) Effect of planting dates on the growth, yield and quality of three cultivars of Sorghum (*Sorghum bicolor* L. Moench). IOP Conf Ser Earth Environ Sci 904:012019. https://doi.org/10.1088/1755-1315/904/1/012019

Amare K, Zeleke H, Bultosa G (2015) Variability for yield, yield related traits and association among traits of Sorghum (Sorghum Bicolor (L.) Moench) varieties in Wollo, Ethiopia. J Plant Breed Crop Sci 7:125–133. https://doi.org/10.5897/JPBCS2014.0469

Assefa A, Bezabih A, Girmay G, Alemayehu T, Lakew A (2020) Evaluation of sorghum (*Sorghum bicolor* (L.) Moench) variety performance in the lowlands area of wag lasta, north eastern Ethiopia. Cogent Food Agric 6:1778603. https://doi.org/10.1080/23311932.2020.1778603

Bhagavatula S, Parthasarathy Rao P, Basavaraj G, Nagaraj N (2013) Sorghum and millet economies in Asia - facts, trends and outlook. International Crops Research Institute for the Semi-Arid Tropics, Patancheru, p 80

Burke MB, Lobell DB, Guarino L (2009) Shifts in African crop climates by 2050, and the implications for crop improvement and genetic resources conservation. Glob Environ Chang 19:317–325. https://doi.org/10.1016/j.gloenvcha.2009.04.003

Butt TA, McCarl BA, Angerer J, Dyke PT, Stuth JW (2005) The economic and food security implications of climate change in Mali. Clim Chang 68:355–378. https://doi.org/10.1007/s10584-005-6014-0

Carcedo AJP, Cejas E, Gambin BL (2021) Adapting sorghum sowing date and genotype maturity to seasonal rainfall variation in a temperate region. In Silico Plants 3:7. https://doi.org/10.1093/insilicoplants/diab007

Chadalavada K, Gummadi S, Kundeti KR, Kadiyala DM, Deevi KC, Dakhore KK, Bollipo Diana RK, Thiruppathi SK (2022) Simulating potential impacts of future climate change on post-rainy season sorghum yields in India. Sustainability 14:334. https://doi.org/10.3390/su14010334

Chipanshi AC, Chanda R, Totolo O (2003) Vulnerability assessment of the maize and Sorghum crops to climate change in Botswana. Clim Chang 61:339–360. https://doi.org/10.1023/B:CLIM.0000004551.55871.eb

Conley SP, Wiebold WJ (2003) Grain Sorghum response to planting date. Crop Manage 2:1–5. https://doi.org/10.1094/CM-2003-0204-01-RS

Damerau K, Patt AG, van Vliet OPR (2016) Water saving potentials and possible trade-offs for future food and energy supply. Glob Environ Chang 39:15–25. https://doi.org/10.1016/j.gloenvcha.2016.03.014

de Cardoso Morais L et al (2015) Sorghum (*Sorghum bicolor* L.): nutrients, bioactive compounds, and potential impact on human health. Crit Rev Food Sci Nutr 57(2):372–390. https://doi.org/10.1080/10408398.2014.887057

Fakrudin B, Lakshmidevamma TN, Ugalat J, Gunnaiah R, Khan J, Gautham Suresh SP et al (2021) Genomic designing for biotic stress resistance in Sorghum. In: Kole C (ed) Genomic designing for biotic stress resistant cereal crops. Springer, Cham, pp 213–255

FAO and ICRISAT (1996) The world sorghum and millet economies: facts, trends and outlook.

FAOSTAT (2023). https://www.fao.org/faostat/en/#home

Gao F-C, Yan H-D, Gao Y, Huang Y, Li M, Song G-L et al (2022) Interpretation of genotype-environment-sowing date/plant density interaction in sorghum [*Sorghum bicolor* (L.) Moench] in early mature regions of China. Front Plant Sci 13:1008198. https://doi.org/10.3389/fpls.2022.1008198/full

Graham-Rowe D (2011) Agriculture: beyond food versus fuel. Nature 474:S6–S8. https://doi.org/10.1038/474S06a

Habyarimana E, Dall'Agata M, De Franceschi P, Baloch FS (2019) Genome-wide association mapping of total antioxidant capacity, phenols, tannins, and flavonoids in a panel of *Sorghum bicolor* and *S. bicolor* × *S. halepense* populations using multi-locus models. PLoS ONE 14(12):e0225979. https://doi.org/10.1371/journal.pone.0225979

Hao H, Li Z, Leng C, Lu C, Luo H, Liu Y et al (2021) Sorghum breeding in the genomic era: opportunities and challenges. Theor Appl Genet 134:1899–1924. https://doi.org/10.1007/s00122-021-03789-z

Hariprasanna K, Rakshit S (2016) Economic importance of sorghum. In: The sorghum genome. Springer, Cham, pp 1–25

Hochman G, Kaplan S, Rajagopal D, Zilberman D (2012) Biofuel and food-commodity prices. Agriculture 2:272–281. https://doi.org/10.3390/agriculture2030272

Huang R (2018) Research progress on plant tolerance to soil salinity and alkalinity in sorghum. J Integr Agric 17:739–746. https://doi.org/10.1016/S2095-3119(17)61728-3

Iqbal A, Sadia B, Khan AI, Awan FS, Kainth RA, Sadaqat HA (2010) Biodiversity in the sorghum (*Sorghum bicolor* L. Moench) germplasm of Pakistan. Genet Mol Res 9:756–764. https://doi.org/10.4238/vol9-2gmr741

Kadam S, Abril A, Dhanapal AP, Koester RP, Vermerris W, Jose S et al (2017) Characterization and regulation of aquaporin genes of Sorghum [*Sorghum bicolor* (L.) Moench] in response to waterlogging stress. Front Plant Sci 8:862. https://doi.org/10.3389/fpls.2017.00862

Khalifa M, Eltahir EAB (2023) Assessment of global sorghum production, tolerance, and climate risk. Front Sustain Food Syst 7:1184373. https://doi.org/10.3389/fsufs.2023.1184373

Khalifa M, Elagib NA, Ahmed BM, Ribbe L, Schneider K (2020) Exploring socio-hydrological determinants of crop yield in under-performing irrigation schemes: pathways for sustainable intensification. Hydrol Sci J 65:153–168. https://doi.org/10.1080/02626667.2019.1688333

Knox J, Hess T, Daccache A, Wheeler T (2012) Climate change impacts on crop productivity in Africa and South Asia. Environ Res Lett 7:034032. https://doi.org/10.1088/1748-9326/7/3/034032

Monk R, Franks C, Dahlberg J (2014) Sorghum. In: Yield gains in major U.S. Field crops. Wiley, Hoboken, pp 293–310

Müller M, Dembélé S, Zougmoré RB, Gaiser T, Partey ST (2020) Performance of three Sorghum cultivars under excessive rainfall and waterlogged conditions in the Sudano-Sahelian zone of West Africa: a case study at the climate-Smart Village of Cinzana in Mali. Water 12:2655. https://doi.org/10.3390/w12102655

Naoura G, Emendack Y, Sawadogo N, Djirabaye N, Tabo R, Laza H et al (2023) Assessment of photoperiod sensitivity and the effect of sowing date on dry-season Sorghum cultivars in southern Chad. Agronomy 13:932. https://doi.org/10.3390/agronomy13030932

Ndlovu E, van Staden J, Maphosa M (2021) Morpho-physiological effects of moisture, heat and combined stresses on *Sorghum bicolor* [Moench (L.)] and its acclimation mechanisms. Plant Stress 2:100018. https://doi.org/10.1016/j.stress.2021.100018

Ogbaga CC, Stepien P, Dyson BC, Rattray NJW, Ellis DI, Goodacre R et al (2016) Biochemical analyses of Sorghum varieties reveal differential responses to drought. PLoS One 11:e0154423. https://doi.org/10.1371/journal.pone.0154423

Parthasarathy Rao P, Hall AJ (2003) Importance of crop residues in crop-livestock systems in India and farmers perceptions of fodder quality in coarse cereals. Field Crop Res 84:189–198

Peñuelas J, Prieto P, Beier C, Cesaraccio C, De Angelis P (2007) Response of plant species richness and primary productivity in shrublands along a north–south gradient in Europe to seven years of experimental warming and drought: reductions in primary productivity in the heat and drought year of 2003. Glob Chang Biol 13:2563–2581. https://doi.org/10.1111/j.1365-2486.2007.01464.x

Perumal R, Hays ARC, Magill UCW, Center E, Prom ULK (2018) Sorghum breeding for biotic stress tolerance Texas a&M University, USA. In: Peterson GC (ed) Achieving sustainable cultivation of sorghum, vol 1. Burleigh Dodds Science Publishing, Cambridge

Pfeiffer BK, Pietsch D, Schnell RW, Rooney WL (2019) Long-term selection in hybrid Sorghum breeding programs. Crop Sci 59:150–164. https://doi.org/10.2135/cropsci2018.05.0345
Rakshit S et al (2014) Changes in area, yield gains, and yield stability of sorghum in major sorghum-producing countries, 1970 to 2009. Crop Sci 54(4):1571–1584. https://doi.org/10.2135/cropsci2012.12.0697
Rao PS et al (2015) Sorghum as a sustainable feedstock for biofuels. In: Biomass and biofuels-advanced biorefineries for sustainable production and distribution. CRC Press, Boca Raton, pp 27–47
Ringler C et al (2010) Climate change impacts on food security in sub-Saharan Africa. In: Insights from comprehensive climate change scenarios, vol 2. IFPRI, Washington
Rosegrant MW, Msangi S (2014) Consensus and contention in the food-versus-fuel debate. Annu Rev Environ Resour 39:271–294. https://doi.org/10.1146/annurev-environ-031813-132233
Shrestha DS, Staab BD, Duffield JA (2019) Biofuel impact on food prices index and land use change. Biomass Bioenergy 124:43–53. https://doi.org/10.1016/j.biombioe.2019.03.003
Srivastava A, Naresh Kumar S, Aggarwal PK (2010) Assessment on vulnerability of sorghum to climate change in India. Agric Ecosyst Environ 138:160–169. https://doi.org/10.1016/j.agee.2010.04.012
Stein K (2007) Food vs biofuel. J Am Diet Assoc 107:1870–1878. https://doi.org/10.1016/j.jada.2007.09.014
Sultan B, Roudier P, Quirion P, Alhassane A, Muller B, Dingkuhn M et al (2013) Assessing climate change impacts on sorghum and millet yields in the Sudanian and Sahelian savannas of West Africa. Environ Res Lett 8:014040. https://doi.org/10.1088/1748-9326/8/1/014040
Tari I, Laskay G, Takács Z, Poór P (2013) Response of sorghum to abiotic stresses: a review. J Agron Crop Sci 199:264–274. https://doi.org/10.1111/jac.12017
Tasie MM, Gebreyes BG (2020) Characterization of nutritional, antinutritional, and mineral contents of thirty-five Sorghum varieties grown in Ethiopia. Int J Food Sci 2020:e8243617. https://doi.org/10.1155/2020/8243617
Tingem M, Rivington M, Bellocchi G, Azam-Ali S, Colls J (2008) Effects of climate change on crop production in Cameroon. Clim Res 36:65–77. https://doi.org/10.3354/cr00733
Upadhyaya HD, Vetriventhan M, Asiri AM, Azevedo CR (2019) Multi-trait diverse germplasm sources from mini core collection for sorghum improvement. Agriculture 9:121. https://doi.org/10.3390/agriculture9060121
Wendmu TA, Cuni-Sanchez A, Abebe HT, de Boer HJ, Abera FA, Westengen OT (2022) Cultural effects on Sorghum varieties grown, traits preferred, and seed management practices in northern Ethiopia. Econ Bot 76:233–249. https://doi.org/10.1007/s12231-022-09555-6
Zander A, Lofton J, Harris C, Kezar S (2021) Grain sorghum production: influence of planting date, hybrid selection, and insecticide application. Agrosyst Geosci Environ 4:e20162. https://doi.org/10.1002/agg2.20162
Zhang R, Zhou Y, Yue Z, Chen X, Cao X, Ai X et al (2019) The leaf-air temperature difference reflects the variation in water status and photosynthesis of sorghum under waterlogged conditions. PLoS One 14:e0219209. https://doi.org/10.1371/journal.pone.0219209
Zhuang D, Jiang D, Liu L, Huang Y (2011) Assessment of bioenergy potential on marginal land in China. Renew Sust Energ Rev 15:1050–1056. https://doi.org/10.1016/j.rser.2010.11.041

Agronomic Practices for Sorghum Production

2

Recep İrfan Nazlı and Mihrab Polat

Abstract

Sorghum is a C_4 crop that belongs to Poaceae family. Sorghum is a multifunctional crop used as forage, food, sugar, fiber, and bioenergy crops in a wide range of environments. Grain sorghum is the fifth most important cereal worldwide after wheat, rice, maize, and barley. Forage sorghum varieties produce greater biomass yield than grain sorghum varieties but have lower grain yield. Sweet sorghum contains high amount of fermentable sugar (10–20%) in its juicy stems, which can easily be fermented to ethanol like sugar cane. Sorghum is a warm-season crop; therefore, it needs warm temperature for good establishment and high productivity. Sorghum, also known as the camel crop, is very resistant to drought and can grow well in semi-arid areas. Sorghum removes a large amount of nutrients from the soil, thanks to its well-developed fibrous root system. Fertilization should be done at the optimum rate and time, taking into account the soil test results and the expected yield average. Yield losses in sorghum cultivation associated with infestation of weeds, insects, and diseases can be minimized by a combination of integrated pest and weed management components including chemical control, natural enemies, host plant resistance, and proper agronomic practices.

Keywords

Drought · Fertilization · Forage sorghum · Grain sorghum · Pest management · Sweet sorghum

R. İ. Nazlı (✉) · M. Polat
Department of Field Crops, Faculty of Agriculture, Cukurova University, Adana, Turkey
e-mail: inazli@cu.edu.tr

E. Habyarimana et al. (eds.), *Omics and Biotechnological Approaches for Product Profile-Driven Sorghum Improvement*,
https://doi.org/10.1007/978-981-97-4347-6_2

2.1 Introduction

Sorghum is a C_4 crop that belongs to Poaceae family and is indigenous to Africa (Oyier et al. 2017). The cultivated sorghum species (*Sorghum bicolor* (L.) Moench) have been currently distributed in almost all tropical and subtropical regions of the world with a warm or temperate climate (Avcioglu et al. 2009). Sorghum is a multipurpose crop that can provide food, forage, fiber, sugar, pulp, and biofuel across a wide range of environments (Oyier et al. 2017). Sorghum genus contains almost eight to ten different types or species depending on the use and morphological characteristics.

Grain sorghum is the fifth most important cereal worldwide after wheat, rice, maize, and barley. In this context, it is the main food source for approximately 750 million people living in semi-arid tropical regions of Africa, Asia, and Latin America (CCCF 2011; Ramatoulaye et al. 2016). Recently bred varieties have 0.6–1.5 m height to facilitate machine harvesting and to minimize harvest losses. Worldwide, approximately 40% of sorghum grain is used for human nutrition, while the remainder is devoted to bioethanol production, industrial use, brewing, and mostly for livestock feed. Grain sorghum has a higher crude protein (9–15%), but lower crude oil content (3%) compared to maize, and has an energy value of 97–100% of maize (Avcioglu et al. 2009). Since sorghum flour is free of gluten, it is frequently used in the diets of celiac patients. On the other hand, researchers and bioethanol producers in the USA consider grain sorghum to be a good maize-like feedstock for ethanol production and to contribute greatly to the country's biofuel needs. Sorghum has starch-rich grains like maize (approximately 60–77%), and its starch content varies considerably depending on the genetic and environmental factors. For this reason, sorghum grain can be easily used as a feedstock for bioethanol production, especially in arid and semi-arid climates where maize does not grow well, passing through the same production processes as maize (Dweikat 2012).

In countries where other cereals (wheat, maize, and rice) are generally preferred for carbohydrate source, sorghum is grown mainly for forage production. Forage sorghum varieties have higher plant height (2–5 m), more tillering, and produce greater biomass yield than grain sorghum varieties but have lower grain yield and finer stems. Varieties suitable for multi-cut system generally belong to Sudan grass or sorghum × Sudan grass hybrid species. Commonly, forage sorghum is grown for direct grazing, but primarily for hay and silage production (Undersander et al. 2010). However, in recent years, there has been an increasing interest in the use of forage sorghum as a biogas substrate as an alternative to maize, especially in Europe. Forage sorghum is the most promising alternative to silage maize for both forage and biogas production due to its high yield potential, drought tolerance, resource use efficiency (water, nutrient, and radiation), and low input requirement, especially in drought-prone areas where water scarcity limits maize growth (Amaducci et al. 2016; Samarappuli and Berti 2018). Additionally, forage sorghum can also effectively avoid environmental problems arising from long-term monocropping of maize such as decreasing biodiversity, increasing pest/disease intensity, nitrate pollution in groundwater, nutrient losses, and soil degradation (Mahmood et al. 2013;

Schulz et al. 2018; Jankowski et al. 2020). Forage sorghum has a similar crude protein content to silage maize, but generally has a lower digestibility and energy value due to its higher fiber content (Mahmood et al. 2013).

Sweet sorghum differs from other sorghum types as it contains high amount of fermentable sugar (10–20%) in its juicy stems, which can easily be fermented to ethanol like sugar cane. Sweet sorghum has gained particular interest as a bioenergy crop since last decades because it can produce satisfactory bioethanol yield in low input cultivation systems due to its superior drought tolerance, high water and fertilizer use efficiency, and wide adaptability to constraint environments (Almodares et al. 2007; Tew et al. 2008). Additionally, sweet sorghum does not lead to food–fuel competition as it is a non-food crop (Hu et al. 2017). On the other hand, sweet sorghum bagasse (remaining stem after sugar extraction) is considered as a valuable feedstock for both second-generation bioethanol and solid biofuel production because it comprises of considerable amount of lignocellulosic material (Singh et al. 2012a). Sweet sorghum is also known as camel crop or sugar cane of the desert thanks to its high fermentable sugar content and drought tolerance (Dar et al. 2018). Sugar accumulation in stems begins with flowering and usually peaks between the soft dough and physiological maturity stages (Teixeira et al. 2017; Oyier et al. 2017; Nazli 2022). Apart from these, sweet sorghum may be regarded as an important silage crop since it can be easily ensiled and produce high amount of lactic acid thanks to its high sugar content and low buffering capacity.

2.2 Adaptation

Sorghum has a C_4 photosynthetic pathway like maize and sugar cane meaning that it is well adapted to environments exposed to high temperatures (>25 °C) during summer with sufficient water and nutrient availability (Edwards et al. 2004; Dweikat 2012). Temperature, day length, water availability, and soil fertility are the most critical factors affecting the adaptation and growth of sorghum in a certain environment (Dweikat 2012).

2.2.1 Soil

Sorghum can grow well in a wide range of soils, from heavy clay soils to light sand, with pH varying from 5.5 to 8.5, but it performs best on deep fertile, well-drained loamy soils having a good water-holding capacity and pH between 6 and 7.5 (Dar et al. 2018; Lubadde et al. 2019; Ajeigbe et al. 2020). Although it generally grows poorly on shallow and sandy soils with a high clay content, it adapts better to these areas than maize (Ajeigbe et al. 2020). Additionally, sorghum can tolerate alkaline salts, aluminum toxicity, and short periods of waterlogging better than other cereals (Lubadde et al. 2019; Ajeigbe et al. 2020).

2.2.2 Temperature

Sorghum is a warm-season crop; therefore, it needs warm temperature for good establishment and high productivity (Dweikat 2012). While a temperature of 17–25 °C is needed for optimum germination, it can be delayed for up to 7–10 days at a temperature below 15 °C (Dweikat 2012; Lubadde et al. 2019; Ajeigbe et al. 2020). For the optimum growth and photosynthesis, sorghum requires a temperature between 32 and 37 °C, while the growth normally ceases at a temperature below 15 °C (Dweikat 2012; Dar et al. 2018). Sorghum has a superior tolerance to high temperature (>37 °C) stress than most warm-season crops, especially maize. However, high temperature conditions occurring within the flowering stage can exceptionally cause destruction of the pollens, which results in a poor grain formation. Since sorghum is susceptible to freezing, prolonged exposure to frost may cause death of the plant. At early seedling stage (1–3 weeks), plants may recover when exposed to temperature of −5 °C for a short time, while older plants may die even at a freezing point (Dweikat 2012; Vanya 2012). Sweet sorghum, an annual sorghum type, can act as a short-lived perennial crop in frost-free tropical climates thanks to its ratoon-growing capability (Kim and Day 2011).

2.2.3 Photoperiod

Since sorghum is a short-day plant, it needs short days, i.e., long nights, to get through the generative stage. Required day length for initiation of reproductive growth normally ranges from 11.5 to 13.5 in photoperiod sensitive varieties, while the optimum photoperiod, which induces flower formation, is between 10 and 11 h (Dar et al. 2018; Kumar and Pathak 2018; Ajeigbe et al. 2020). Sorghum is most sensitive to photoperiod during flower initiation (Ajeigbe et al. 2020). On the other hand, some sorghum varieties are photoperiod insensitive (Folliard et al. 2004). Biomass sorghum varieties are usually photoperiod sensitive; therefore, they have longer vegetative period, thus producing greater biomass yield compared to other sorghum types.

2.2.4 Water

Sorghum, also known as the camel crop, is very resistant to drought and can grow well even in areas receiving little rainfall (<300 mm) during a growing season (Youngquist et al. 1990). However, it requires nearly 450–800 mm of water for the optimum yield depending on the variety and ecological conditions (Ajeigbe et al. 2020). The water requirement is low during the seedling and tillering, but increases toward the booting, and finally reaches peak within the flowering stage (Dweikat 2012). Severe drought stress that occurs during boot and flowering stages can cause poor grain yield because of incomplete pollination. Sorghum usually continues high water use till the end of the soft dough stage, but the water requirement gradually decreases during hard dough and physiological maturity stages (Dweikat 2012).

Sorghum owes its superior drought tolerance compared to maize to the following factors:

- Sorghum absorbs water efficiently thanks to its fibrous root system that can develop up to 2.5 m deep and spread laterally 30–60 cm depending on the texture and water status of soil (Avcioglu et al. 2009).
- Sorghum has a smaller leaf area than maize, which limits transpiration (Kakani et al. 2011). Additionally, its leaves tend to be folded up unlike in maize that are rounded under water-stressed conditions.
- Varieties resistant to post-flowering drought stress, also called "stay-green," have fewer tillering, leaf area per plant, and number of leaves per culm; higher soluble sugar content; and they maintain green leaves under terminal drought conditions, compared to "senescent" varieties. This trait enables them to retain chlorophyll in their leaves and maintain the ability to carry out photosynthesis for a long time when subjected to post-flowering drought, consequently having significantly high yield potential (Harris et al. 2007; Burke et al. 2010; Borrell et al. 2014; Tovignan et al. 2020).
- Sorghum accumulates epicuticular wax on its surface (leaves, sheath, and culm) increasing the leaf radiate reflection and prevents non-stomatal water loss under water-deficit conditions (Punnuri et al. 2017; Sanjari et al. 2021).
- Sorghum has relatively low amounts of stomata that rapidly close to minimize water loss when exposed to warm and drought conditions (Dweikat 2012; Lubadde et al. 2019).
- Sorghum goes into dormancy when exposed to drought during vegetative stage, but resumes growth as soon as soil moisture conditions improve (Dweikat 2012).

2.3 Planting

2.3.1 Land Preparation

Sorghum requires a fine-grained, sufficiently moist and well-drained soil for good emergence, root growth, and seedling development. In order to prepare a seedbed that is free of weeds and ensures a good soil–seed contact, the field should be ploughed once or twice at a depth of 20–25 cm, then harrowed with a field cultivator about 1–2 weeks before planting (Avcioglu et al. 2009). However, tillage operations should be minimized to maintain soil moisture in dry areas (Ajeigbe et al. 2020).

2.3.2 Planting Time

Soil moisture and temperature are the key elements that determine the planting time of sorghum. For adequate seed germination and emergence, the field water capacity should be between 25% and 50% and temperature should be at least 15 °C at planting time (Fawusi and Agboola 1980; Ajeigbe et al. 2020). When the soil moisture is too low, the seeds fail to germinate sufficiently, or, even if they germinate, the coleoptile dies after reaching the soil surface as moisture runs out. If planting is done soon after heavy rains or flooding, excess moisture can cause the seeds to rot (Anonymous 2019). Soil temperature below 15 °C decreases germination and nutrient uptake, delays emergence, and finally leads to poor establishment. Moreover, sorghum seeds and seedlings are more susceptible to insect attack and disease infection under cold and moist conditions (Anonymous 2019).

Sorghum plantings may be delayed till late May in humid continental climates to protect the seeds from destructive effects of last spring frosts under main cropping conditions. However, sorghum can be planted in the first week of April at the earliest in temperate climates like Mediterranean. On the other hand, sowing can be done between first weeks of June and July if sorghum is grown as a second crop. Early cultivar should be chosen as much as possible in second cropping systems. In semi-arid tropics, sorghum has been grown during two distinct growing seasons: *kharif* (rainy season; June–October) and *rabi* (post-rainy season; October–January) (Reddy et al. 2012). Although sowing can be done in a wide range from May to July in the *kharif* season, it should be completed between September 25 and October 15 in *rabi* season depending on the rainfall occurrence (Ratnavathi et al. 2011; Kholova et al. 2013; Kumar et al. 2013; Karhale et al. 2014; Reddy et al. 2014). As grain and sweet sorghum have been principally grown during off-season (January–April) as succession crops to soybean and sugar cane, respectively, in Brazil, they usually have been *sown* within January and February (Júnnyor et al. 2015; Menezes et al. 2022; Vieira et al. 2021).

2.3.3 Row Configuration

Inter-row spacing applied for sorghum cultivation varies greatly by ecological conditions, especially soil moisture level (Whish et al. 2005; Abunyewa et al. 2010; Karhale et al. 2014; Nielsen et al. 2018; Menezes et al. 2022; Mubarik et al. 2022). In drought-prone environments, wider rows improve grain yield because they enable sorghum plants to use available water more efficiently. For this reason, while grain sorghum is usually planted in 45 cm row spacing in *kharif* season, row spacing is increased up to 60 cm in *rabi* season in India (Madhusudhana et al. 2007; Elangovan et al. 2010; Talwar et al. 2010; Karhale et al. 2014; Choudhary et al. 2017; Kishor et al. 2017). On the other hand, in the USA and Australia, sorghum has been conventionally planted in row spacings of 0.76 and 1 m, respectively (Whish et al. 2005; Abunyewa et al. 2010; Vigil et al. 2015; Nielsen et al. 2018). As a result, growers must choose the row spacing best suited to their particular growing conditions for the sustainable production.

Skip-row configuration, which includes widening the row spacing while maintaining plant density, promotes grain yield of sorghum in water-limiting environments (Nielsen et al. 2018). There are three skip-row configurations, P1S1

(Plant1–Skip1), P2S1 (Plant2–Skip 1), and P2S2 (Plant2–Skip2), which are commonly used in the USA and Australia. In P1S1 (Plant1–Skip1), seeds are sown in alternating rows, essentially creating a spacing of 1.52 or 2 m; in P2S1 (Plant2–Skip1), two rows are planted at 0.76 or 1 m spacing and then the other rows are skipped, leaving the same row spacing as P1S1; and in P2S2 (Plant2–Skip2), rows are planted in the same way as in P2S1, but every third and fourth row is skipped leaving a 2.29 or 3 m row spacing (Whish et al. 2005; Abunyewa et al. 2010). In the skip-row planting system, sorghum plants cannot reach the soil water stored under the unplanted row areas during vegetative growth period, as their roots do not develop very well. In this way, a significant amount of water is conserved in soils with high water-holding capacity for use during flowering and grain filling periods when the plants are most sensitive to water stress (Whish et al. 2005; Abunyewa et al. 2010; Nielsen et al. 2018). Abunyewa et al. (2010) reported that P2S2 increased grain yield of sorghum by 123% compared to conventional row configuration (0.76 m) in an experiment site receiving annual rainfall of 379 mm located in Nebraska, USA. However, conventional or narrow row spacing outperforms skip-row plantings in areas where water stress is not a concern due to maximized light interception, increased nutrient uptake, and reduced weed emergence (Whish et al. 2005; Abunyewa et al. 2010).

As with the inter-row spacing, the optimum intra-row spacing for grain sorghum is highly dependent on the soil water availability, which is ranged from 7 to 30 cm worldwide (Emeklier and Köksoy 1997; Assefa and Staggenborg 2010; Karhale et al. 2014; Lubadde et al. 2019; Ajeigbe et al. 2020; Menezes et al. 2022). Wider spacings within rows allow the plant roots to absorb larger volume of water from the soil during critical growth periods, thereby increasing the grain yield under dryland conditions. However, narrower intra-row spacings that increase resource use efficiency and reduce weed competition of the crop should be preferred for high grain yield potential under wet conditions. For example, Assefa and Staggenborg (2010) reported that intra-row spacings used in grain sorghum cultivation in Kansas, USA, between 1957 and 2008 ranged from 17 to 26.9 cm and 7 to 8 cm under dryland and irrigated conditions, respectively.

Since green and dry biomass production per unit area are the two most important components representing the yield in forage sorghum, water stress during the vegetative stage restricts sustainable production more than what occurs at flowering and grain filling stages. Therefore, the forage sorghum varieties are generally planted in narrower inter- (15–60 cm) and intra-row spacings (5–15 cm) (Causley 1990; Iptaş and Yılmaz 1995; Turgut et al. 2005; Oten 2017; Zhou et al. 2017).

Like grain sorghum, sweet sorghum also has been conventionally planted in 76 cm inter-row spacing in the USA (Erickson et al. 2011; Houx III and Fritschi 2013; Maw et al. 2016; Briand et al. 2018; Burnham et al. 2022). However, recent reports achieved in Oklahoma (Godsey et al. 2012) and Iowa (Bonin et al. 2016) showed that narrow row spacings (20 and 38 cm) significantly increased dry matter and ethanol yields of sweet sorghum compared to conventional row spacing (76 cm) in fertile soils with high water availability, despite having no evident effect on brix degree or fermentable sugar content. Additionally, Adams et al. (2015) concluded that inter-row spacing of 61 cm significantly enhanced biomass and sugar yields of sweet sorghum in comparison to 76 cm in Florida. Inter-row spacing applied in sweet sorghum cultivation in India varies between 45 and 75 cm, but mostly 60 cm is preferred (Reddy et al. 2008; Ratnavathi et al. 2011; Sawargaonkar et al. 2013; Rao et al. 2013). Sawargaonkar et al. (2013) indicated that there was no significant difference between 45 and 60 cm inter-row spacings in terms of brix degree and stalk, juice, and potential ethanol yields in Patancheru, India. In the rest of the world, mainly in the Mediterranean region, Brazil, and China, sweet sorghum has been usually planted in row spacings between 50 and 70 cm (Petrini et al. 1993; Cosentino et al. 2012; Júnnyor et al. 2015; Tang et al. 2018; Diallo et al. 2019; Da Silva et al. 2019; Nazli 2020; De Oliveira et al. 2022). Optimal intra-row spacing has been commonly ranged from 5 to 15 cm for sweet sorghum cultivation worldwide (Turgut et al. 2005; Erickson et al. 2011; Rao et al. 2013; Reddy et al. 2014). On the other hand, the optimal seeding rate is ranged from 10 to 50 kg/ha depending on the type and seed size (Avcioglu et al. 2009).

2.3.4 Planting Depth

Sorghum seeds should be planted at 2–5 cm soil depth, depending on the seed size, soil type, and moisture conditions. In sandy and dry soils, seeds should be planted deeper than in wet and heavy soils (Avcioglu et al. 2009; Erickson et al. 2011; Ajeigbe et al. 2020). Planting deeper than 5 cm causes a poor establishment in two ways: (1) it slows emergence that may increase seed losses due to the increased risk of pest and disease attacks; and (2) it may avoid the coleoptiles to reach the soil surface, thus causing the emergence to fail. On the other hand, warm and windy weather conditions may lead to rapid seed bed drying, which result in considerable decrease in germination rate when the seeds are planted at levels shallower than 2–3 cm. Pressing the soil with a roller immediately after planting increases the seed–soil contact, thereby encouraging the germination.

2.4 Fertilization

Sorghum removes a large amount of nutrients from the soil thanks to its well-developed fibrous root system. For the sustainable production, fertilization should be done at the optimum rate and time, taking into account the soil test results and the expected yield average. While insufficient fertilization reduces yield and quality, its overapplication causes environmental pollution, nitrous oxide (NOx) emission, and deterioration in soil structure (Khan et al. 2008; Nazli et al. 2014).

2.4.1 Phosphorus (P)

Phosphorus is the second most required nutrient for sorghum development after nitrogen (N) due to its vital roles in the respiration, photosynthesis, cell enlargement, cell division, seed formation, and root development (Mengel et al. 2001; Gupta et al. 2016; Asif et al. 2022). P fertilization improves yield and quality of forage sorghum by increasing leaf area, stem diameter, plant height, and number of leaves (Mahmud et al. 2003). On the other hand, P deficiency in sorghum is characterized by symptoms like poor seedling vigor and root development, young leaves with dark green and reddish-purple margins, older leaves with brown and irregular stains, and delayed or uneven maturity (Camacho et al. 2002; Anonymous 2005; Mattos et al. 2020). Long-term fallows make sorghum more susceptible to P deficiency because it reduces the presence of *Arbuscular mycorrhizae* (AM) fungi, which facilitate the uptake of P from the soil by plant roots (QDAF 2011).

Sorghum is more tolerant to low soil P availability than winter cereals. The crop responds well to P fertilization in soils with less than 10 mg kg^{-1} P (52.7 kg ha^{-1} P_2O_5) (Anonymous 2005). Amount of P fertilizer required for optimum growth of sorghum depends on soil texture and organic matter and available P contents in soil (Ogunlela 1988; Ajeigbe et al. 2020; Asif et al. 2022). Earlier studies clearly showed that optimum P fertilization rate for both grain and forage sorghum in P-deficient soils (<10 mg kg^{-1}) varied between 40 and 80 kg ha^{-1} (Ogunlela 1988; Sahrawat et al. 1995; Ajeigbe et al. 2020; Asif et al. 2022; Rehman et al. 2022).

P should either be broadcast prior to planting or placed in bands with seeds during planting in the form of commercial preplant fertilizers containing at least one of N and K (potassium) together with P (Diammonium Phosphate, Mono Ammonium Phosphate, NPK 15:15:15, etc.) (Wortmann et al. 2013).

2.4.2 Nitrogen (N)

Nitrogen (N) is the major nutrient limiting the growth of sorghum; therefore, its adequate supply to crop is essential for sustainable production (Reddy et al. 2008). N has a significant role on numerous plant functions including constituent of chlorophyll, protoplasm, and enzymes; cell division; and synthesis of sucrose and protein (Almodares et al. 2008; Sune et al. 2017). Deficient N supply limits vegetative growth, reduces protein content, delays flowering, and leads to incomplete grain fill, thus decreasing biomass and grain yields of sorghum. Deficiency symptoms are generally evident on older leaves that become mainly pale green with pale yellow chlorosis and pale brown necrosis (Anonymous 2005). On the other hand, excessive N supply must be avoided because it raises production costs, damages soil structure, increases nitrous oxide (NOx) emissions, and leads to groundwater contamination (Khan et al. 2008; Sowiński and Głąb 2018; Tang et al. 2018). NOx is one of the three major greenhouse gasses and has 265–298 times greater global warming potential (GWP) than CO_2 (Sardá et al. 2020). Therefore, determination of N fertilizer amount required for yield optimization is very crucial to achieving both economically and environmentally viable food, feed, and biofuel production from sorghum (Maw et al. 2016; Nazli et al. 2020). Soil analysis to determine the available N content in the soil can prevent excessive and insufficient N fertilization. Total N content in soil usually is not considered because only 1–2% of organic N is in a readily available form to plants; therefore, it is not reliable data for fertilization recommendations (Liu et al. 2003; Liang et al. 2005; Bao et al. 2006). The N fertilization requirement in sorghum cultivation greatly depends on expected yield, previous crop, the contents of residual nitrate-nitrogen (NO_3–N) and organic matter in soil, and climatic conditions (Wortmann et al. 2013; Sainju et al. 2018). Grain sorghum generally requires similar amount of N supply with maize for target yield in most environments. In contrast, sweet sorghum needs between 37% and 50% less N fertilizer to produce similar bioethanol yield as maize (Anderson et al. 1995; Kering et al. 2017; Dar et al. 2018). The optimal N fertilization rates required to produce sustainable yield from grain, forage, and sweet sorghums are presented in Tables 2.1, 2.2, and 2.3.

Compound fertilizers (DAP, 20-20-0, 15-15-15, etc.), urea, ammonium nitrate, and ammonium sulfate are the most common N fertilizers applied in sorghum cultivation as a preplant and side-dress. In dry conditions, total fertilizer N is generally applied as a preplant due to unreliability that sufficient rainfall will occur to

Table 2.1 Yield optimization of grain sorghum in diverse ecological conditions under different N fertilization regimes

Region	Rainfed/ irrigated	N fertilization rates (kg ha^{-1})	Optimum grain yield (kg ha^{-1})	Optimum N rate (kg ha^{-1})	Reference
Adana, Turkey	Irrigated	0, 50, 100, 150, 200	6585	150	Arslangiray (1998)
Esfahan, Iran	Irrigated	0, 80, 160, 200	4900	0	Soleymani et al. (2011)
Lake Kyoga Basin, Uganda	Rainfed	0, 30, 60, 90	2310	90	Kaizzi et al. (2012)
Kansas, USA	Rainfed	0, 33, 66, 100	7764	66	Kaufman et al. (2013)
Tabriz, Iran	Irrigated	40, 80, 120	7942	120	Zand et al. (2014)
Northern Ethiopia	Rainfed	0, 50, 100, 150	2083	100	Gebrelibanos and Dereje (2015)
Lake Tana Basin, Ethiopia	Rainfed	0, 23, 41, 64, 87	4046	87	Melaku et al. (2018)
Guinea Savannah, Ghana	Rainfed	0, 30, 60, 90, 120, 150	418–1276	90	Kugbe et al. (2019)
Sotuba, Mali	Rainfed	0, 89, 178	4577–6611	178	Dembele et al. (2021)
Kansas, USA	Rainfed	0, 22, 45, 67	4461	67	Obour et al. (2022)

Table 2.2 Yield optimization of forage sorghum in diverse ecological conditions under different N fertilization regimes

Region	Rainfed/ irrigated	N fertilization rates (kg ha^{-1})	Optimum Dry Matter Yield (t ha^{-1})	Optimum N rate (kg ha^{-1})	Reference
Adana, Turkey	Irrigated	0, 60, 90, 120, 180	13.16	120	Arslangiray (1998)
Faisalabad, Pakistan	Rainfed	0, 50, 100, 150	22.5	150	Ayub et al. (2002)
Khorramabad, Iran	Irrigated	0, 100, 150	13.71	100	Joorabi et al. (2015)
Mandya, India	Irrigated	15, 30, 45, 60	17.04	30	Somashekar et al. (2015)
Islamabad, Pakistan	Irrigated	0, 60, 120	6.9–23.1	120	Sher et al. (2016)
Oklahoma, USA	Rainfed	0, 84, 168, 252	11.1–16.3	84	Haankuku et al. (2014)
Edirne, Turkey	Irrigated	0, 40, 80, 120, 160	41.86	160	Atalay (2019)
Gujarat, India	Rainfed	60, 80, 100	11.28	80	Chaudhary et al. (2018)
North China	Irrigated	60, 120, 240	8.6	60	Tang et al. (2018)
Kansas, USA	Rainfed	0, 28, 56, 84, 112, 140	6.01–7.11	84	Holman et al. (2019)
Sharkia, Egypt	Irrigated	35, 70, 105	19.3–32.3	105	Ziki et al. (2019)

promote plant growth in later stages. In irrigated conditions, maximum 50% of the total fertilizer N is broadcasted preplant or at planting, then the remainder is applied as side-dress by splitting it once or several times in the form of urea, ammonium nitrate, or ammonium sulfate. Finally, this N fertilization regime not only increases the yield and protein content, but also limits NOx emission, nitrate leaching, and ammonia volatilization (Junejo et al. 2013; Pan et al. 2016).

N utilization is maximum in sorghum between five-leaf and booting stages when approximately 65–70% of total N uptake is completed. For this reason, sufficient N fertilizer should be supplied to the crop during this period for both supporting the rapid growth and a good pollination that will ensure adequate grain formation and filling (Anonymous 2019). In this case, two split top N applications after five-leaf stage and prior to booting stage may be appropriate to achieve high grain yields from sorghum. However, top dressing of N fertilizer in equal amounts shortly after each cut considerably increases forage yield and nutritive value of multi-cut forage sorghum varieties (Iptas and Brohi 2003; Turgut et al. 2005).

Table 2.3 Yield optimization of sweet sorghum in diverse ecological conditions under different N fertilization regimes

Region	Rainfed/ irrigated	N fertilization rates (kg ha^{-1})	Optimum sugar yield (kg ha^{-1})	Optimum N rate (kg ha^{-1})	Reference
Midwestern USA	Rainfed	0, 56, 112, 168, 224	1217	112	Maw et al. (2016)
Uttarakhand, India	Rainfed	0, 60, 120, 180	2130	120	Singh et al. (2014)
Patancheru, India	Irrigated	0, 63, 90, 150	5300	90	Kurai et al. (2015)
North China	Irrigated	60, 120, 240	3089	120	Tang et al. (2018)
Quebec, Canada	Rainfed	0, 40, 80, 120, 160	2940	80	Thivierge et al. (2015)
Adana, Turkey	Irrigated	100, 150, 200	1402	150	Nazli et al. (2020)
Isfahan, Iran	Rainfed	0, 42, 84	2980	84	Almodares et al. (2009)
Florida, USA	Irrigated	45, 90, 135, 180	5820	45	Erickson et al. (2012)
Missouri, USA	Irrigated	0, 22, 45, 67, 90, 112, 134	2900–9600	45	Holou and Stevens (2012)

2.4.3 Potassium (K)

Potassium (K) is an essential element for plant growth and development and is the most abundant cation in plants, making up 3–5% of a plant's total dry weight (Marschner 1995; Amasiab et al. 2023). K plays a vital role in numerous metabolic processes including water regulation, enzyme activation, nutrient translocation, root growth, cell expansion, stomatal conductance, photosynthesis, osmotic adjustment, and protein synthesis (Dahmardeh et al. 2015; Chaudhary et al. 2018; Rehman et al. 2022; El-Mageed et al. 2023). K deficiency in sorghum is characterized by yellow or necrotic leaf margins beginning at the lower leaves and spreading to the upper leaves (Bankole 2022; Nyoni et al. 2022).

On the other hand, K significantly increases plant resistance to abiotic stress. Under water-deficit conditions, K fertilization increases plant resistance to drought via regulation of osmotic pressure, maintenance of turgor potential, stimulation of root growth, reduction of transpiration, and enhancement of photosynthetic rate (Umar 2006; Rohbakhsh 2013; Abdelhameid 2020). Additionally, a sufficient supply of K can protect the crop from oxidative damage of salinity stress (El-Mageed et al. 2023).

K is usually applied preplant—broadcast or preplant—banded, at the rate of 10–25 kg/ha based on the available potassium content in soil. Several potassium sources commonly used in sorghum cultivation are potassium chloride, potassium sulfate, potassium nitrate, and potassium oxide (Whitney 1998). As the most K sources have a high salt index, their direct contact with sorghum seed should be avoided.

2.4.4 Other Minerals

Adequate sulfur (S) supply is essential for high yield potential in grain sorghum because it allows N to be used more effectively by the plant (Bean 2019). Sulfur is taken up by the plants in the form of sulfate, which is mostly accumulated in subsoils. Therefore, the plants usually contain low levels of sulfur in the early vegetative stage, while the sulfur level increases concurrent with the root extension and elongation (Espinoza and Ross 2003). Sorghum plants grown on soils with low organic matter may suffer from S deficiency. On these soils, sufficient S should be supplied to the plants as preplant—broadcast or preplant—banded along with the essential nutrients (N, P, and K). For the optimum grain yield, fertilization rates of N to S ratio should be kept at approximately 15 to 1 (Bean 2019).

Iron (Fe) deficiency is most likely to occur on highly calcareous or saline-sodic soils with low organic matter and high pH level (above 7.8) (Whitney 1998; Wortmann et al. 2013). Fe chlorosis in sorghum, which is evident as yellow stripes on leaves, begins from the upper leaves and continue to the lower leaves (Wortmann et al. 2013). Foliar sprays of Fe and manure application are the most common and effective methods used to ameliorate the Fe chlorosis in sorghum (Wortmann et al. 2013).

Zinc plays a critical role in many metabolic processes that are vital for plant development, such as enzyme activation, protein and chlorophyll synthesis, ion transport, and carbohydrate formation. Therefore, poor Zn nutrition is one of the major considerations leading to significant yield and quality losses in grain sorghum (Saleem et al. 2022). Zn deficiency in sorghum has been generally observed on sandy and heavy clay soils with low organic matter content and high pH/or high P levels (Anonymous 2017). Additionally, soil erosion, root diseases, removing of topsoil, and flooding can also lead to reduction of Zn availability in soil. In contrast, the presence of *Arbuscular mycorrhizae* (AM) fungi in the soil increases uptake of Zn by plants (Wortmann et al. 2013; Anonymous 2017). As with S, Zn sources such as zinc chelates, zinc oxides, zinc sulfate, or ammoniated zinc solutions are generally applied as a basal fertilizer in conjunction with P and K by broadcasting or banding (Wortmann et al. 2013; Anonymous 2017).

Soil acidification is a natural process, but it can be accelerated by several agricultural practices including application of N- and S-based chemical fertilizers, monoculture farming, nitrate leaching, and growth of legumes (Wortmann et al. 2013; Goulding 2016). Liming is the most common and oldest management practice used to eliminate the soil acidification and related problems in agriculture (Prescott et al. 2021). As sorghum grows well in the pH range 5.5–8.5, liming application is recommended when the soil pH is below 5.5 to maintain the productivity (Wortmann et al. 2013; Ajeigbe et al. 2020).

2.5 Weed Management

Weeds compete sorghum plants for light, water, and nutrients, mostly in the early development stages, that may result in yield losses up to 70% (Idziak et al. 2013; Pimentel et al. 2019). Besides, some weed species serve as hosts of pest and diseases, thus leading to reduction in crop quality as well as the yield (Idziak et al. 2013; Nyoni et al. 2022). Weed control should be started immediately after the harvest of the previous crop. At least 1 month before sorghum planting, a control plan should be created by evaluating weed species present and their relative abundance in the field (Regehr 1998). The efficiency of a weed control is highly dependent on environmental conditions, cultivar, preferred method (cultural or chemical), and intensity and biology of weeds (Idziak et al. 2013; Kaczmarek 2017).

There are a wide range of broadleaf (pigweeds, common milkweed, prickly burweed, little hogweed, witchweed, etc.) and grass (foxtails, shattercane, proso millet, Bermuda grass, johnsongrass, etc.) weed species that infest sorghum worldwide (Pandian et al. 2022). Among these, johnsongrass (*Sorghum halepense* L.) and witchweed (*Striga hermonthica* L.) are the most troublesome weeds due to their high invasiveness. As johnsongrass has been genetically very similar to cultivated sorghum varieties, there have been currently no registered herbicide that selectively kills johnsongrass in sorghum field (Anonymous 2017). Therefore, post-emergence control of johnsongrass must be based on only mechanical or hand hoeing. On the other hand, crop rotation and repeated tillage are the most effective methods to control johnsongrass before planting. Witchweed is a very common parasitic weed in Sub-Saharan Africa that is usually carried by water, wind, and animals, and germinates in the roots of the host plant (Mbuvi et al. 2017). Witchweed consumes water and nutrients of infested sorghum plant, thus resulting in yield losses of up to 100% (Muchira et al. 2021). Sorghum farmers in Sub-Saharan Africa traditionally control witchweed by hoeing, crop rotation, and intercropping (Muchira et al. 2021).

An integrated approach including crop rotation, herbicide application, and tillage should be adopted for the effective weed management in sorghum cultivation. Although preplant and post-emergence tillage operations control weeds to a large extent, it does not provide sufficient protection compared to herbicide application (Regehr 1998). Herbicides are divided into three groups according to time they are applied: preplant, pre-emergence, and post-emergence (Nyoni et al. 2022).

Planting sorghum on a weed-free soil not only reduces weed competition in early development stage, but also decreases costs by reducing pre-emergence and post-emergence herbicide use. If the field is invaded by particularly perennial weeds such as johnsongrass and hemp dogbane prior to planting, tillage alone will likely be insufficient to remove them effectively. In this case, application of glyphosate or its combination with 2.4–D or Dicamba about 2 weeks before planting can provide a very effective weed control (Barber et al. 2015).

Atrazine is the most widely used herbicide in sorghum fields around the world, both as pre-emergence and post-emergence, that effectively controls most of the broadleaf weeds, but has a slight activity against grass weeds (Pimentel et al. 2019). Therefore, atrazine is often the only herbicide applied in sites where grass weeds are poorly existing. However, increasing resistance of weed species to atrazine is an emerging problem in most sorghum cultivation areas. On the other hand, high pre-emergence application of atrazine (>1.2 kg/ha) can cause significant injury and delay seedling development, especially in cool and wet conditions (Barber et al. 2015). In this context, combined application of atrazine with S-metolachlor at planting may reduce phytotoxicity risk caused by atrazine and manage weeds more effectively, particularly in sites where annual grass species are dense (Anonymous 2017; Pimentel et al. 2019). If S-metolachlor is to be applied pre-emergence, seeds should be treated with a safener containing active ingredient of fluxofenim to avoid injuries associated with the phytotoxic effects of S-metolachlor (Głab and Sowinski 2019).

If weeds are still dense in the field despite the preplant and pre-emergence herbicide applications, they must be controlled till three-leaf stage in order to prevent significant yield losses (Anonymous 2017; Kaczmarek 2017). Since herbicides used safely in sorghum fields to control grass species have not evident activity on emerged weeds, post-emergence herbicides have been generally applied to control broadleaf weed species. Atrazine, 2,4-D, dicamba, prosulfuron, and Simazine are widely used post-emergence broadleaf herbicides in sorghum fields (Barber et al. 2015).

2.6 Insect Management

Sorghum is infested by nearly 150 insect pests from emergence through the harvest, which cause an estimated loss of 1 billion US$ annually (ICRISAT 1992; Sharma et al. 2006; Aruna and Padmaja 2009). The amount of economic loss varies depending on the insect species, density of the infestation, stage of development, and environmental factors (Regehr 1998). In order to minimize losses in sorghum caused by insect attack, an integrated pest management should be adopted including chemical control, natural enemies, host plant resistance, and proper agronomic practices (Sharma et al. 2006; Aruna et al. 2011).

2.6.1 Sorghum Shoot Fly

Sorghum shoot fly (*Atherigona soccata*) is one of the most destructive pests in Asia, Africa, and the Mediterranean Europe causing economic losses of approximately US$ 274 million annually (Dhillon et al. 2005; Sharma et al. 2006; Chamarthi 2008; Mohammed et al. 2016). In India, estimated yield losses caused by shoot fly damage ranged between 50% and 100% in grain sorghum, and 45% and 68% in forage sorghum (Satish et al. 2009; Aruna et al. 2011; Mohammed et al. 2016). The female fly lays white elongated cigar-shaped eggs on the lower surface of the leaf, parallel to leaf midrib (Padmaja et al. 2010; Aruna et al. 2011; Mohammed et al. 2016). After egg hatching, the larvae crawl to the plant whorl and move downward between the folds of the young leaves until they reach the growing point (Dhillon et al. 2005). After the larvae eat the growing point, the central leaf is wilted and later dried up, which is called as "deadheart syndrome" (Chamarthi et al. 2011; Ajeigbe et al. 2020). When the central leaf dies, the plant may produce axial tillers to survive having lower productivity, but if infestation continues, they are also damaged and cannot produce any grain (Satish et al. 2009). The early-sown crops can escape from the shoot fly infestation, but late-sown crops are mostly damaged (Kumar et al. 2008). The shoot fly attacks sorghum seedlings during 5–30 days after emergence (Aruna et al. 2011; Prasad et al. 2015; Mohammed et al. 2016). Although sorghum is the most important host of shoot fly, it may also infest several grass species such as johnsongrass (*Sorghum halepense* L.), barnyard grass (*Echinochloa colonum* L.), maize (*Zea mays* L.), kodo millet (*Paspalum scrobiculatum* L.), pearl millet (*Pennisetum glaucum* L.), and lemongrass (*Cymbopogon* spp.) (Baliddawa and Lyon 1974; Malipatil 2008). Yield losses associated with shoot fly infestation can be minimized through combination of integrated pest management components involving high seeding rate, timely planting, using resistant cultivars, effective control of host weeds, natural enemies, and application of proper insecticides (Sharma et al. 2005; Riyazaddin et al. 2015; Mohammed et al. 2016). It is not always possible to adjust the planting time to avoid shoot fly damage because it is largely dictated by rainfall situation (Kumar et al. 2008; Chamarthi et al. 2011). Although several insecticides such as cypermethrin, carbofuran, endosulfan, thionex, and carbaryl have currently been applied a to limited extent to control shoot fly, their high cost and toxicity hazards limit their widespread use in sorghum fields (Aruna and Padmaja 2009; Lubadde et al. 2019; Ajeigbe et al. 2020; Nyoni et al. 2022). Therefore, host plant resistance stands out as the most effective component in minimizing the economic losses caused by shoot fly because it does not involve any extra cost and has no negative effect on environment (Dhillon et al. 2005). A number of genotypes resistant to this pest individually have been identified, but the levels of resistance are low to moderate (Prasad et al. 2015).

2.6.2 Stem Borers

A number of stem borer species have been identified to cause great yield losses in sorghum so far such as African maize borer (*Busseola fusca* (Fuller)), European maize borer (*Ostrinia nubilalis* Hbn.), pink stem borer (*Sesamia nonagrioides* Lef. Badolato), and spotted stem borer (*Chilo partellus* Swinhoe). Of these, the spotted stem borer is the predominant species in Asia and Africa causing the economic losses of nearly US$ 334 million annually (Sharma et al. 2007; Kamala et al. 2012; Vashisth et al. 2022). Early instar larvae of the spotted stem borer begin to feed in whorl soon after hatching and lead to irregularly shaped pinholes that later develop into elongated lesions on the leaves (Prasad et al. 2015). The older larvae enter the apical meristem and reach the growth point and then cut it off, causing deadheart syndrome, as the shoot fly does (Singh et al. 2012b). Third instar larvae feed inside the stem and cause stem tunneling (Singh et al. 2012b). If the tunnels reach the peduncle, heading may be prevented. As a result, the injuries of the stem borers on the plant occur as leaf feeding, deadhearts, stem tunneling, and exit holes (Bardner and Fletcher 1974; Singh et al. 2012b). Extensive damage can reduce plant vigor and photosynthetic efficiency, and juice quality; delay or prevent flowering completely; make the plant susceptible to lodging and diseases; and eventually cause considerable yield losses (Prasad et al. 2015; Nyoni et al. 2022; Vashisth et al. 2022). Stem borer infestation commences nearly 2 weeks after seedling emergence and continues until the harvest (Singh et al. 2012b; Kamala et al. 2012). Control of stem borers can be achieved by destruction of stover to kill overwintering larvae; deep ploughing; rotating or intercropping of sorghum with non-host crops; using biocontrol agents such as parasitoids, predators, and pathogens; spraying insecticides; and planting resistant cultivars (Lubadde et al. 2019; Nyoni et al. 2022; Vashisth et al. 2022). Several insecticides including active ingredients of endosulfan, lambda-cyhalothrin, carbaryl, and trichlorfon have been commonly used to chemically control stem borers in sorghum (Ajeigbe et al. 2020; Nyoni et al. 2022). However, insecticides unfortunately do not provide adequate protection because they mostly cannot reach the larvae that live inside the leaf whorls and the stem (Vashisth et al. 2022). As with sorghum shoot fly, host plant resistance is considered the most effective method to control stem borers. Several genotypes that are resistant to the spotted stem borer have been improved, but the levels of resistance are low to moderate (Prasad et al. 2015). Nonetheless, wild relatives of sorghum, which have proven to show high levels of resistance to spotted stem borer, are promising for breeding programs to develop resistant varieties (Kamala et al. 2012). The host plant resistance is associated with the production of secondary plant metabolites and defensive proteins as soon as insect oral secretion is started and plant cell is injured (Vashisth et al. 2022). Fast-growing, large internode length, short peduncles, long and thin stems, early panicle initiation, and yellowish-green leaves with high trichome density have been reported to be the major plant characters associated with resistance to stem borer (Vashisth et al. 2022).

2.6.3 Sorghum Midge

Sorghum midge (*Stenodiplosis sorghicola* Coquillett) is widely distributed in grain sorghum cultivation areas of Uganda, Sudan, Nigeria, and Australia (Jordan et al. 1998; Sharma et al. 1999; Lubadde et al. 2019; Ajeigbe et al. 2020). The 30–120 eggs laid by the female sorghum midge in the flower spikelets hatch in just 2–3 days (Sharma 1997; Lubadde et al. 2019). The white larvae move on the developing ovary and feed immature seeds, thus preventing seed kernel development especially in late-sown crops (Anonymous 2017; Lubadde et al. 2019). Its rapid development cycle allows 9–12 generations to occur in a growing season, which may cause up to 100% grain loss because of high populations and infestations (Anonymous 2017; Lubadde et al. 2019; Ajeigbe et al. 2020). Early planting of resistant varieties is most effective method to control sorghum midge. Additionally, crop rotation, fallowing, and controlling the host weeds such as johnsongrass and Sudan grass are considered the alternative options to reduce sorghum midge damage (Lubadde et al.

2019). Several insecticide applications are needed to reduce sorghum midge populations in highly infected late-sown crops (Ajeigbe et al. 2020).

2.6.4 Cutworms

Cutworm species (*Agrostis* spp.) damage the sorghum plant by cutting young seedlings and feeding on the leaves (Anonymous 2017; Nyoni et al. 2022). Adults lay their eggs on the stems and leaves of the plants or soil singly or in groups. A female cutworm can lay 1500–2800 eggs in her life cycle. The larvae hatching from eggs cause damage by eating the fresh leaves and shoots of the plants in the early periods (Anonymous 2008). In the later periods, they sever the stems of young seedlings at or near soil surface, causing the plants to dry and collapse (Anonymous 2017; Nyoni et al. 2022). The larvae stay curled below the soil surface or at the bottom of the plants during the day, but feed at night by moving to the ground level (Anonymous 2008). Cutworms can be controlled by weed management, seed coating, and insecticide application (Nyoni et al. 2022).

2.6.5 Armyworms

The larvae of armyworm species (African armyworm, fall armyworm, beet armyworm, etc.) damage the sorghum plants by feeding on leaves generally in the early vegetative stage. Although they are occasional pests of sorghum, high infestation rates associated with heavily rains following drought periods may cause serious economic damage (Ajeigbe et al. 2020). Except for day feeding once, most armyworm species feed at night but hide under clods of soil or at the base of plants during the day (Anonymous 2017).

African armyworms (*Spodoptera exempta* Walker) attack the leaf in swarms and leave the midrib on the lamina by cutting and chewing (Nyoni et al. 2022).

Since fall armyworms (*Spodoptera frugiperda*) feed both on the whorl and panicle, it causes damage to sorghum not only during the vegetative period but also during the grain filling stage. Young larvae cause irregularly shaped holes in emerging leaves, while older larvae feed on newly developing grains (Nyoni et al. 2022).

Early instar larvae of beet armyworm (*Spodoptera exigua*) eat the upper and lower epidermis tissues of leaf and finally leave a membrane on it. As the larvae develop, they also consume the membrane, leaving only the leaf veins. If they are not controlled, they may consume the entire leaf in later periods, which results in grain yield losses up to 100% (Anonymous 2008).

Since the pupae of armyworm species overwinter under the ground, deep ploughing prior to planting is considered an effective control method. Additionally, crop rotation, intercropping, removal of host plants, and conservative agriculture can reduce risk of infestation by armyworms. If the plant is largely infested despite all these precautions, one of recommended insecticides (lufenuron, carbaryl, malathion, permethrin, and lambda-cyhalothrin) must be applied immediately to the field (Nyoni et al. 2022).

2.6.6 Aphid

Corn leaf aphid (*Rhopalosiphum maidis*), greenbug (*Schizaphis graminum*), and sugarcane aphid (*Melanaphis sacchari*) are the major aphid species damaging to sorghum. Corn leaf aphid has a dark green color and usually develops in the whorl of the plant but may also be on the underside of leaves, stems, or in panicles (Wright et al. 2006; Mostafa et al. 2022). Corn leaf aphid sucks plant juice, causing yellow spots on the leaves. Besides, they rarely produce honeydew, which causes the grain to become sticky, resulting in blockages and breakdowns of harvesters (Anonymous 2017). Nonetheless, control is seldom necessary as sorghum varieties are generally highly tolerant to corn leaf aphid. However, if the population is high enough to cause significant yield losses, insecticides such as dimethoate, mercaptothion, and acetamiprid must be sprayed (Nyoni et al. 2022).

Greenbugs are lime-green or greenish-yellow aphids that generally colonize underside of the leaves from seedling stage to maturity (Wright et al. 2006). Greenbug secretes saliva that is very toxic to sorghum, as they feed on leaves, resulting in breakdown of the tissues and severe chlorosis (Michauld 2014). Besides, uncontrolled colonies may also cause flower sterility. However, planting insecticide-treated seeds largely prevents greenbug damage.

Sugarcane aphid has been one of the most destructive pests of sorghum in the USA (Brewer et al. 2016; Knutson et al. 2016; Filbert 2021). Similar to greenbugs, they also colonize underside of leaves and suck the plant sap but do not inject a toxin (Knutson et al. 2016). Feeding causes symptoms such as purpling of the leaves, chlorosis, and necrosis as the leaf tissue dies (Knutson et al. 2016; Filbert 2021). Infestations before the booting stage can cause plant death, while infestations during pollination can cause significant yield loss due to insufficient grain filling (Filbert 2021). Infestations in forage sorghum kill leaves, stunt plant growth, and decrease forage yield and quality (Knutson et al. 2016). On the other hand, sugarcane aphid largely excretes sticky honeydew, as corn leaf aphid does, that complicates separation of grain from rest of the plant during the harvest (Filbert 2021). Additionally, black sooty mold, a saprophytic fungus, grows on the honeydew that inhibits light absorption needed for photosynthesis and reduces the effectiveness of insecticides and harvest aid chemicals (Knutson et al. 2016; Brewer et al. 2016). Control of sugarcane aphid can be achieved by treatment of seed with insecticide using biocontrol agents such as lady beetles, syrphid flies, green and brown lacewings, and aphelinid parasitoids; application of selective insecticides; and planting resistant cultivars (Knutson et al. 2016; Brewer et al. 2016). Resistant genotypes have genes that effectively protect them against sugarcane damage or rapid population increases. Resistant cultivars can withstand or recover quickly from insect attack thanks to their genes that effectively protect them against sugarcane aphid damage or rapid population increases (Knutson et al. 2016).

2.7 Disease Management

2.7.1 Anthracnose

Anthracnose is the most common and destructive disease of sorghum worldwide, caused by the fungus *Colletotrichum sublineola* (Xavier et al. 2017; Ajeigbe et al. 2020). The fungus can attack all parts of the plant including the stem, panicle, and leaves, and grain yield losses associated with anthracnose can reach up to 50%, especially in warm and humid climates (Xavier et al. 2017; Stutts and Vermerris 2020). The fungus leads to both leaf spot and stalk rot, which are called leaf anthracnose and stalk red root, respectively (LeBeau et al. 1951; Xavier et al. 2017). Leaf anthracnose appears as small, circular to elliptical, dark reddish-purple to tan-colored spots on both surfaces of the leaves. As the disease progresses, the spots increase in number and size, their centers turn grayish to dark straw, and numerous acervuli with prominent hairs appear (LeBeau et al. 1951; Anitha et al. 2020). Stalk red root appears with the development of cankers on the panicle, which will later turn tan to blackish-purple (Isakeit 2020). Infected stem when split open shows discoloration, which may be continuous over a large area or, more generally, discontinuous giving the stem a marbled appearance (Nyoni et al. 2022). Discoloration type seems to be largely a varietal characteristic and may vary from yellowish tan to reddish-orange and from reddish-purple to blackish-purple (LeBeau et al. 1951). The main strategies for the control of anthracnose in sorghum are crop rotation with legume crops, removal of crop residues and alternate hosts, host plant resistance, natural enemies, and adoption of chemical control (Xavier et al. 2017; Anitha et al. 2020). Chemical control includes seed treatment prior to planting with fungicides such captan, thiram, or mefenoxam, and foliar application of fungicides such as carbendazim, captafol, or mancozeb during the vegetative growth period (Anitha et al. 2020).

2.7.2 Downy Mildew

Downy mildew is one of the most devastating sorghum diseases caused by the fungus *Peronosclerospora sorghi*. It is generally prevalent in tropical and subtropical regions that can cause grain yield losses between 12% and 78% in India, the USA, and Africa (Anitha et al. 2020). Infected seedlings seem stunted and chlorotic and have pale yellow or light-colored striping on their leaves (Anitha et al. 2020; Nyoni et al. 2022). Under cool, humid conditions the fungus may produce a white downy growth on the lower leaf surface (Anitha et al. 2020). As the plant ages, the chlorotic tissue stripes and leaves become shredded that may cause plant death prematurely (Anitha et al. 2020; Nyoni et al. 2022). Since downy mildew is a soil-borne disease, several cultural control methods such as deep tillage, crop rotation, destruction of crop residues, and removal of host weeds markedly reduce economic losses arising from fungus infection (Anitha et al. 2020). However, these measures are generally inadequate in cases where infection is intense. In this case, seed treatment with fludioxonil or mefenoxam followed by foliar spray of metalaxyl can provide a sufficient protection (Anitha et al. 2020).

2.7.3 Seedling Blight

Seedling blight, also known as damping off, is mostly caused by soil-borne pathogens such as *Pythium*, *Fusarium*, *Aspergillus*, and *Rhizoctonia* independently or in their combinations (Ajeigbe et al. 2020). The likelihood appearance of these pathogens depends largely on weather conditions at planting. *Fusarium* infections usually occur in warm and humid conditions at, or shortly after, planting. *Fusarium* causes root rot during seedling stage, which reduces plant growth and the overall plant stand (Anonymous 2019). In cool and wet conditions, sorghum may be more vulnerable to *Pythium* spp. because of slower seed germination, delayed emergence, and retarded root growth (Anonymous 2005). Infection of *Pythium* spp. may prevent seed germination and cause root damage that result in poor seedling vigor, death, and sparse or irregular stands (Anonymous 2005; Isakeit 2020). In order to minimize seedling diseases in a sorghum field, early planting should be avoided and the seeds should be treated with fungicides before planting (Anonymous 2005).

2.7.4 Ergot

Ergot is caused by the fungus *Sphacelia sorghi*, which is also known as sugary disease producing honeydew on the affected panicle (Lubadde et al. 2019). The infection occurs during the flowering stage and is encouraged by warm temperatures (15–30 °C) and high humidity conditions (Anonymous 2005). The fungus attacks flower ovaries, turning them into a whitish, fungal mass (Anitha et al. 2020). Infected flowers release an amber-colored, sticky fluid known as "honeydew," which drips onto the leaves and ground (Nyoni et al. 2022). Ergot may significantly decrease grain yield of sorghum because of poor grain set and cause harvesting difficulties with the sticky honeydew clogging machinery (Anonymous 2005). Additionally, ergots are toxic to livestock, especially cows, sows, and beef cattle (Anonymous 2017). Control of ergot disease involves: adjusting planting so that flowering occurs in low humidity conditions, removal of crop residues, deep ploughing, crop rotation, planting resistant varieties, seed treatment with thiram or captan, and foliar application of fungicides such as mancozeb and carbendazim (Lubadde et al. 2019).

2.7.5 Smut

Smut is the most devastating group of sorghum diseases around the world, causing serious economic losses in both local and improved varieties. There are four types of smut diseases that have been identified and are listed as follows: covered kernel smut (*Sporisorium sorghi*), head smut (*Sporisorium reilianum*), loose smut (*Sporisorium cruentum*), and long smut *(Sporisorium ehrenbergii*) (Ajeigbe et al. 2020; Anitha et al. 2020).

Kernel smut affects the panicle. Plants with coated kernel smut can be identified before the head emerges. It shows up as panicles that have formed cone-shaped capsules. Spore-containing capsules may cover a portion or the entire panicle. The damaged panicles lose all of their kernels, which are then replaced with a dark brown powdery substance (Lubadde et al. 2019; Nyoni et al. 2022).

Symptoms of head smut appear at the booting stage when the head is replaced by a large smut gall covered by a whitish-gray membrane. Before the head emergence, the membrane ruptures and then releases a mass of dark brown to black smut spores (Anitha et al. 2020; Nyoni et al. 2022). Affected plants may tiller excessively and become stunted (Isakeit 2020). On the other hand, the disease can lead to partial infertility (Anonymous 2005). The spores of the disease are easily dispersed by wind and rain, infecting more plants (Nyoni et al. 2022).

Loose smut manifests with smut sori or galls covered by a thin, gray membrane that replaces individual kernels (Anitha et al. 2020). Infected plants differ from healthy plants in that they are stunted, have weaker stems, frequently tiller more, and can flower up to 2 weeks earlier (Isakeit 2020).

Long smut is a widespread disease in Africa and India. The disease appears with the development of elongated, cylindrical, and slightly curved spore sacs (sori), which is covered by thick creamy brown covering membrane, called peridium. Infection occurs only during the booting stage (Prom et al. 2017; Anitha et al. 2020).

In order to effectively control smut, certified seeds that are disease-free and dressed with fungicides (captan, thiram, mefenoxam, etc.) must be primarily preferred (Anitha et al. 2020; Nyoni et al. 2022). Crop residues carrying smut galls should be burned or destroyed immediately before the spores are spread around. Since the smut is a soil-borne disease, sorghum should not be planted in the infested field before 4 years (Ajeigbe et al. 2020). During this period, crop rotation with non-host crops, especially legumes, can be effective in removing the disease agents from the field (Lubadde et al. 2019).

2.7.6 Rust

Rust is caused by the fungus *Puccinia purpurea* and first appears as small purple-red or tan spots that subsequently grow to form elongated, raised pustules on both the surfaces of leaves (Anonymous 2005; Lubadde et al. 2019). These pustules rupture and release the powdery mass of reddish-brown spores that are commonly dispersed by wind (Lubadde et al. 2019). As a result of excessive rust infections, leaves are destroyed and lodging occurs (Anonymous 2005; Anitha et al. 2020). The incidence of rust is particularly prevalent in late-sown crops under cool and humid conditions (Anonymous 2005). It is widely spread over all sorghum-growing areas of the world and can cause yield losses up to 65% (Anitha et al. 2020). Control of rust can be achieved by early sowing, field sanitation including removal of crop residues and weed hosts before the rains begin, crop rotation with legume crops to break the disease lifecycle, the use of disease-free certified seeds and tolerant varieties, seed treatment with contact fungicides, and foliar sprays of fungicides such as hexaconazole, oxycarboxin, difenoconazole, and mancozeb (Lubadde et al. 2019; Anitha et al. 2020).

2.7.7 Leaf Blight

Leaf blight, caused by the fungus *Exserohilum turcicum*, first appears as small flecks that later grow into long, thin lesions with a deep red edge and a brownish center. Severe infections may result in complete wilting of the leaves (Anitha et al. 2020). The disease is common in India and South Africa and may cause yield losses as high as 50% under moderate temperatures and wet conditions (Anitha et al. 2020; Nyoni et al. 2022). Leaf blight can be effectively controlled by planting resistant hybrids, crop rotation with non-host crops, seasonal tillage, destruction of crop residues and host weeds, seed dressing with captan or thiram, and foliar sprays of fungicides such as hexaconazole, azoxystrobin, difenoconazole, and tebuconazole (Anonymous 2005; Anitha et al. 2020; Nyoni et al. 2022).

2.8 Harvest

Depending on the variety and environmental conditions, sorghum becomes ready for harvesting in 90–140 days after planting (Lubadde et al. 2019). For the long-term and safe storage, harvesting grain sorghum should be done when the moisture content of grains falls to 12–14%. But if drying facilities are available, harvesting may commence at 20–25% grain moisture, followed by drying to 12–14% moisture (Avcioglu et al. 2009; Anonymous 2017). Application of preharvest desiccants such as diquat and glyphosate allows earlier harvest by accelerating drying of grains because it stops further water use by the crop (Taylor 1998; Anonymous 2005). Besides, this application kills immature tillers, weeds, and green foliage, thereby decreasing the quantity of green vegetation passing through the combined harvester. Crop desiccants should be applied at physiological maturity when the grain moisture is between 25% and 35% (Nyoni et al. 2022).

Forage sorghum can be grazed directly or baled for hay and ensiled for silage. Toxicity of hydrocyanic acid (HCN) is the most serious problem in grazing. The HCN content of the young plants increases until they reach 50–60 cm in height, then it tends to decrease. When the plants grow to 100 cm tall, the HCN poisoning generally becomes less important (Avcioglu et al. 2009). HCN poisoning can be also triggered by abiotic stress factors such as drought, waterlogging, and nutrient deficiency (Anonymous 2005). Forage sorghum should be harvested before early head emergence and within dough stage for the production of hay and silage, respectively (Anonymous 2005; Avcioglu et al. 2009; Lyons et al. 2019). On the other hand, multiple harvest per growing season is generally possible when it is grown in warm and temperate regions.

Sweet sorghum has usually been grown for the production of sugar, bioethanol, and silage around the world. Sweet sorghum should be harvested when the sugar yield reaches the maximum level, which is mainly determined by sugar content and juice yield of the crop (Nazli 2020). Optimum harvest time for sweet sorghum varies from soft dough to physiological maturity stages depending on the variety and ecological conditions (Subramanian 2013; Oyier et al. 2017; Teixeira et al. 2017; Nazli 2022).

References

Abdelhameid NM (2020) Effect of mycorrhizal inoculation and potassium fertilization on grain yield and nutrient uptake of sweet sorghum cultivated under water stress in calcareous soil. Egypt J Soil Sci 60(1):17–29

Abunyewa AA, Ferguson RB, Wortmann CS, Lyon DJ, Mason SC, Klein RN (2010) Skip-row and plant population effects on sorghum grain yield. Agron J 102(1):296–302

Adams CB, Erickson JE, Campbell DN, Singh MP, Rebolledo JP (2015) Effects of row spacing and population density on yield of sweet sorghum: applications for harvesting as billets. Agron J 107(5):1831–1836

Ajeigbe HA, Angarawai II, Akinseye F, Inuwa AH, AbdulAzeez T, Vabi MB (2020) Handbook on improved agronomic practices for sorghum production in north east Nigeria. Manual. The U.S. Government's Global Hunger & Food Security Initiative (Feed the Future), Nigeria:1–15

Almodares A, Hadi MR, Ranjbar M, Taheri R (2007) The effects of nitrogen treatments, cultivars and harvest stages on stalk yield and sugar content in sweet sorghum. Asian J Plant Sci 6(2):423–426

Almodares A, Taheri R, Chung M, Fathi M (2008) The effect of nitrogen and potassium fertilizers on growth parameters and carbohydrate contents of sweet sorghum cultivars. J Environ Biol 29(6):849–852

Almodares A, Jafarinia M, Hadi MR (2009) The effects of nitrogen fertilizer on chemical compositions in corn and sweet sorghum. Am Eurasian J Agric Environ Sci 6(4):441–446

Amaducci S, Colauzzi M, Battini F, Fracasso A, Perego A (2016) Effect of irrigation and nitrogen fertilization on the production of biogas from maize and sorghum in a water limited environment. Eur J Agron 76:54–65

Amasiab EO, Eltaib MZ, Abdalla AH (2023) Effect of potassium sulfate fertilizer on the mineral profile of sorghum forage and their uses as ruminant feed. Int J Plant Soil Sci 35(5):164–173

Anderson IC, Buxton DB, Hallam A, Hunter E (1995) Biomass production and ethanol potential from sweet sorghum. In: Leopold Center for Sustainable Agriculture 4. Iowa State University, Ames, IA, pp 97–101

Anitha K, Das IK, Holajjer P, Sivaraj N, Reddy CR, Balijepalli SB (2020) Sorghum diseases: diagnosis and management. In: Sorghum in the 21st century: food–fodder–feed–fuel for a rapidly changing world, pp 565–619

Anonymous (2005) Grain sorghum. Agfact P3.3.5 First Edition. NSW Department of Primary Industries. www.dpi.nsw.gov.au

Anonymous (2008) Sebzelerde bozkurt (*Agrostis* spp.). Zirai Mücadele Teknik Talimatları, Cilt 3, Tarımsal Araştırmalar ve Politikalar Genel Müdürlüğü. pp 263–265. https://www.tarimorman.gov.tr/TAGEM/Belgeler/Teknik%20tal%C4%B1matlar%202008/C%C4%B0LT%203.pdf

Anonymous (2017) Sorghum Northern Region—GrowNotes™. https://grdc.com.au/resources-and-publications/grownotes/crop-agronomy/sorghumgrownotes

Anonymous (2019) Best practices for planting grain sorghum. Pasific seeds. https://www.pacific-seeds.com.au
Arslangiray C (1998) Çukurova Koşullarında İkinci Ürün Olarak Yetiştirilen Bazı Tane Sorgum (*Sorgum bicolor* L.) ve Sorgum x Sudanotu (*Sorgum sudanense* L.) Melez Çeşitlerinde Azot Gübrelemesinin Dane ve Hasıl Verimine Etkisi Üzerinde Araştırmalar. Doktora Tezi, Çukurova Üniversitesi, Fen Bilimleri Enstitüsü, Tarla Bitkileri Anabilim Dalı, Adana
Aruna C, Padmaja PG (2009) Evaluation of genetic potential of shoot fly resistant sources in sorghum (*Sorghum bicolor* (L.) Moench). J Agric Sci 147(1):71–80
Aruna C, Bhagwat VR, Madhusudhana R, Sharma V, Hussain T, Ghorade RB, Khandalkar HG, Audilakshmi S, Seetharama N (2011) Identification and validation of genomic regions that affect shoot fly resistance in sorghum [*Sorghum bicolor* (L.) Moench]. Theor Appl Genet 122:1617–1630
Asif M, Adnan M, Safdar ME, Akhtar N, Khalofah A, Alzuaibr FM (2022) Integrated use of phosphorus and growth stimulant (actibion) improves yield and quality of forage sorghum (*Sorghum bicolor* L.). J King Saud Univ Sci 34(7):102236
Assefa Y, Staggenborg SA (2010) Grain sorghum yield with hybrid advancement and changes in agronomic practices from 1957 through 2008. Agron J 102(2):703–706
Atalay M (2019) Farklı azot dozu uygulamalarının sorgum x Sudan otu (*Sorghum bicolor* (L.) Moench x Sorhgum sudanense (Piper) Stapf) melez çeşitlerinin verim ve bazı kalite özelliklerine etkileri. Tekirdağ Namık Kemal Üniversitesi Fen Bilimleri Enstitüsü, Yayınlanmamış Yüksek Lisans Tezi, Tekirdağ
Avcioglu R, Geren H, Kavut YT (2009) Sorgum sudanotu ve SorgumxSudanotu Melezi. Yem Bitkileri. Buğdaygil ve Diğer Familyalardan Yem bitkileri, Tarım ve Köy işleri Bakanlığı, Tarımsal Üretim ve Geliştirme Genel Müdürlüğü, İzmir, Cilt III, 843s
Ayub M, Nadeem MA, Tanveer A, Husnain A (2002) Effect of different levels of nitrogen and harvesting times on the growth, yield, and quality of sorghum fodder. Asian J Plant Sci 1(4):304–307
Baliddawa CW, Lyon WF (1974) Sorghum shoot fly species and their control in Uganda. PANS Pest Articles News Summaries 20(1):20–22
Bankole JA (2022) Growth and yield of sorghum cultivars as influenced by population density of components of soya bean. Doctoral dissertation, Kwara State University, Nigeria
Bao X, Watanabe M, Wang Q, Hayashi S, Liu J (2006) Nitrogen budgets of agricultural fields of the Changjiang River basin from 1980 to 1990. Sci Total Environ 363(1–3):136–148
Barber T, Scott B, Norsworthy JK (2015) Weed control in grain sorghum. In: Arkansas grain sorghum production handbook, vol 8, pp 1–14
Bardner R, Fletcher KE (1974) Insect infestations and their effects on the growth and yield of field crops: a review. Bull Entomol Res 64(1):141–160
Bean (2019) Fertilizing grain sorghum. Sorghum checkoff. https://www.sorghumcheckoff.com/agronomy-nsghts/fertlzng-gran-sorghum
Bonin CL, Heaton EA, Cogdill TJ, Moore KJ (2016) Management of sweet sorghum for biomass production. Sugar Tech 18:150–159
Borrell AK, Mullet JE, George-Jaeggli B, van Oosterom EJ, Hammer GL, Klein PE, Jordan DR (2014) Drought adaptation of stay-green sorghum is associated with canopy development, leaf anatomy, root growth, and water uptake. J Exp Bot 65(21):6251–6263
Brewer MJ, Bowling R, Michaud JP, Jacobson AL (2016) Sugarcane aphid: a new sorghum pest in North America. Texas A&M Agrilife Extension ENTO-056
Briand CH, Geleta SB, Kratochvil RJ (2018) Sweet sorghum (*Sorghum bicolor* [L.] Moench) a potential biofuel feedstock: analysis of cultivar performance in the Mid-Atlantic. Renew Energy 129:328–333
Burke JJ, Franks CD, Burow G, Xin Z (2010) Selection system for the stay-green drought tolerance trait in sorghum germplasm. Agron J 102(4):1118–1122
Burnham MB, Simon SJ, Lee D, Kent AD, DeLucia EH, Yang WH (2022) Intra-and inter-annual variability of nitrification in the rhizosphere of field-grown bioenergy sorghum. GCB Bioenergy 14(3):393–410

Camacho R, Malavolta E, Guerrero-Alves J, Camacho T (2002) Vegetative growth of grain sorghum in response to phosphorus nutrition. Sci Agric 59:771–776

Causley DC (1990) Effect of minimum tillage, sowing rate, and sowing time on the yield of a sorghum-sudangrass hybrid in the Manawatu, New Zealand. N Z J Agric Res 33(1):15–20

CCCF (2011) Working paper on mycotoxins in sorghum. Joint FAO/WHO Food Standards Codex Committee on Contaminants in Foods (CCCF)

Chamarthi SK (2008) Biochemical mechanisms of resistance to shoot fly, *Atherigona soccata* (Rondani) in sorghum, *Sorghum bicolor* (L.) Moench. Doctoral dissertation, Jawaharlal Nehru Technological University

Chamarthi SK, Sharma HC, Sahrawat KL, Narasu LM, Dhillon MK (2011) Physico-chemical mechanisms of resistance to shoot fly, Atherigona soccata in sorghum, Sorghum bicolor. J Appl Entomol 135(6):446–455

Chaudhary JD, Pavaya RP, Malav JK, Dipika G, Chaudhary N, Kuniya NK, Vina A, Patel I, Jat JR (2018) Effect of nitrogen and potassium on yield, nutrient content and uptake by forage sorghum (*Sorghum bicolor* (L.) Moench) on loamy sand. Int J Chem Stud 6(2):761–765

Choudhary SK, Jat MK, Mathur AK (2017) Effect of micronutrient on yield and nutrient uptake in sorghum. J Pharmacogn Phytochem 6(2):105–108

Cosentino SL, Mantineo M, Testa G (2012) Water and nitrogen balance of sweet sorghum (Sorghum bicolor Moench (L.)) cv. Keller under semi-arid conditions. Ind Crop Prod 36(1):329–342

Da Silva TM, de Oliveira AB, de Moura JG, da Trindade Lessa BF, de Oliveira LSB (2019) Potential of sweet sorghum juice as a source of ethanol for semi-arid regions: cultivars and spacing arrangement effects. Sugar Tech 21:145–152

Dahmardeh K, Rad MRP, Rad MRN, Hadizadeh M (2015) Effects of potassium rates and irrigation regimes on yield of forage sorghum in arid regions. Int J Agron Agric Res 6(4):207–212

Dar RA, Dar EA, Kaur A, Phutela UG (2018) Sweet sorghum-a promising alternative feedstock for biofuel production. Renew Sust Energ Rev 82:4070–4090

De Oliveira LP, Cabral PDS, e Silva FHDL, Neto AR, Silva FG, Pereira LD (2022) Performance and genetic diversity of pre-commercial sweet sorghum hybrids in Central-Western and Southern Brazil. Renew Energy 182:992–997

Dembele JSB, Gano B, Kouressy M, Dembele LL, Doumbia M, Ganyo KK, Sanogo S, Togola A, Traore K, Vaksman M, Teme N (2021) Plant density and nitrogen fertilization optimization on sorghum grain yield in Mali. Agron J 113(6):4705–4720

Dhillon MK, Sharma HC, Singh R, Naresh JS (2005) Mechanisms of resistance to shoot fly, Atherigona soccata in sorghum. Euphytica 144:301–312

Diallo B, Li M, Tang C, Ameen A, Zhang W, Xie GH (2019) Biomass yield, chemical composition and theoretical ethanol yield for different genotypes of energy sorghum cultivated on marginal land in China. Ind Crop Prod 137:221–230

Dweikat (2012) Chapter 19: Sorghum. In: Kole C, Joshi CP, Shonnard DR (eds) Handbook of bioenergy crop plants, pp 524–535. ISBN: 978-1-4398-1684-4

Edwards GE, Franceschi VR, Voznesenskaya EV (2004) Single-cell C4 photosynthesis versus the dual-cell (Kranz) paradigm. Annu Rev Plant Biol 55:173–196

Elangovan M, Prabhakar P, Chandra Sekara Reddy D (2010) Characterization and evaluation of sorghum [Sorghum bicolor (L.) Moench] germplasm from Karnataka, India. Karnataka J Agric Sci 20(4):840–842

El-Mageed TAA, Semida WM, Abdou NM, El-Mageed SAA (2023) Coupling effects of potassium fertilization rate and application time on growth and grain yield of wheat (*Triticum aestivum* L.) plants grown under Cd-contaminated saline soil. J Soil Sci Plant Nutr 23(1):1070–1084

Emeklier HY, Köksoy NF (1997) Sorgum (*Sorgum bicolor* L, Moench)'da Ekim Zamanı ve Bitki Sıklığının Verim Öğelerine Etkisi. J Agric Sci 3(03):20–28

Erickson JE, Helsel ZR, Woodard KR, Vendramini JMB, Wang Y, Sollenberger LE, Gilbert RA (2011) Planting date affects biomass and brix of sweet sorghum grown for biofuel across Florida. Agron J 103(6):1827–1833

Erickson JE, Woodard KR, Sollenberger LE (2012) Optimizing sweet sorghum production for biofuel in the southeastern USA through nitrogen fertilization and top removal. Bioenergy Res 5:86–94

Espinoza L, Ross J (2003) Fertilization and liming. In: Corn production handbook, pp 23–27

Fawusi MOA, Agboola AA (1980) Soil moisture requirements for germination of sorghum, millet, tomato, and celosia 1. Agron J 72(2):353–357

Filbert S (2021) Sugarcane aphids in grain sorghum. Nat Sci Educ 50(1):e20050

Folliard A, Traoré PCS, Vaksmann M, Kouressy M (2004) Modeling of sorghum response to photoperiod: a threshold–hyperbolic approach. Field Crop Res 89(1):59–70

Gebrelibanos G, Dereje A (2015) Nitrogen fertilization effect on grain sorghum (Sorghum bicolor L. Moench) yield, yield components and witchweed (*Striga hermonthica* (Del.) Benth) infestation in Northern Ethiopia. Int J Agric Res 10(1):14–23

Głab L, Sowinski J (2019) Sustainable production of sweet sorghum as a bioenergy crop using biosolids taking into account greenhouse gas emissions. Sustain For 11:3033–3050

Godsey CB, Linneman J, Bellmer D, Huhnke R (2012) Developing row spacing and planting density recommendations for rainfed sweet sorghum production in the southern plains. Agron J 104(2):280–286

Goulding KWT (2016) Soil acidification and the importance of liming agricultural soils with particular reference to the United Kingdom. Soil Use Manag 32(3):390–399

Gupta RK, Rathore N, Singh Y, Singh B (2016) Effect of phosphorus fertilization on its transformations in different soils under dry direct-seeded rice. J Indian Soc Soil Sci 64(3):230–234

Haankuku C, Epplin FM, Kakani VG (2014) Forage sorghum response to nitrogen fertilization and estimation of production cost. Agron J 106(5):1659–1666

Harris K, Subudhi PK, Borrell A, Jordan D, Rosenow D, Nguyen H, Klein P, Klein R, Mullet J (2007) Sorghum stay-green QTL individually reduce post-flowering drought-induced leaf senescence. J Exp Bot 58(2):327–338

Holman JD, Obour AK, Mengel DB (2019) Nitrogen application effects on forage sorghum production and nitrate concentration. J Plant Nutr 42(20):2794–2804

Holou RA, Stevens G (2012) Juice, sugar, and bagasse response of sweet sorghum (*Sorghum bicolor* (L.) M oench cv. M81E) to N fertilization and soil type. GCB Bioenergy 4(3):302–310

Houx JH III, Fritschi FB (2013) Influence of midsummer planting dates on ethanol production potential of sweet sorghum. Agron J 105(6):1761–1768

Hu SW, Wu LM, Persson S, Peng LC, Feng SQ (2017) Sweet sorghum and Miscanthus: two potential dedicated bioenergy crops in China. J Integr Agric 16(6):1236–1243

ICRISAT (1992) Annual report 1991. International Crop Research Institute for Semi-arid Tropics, Patancheru

Idziak R, Skrzypczak W, Waligora H, Woznica Z (2013) The effect of mesotrione applied with adjuvants on weed control efficacy and forage sorghum tolerance. Turk J Agric For 37(3):265–270

Iptas S, Brohi AR (2003) Effect of nitrogen rate and stubble height on dry matter yield, crude protein content and crude protein yield of a sorghum–sudangrass hybrid [*Sorghum bicolor* (L.) Moench× Sorghum sudanense (Piper) Stapf.] in the three-cutting system. J Agron Crop Sci 189(4):227–232

Iptaş S, Yılmaz M (1995) Silajlık Sorgum (*Sorghum Bicolor* (L.) Moench) Ve Sorghum-Sudanotu Melezleri (*Sorghum Bicolor X Sorghum Sudanense* Stapf)'de Farklı Sıra Aralıklarının Bazı Morfolojik Ve Tarımsal Özelliklere Etkisi Üzerine Bir Araştırma. J Agric Faculty Gaziosmanpaşa Univ (JAFAG) 1995(1)

Isakeit T (2020) Diagnosis and management of sorghum diseases. Texas A&M AgriLife Center, Corpus Christi. https://ccag.tamu.edu/wpcontent/uploads/sites/10/2020/12/2010_SORGHUM_DISEASES.pdf

Jankowski KJ, Sokólski MM, Dubis B, Załuski D, Szempliński W (2020) Sweet sorghum-biomass production and energy balance at different levels of agricultural inputs. A six-year field experiment in north-eastern Poland. Eur J Agron 119:126119

Joorabi S, Akbari N, Chaichi MR, Azizi K (2015) Effect of sowing date and nitrogen fertilizer on sorghum (Sorghum bicolor L. var. speed feed) forage production in a summer intercropping system. Cercetări Agronomice în Moldova 48(3):63–72

Jordan DR, Tao YZ, Godwin ID, Henzell RG, Cooper M, McIntyre CL (1998) Loss of genetic diversity associated with selection for resistance to sorghum midge in Australian sorghum. Euphytica 102:1–7

Junejo N, Khanif M, Hanfi M, Yunus W, Dharejo K (2013) Role of inhibitors and biodegradable material in mitigation of nitrogen losses from fertilized lands. Afr J Biotechnol 10:3504–3514

Júnnyor WDSG, Severiano EDC, Silva AGD, Gonçalves WG, Andrade R, Martins BRR, Custódio GD (2015) Sweet sorghum performance affected by soil compaction and sowing time as a second crop in the Brazilian Cerrado. Rev Bras Ciênc Solo 39:1744–1754

Kaczmarek S (2017) A study on *Sorghum bicolor* (L.) Moench response to split application of herbicides. J Plant Protect Res 57:156–157

Kaizzi KC, Byalebeka J, Semalulu O, Alou I, Zimwanguyizza W, Nansamba A, Musinguzi P, Ebenyat P, Hyuha T, Wortmann CS (2012) Sorghum response to fertilizer and nitrogen use efficiency in Uganda. Agron J 104(1):83–90

Kakani VG, Vu JC, Allen LH Jr, Boote KJ (2011) Leaf photosynthesis and carbohydrates of CO_2-enriched maize and grain sorghum exposed to a short period of soil water deficit during vegetative development. J Plant Physiol 168(18):2169–2176

Kamala V, Sharma HC, Manohar Rao D, Varaprasad KS, Bramel PJ, Chandra S (2012) Interactions of spotted stem borer Chilo partellus with wild relatives of sorghum. Plant Breed 131(4):511–521

Karhale MB, Jaybhaye PR, Asewar BV, Shinde PB (2014) Effect of different sowing dates on growth and yield of kharif sorghum hybrids. Karnataka J Agric Sci 24(3):289–293

Kaufman RC, Wilson JD, Bean SR, Presley DR, Blanco-Canqui H, Mikha M (2013) Effect of nitrogen fertilization and cover cropping systems on sorghum grain characteristics. J Agric Food Chem 61(24):5715–5719

Kering MK, Temu VW, Rutto LK (2017) Nitrogen fertilizer and panicle removal in sweet sorghum production: effect on biomass, juice yield and soluble sugar content. J Sustain Bioenergy Syst 7(1):14–26

Khan HZ, Malik MA, Saleem MF (2008) Effect of rate and source of organic material on the production potential of spring maize (*Zea mays* L.). Pak J Agric Sci 45(1):40–43

Kholova J, McLean G, Vadez V, Craufurd P, Hammer GL (2013) Drought stress characterization of post-rainy season (rabi) sorghum in India. Field Crop Res 141:38–46

Kim M, Day DF (2011) Composition of sugar cane, energy cane, and sweet sorghum suitable for ethanol production at Louisiana sugar mills. J Ind Microbiol Biotechnol 38(7):803–807

Kishor K, Kaushik MK, Yadav VK, Gautam P, Chugh A (2017) Effect of fertility levels on yield and yield attribute of different sorghum [*Sorghum bicolar* (L.) Moench] genotypes. J Pharmacogn Phytochem 6(4):541–543

Knutson A, Bowling R, Brewer MJ, Bynum E, Porter P (2016) The sugarcane aphid: management guidelines for grain and forage sorghum in Texas. ENTO-035. Texas A&M AgriLife Extension, College Station, TX

Kugbe JX, Mbawuni AM, Wisdom A (2019) Growth, yield and yield components response of local sorghum varieties to nitrogenous fertilizer rates in northern Ghana. World J Agric Soil Sci 3(5):1–10

Kumar P, Pathak S (2018) Short-term response of plants grown under heavy metal toxicity. Heavy Metals 69(75722):10–5772

Kumar A, Reddy BV, Sharma HC, Ramaiah B (2008) Shoot fly (*Atherigona soccata*) resistance in improved grain sorghum hybrids. J SAT Agric Res 6:1–4

Kumar S, Arya RK, Dahiya KK (2013) Optimising sowing time of multi-cut sorghum for maximising fodder yield. Forage Res 39(2):102–103

Kurai T, Morey SR, Wani SP, Watanabe T (2015) Efficient rates of nitrogenous fertiliser for irrigated sweet sorghum cultivation during the post-rainy season in the semi-arid tropics. Eur J Agron 71:63–72

LeBeau FJ, Stokes LE, Coleman OH (1951) Anthracnose and red rot of sorghum. Technical Bulletin No. 1035, USDA

Liang L, Nagumo T, Hatano R (2005) Nitrogen cycling with respect to environmental load in farm systems in Southwest China. Nutr Cycl Agroecosyst 73:119–134

Liu X, Ju X, Zhang F, Pan J, Christie P (2003) Nitrogen dynamics and budgets in a winter wheat–maize cropping system in the North China Plain. Field Crop Res 83(2):111–124

Lubadde G, Ebiyau J, Aru JC, Andiku C, Wandulu JA, Ugen MA (2019) Sorghum production handbook for Uganda. In: National Semi Arid Resources Research Institute of the National Agricultural Research Organisation (NaSARRI-NARO), Uganda

Lyons SE, Ketterings QM, Godwin GS, Cherney DJ, Cherney JH, Van Amburgh ME, Meisinger JJ, Kilcer TF (2019) Optimal harvest timing for brown midrib forage sorghum yield, nutritive value, and ration performance. J Dairy Sci 102(8):7134–7149

Madhusudhana R, Padmaja PO, Satish K, Mohan ŚM, Srinivas G, Reddy RN (2007) Heterosis for shoot fly [Atherigona soccata (Rondani)] resistance in sorghum. Indian J Genet Plant Breed 67(03):299–300

Mahmood A, Ullah H, Ijaz M, Javaid MM, Shahzad AN, Honermeier B (2013) Evaluation of sorghum hybrids for biomass and biogas production. Aust J Crop Sci 7(10):1456–1462

Mahmud K, Ahmad I, Ayub M (2003) Effect of nitrogen and phosphorus on the fodder yield and quality of two sorghum cultivars (*Sorghum bicolor* L.). Int J Agric Biol 5(1):61–63

Malipatil M (2008) Industry biosecurity plan for the grains industry threat specific contingency plan: sorghum shoot fly Atherigona soccata. Grain Research & Development Corporation, pp 1–22

Marschner H (1995) Mineral nutrition of higher plants. Academic Press, San Diego

Mattos BB, Marriel IE, de Sousa SM, Lana UDP, Schaffert RE, Gomes EA, Oliveira-Paiva CA (2020) Sorghum genotypes response to inoculation with phosphate solubilizing bacteria. Revista Brasileira de Milho e Sorgo 19(e1177):2020

Maw MJ, Houx JH III, Fritschi FB (2016) Sweet sorghum ethanol yield component response to nitrogen fertilization. Ind Crop Prod 84:43–49

Mbuvi DA, Masiga CW, Kuria E, Masanga J, Wamalwa M, Mohamed A, Odeny DA, Hamza N, Timko MP, Runo S (2017) Novel sources of witchweed (Striga) resistance from wild sorghum accessions. Front Plant Sci 8:116

Melaku ND, Bayu W, Ziadat F, Strohmeier S, Zucca C, Tefera ML, Ayalew B, Klik A (2018) Effect of nitrogen fertilizer rate and timing on sorghum productivity in Ethiopian highland Vertisols. Arch Agron Soil Sci 64(4):480–491

Menezes CBD, Silva KJD, Teodoro LPR, Santos CVD, Julio BHM, Portugal AF, Batista PSC, Carvalho AJ, Teodoro PE (2022) Grain sorghum hybrids under drought stress and full-irrigation conditions in the Brazilian semiarid. J Agron Crop Sci 208(6):868–875

Mengel K, Kirkby EA, Kosegarten H, Appel T (2001) Principles of plant nutrition. Springer, Dordrecht

Michauld (2014) Greenbug. Kansas State University Agricultural Experiment Station and Cooperative Extension Service, Manhattan, KS

Mohammed R, Are AK, Munghate RS, Bhavanasi R, Polavarapu KKB, Sharma HC (2016) Inheritance of resistance to sorghum shoot fly, Atherigona soccata in sorghum, *Sorghum bicolor* (L.) Moench. Front Plant Sci 7:543

Mostafa AM, Harrington K, Burayu W (2022) Sugarcane Aphid, Melanaphis sacchari (Zehntner), (Hemiptera: Aphididae) in Arizona Sorghum. The University of Arizona Cooperative Extension

Mubarik MK, Hussain K, Abbas G, Altaf MT, Baloch FS, Ahmad S (2022) Productivity of sorghum (Sorghum bicolar L.) at diverse irrigation regimes and sowing dates in semi-arid and arid environment. Turk J Agric For 46(1):1–18

Muchira N, Ngugi K, Wamalwa LN, Avosa M, Chepkorir W, Manyasa E, Nyamongo D, Odeny DA (2021) Genotypic variation in cultivated and wild sorghum genotypes in response to Striga hermonthica infestation. Front Plant Sci 12:671984

Nazli RI (2020) Evaluation of different sweet sorghum cultivars for bioethanol yield potential and bagasse combustion characteristics in a semiarid Mediterranean environment. Biomass Bioenergy 139:105624

Nazli RI (2022) Effects of harvest time on bioethanol production and bagasse characteristics for combustion and forage in sweet sorghum. Field Crop Res 280:108464

Nazli RI, Tansi V, Gulnaz O, Kafkas E, Kusvuran A, Ozturk HH, Bostan Budak D (2020) Interactive effects of nitrogen and humic substances applications on bioethanol production from sweet sorghum and combustion characteristics of its bagasse. Agronomy 10(9):1397

Nazli RI, Kuşvuran A, Inal I, Demirbaş A, Tansi V (2014) Effects of different organic materials on forage yield and quality of silage maize (Zea mays L.). Turk J Agric For 38(1):23–31

Nielsen DC, Vigil MF, Henry WB (2018) Skip row planting configuration shifts grain sorghum water use under dry conditions. Field Crop Res 223:66–74

Nyoni N, Dube M, Bhebhe S, Sibanda B, Maphosa M, Bombom A (2022) Sorghum production manual: a guide to agronomic practices for sorghum production in Zimbabwe. Lupane State University, Zimbabwe, p 45

Obour AK, Holman JD, Assefa Y (2022) Grain sorghum productivity as affected by nitrogen rates and available soil water. Crop Sci 62(3):1360–1372

Ogunlela VB (1988) Growth and yield responses of dryland grain sorghum to nitrogen and phosphorus fertilization in a ferruginous tropical soil (Haplustalf). Fertil Res 17:125–135

Oten M (2017) The effects of different sowing time and harvesting height on hydrocyanic acid content in some silage sorghum (*Sorghum bicolor* L.) varieties. Turk J Field Crops 22(2):211–217

Oyier MO, Owuoche JO, Oyoo ME, Cheruiyot E, Mulianga B, Rono J (2017) Effect of harvesting stage on sweet sorghum (*Sorghum bicolor* L.) genotypes in western Kenya. ScientificWorldJournal 2017:8249532

Padmaja PG, Madhusudhana R, Seetharama N (2010) Sorghum shoot fly. Directorate of Sorghum Research, Hyderabad

Pan B, Lam SK, Mosier A, Luo Y, Chen D (2016) Ammonia volatilization from synthetic fertilizers and its mitigation strategies: a global synthesis. Agric Ecosyst Environ 232:283–289

Pandian BA, Sexton-Bowser S, Prasad PV, Jugulam M (2022) Current status and prospects of herbicide-resistant grain sorghum (*Sorghum bicolor*). Pest Manag Sci 78(2):409–415

Petrini C, Belletti A, Salamini F (1993) Accumulation and distribution of dry matter and soluble carbohydrates in two sweet sorghum cultivars influence of sowing date and harvesting time. Eur J Agron 2(3):185–192

Pimentel GV, Guimarães DF, Moreira SG, Ávila MOT, Martins IA, Bruzi AT (2019) Selectivity and effectiveness of herbicides in the grain sorghum crop. Planta Daninha 37

Prasad G, Babu KS, Subbarayudu B, Bhagwat VR, Patil JV (2015) Identification of sweet sorghum accessions possessing multiple resistance to shoot fly (Atherigona soccata Rondani) and spotted stem borer (Chilo partellus Swinhoe). Sugar Tech 17(2):173–180

Prescott CE, Katzensteiner K, Weston C (2021) Soils and restoration of forested landscapes. In: Soils and landscape restoration. Academic Press, London, pp 299–331

Prom LK, Cisse N, Perumal R, Cuevas H (2017) Screening of sorghum lines against long smut and grain mold pathogens. Int J Plant Pathol 8(2):23–27

Punnuri GS, Harris-Shultz K, Knoll J, Ni X, Wang H (2017) The genes bm2 and blmc that affect epicuticular wax deposition in sorghum are allelic. Crop Sci 57(3):1552–1556

QDAF (2011) Sorghum—nutrition, irrigation and harvest issues. Department of Agriculture, Fisheries and Forestry Queensland. http://www.daff.qld.gov.au/plants/field-crops-and-pastures/broadacre-field-crops/sorghum/nutrition,-irrigation-and-harvest

Ramatoulaye F, Mady C, Fallou S (2016) Production and use sorghum: a literature review. J Nutr Health Food Sci 4(1):1–4

Rao SS, Patil JV, Umakanth AV, Mishra JS, Ratnavathi CV, Prasad GS, Rao BD (2013) Comparative performance of sweet sorghum hybrids and open pollinated varieties for millable stalk yield, biomass, sugar quality traits, grain yield and bioethanol production in tropical Indian condition. Sugar Tech 15:250–257

Ratnavathi CV, Chakravarthy SK, Komala VV, Chavan UD, Patil JV (2011) Sweet sorghum as feedstock for biofuel production: a review. Sugar Tech 13(4):399–407

Reddy BV, Ramesh S, Ashok Kumar A, Gowda CLL (2008) Sorghum improvement in the new millennium. International Crops Research Institute for the Semi-Arid Tropics, Patancheru

Reddy BV, Reddy PS, Sadananda AR, Dinakaran E, Ashok Kumar A, Deshpande SP, Srinivasa Rao P, Sharma HC, Sharma R, Krishnamurthy L, Patil JV (2012) Postrainy season sorghum: constraints and breeding approaches. J SAT Agric Res 10(1):1–12
Reddy PS, Reddy BVS, Rao PS (2014) Genotype by sowing date interaction effects on sugar yield components in sweet sorghum (*Sorghum bicolor* L. Moench). SABRAO J Breed Genet 46(2):241–255
Regehr (1998) Weed control. In: Grain sorghum production handbook. Kansas State University Agricultural Experiment Station and Cooperative Extension Service, Manhattan, KS, pp 10–12
Rehman A, Qamar R, Altaf MM, Alwahibi MS, Al-Yahyai R, Hussain M (2022) Phosphorus and potassium application improves fodder yield and quality of sorghum in Aridisol under diverse climatic conditions. Agriculture 12(5):593
Riyazaddin M, Kavi Kishor PB, Ashok Kumar A, Reddy BV, Munghate RS, Sharma HC (2015) Mechanisms and diversity of resistance to sorghum shoot fly, Atherigona soccata. Plant Breed 134(4):423–436
Rohbakhsh H (2013) Alleviating adverse effects of water stress on growth and yield of forage sorghum by potassium application. Adv Environ Biol 7(1):40–46
Sahrawat KL, Rego TJ, Burford JR, Rahman MH, Rao JK, Adam A (1995) Response of sorghum to fertilizer phosphorus and its residual value in a Vertisol. Fertil Res 41:41–47
Sainju UM, Singh HP, Singh BP, Whitehead WF, Chiluwal A, Paudel R (2018) Cover crop and nitrogen fertilization influence soil carbon and nitrogen under bioenergy sweet sorghum. Agron J 110(2):463–471
Saleem MH, Usman K, Rizwan M, Al Jabri H, Alsafran M (2022) Functions and strategies for enhancing zinc availability in plants for sustainable agriculture. Front Plant Sci 13:1033092
Samarappuli D, Berti MT (2018) Intercropping forage sorghum with maize is a promising alternative to maize silage for biogas production. J Clean Prod 194:515–524
Sanjari S, Shobbar ZS, Ghanati F, Afshari-Behbahanizadeh S, Farajpour M, Jokar M, Khazaei A, Shahbazi M (2021) Molecular, chemical, and physiological analyses of sorghum leaf wax under post-flowering drought stress. Plant Physiol Biochem 159:383–391
Sardá LG, Higarashi MM, Nicoloso RS, Falkoski C, Ribeiro SM, Silveira CA, Soares HM (2020) Effects of dicyandiamide and Mg/P on the global warming potential of swine slurry and sawdust cocomposting. Environ Sci Pollut Res 27:30405–30418
Satish K, Srinivas G, Madhusudhana R, Padmaja PG, Nagaraja Reddy R, Murali Mohan S, Seetharama N (2009) Identification of quantitative trait loci for resistance to shoot fly in sorghum [Sorghum bicolor (L.) Moench]. Theor Appl Genet 119:1425–1439
Sawargaonkar GL, Patil MD, Wani SP, Pavani E, Reddy BVSR, Marimuthu S (2013) Nitrogen response and water use efficiency of sweet sorghum cultivars. Field Crop Res 149:245–251
Schulz VS, Munz S, Stolzenburg K, Hartung J, Weisenburger S, Mastel K, Möller K, Claupein W, Graeff-Hönninger S (2018) Biomass and biogas yield of maize (*Zea mays* L.) grown under artificial shading. Agriculture 8(11):178
Sharma HC (1997) Screening for resistance to sorghum midge. In: Sharma HC, Singh F, Nwanze KF (eds) Plant resistance to insects in sorghum. International Crops Research Institute for the Semi-Arid Tropics, Patancheru, pp 72–78. ISBN 92-9066-382-0
Sharma HC, Mukuru SZ, Manyasa E, Were JW (1999) Breakdown of resistance to sorghum midge, Stenodiplosis sorghicola. Euphytica 109:131–140
Sharma HC, Reddy BV, Dhillon MK, Venkateswaran K, Singh BU, Pampapathy G, Folkertsma RT, Hash CT, Sharma KK (2005) Host plant resistance to insects in sorghum: present status and need for future research. Int Sorghum Millets Newsl 46:36–43
Sharma HC, Dhillon MK, Pampapathy G (2006) Multiple-resistance to sorghum shoot fly, spotted stemborer and sugarcane aphid in sorghum. Int J Trop Insect Sci 26(4):239–245
Sharma HC, Dhillon MK, Pampapathy G, Reddy BVS (2007) Inheritance of resistance to spotted stem borer, Chilo partellus, in sorghum, Sorghum bicolor. Euphytica 156:117–128
Sher A, Hassan FU, Ali H, Hassan W (2016) Seed rate and nitrogen application effects on production and brix value of forage sorghum cultivars. Grassl Sci 62(2):119–127

Singh MP, Erickson JE, Sollenberger LE, Woodard KR, Vendramini JMB, Fedenko JR (2012a) Mineral composition and biomass partitioning of sweet sorghum grown for bioenergy in the southeastern USA. Biomass Bioenergy 47:1–8

Singh BU, Sharma HC, Rao KV (2012b) Mechanisms and genetic diversity for host plant resistance to spotted stem borer, Chilo partellus in sorghum, Sorghum bicolor. J Appl Entomol 136(5):386–400

Singh SP, Joshi YP, Singh RK, Meena VS (2014) Influence of nitrogen levels and seed rate on growth, yield and quality of sweet sorghum. Ann Biol 30(1):89–93

Soleymani A, Shahrajabian MH, Naranjani L (2011) The effect of plant density and nitrogen fertilization on yield, yield components and grain protein of grain sorghum. J Food Agric Environ 9(3):244–246

Somashekar KS, Shekara BG, Kalyanamurthy KN, Lohithaswa HC (2015) Growth, yield and economics of multi-cut fodder sorghum (*Sorghum sudanense* L.) as influenced by different seed rates and nitrogen levels. Forage Res 40(4):247–250

Sowiński J, Głąb L (2018) The effect of nitrogen fertilization management on yield and nitrate contents in sorghum biomass and bagasse. Field Crop Res 227:132–143

Stutts LR, Vermerris W (2020) Elucidating anthracnose resistance mechanisms in sorghum—a review. Phytopathology 110(12):1863–1876

Subramanian SK (2013) Agronomical, physiological and biochemical approaches to characterize sweet sorghum genotypes for biofuel production. Kansas State University, Manhattan, KS

Sune AD, Hadole SS, Rathod GLESV, Ghatol PU (2017) Long term effect of nutrient management on yield, nutrient uptake and residual fertility after harvest of sorghum. J Sci Agric Eng 6:204–208

Talwar HS, Prabhakar ME, Kumabi A, Patil JMJ (2010) Strategies to improve post flowering drought tolerance in rabi sorghum for predicted climate change scenario. Crop Improv 37(2):93–99

Tang C, Yang X, Chen X, Ameen A, Xie G (2018) Sorghum biomass and quality and soil nitrogen balance response to nitrogen rate on semiarid marginal land. Field Crop Res 215:12–22

Taylor (1998) Harvesting grain sorghum. In: Grain sorghum production handbook. Kansas State University Agricultural Experiment Station and Cooperative Extension Service, Manhattan, KS, pp 24–25

Teixeira TPM, Pimentel LD, dos Santos Dias LA, da Costa Parrella RA, da Paixão MQ, Biesdorf EM (2017) Redefinition of sweet sorghum harvest time: new approach for sampling and decision-making in field. Ind Crop Prod 109:579–586

Tew TL, Cobill RM, Richard EP (2008) Evaluation of sweet sorghum and sorghum× sudangrass hybrids as feedstocks for ethanol production. Bioenergy Res 1:147–152

Thivierge MN, Chantigny MH, Bélanger G, Seguin P, Bertrand A, Vanasse A (2015) Response to nitrogen of sweet pearl millet and sweet sorghum grown for ethanol in eastern Canada. Bioenergy Res 8:807–820

Tovignan TK, Adoukonou-Sagbadja H, Diatta C, Clément-Vidal A, Soutiras A, Cisse N, Luquet D (2020) Terminal drought effect on sugar partitioning and metabolism is modulated by leaf stay-green and panicle size in the stem of sweet sorghum (*Sorghum bicolor* L. Moench). CABI Agric Biosci 1:1–11

Turgut I, Bilgili U, Duman A, Acikgoz E (2005) Production of sweet sorghum (*Sorghum bicolor* L. Moench) increases with increased plant densities and nitrogen fertilizer levels. Acta Agric Scand Sect B Soil Plant 55(3):236–240

Umar S (2006) Alleviating adverse effects of water stress on yield of sorghum, mustard and groundnut by potassium application. Pak J Bot 38(5):1373–1380

Undersander DJ, Smith LH, Kaminski AR, Kelling KA, Doll JD (2010) Sorghum-forage. Alternative field crops manual. Univ. of Wisconsin-Ext. Wisconsin Cooperative Ext, Madison, WI

Vanya CL (2012) Impact of climate variability and change on sorghum production in the lower shire valley in Malawi. Doctoral dissertation, University of Nairobi

Vashisth S, Jagdish J, Sharma SP, Sharma HC (2022) Biochemical mechanisms of induced resistance to Chilo partellus in sorghum. Int J Pest Manag:1–12

Vieira PVD, Freitas PSLD, Silva ALBRD, Souza ACS, Voroniak J (2021) Agronomic characteristics and grain yield of sorghum and maize hybrids grown with different sowing times. Revista Caatinga 34:780–790

Vigil MF, Nielsen DC, Poss D, Calderon F, Henry B (2015) Skip-row planting as a strategy for drought mitigation in the west central Great Plains. In: 2015 annual meeting ASA-CSSA-SSSA. 18 Nov 2015, Minneapolis, MN. https://dl.sciencesocieties.org/publications/meetings/download/pdf/2015am/93380. Accessed 28 Dec 2017

Whish J, Butler G, Castor M, Cawthray S, Broad I, Carberry P, Hammer G, McLean G, Routley R, Yeates S (2005) Modelling the effects of row configuration on sorghum yield reliability in north-eastern Australia. Aust J Agric Res 56(1):11–23

Whitney D (1998) Fertilizer requirements. In: Grain sorghum production handbook. Kansas State University Agricultural Experiment Station and Cooperative Extension Service, Manhattan, KS, pp 12–14

Wortmann CS, Ferguson RB, Hergert GW, Shapiro CA, Shaver TM (2013) Nutrient management suggestions for grain sorghum. Bulletin of Nebraska University, pp 206–215

Wright R, Danielson S, Mayo ZB (2006) Management of greenbugs in sorghum. NebGuide, Univesity of Nebrabka-Lincoln Extension

Xavier KV, Pfeiffer T, Parreira DF, Chopra S, Vaillancourt L (2017) Aggressiveness of Colletotrichum sublineola strains from Sorghum bicolor and S. halepense to sweet sorghum variety Sugar Drip, and their impact on yield. Plant Dis 101(9):1578–1587

Youngquist JB, Carter DC, Clegg MD (1990) Grain and forage yield and stover quality of sorghum and millet in low rainfall environments. Exp Agric 26(3):279–286

Zand N, Shakiba MR, Vahed MM, Nasab ADM (2014) Response of sorghum to nitrogen fertilizer at different plant densities. Int J Farm Allied Sci 3(1):71–74

Zhou H, Liu H, Zhou X, Jia H, Wei Z, Li S, Hou S (2017) Effects of sowing time on biological yield of forage sorghum [*Sorghum bicolor* (L.) Moench] in autumn idle land. Agric Biotechnol 6(5):38–42

Ziki SJL, Zeidan EMI, El-Banna AYA, Omar AEA (2019) Influence of cutting date and nitrogen fertilizer levels on growth, forage yield, and quality of Sudan grass in a semiarid environment. Int J Agron 2019:1–9

Adaptability and Cultivation of Sorghum Under Climate Change Scenarios

3

Kaliamoorthy Sivasakthi, Tharanya Murugesan, and Sunita Choudhary

Abstract

This chapter thoroughly investigates the vital role of sorghum (*Sorghum bicolor*) in arid agroecosystems, specifically emphasizing its significance in South Asia and sub-Saharan Africa. It highlights significant challenges in post-rainy sorghum production, attributing them to the complexities of genotype–environment interactions and the limited integration of essential physiological traits into sorghum breeding, further complicated by the absence of sophisticated, breeder-friendly phenotyping tools. To address these challenges, the chapter meticulously characterizes the post-rainy sorghum production areas in India, utilizing advanced modeling tools, revealing distinct production zones and drought stress patterns that impact yields, necessitating optimization strategies. Innovative approaches, like homogeneous system units, are introduced to augment sorghum production through well-planned genotype × management combinations.

The chapter explores the modification of plant water-use traits as a potential solution to balance trade-offs between grain and stover production. It examines component traits crucial for drought adaptation, extending beyond grain yield considerations to factors such as limited transpiration, soil drought, and canopy development. Specific traits, including leaf size, leaf angle, transpiration efficiency, and the stay-green trait, are underscored for their pivotal roles in enhancing crop resilience to water constraints.

The chapter also investigates potential molecular and genetic approaches to fortify sorghum drought adaptation. It concludes by emphasizing the complexity of enhancing drought adaptation in sorghum and advocates for precise phenotyping techniques, illustrated through sensor-based high-throughput phenotyping.

K. Sivasakthi · T. Murugesan · S. Choudhary (✉)
International Crop Research Institute for the Semi-Arid Tropics, Patancheru, Telangana, India
e-mail: Sunita.choudhary@icrisat.org

E. Habyarimana et al. (eds.), *Omics and Biotechnological Approaches for Product Profile-Driven Sorghum Improvement*,
https://doi.org/10.1007/978-981-97-4347-6_3

Keywords

Sorghum · Climate change · Drought adaptation · Phenotyping · Aquaporins · Quantitative trait loci

3.1 Introduction

Sorghum (*Sorghum bicolor* (L.) Moench) holds significant agricultural importance in the semi-arid and arid regions of South Asia and sub-Saharan Africa, where it frequently faces challenges such as drought, low and unpredictable rainfall, and high average temperatures (Nagesh Kumar et al. 2022).

Sorghum Cultivation and Production Status Globally, sorghum is cultivated across 40.25 million hectares, producing approximately 58.70 million tons with a productivity rate of 1458 kg/ha (FAO 2022). Notably, the top 10 sorghum producers, including the United States, Sudan, Mexico, Nigeria, India, Niger, Ethiopia, Australia, Brazil, and China, collectively contribute to around 77% of the global sorghum production (Aruna and Cheruku 2019).

Globally, sorghum production reached approximately 62.51 million tons in the 2020–2021 season. The foremost producer was the United States yielding 9.4 million tons, trailed by Nigeria, Ethiopia, and Sudan. India ranked fifth with 4.78 million tons of sorghum produced on 4.39 million hectares in 2020–2021. During the kharif season of 2021–2022, India produced an estimated 1.54 million tons of sorghum. Andhra Pradesh significantly contributed to India's production, yielding 4.10 lakh tons in 2020–2021 from 1.20 lakh hectares with a productivity of 3428 kg/hectare (https://angrau.ac.in/downloads/AMIC/OutlookReports/2021/9-SORGHUM_January%20to%20December%202021.pdf).

Sorghum Cultivation in India Sorghum holds a prominent position in India as the fourth most crucial food crop, cultivated in two distinct cropping seasons: the rainy season (June–October) and the post-rainy season (October–January). Traditionally, sorghum harvested after the rainy season plays a crucial role in securing food and fodder for millions of rural families in India (Upadhyaya et al. 2017), while rainy season sorghum is primarily used for feed and fodder due to its susceptibility to grain mold infections and leaf diseases.

India, the world's second-largest sorghum producer after Nigeria, faces low productivity at 1250 kg/ha, significantly lower than countries like China and the United States, where yields reach 4854 and 4594.6 kg/ha, respectively. India's below-par sorghum yield is attributed to several factors, including the lack of appropriate varieties for different agro-ecological zones, abiotic stresses like drought, cold, and soil problems, and losses due to pests, which account for approximately 32% during the rainy season and 26% during the post-rainy season (Nagesh Kumar et al. 2022).

India produces approximately 3 million tons of sorghum from 5.7 million hectares during the post-rainy season (Kholova et al. 2013).

The clean, lustrous, and bold sorghum grains harvested after the rainy season dominate food usage in India, making post-rainy sorghum cultivation crucial for sustaining both human and livestock populations. However, the area dedicated to post-rainy sorghum cultivation in India has significantly decreased from 10.25 million hectares in 1999–2000 to 5.82 million hectares in 2014–2015, resulting in a decline in total production from 8.68 million tons to 5.39 million tons (Chapke et al. 2017).

3.2 Predicted Weather Patterns and Their Effects on Sorghum Yield

According to the findings of Boyle (2018), climate projections based on historical data suggest potential adverse impacts on agricultural crop production in forthcoming decades (Fig. 3.1). These impacts are of particular concern in regions with high levels of food insecurity (Lobell and Burke 2008; Rosenzweig et al. 2014). Notably, areas at the greatest risk of declining agricultural productivity overlap with regions heavily dependent on sorghum as a staple food source, such as Sub-Saharan Africa and Southern Asia. For instance, the Sahel in Africa faces significant challenges, with hundreds of thousands of child deaths due to hunger-related causes (Battisti and Naylor 2009). This biogeographic zone is especially vulnerable to rising temperatures. Additionally, other key sorghum-producing areas such as the regions of Australia, North America, and South America, are expected to experience reduced crop productivity by 2080 (Cline's 2007).

3.3 Environmental Characterization of Post-rainy Sorghum in India and Simulation Study Under Changing Climate Scenario

The most prone environment of sorghum under climate change scenario in Asia is post-rainy sorghum cultivation regions under heavy reliance on remaining soil moisture of rainy season. This cultivation is mostly taken in India's Deccan Plateau, where soil moisture availability is limited due to early cessation of rain and crop often suffer terminal drought stress. It is crucial to grasp these environmental dynamics for sustainable sorghum production which are early indicators of future climate change environment. To achieve this, researchers employed crop simulation modeling, as outlined in Kholova et al. (2013), to delineate four unique production zones within the "rabi sorghum belt" of India. This research identified common drought patterns, quantified the occurrences of stress, and evaluated their impact on both grain and stover yields. Here are the key findings and analyses:

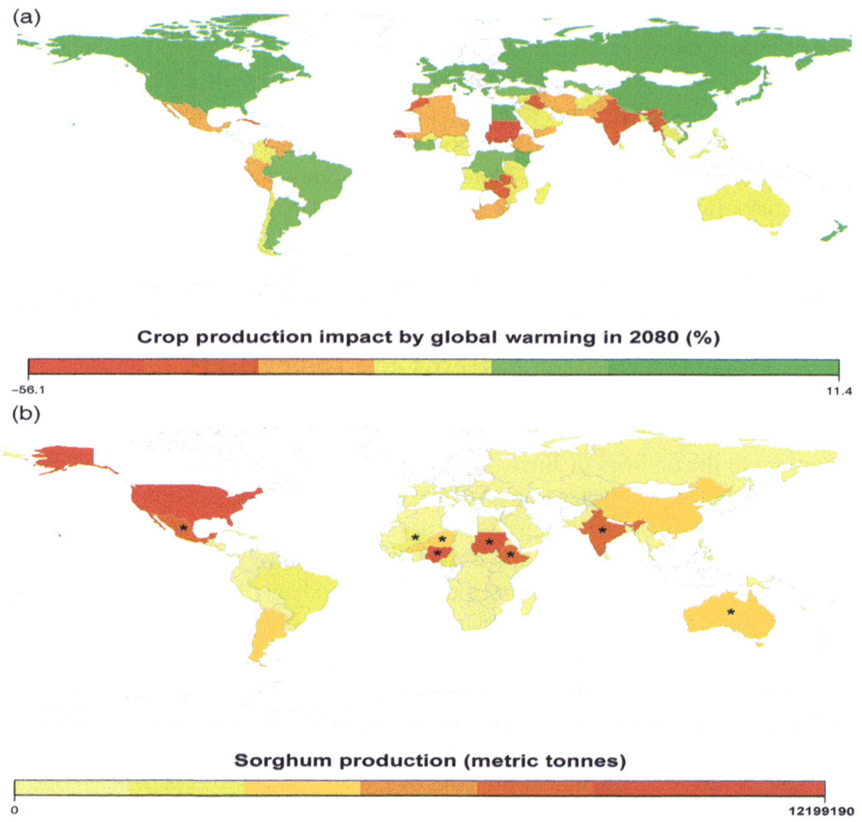

Fig. 3.1 The predicted consequences of global warming on crop yields have profound implications for regions that are major sorghum producers. (**a**) The nations poised to experience the most substantial declines in crop production are mainly located in semi-arid tropical areas, a fact underscored in Cline's 2007 research. (**b**) Predominantly, the countries renowned for their sorghum production face the prospect of significant temperature increases, with Sub-Saharan Africa taking the lead. Other regions, such as India, Mexico, and Australia, are also expected to be affected. In the document, asterisks denote countries where sorghum production is substantial and where adverse effects on crop yield by climate change are projected. The yearly sorghum yield data is based on figures from 2016, obtained from the CGIAR Agricultural Science and Technology Indicators (www.asti.cgiar.org). This graph retrieved from Boyles et al. (2019); The Plant Journal, Volume: 97, Issue: 1, Pages: 19–39, First published: 27 September 2018, https://doi.org/10.1111/tpj.14113

Characterizing Sorghum Production Zones The study classified the region into four distinct production zones (Central, South, North, and Far South) based on location, yield, and environmental conditions. Each zone displayed unique characteristics, with variations in average yields attributed to differing soil quality and rainfall patterns.

Drought Stress Patterns The research pinpointed five distinct drought stress patterns (vegetative stress, pre-flowering stress, post-flowering stress, post-flowering relived stress and mild stress) affecting sorghum production during the rabi season. These patterns encompassed terminal drought stress across different growth stages, ranging from severe to mild stress conditions. Notably, severe stress types, such as pre-flowering terminal drought, substantially diminished yields.

Frequency of Drought Stress Patterns An in-depth analysis within each production zone uncovered the prevalence of specific stress patterns and their impact on both grain and stover yields. For instance, despite being a significant producer, the central production zone experienced severe stress types, resulting in substantial yield reductions.

Comparative Analysis Across the Rabi Belt A comprehensive examination spanning the entire rabi sorghum belt revealed that approximately 60% of cases faced water limitations. Severe stress patterns, like vegetative and pre-flowering terminal drought, led to significant reductions in grain yields, although stover yields were less affected in specific drought conditions.

Sorghum Production Potential The research shed light on the potential for achieving higher grain and biomass yields in different zones under ideal conditions with adequate water and nitrogen supply. This potential varied among the major zones, influenced by factors like radiation levels and temperature conditions.

Enhancing Sorghum Cultivation The study emphasized the importance of understanding drought patterns and the potential for crop improvement. The research aims to minimize trade-offs between grain and stover production while considering factors such as profitability and risk.

Innovative Approaches for Optimization In 2022, Ronanki and colleagues showcased an innovative method by establishing "homogeneous system units" (HSUs) rooted in geographic attributes. These HSUs streamline the process of pinpointing the most effective genotype × management (GxM) combinations to improve and stabilize post-rainy sorghum production (Fig. 3.2).

Modifying Plant Water-Use Traits Through simulation experiments, Kholová et al. (2014) explored modifications to plant traits related to drought adaptation. These alterations impacted grain yield and the expression of the stay-green (SG) phenotype, often accompanied by trade-offs between grain and stover production in specific scenarios.

In summary, this research offers a comprehensive understanding of the challenges and opportunities in post-monsoon sorghum cultivation in India's Deccan Plateau. The insights into drought patterns, production potential, and strategies for

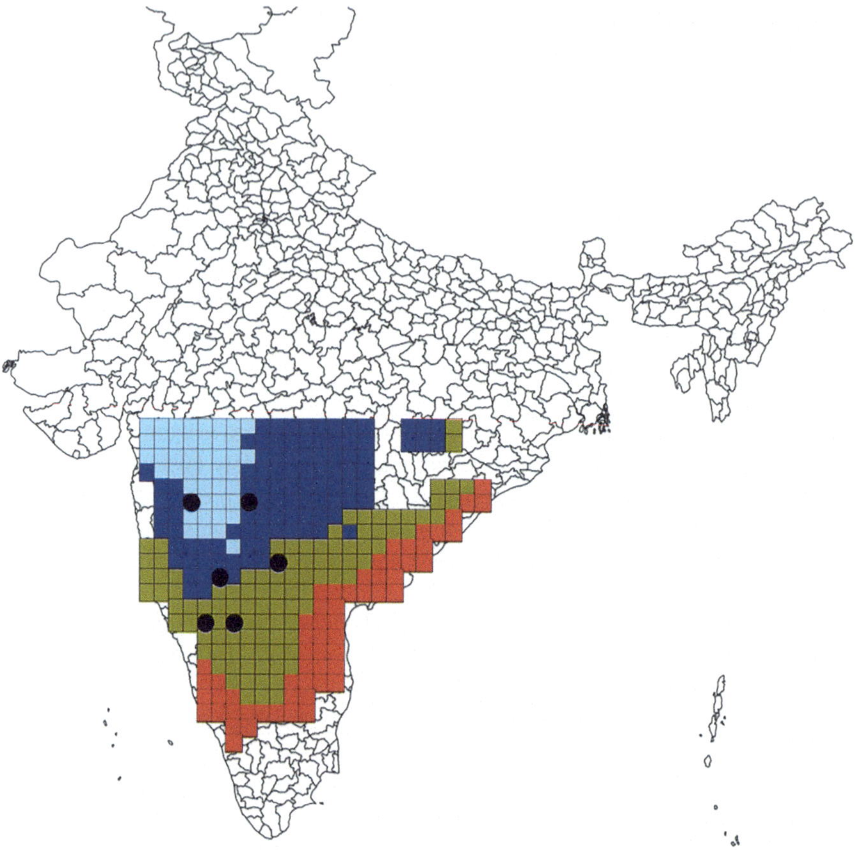

Fig. 3.2 Map of the sorghum rabi production regions in India featuring four major production zones represented through homogeneous units (HSUs)—Central, Northern, Southern, and Far South. The inset map indicates the geographical placement of the major rabi tract and the location of the AICRP testing sites

improvement are valuable and have the potential to influence future agricultural practices, ultimately enhancing sorghum production in the region.

3.4 Water-Saving Traits in Sorghum to Combat Drought

The breeder primarily focused on grain yield, which is a complex trait greatly influenced by the interplay of genetics (G), environment (E), and management (M). However, it is essential to recognize that grain yield is the result of several component traits (Fig. 3.3), each possessing its unique value and impacting yield formation through interactions with the environment. Consequently, it becomes imperative to concentrate on these component traits rather than solely fixating on grain yield. In this context, our primary focus revolves around plant water-use-related traits that have a direct connection to drought adaptation, although there are numerous other component traits to consider.

The water-conserving traits for drought adaptation in sorghum encompass a range of traits. These include the ability to limit transpiration rates under atmospheric drought conditions, reducing plant transpiration when soil moisture is high (soil drought management), optimizing plant hydraulics for efficient water transport, influencing canopy development through factors like leaf area, leaf angle, and leaf width, adjusting water extraction patterns during critical crop growth stages, improving transpiration efficiency (TE), and promoting the SG phenotype. When combined with appropriate crop management practices, these traits collectively enhance the crop's resilience in water-limited environments.

3.4.1 Limited Transpiration Rate (Atmospheric Drought)

The limited transpiration rate trait has been reported in various crops, including sorghum (*Sorghum bicolor* (L.) Moench). This trait restricts the transpiration rate, particularly when plant is exposed to high vapor pressure deficit (VPDs) during the midday period. The advantage of the limited transpiration rate trait lies in its ability to conserve water under conditions of elevated midday VPD. This conservation of water resources positions the crop more effectively to cope with drought conditions

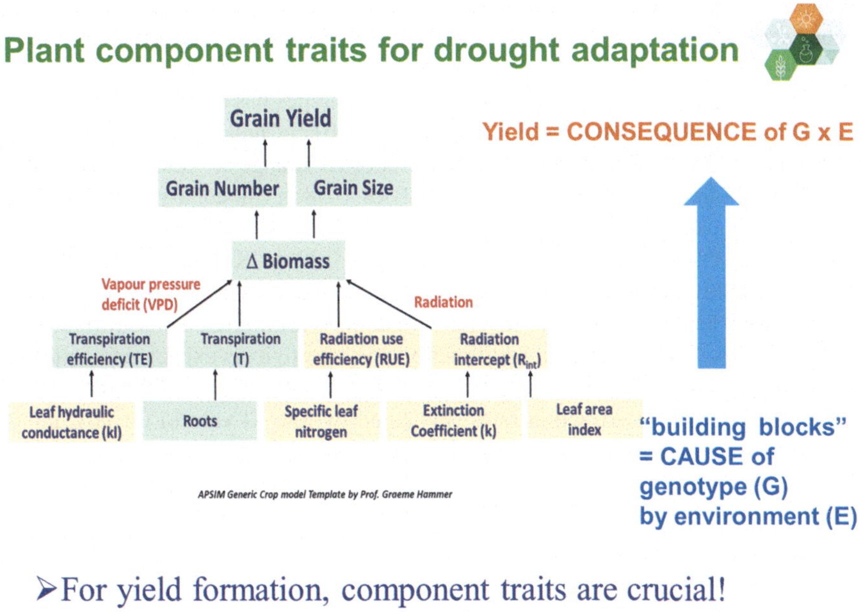

Fig. 3.3 Plant component traits related to drought adaptation in crops. This image is a modification of the APSIM Generic Crop Model template by Professor Graeme Hammer

later in the growing season. Significant variation in the limited transpiration rate (TRlim) trait in response to increasing VPD has been observed in sorghum. Choudhary et al. (2013a) reported that among twelve genotypes studied, seven exhibited a TRlim response. Choudhary et al. (2013b) documented findings from two distinct sorghum genotypes, one with a limited transpiration rate (SC15) and the other with an unrestricted transpiration rate (SC1205). Similarly, Gholipoor et al. (2010) documented a breakpoint VPD response in 17 sorghum genotypes, ranging from 1.62 to 2.72 kPa. Additionally, Karthika et al. (2019) observed a breakpoint VPD response in 16 different sorghum genotypes, ranging from 2.4 to 3.4 kPa. Several simulation studies, including those by Sinclair et al. (2005) as well as Kholová et al. (2014), have highlighted the potential for enhancing sorghum or soybean crop yields in semi-arid environments through the utilization of the physiological mechanism associated with a limited transpiration trait. Reducing the transpiration rate can lead to the conservation of soil moisture during crucial post-flowering stages, such as grain filling, and contribute to improved adaptation to terminal drought, as discussed in studies like Richards and Passioura (1989) and Sinclair et al. (2005).

3.4.2 Transpiration Response to Soil Drying (Soil Drought)

In the context of progressively drying soil, a method for conserving soil water involves managing plant water usage by reducing transpiration when soil moisture is relatively high (fraction of transpirable soil water (FTSW)). Genotypes exhibiting higher sensitivity to soil drying initiate stomatal closure at relatively higher soil moisture levels, aiding in soil water conservation (Sinclair and Rufty 2012; Vadez et al. 2014). Variability in this response across genotypes has been noted in various crop species, sorghum included (Gholipoor et al. 2012; Choudhary et al. 2013a, b; Karthika et al. 2019). Different studies have reported a range of variability in NTR-FTSW threshold points in sorghum, such as 0.48–0.38 in Choudhary et al. (2013a), 0.47–0.41 in Choudhary et al. (2013b), 0.53–0.32 in Choudhary et al. (2020), 0.48–0.32 in Gholipoor et al. (2012), and 0.42–0.22 in Karthika et al. (2019). Stomatal sensitivity to soil drying can potentially aid in conserving soil water and, consequently, lead to improved yields in situations with limited water availability, as discussed by Sinclair (2012). Additionally, this sensitivity is expected to enhance TE, as suggested by Vadez et al. (2014). In sorghum, Kholová et al. (2014) simulated lower soil water extraction rates (indicated by low Kl values) for post-rainy sorghum cultivars. This resulted in a moderate stover yield increment in majority of stress scenarios while maintaining similar grain yields. Lower Kl values led to earlier stress onset but extended the growth period, enhancing crop resilience and the potential for expressing the SG phenotype, particularly when stress occurred after flowering and was subsequently relieved.

3.4.3 Canopy Dynamics

Plants utilize water in the process of biomass production for the growth of their leaf area, known as the canopy. Variations in canopy development were observed among different species or genotypes within species, which was influenced by their distinct patterns of water usage (Vadez et al. 2013). This includes aspects such as leaf area, leaf angle, leaf width, and the overall structure of the canopy. The progress of canopy development significantly impacts the overall growth, light absorption, and photosynthetic efficiency of a sorghum crop.

Leaf Size Leaf size plays a crucial role in determining canopy leaf area development, thereby impacting crop photosynthesis, transpiration, and yield (Duncan et al. 1967; Kholová et al. 2014; Zhu et al. 2012). In a study by Kholová et al. (2014), the research focused on exploring the variability in leaf size within sorghum stay-green introgression lines (ILs) when compared to their respective recurrent parent. The findings showed that the average leaf size exhibited a range of variation: from 110 to 166 during 2010–2011, 188 to 237 in 2011–2012, and 125 to 197 in 2012–2013. Compared to the recurrent parent, certain stay-green ILs exhibited larger leaves size, while others had smaller leaves. These variations in leaf size were ultimately associated with differences in canopy development.

Leaf Numbers In the study conducted by Kholová et al. (2014), it was documented that the mean count of fully expanded leaves exhibited variability across the different sorghum lines. Within the stay-green ILs, this range extended from a minimum of 9 to a maximum of 10.2. Notably, among all the materials tested, the maximum number of leaves observed was 14. These disparities in leaf expansion and leaf count also had a significant impact on the development of the canopy.

Leaf Area Total plant leaf area in sorghum is essential for capturing light and performing photosynthesis, impacting the plant's growth and overall health. It is a valuable metric for researchers and farmers, influencing crop yield and agricultural practices. In a 2014 study conducted by Kholova et al. the research aimed to investigate the variation in leaf size among sorghum stay-green ILs. The results revealed that the total plant leaf area (cm^2) displayed a considerable range of variability. In the years 2010–2011, it ranged from 1190 to 1807, in 2011–2012, it spanned from 2136 to 2728, and in 2012–2013, it varied between 1367 and 2134. These changes in total plant leaf area were ultimately linked to differences in canopy development.

In 2019, Karthika and their team conducted two experiments, which involved 14 sorghum genotypes from eight established recombinant inbred lines (RILs) mapping populations. Their primary focus was on examining various aspects of canopy development. The results of their research indicated that the total plant leaf area (cm^2) ranged from 364 to 591 in the first experiment and from 289 to 674 in the second experiment. Additionally, the study also involved an investigation of several other canopy development traits, including digital biomass, leaf area index (LAI), projected leaf area, and plant height, utilizing a high-throughput phenotyping platform known as LeasyScan (for more details, see Vadez et al. 2015). Remarkably, the research unveiled a spectrum of fold differences in various traits: a 2.7-fold variance in digital biomass (ranging from 3,219,018 to 8,749,690), a 2-fold variation in LAI (from 0.08 to 0.16), a 2-fold diversity in projected leaf area (spanning from 145 to 299), and a 1.45-fold difference in plant height (ranging from 141 to 204). These variations in canopy development parameters among the genotypes or RILs had a profound impact on overall plant canopy development.

Leaf Angle Leaf angle constitutes a fundamental aspect of canopy architecture and plays a pivotal role in influencing the distribution of light within the canopy. This characteristic has been associated with variations in yield in densely planted crops (Hammer et al. 2009; Wang et al. 2007; Mantilla-Perez and Salas Fernandez 2017). In canopies with a high LAI, upper leaves often get too much light, while lower leaves do not get enough. When upper leaves are more erect, it reduces mutual shading in high LAI canopies, leading to a more even light distribution (Duncan et al. 1967; Duncan 1971; Clegg et al. 1974; McMillen and McClendon 1978). This benefits the top leaves, as they operate below light saturation, and the lower leaves, which receive more light. This boosts individual leaf photosynthesis, enhancing canopy radiation use efficiency. It also improves nitrogen distribution and delays lower leaf senescence, aiding carbohydrate storage and grain filling (Hammer et al. 2009; Murchie and Niyogi 2011).

Developing sorghum genotypes tailored to diverse light conditions necessitates understanding a plant's capacity to capture light, particularly influenced by leaf angle. Zhi et al. (2022a, b) study to identify the genetic underpinnings of leaf angle by conducting three years of field trials at two locations, utilizing a sorghum diversity panel consisting of 729 accessions. The findings demonstrated significant variability in leaf angle among the genotypes within the diversity panel, spanning from 20° to 82°. Leaf angle accounted for 36% of the variation in the canopy light extinction coefficient, underscoring its significant impact on light interception across the entire canopy. Additionally, the study revealed a consistent pattern: Guinea sorghum races typically had the largest leaf angle, while Durra sorghum races had the smallest leaf angle. This emphasizes the potential influence of leaf angle in adapting to different environments. Through a genome-wide association study, 33 quantitative trait loci (QTLs) relevant with leaf angle were identified.

Leaf Width The dimensions of the largest leaf, both in terms of length and width, serve as indicators of the total plant leaf area, which, in turn, correlates with the water requirements of plants through canopy transpiration (Borrell et al. 2014; Kholová et al. 2014; George-Jaeggli et al. 2017). The recent findings indicate that in C4 crops, wider leaves are positively linked to stomatal conductance but inversely related to mesophyll conductance and intrinsic water use efficiency. This relationship may be due to the inverse correlation between leaf width and vein density, which is usually positively correlated with stomatal density (Zhi et al. 2022a, b). Wider leaves result in higher boundary layer resistance, leading to increased stomatal conductance. This ultimately contributes to a decrease in intrinsic water use efficiency (Baird et al. 2021; Zhi et al. 2022a, b). Therefore, widening sorghum leaves may be a valuable breeding goal to enhance its adaptability to hotter and drier climates in the future.

In a 2022 study by Zhi and colleagues, the primary objectives were to assess the connection between leaf width and whole-plant transpiration efficiency (TEwp) and to explore the genetic factors influencing leaf width in sorghum. The research encompassed two lysimetry experiments involving 16 different genotypes with varying stomatal conductance and three field trials utilizing a diverse sorghum panel comprising 701 lines. The results demonstrated a notable range in leaf width, spanning from 43.1 to 116.8 mm across the three trials. Notably, the lysimetry experiments revealed a negative correlation between leaf width and plant TE. This suggests that narrower leaves may lead to reduced plant transpiration without compromising biomass accumulation. Furthermore, the study found that approximately 70% of the observed leaf width variation could be attributed to racial groups. The guinea racial types consistently exhibited narrower leaves in all trials compared to the other four racial groups (Bicolor, Durra, Kafir, Caudautum), while the East African Durra types displayed wider leaves in comparison to the other racial types. Importantly, 19 QTLs were discovered through genome-wide association studies focusing on leaf width after accounting for flowering time.

3.4.4 Water Extraction Capacity and Pattern

Crops thriving under terminal drought conditions are marked by improved water availability and access during grain filling (Vadez et al. 2007). Enhancing post-anthesis water availability can be achieved by optimizing soil moisture management to conserve water for grain filling and by reducing water use before anthesis or improving transpiration efficiency (Kholová et al. 2010a, b).

Vadez et al. (2011a) conducted a study to assess yield, TE, and water extraction across 149 germplasm entries from the sorghum reference set, which encompassed three control cultivars. They used a lysimetric system in outdoor conditions under both terminal water stress and full irrigation. The study's findings unveiled significant variability in the total water extracted under drought stress among the 149 entries, with values spanning from 10,600 to 15,200 g/plant. Similarly, in fully irrigated conditions, the amount of water extracted by the plants exhibited notable variation, ranging from 10,500 to 42,300 g/plant. Furthermore, the research scrutinized the total water uptake under drought stress conditions for each distinct racial group. The Durra race entries displayed the greatest water extraction capacity, while the Caudatum–Bicolor and Caudatum–Durra intermediate races exhibited lower water extraction capabilities. In terms of TE, the Durra, Caudatum, and Caudatum–Guinea races had the highest TE, with the Guinea race showing the lowest TE.

In a study by Vadez et al. (2011b), the research focused on evaluating variations in water extraction under conditions of both terminal drought and well-irrigated conditions. This assessment involved 29 ILs that were developed to target SG QTLs Stg1, Stg2, Stg3, Stg4, StgA, and StgB in the S35 background, as well as 16 ILs targeting Stg1, Stg3, Stg4, and StgB in the R16 background. The study's findings indicated that, overall, entries in the R16 background exhibited a higher total water extraction (13,900 g kg^{-1}) compared to the S35 background (13,140 g kg^{-1}). In the S35 background, the total water extracted from the soil profile between 30 days after sowing (DAS) and maturity varied from 10,880 to 14,640 g/plant. Stg1 had a positive impact on water extraction in the S35 background, but it had no effect in the R16 background. Furthermore, an inverse correlation between pre- and post-anthesis water utilization were reported in both the S35 background ($R^2 = 0.51$) and the R16 background ($R^2 = 0.87$). Among the local checks, two Durra race entries, Phule Maulee and CRS4, demonstrated the highest water extraction capacities, measuring 15,180 and 15,260 g/plant, respectively. In contrast, the Caudatum race NTJ-2, which is a popular post-rainy season-adapted sweet and grain + stover sorghum variety, exhibited a lower water extraction capacity.

3.4.5 Transpiration Efficiency

Increased TE, often defined as the ratio of mass accumulation to transpiration, is frequently proposed as a key avenue for genetic enhancements aimed at boosting crop yields in conditions with limited water resources. Nevertheless, a detailed examination of TE reveals its complexity, as it is complicatedly linked to both physiological and environmental factors. Physiological factors encompass elements like leaf photosynthetic capacity, the biochemical composition of plant products, and potential constraints on water flow in the plant due to hydraulic limitations. Environmental factors involve variables such as atmospheric CO2 concentration and atmospheric VPD (Sinclair and Rufty 2012).

3.4.5.1 TE Insights from Case Studies

Simulation of TE Simulation studies indicate that increased TE significantly enhances grain yield, especially in drought conditions. In long-term simulations involving sorghum and wheat, it was found that heightened TE was typically linked with enhanced grain yield during post-anthesis drought stress. However, the relationship between TE and yield is not direct, as TE can be influenced by factors such as photosynthetic capacity and transpiration rates. For instance, reducing maximum transpiration rates can lead to higher TE. This can be advantageous in environments where crops depend on stored soil moisture, as it delays the onset of drought stress (Sinclair et al. 2005; Chenu et al. 2018).

TE Assessment in Diversity Panel In the study by Vadez et al. (2011a), they assessed the yield, TE, and water extraction in 149 germplasm entries from the sorghum reference set, alongside three control cultivars. This assessment was conducted using a lysimetric system under both terminal water stress and fully irrigated outdoor conditions. The results indicated significant variation in TE across different entries, with values varying from 3.21 to 6.09 g kg^{-1} of water transpired under drought stress conditions. Similarly, under fully irrigated conditions, TE also displayed considerable variability, spanning from 2.95 to 5.59 g kg^{-1}. For drought stress conditions, TE was evaluated for each specific race. The Guinea race had the lowest average TE values at 4.29 g kg^{-1} ($n = 13$), followed by the Kafir race at 4.58 g kg^{-1} ($n = 6$). In contrast, the Guinea–Caudatum, Durra, and Caudatum races exhibited the highest mean TE values, measuring 5.09, 5.05, and 4.98 g kg^{-1}, respectively, with sample sizes of 25, 20, and 32, respectively. In summary, the variation in TE was primarily attributed to genetic factors rather than genotype-environment interactions, and notably, high TE variants were identified, particularly within the Guinea-Caudatum, Durra, and Caudatum races.

TE Assessment in Stay-Green ILs In the study by Vadez et al. (2011b), they investigated TE, water extraction, harvest index (HI), and yield under both terminal drought and fully irrigated conditions. Two sets of ILs were examined: one with 29 ILs targeting SG QTLs in the S35 background, and another with 16 ILs targeting similar QTLs in the R16 background. In the S35 background, TE ranged from 4.35 to 6.08 g kg^{-1}, with Stg3 and StgB showing the highest mean values. In the R16 background, TE had a broader range, from 3.69 to 5.49 g kg^{-1}, with the StgB ILs exhibiting notably higher TE than the recurrent R16 background.

Link Between TE and Soil Water Conservation Traits In 2015, Vadez and colleagues put forward a significant proposition. They emphasized a robust correlation that has been established between notable disparities in TE among different crops and particular plant traits that efficiently mitigate water loss under high vapor-pressure deficits. This breakthrough not only advances our understanding of the genetic underpinnings of TE, particularly in terms of plant hydraulics and the potential involvement of aquaporins but also opens promising avenues for genetic improvements through targeted breeding. It is crucial to emphasize that even minimal water usage during critical crop stages, such as grain filling, holds immense importance. The research delves into the efficiency of water utilization during these pivotal growth phases.

Link Between TE and Other Factors (Species, Soil Type, and Source–Sink Relationships) In their 2021 study, Vadez and colleagues investigated factors affecting TE in major C4 cereals, including species differences, soil type, and source–sink relationships. They found that maize generally exhibited higher TE than pearl millet and sorghum. High-clay soil was associated with higher TE under high VPD, though the effect depended on the species. Species fitness appeared specific to soil type.

Interestingly, removing cobs significantly reduced TE in maize under high VPD, while panicle removal had no such effect in pearl millet. This suggests that source–sink balance played a role in TE variation. Species differences in TE were attributed to their capacity to limit transpiration under high VPD, possibly influenced by breeding history, favoring source–sink balance in maize. This implies potential for improving TE in pearl millet and sorghum through breeding.

Regarding soil conditions, the study emphasized the need to consider hydraulic characteristics and root systems together for a better understanding of stomatal regulation and transpiration restriction under high VPD. Lastly, the research underscored the significance of sink strength in regulating transpiration and photosynthesis, thus influencing TE.

Difference in TE Linked with Water-Saving Physiological Mechanisms In their 2019 study, Geetika and colleagues examined the variability in TEwp in 27 sorghum genotypes. Their findings demonstrated significant genotypic differences in TEwp for sorghum under well-irrigated conditions. However, this variation was only partly linked to how different genotypes responded at the plant level to VPD. The study unveiled a close correlation between plant-level transpiration rates and leaf stomatal conductance (gsleaf). Consequently, variations in TEwp were attributed to differences in the underlying physiological mechanisms operating at the organ level. Interestingly, genotypes with a shared genetic background typically exhibited similar combinations of transpiration rates and photosynthetic capacity. This insight can be valuable for identifying germplasm with attributes that enhance TEwp under specific environmental conditions.

Blum (2009) argues that the importance of TE becomes evident in water-scarce conditions, emphasizing the need to optimize water use when it is limited. In this context, traits that improve water utilization are crucial. Notably, two water-saving traits discussed earlier, increased sensitivity to VPD and soil drying, contribute to more effective water use. While these traits enhance TE, their primary impact lies in conserving soil water, allowing plants to remain active for longer during water-deficit periods. Consequently, breeding efforts should prioritize traits that enhance a crop's efficient water use rather than solely focusing on increasing TE (Sinclair and Rufty 2012).

3.4.6 Stay-Green Trait in Sorghum

The "stay-green" (SG) trait is a secondary attribute that allows crop plants to retain their green leaves and photosynthetic capacity for an extended period post-anthesis, particularly in challenging conditions like drought and heat stress. Consequently, SG plants exhibit a prolonged grain-filling period and, as a result, achieve higher yields compared to non-SG plants. The SG trait has been acknowledged as a valuable feature for selecting commercially bred cereals, as it plays a role in addressing the current yield plateau, along with offering adaptability and stability in yields. Breeding for functional SG has made significant contributions to improving crop yields, especially when combined with other advantageous traits (Kamal et al. 2019, 2018; Borrell et al. 2014; Christopher et al. 2016).

Recent research has highlighted that the adaptation of SG sorghum to drought conditions is influenced by various factors, including canopy development, leaf anatomy, root growth, and water uptake and utilization (Borrell et al. 2014; George-Jaeggli et al. 2017; Kamal et al. 2018; Vadez et al. 2011a, b). For instance, studies by Kassahun et al. (2010) and Vadez et al. (2011a, b) have shown that SG QTLs have an impact on water uptake, TE, and grain yield, although the extent of this impact varies depending on the genetic background and environmental conditions.

Moreover, during the flowering stage, SG QTLs alter characteristics like tillering, leaf number, and leaf size, resulting in a smaller canopy size. This reduction in canopy size at flowering reduces water usage before anthesis and increases water availability during grain filling, ultimately enhancing grain yield under post-flowering water stress (Borrell et al. 2014; George-Jaeggli et al. 2017). Simulation studies in sorghum have suggested that adding just 1 mm of transpired water during grain filling could increase grain yield by approximately 30 kg/hectare (Hammer et al. 2009).

Recent studies, such as the one conducted by Kamal et al. (2018), have suggested that SG QTLs transferred from "B35" into the Sudanese background "Tabat" possess the capability to modulate transpiration rates and water usage in response to varying drought intensities. This information collectively implies that the SG trait enhances water availability after anthesis, delaying leaf senescence and consequently improving sorghum yield. The SG phenotype seems to arise from the interplay between SG loci, primarily governing plant size and consequently water demand and utilization, and environmental factors influencing water supply from the soil.

Considering the extensive research efforts, SG can be viewed as a coordinated system that ensures an adequate water supply for growth and production under the conditions of terminal drought stress. Notably, SG phenotypes increase water availability, leading to better photosynthesis rates, HI, and biomass in SG ILs compared to non-SG phenotypes (Kamal et al. 2018). These recent investigations highlight the high productivity of SG genotypes under terminal drought stress in sorghum.

Interestingly, the positive influence of SG on yield has been reported across various genetic backgrounds (Jordan et al. 2012), and all four SG QTLs have been found to increase grain yield under both drought and well-watered conditions without negatively affecting yield under adequate water conditions (Kamal et al. 2018; Borrell et al. 2014). In sorghum, only enhanced water productivity (TE) has the potential to simultaneously improve grain and stover yield while enhancing the crop's resilience (Kholová et al. 2014).

3.4.7 Root System Architecture Traits for Drought Adaptation

The architecture of the root system (RSA) plays a crucial role in determining water uptake during periods of moisture stress, making it a valuable trait for enhancing drought tolerance in sorghum through breeding efforts (Joshi et al. 2016). In sorghum, three types of roots are typically observed: primary roots, secondary or adventitious roots, and brace or buttress roots. Primary roots come from the radicle but do not last. Secondary or adventitious roots grow from the first node near the mesocotyle and higher nodes and provide nutrients. Brace or buttress roots develop from basal node root primordia and are short, non-growing roots (Li et al. 2014a, b). Different aspects of RSA, such as the numbers, length, diameter, weight, volume, density, surface area, and elongation rate of roots, are the factors that influence the ability of sorghum varieties to thrive and develop in regions prone to drought (Joshi et al. 2017). Sorghum varieties that are well-suited for environments with limited moisture typically feature deep and robust root systems, as noted by Blum et al. (1997). Given the intricate nature of drought, there is no singular trait or mechanism that can entirely mitigate the issue. Instead, all traits contributing to drought adaptation should be leveraged to their fullest extent possible (Duresso et al. 2023; Teshome and Zhang 2019). The highly variable timing and intensity of drought stress can lead to intricate interactions between a genotype and its environment, ultimately impacting grain yield. Grain yield during drought is closely tied to post-flowering water use, which can be improved by increasing water accessibility or reducing water use before flowering. RSA, the layout of roots in the soil, is pivotal for water acquisition and drought adaptation. However, it has not been widely integrated into breeding programs due to measurement challenges and variable soil conditions (Singh et al. 2012). Recently, there has been increased focus on root growth angle's role in determining shallow or deep root systems, a critical trait for drought-tolerant breeding. While understanding the importance of roots in acquiring water and nutrients is clear, getting accurate data on root traits in their natural environment remains a challenge. Controlled studies also show that root architecture influences above-ground plant performance. Nodal root angle is a key part of RSA, impacting how mature plant roots are distributed in the soil and their ability to adapt to drought. Narrower root angles promote deeper water access in crops like sorghum (Duresso et al. 2023; Singh et al. 2012; Joshi et al. 2017). Despite the rapid progress in sorghum genomics, the absence of cost-effective, high-throughput phenotyping platforms for evaluating RSA parameters poses a significant obstacle. This limitation complicates efforts to comprehend the genetic mechanisms governing these traits and to integrate them into breeding programs through marker-assisted selection (MAS). The study by Joshi et al. (2017) provides a comprehensive exploration of diverse phenotyping methods employed to explore the genetic and physiological underpinnings of root traits in cereal crops.

This section provides an overview of four case studies that examine the diversity of root traits in different germplasm collections, such as diversity panels, RILs, advanced breeding lines, and cultivars, and how these traits are associated with sorghum's ability to adapt drought. Additional details are available in the following content.

3.4.7.1 Insights from Case Studies

Case Study 1 (Sorghum Inbred and Hybrid)

Singh et al. (2011) conducted a study involving 44 inbred lines and 30 hybrid lines to investigate various root traits, such as nodal root angle, root length, root diameter, and root dry weight. The findings demonstrated significant variability in both inbred and hybrid lines. Among the 44 inbred lines, nodal root angles ranged from 14.4° to 50.4°, root lengths varied between 214 m and 452 m, root diameters ranged from 1.24 mm to 2.68 mm, and root dry weights spanned from 0.04 g to 0.18 g. Similarly, within the 30 hybrid lines, nodal root angles showed variability between 14.3° and 42.8°, root lengths ranged from 200 m to 540 m, root diameters varied from 1.16 mm to 2.53 mm, and root dry weights ranged from 0.08 g to 0.19 g. In comparison to the hybrids, the inbred lines exhibited greater diversity in nodal root angles. The findings of this study indicate the possibility of utilizing genetic diversity in nodal root angles to improve sorghum traits. The study identified lines with both extremely narrow and wide nodal root angles, which could potentially serve as donors for this trait. Importantly, the screening and selection of nodal root angles should not be hindered by potential variations in plant size between inbred and hybrid lines.

Case Study 2 (Recombinant Inbred Lines)

In 2012, Mace and colleagues conducted a study focusing on root traits, specifically nodal root angles, in sorghum. They investigated 141 F6 RILs that were developed through single seed descent based on a cross between inbred lines B923296 (with a narrow nodal root angle) and SC170-6-8 (with a wide nodal root angle). Genotyping with 377 DArT markers allowed the construction of a genetic linkage map for the RIL population, comprising 337 markers across 17 linkage groups corresponding to sorghum chromosomes. The study revealed that nodal root angles in the 141 RILs ranged from 14.5° to 32.3°, closely matching the angles of the parent lines. Nodal root angle QTLs explained 58.2% of the phenotypic variation and were consistent across diverse inbred lines. Importantly, some of these QTLs shared homology with root angle QTLs in rice and maize and were found to co-locate with QTLs related to the SG trait in sorghum. This discovery of nodal root angle QTLs offers promising opportunities to enhance the mechanisms of drought adaptation in sorghum.

Case Study 3 (Breeding Lines and Cultivars Related to Stay-Green Trait)

In 2023, Demissie and their team conducted a study to assess the impact of root systems, specifically root angle traits, on sorghum's adaptation to drought stress and grain yield performance. They assessed 428 sorghum genotypes sourced from the Ethiopian breeding program across diverse drought-stress environments. This included both SG and non-SG genotypes, along with released sorghum varieties. The study found that root angles varied from 7.8° to 21.3°, with non-SG genotypes having angles ranging from 8.0° to 30.5° and SG genotypes from 12.0° to 29.0°. Additionally, root traits such as length (38–71 cm), number (7–15.7), root hairs

(22–58), fresh weight (1.14–4.7 g), dry weight (0.17–0.83 g), and root-to-shoot ratio (0.09–0.4) displayed substantial variation. Negative correlations were observed between root angle and leaf area, and there was a negative phenotypic correlation between root angle and root length ($r = -0.018$). The research highlights the importance of considering sorghum genotypes with narrow root angles to enhance grain yield in drought-prone areas, as these genotypes demonstrate a strong connection between root angles and grain yield, benefiting both SG and non-SG sorghum types. The SG trait can be a valuable tool for selecting genotypes with narrow root angles, especially in SG sorghum varieties.

Case Study 4 (Sorghum Diversity Panels)

In 2023, Duresso and their team conducted a study examining 289 sorghum germplasm collections from various sources, including research programs and centers. Their goal was to assess the genetic diversity in root traits. The results showed that the nodal root angle ranged from 8.9° to 21.3°, root length varied from 43.5 to 69.67 cm, the number of roots varied from 7.0 to 15.67, and the fresh and dry weights of the roots varied within the ranges of 1.14 to 4.69 g and 0.17 to 0.77 g, respectively. The study revealed significant positive correlations between several root traits, notably between root dry weight (RDW) and root-to-shoot ratio (RS). Additionally, there were significant and positive correlations between root-branch number (RN), root fresh weight (RFW), RDW, root length (RL), and RS. The study highlighted significant genetic diversity in root and shoot traits, vital for improving sorghum's performance in water-stress conditions. To move forward, they plan to validate these findings by studying how plants respond to water-stress and investigating the genes behind root angles and deep rooting for effective use in breeding programs.

3.5 New Perspectives on the Potential Role of Aquaporins in Water Regulation Under Drought

Aquaporins (AQPs) are transmembrane proteins that help maintain cellular water balance. These proteins belong to the diverse plant AQP protein family, which comprises subfamilies such as plasma membrane intrinsic proteins (PIPs), tonoplast intrinsic proteins (TIPs), NOD 26-like intrinsic proteins (NIPs), small basic intrinsic proteins (SIPs), and unclassified X intrinsic proteins (XIPs). AQPs play a vital role in passive transport of water and various solutes, like CO_2, NH_3, and more (Kaldenhoff et al. 2007; Li et al. 2014a, b). AQPs have vital functions in regulating the movement of water in and out of plant cells, essential for water transport within plants and sustaining cell health. The performance of plant AQPs seems to be closely connected to the expression of plant resilience to biotic and abiotic stresses (Li et al. 2014a, b). The regulation of water movement through cell membranes and across successive cells is achieved through the opening and closing of AQP gates (Zhao et al. 2008). AQPs are found in both the plasma membrane of plant cells and the tonoplast membranes surrounding vacuoles, which typically constitute a significant portion of plant cell volume. Consequently, AQPs have a significant impact on aspects such as overall plant water balance, which includes osmoregulation, expansion of individual cells, long-distance transport, and efficient utilization of water resources by plants (Shekoofa and Sinclair 2018). Empirical evidence highlights the pivotal role of AQPs in shaping two vital plant traits: their responsiveness to soil moisture levels and sensitivity to high atmospheric VPD. These traits empower plants to conserve soil water when facing water deficits, ensuring their physiological functions. AQPs, by regulating hydraulic conductance, contribute to conserving crop water resources in early soil drying and under high VPD conditions. Identifying these traits in crop germplasm has led to breeding programs for dryland-adapted cultivars. Further insights into AQP expression will enhance screening for water conservation traits and selecting cultivars well-suited to diverse water-deficit conditions in crops (Shekoofa and Sinclair 2018).

Sorghum AQPs

In Reddy et al.'s (2015) study, a comprehensive investigation of Sorghum AQP genes was conducted at a genome-wide scale (Table 3.1). This study encompassed aspects such as their relationships with other species, chromosomal distribution, sequence characteristics, and expression levels. The research identified 41 unique AQP genes categorized into four subfamilies: PIPs, TIPs, NIPs, and SIPs. Their physical locations within the Sorghum genome were found to be unevenly distributed (Table 3.1). Topological analysis revealed that SbAQP genes had two to seven transmembrane domains, with PIPs typically having four to six domains (Table 3.1). The genes also contained introns, with varying intron numbers from zero to four. Furthermore, in-silico analysis of SbAQP gene promoters indicated diverse functions associated with both plant development and responses to abiotic stress.

Abiotic Stress Inducible Expression

In the study by Reddy et al. (2015), the expression of Sorghum AQP genes was examined using qRT-PCR in various tissues and under different abiotic stress conditions. The research unveiled distinct expression patterns among different members of the AQP gene family. Notably, drought and salt stress had a more profound impact on the levels of SbAQP transcripts compared to cold stress and abscisic acid (ABA) treatment. Specific SbAQP genes exhibited either up-regulation or down-regulation in response to these stressors. For instance, during drought stress, genes like PIP2.4, PIP2.5, and PIP2.9 were significantly up-regulated, while others like PIP1.6, PIP2.1, PIP2.4, PIP2.10, and TIP4.2 were primarily down-regulated. The intricate expression patterns observed in subfamilies like SbPIP, TIP, and NIP suggest that Sorghum AQP genes fulfill diverse roles in assisting plants in adapting to environmental challenges. This study lays the groundwork for future investigations into SbAQP genes under varying environmental conditions.

Tissue Specificity of Sorghum AQPs

Regarding the specificity of Sorghum AQPs in different tissues, including seedling, root, panicle, and seed, the research revealed intricate and distinct expression patterns. These patterns highlight the specific functions of these genes within each

Table 3.1 Sorghum AQP genes: genome-wide analysis and comprehensive detail

S. No.	Gene name	Protein ID	ORF (bp)	Protein (AA)	Molecular weight (kDa)	pI	Chromosome			Intron	5′ upstream region (bp)	Transmembrane domain	Subcellular localization
							Location	Start	End				
1	PIP1;1	XP_002446929	867	288	30.66	9.06	6	54,149,522	54,151,606	3	–	6	Plastid
2	PIP1;2	XP_002454508	870	289	30.79	9.08	4	62,376,719	62,380,912	3	2500	6	Plastid
3	PIP1;5	XP_002453072	873	290	30.79	8.58	4	67,357,489	67,359,050	1	1960	6	Plastid
4	PIP1;6	XP_002438067	891	296	30.86	8.31	10	7,434,249	7,435,606	1	–	6	Plastid
5	PIP2;1	XP_002461930	873	290	30.27	7.97	2	16,901,298	16,904,960	2	2500	6	Plastid
6	PIP2;2	XP_002461932	861	286	29.91	8.43	2	16,954,792	16,955,852	2	2500	7	Plastid
7	PIP2;3	XP_002461931	861	286	29.92	8.74	2	16,940,031	16,941,473	2	2500	7	Plastid
8	PIP2;4	XP_002452483	870	289	30.38	7.02	4	56,555,048	56,558,379	3	2500	6	Plastid
9	PIP2;5	XP_002446796	879	292	30.46	797	6	52,066,384	52,068,417	2	1795	6	Plastid
10	PIP2;6	XP_002461936	861	286	29.91	8.42	2	17,037,274	17,038,686	2	2500	7	Plastid
11	PIP2;7	XP_002461933	861	286	29.9	8.43	2	16,963,907	16,964,967	2	2500	7	Plastid
12	PIP2;8	XP_002489214	849	282	29.78	9.28	–	2,606,099	2,607,154	0	2500	6	Plastid
13	PIP2;9	XP_002461934	594	197	20.71	10.15 2	2	16,987,880	16,988,880	2	–	5	Vacuolar
14	PIP2;10	Sb02g031390	981	326	35.01	10.28 2	2	66,251,124	66,252,975	3	–	4	Plastid
15	TIP1;1	XP_002465859	753	250	25.77	6.79	1	70,246,505	70,248,652	1	2500	6	Plastid
16	TIP1;2	XP_002459183	770	258	26.02	6.5	3	74,371,021	74,372,588	1	2500	7	Cytoplasmic
17	TIP2;1	XP_002452808	750	249	25.4	6.94	4	62,870,930	62,872,148	1	2500	6	Plastid
18	TIP2;2	XP_002438430	747	248	24.98	6.5	10	41,279,125	41,280,840	2	2500	6	Cytoplasmic
19	TIP2;3	XP_002448289	750	249	25.13	6.67	6	53,686,490	53,687,675	1	2500	6	Plastid
20	TIP3;1	XP_002467022	801	266	27.64	8.77	1	19,171,664	19,172,991	1	2500	6	Mitochondria
21	TIP3;2	XP_002446824	807	268	27.71	9.61	6	52,408,577	52,409,875	2	751	6	Cytoplasmic
22	TIP3;3	XP_002468661	816	271	27.63	9.94	1	72,866,523	72,867,550	2	2500	5	Plastid
23	TIP4;2	XP_002439483	945	314	32.77	7.7	9	14,625,543	14,628,414	2	2500	6	Cytoplasmic
24	TIP4;3	XP_002457068	957	318	32.02	8.76	3	561,153	566,716	2	2500	6	Vacuolar

S. No.	Gene name	Protein ID	ORF (bp)	Protein (AA)	Molecular weight (kDa)	pI	Chromosome			Intron	5′ upstream region (bp)	Transmembrane domain	Subcellular localization
							Location	Start	End				
25	TIP4;4	XP_002457071	759	252	25.27	7.32	3	614,502	615,348	1	2029	6	Cytoplasmic
26	TIP5;1	XP_002448288	669	222	22.05	6.91	6	53,684,983	53,685,785	1	364	5	Plastid/ Vacuolar
27	TIP5;2	XP_002445477	912	303	32.1	10.9	7	522,047,703	52,205,718	3	2500	6	Chloroplast
28	NIP1;1	XP_002453573	864	287	30.19	9.04	4	9,508,112	9,511,126	3	2500	6	Cytoplasmic
29	NIP1;2	XP_002454982	816	271	28.34	8.21	3	2,265,494	2,266,785	4	1449	5	Plastid
30	NIP1;3	XP_002440774	852	283	29.5	8.21	9	9,954,486	9,957,794	3	2500	5	Plastid
31	NIP1;4	XP_002437134	423	140	14.73	9.95	10	48,255,831	48,256,510	2	149	3	Chloroplast
32	NIP1;5	XP_002437133	441	146	15.48	4.14	10	48,255,104	48,255,681	1	2500	2	Cytoplasmic
33	NIP2;1	XP_002454286	894	297	32.04	7.32	4	57,971,834	57,975,693	4	2500	6	Plastid
34	NIP2;2	XP_002438105	888	295	31.36	7.93	10	8,110,238	8,115,151	4	2500	6	Plastid
35	NIP3;1	XP_002464380	906	301	31.16	8.99	1	17,661,019	17,666,112	3	2500	5	Plastid
36	NIP3;2	XP_002445047	870	289	30.01	9.01	7	3,939,661	3,940,905	3	2500	6	Vacuolar
37	NIP3;3	XP_002443852	894	297	31.17	7.85	7	3,952,679	3,953,866	3	1438	6	Plastid
38	NIP4;1	XP_002455311	870	289	30.15	7.43	3	8,731,107	8,732,628	4	1730	6	Plastid
39	SIP1;1	XP_002449310	741	246	25.4	8.47	5	13,710,573	13,714,038	2	2500	5	Plastid
40	SIP1;2	XP_002441068	732	243	25.57	9.41	9	48,712,871	48,716,641	2	2500	6	Plastid
41	SIP2;1	XP_002465351	750	249	26.76	10.2	1	60,518,728	60,520,588	2	1602	7	Chloroplast

The table presents the following information: The putative AQP gene name, NCBI Accession number ID, open reading frame (ORF) size, amino acid length, predicted molecular mass for the deduced proteins, isoelectric point (pI), intron number with size, chromosome location and position, 5′ upstream of the translational start site, transmembrane domain helices number, and predicted subcellular localization need to be determined for each AQP gene analyzed. The table data originates from Reddy et al.'s (2015) study titled "Genome-wide identification and characterization of the aquaporin gene family in *Sorghum bicolor* (L.)" was published in the journal *Plant Gene* in 2015, spanning pages 18–28

tissue during different stages of development. Among the tissues studied, seedling, root, and panicle displayed higher AQP gene expression compared to seeds, underlining their importance in these tissues during active growth. Several SbAQP genes, such as PIP2;3, PIP2;5, PIP2;8, PIP2;9, TIP2;1, TIP2;3, TIP3;1, NIP1;1, NIP1;2, and NIP4;1, showed up-regulation in all tested tissues, while PIP2;10 and TIP4;2 exhibited repression. Certain SbAQPs, like TIP3;1 and PIP2.9, demonstrated higher transcript levels in root tissue. Conversely, genes like PIP2;5, PIP2;9, TIP2;1, TIP3;1, and TIP5;2 displayed lower expression in seeds compared to other tissues. Additionally, TIP1;1, TIP4;3, and NIP2;2 were predominantly expressed in the panicle and seed. Notably, specific genes were exclusively detected in particular tissues, such as PIP2;9 in seeds, and PIP1;5 and NIP1;1 in roots. To summarize, all SbAQP subfamily genes exhibited upregulation in panicle, root, and seedling tissues, with some exceptions like PIP2;7, TIP3;3, TIP4;2, TIP4;3, TIP5;1, TIP5;2, NIP1;4, and NIP1;4. In the case of seeds, SbAQP subfamily genes did not follow a uniform pattern, displaying differential expression. These results highlight the tissue-specific functions of SbAQP genes throughout normal plant development and set the stage for additional investigation into the expression patterns of SbAQP gene family members. Such exploration can deepen our comprehension of their distinct roles, facilitating the identification of candidate genes for future targeted functional analyses in specific tissues.

Sorghum AQPs Inhibition

In 2013, Choudhary et al. conducted a study involving two sorghum genotypes, SC15 and SC1205. They found that genotype SC15 displayed significantly lower Kleaf compared to SC1205. This distinction in leaf conductance suggested that manipulating plant hydraulic conductance could be a strategy to conserve soil water, particularly during late-season water deficits. The research involved subjecting these two sorghum genotypes to increasing VPD to assess their transpiration rates. The results showed that SC1205 displayed a consistent increase in transpiration rate with rising VPD, indicating that there was no evident restriction to water transport in the plant across the entire range of VPD. On contrary, SC15 exhibited a distinct change in transpiration rate at a VPD of 2.1 kPa, beyond which the rate of increase in transpiration was significantly reduced. Furthermore, the study explored the effects of AQP inhibitors, including 50 mM HAuCl4, 1 mM ZnCl2, 500 mM HgCl2, AgNO3, and CHX, on transpiration in both leaves and whole plants (roots) (Table 3.2). The results demonstrated that SC15 and SC1205 responded differently to these inhibitors (Table 3.2). Specifically, treatments with gold and zinc inhibitors resulted in a more substantial decrease in transpiration rate in SC15 compared to SC1205. Conversely, with silver, mercury, and CHX inhibitors, the greater decrease in transpiration occurred in SC1205 rather than SC15 (Table 3.2). However, when considering intact plants, the responses to all four aquaporin inhibitors were the same for both genotypes, with SC15 consistently exhibiting a greater decrease in transpiration compared to SC1205 in all cases. These findings shed light on the role

Table 3.2 Reduction in transpiration rate (TR) following treatment with various aquaporin inhibitors in both leaf and intact plant (roots)

Genotypes	$HAuCl_4$	$AgNO_3$	$HgCl_2$	$ZnCl_2$	CHX
Leaf inhibition	50 μM	500 μM	500 μM	1 mM	500 μM
SC15	68.5 ± 4.6	20.7 ± 5.4	33.3 ± 2.7	72.9 ± 6.3	36.3 ± 3.2
SC1205	31.4 ± 3.2	36.7 ± 3.4	51.9 ± 4.4	46.1 ± 4.9	60.5 ± 6.4
p	***	***	***	***	***
Root inhibition	250 μM	500 μM	500 μM	1 mM	–
SC15	24.4 ± 3.1	38.2 ± 3.1	30.5 ± 3.1	23.4 ± 6.1	–
SC1205	17.1 ± 2.4	23.1 ± 5.2	14.9 ± 4.2	4.6 ± 3.8	–
p	**	***	***	***	–

The data presented show the average percentage decrease in TR ± standard error. "CHX" represents cycloheximide, and "*P*" indicates the probability of a difference between the two genotypes. ** denotes a significance level of $P < 0.01$, while *** indicates $P < 0.001$. The information in the table is extracted from Choudhary et al. (2013b) study titled "Hydraulic conductance of intact plants of two contrasting sorghum lines, SC15 and SC1205," published in the journal *Functional Plant Biology*, 40, pp. 730–738. https://doi.org/10.1071/FP12338

of AQP inhibitors and genotypic differences in regulating transpiration in sorghum plants.

3.6 Enhancing Drought Adaptation in Sorghum: A Phenotyping Framework for Breeding Initiatives

Drought adaptation is complex trait, and the pursuit of sustainable drought-adaptation crops demands a thorough understanding of the specific drought stress factors within a given environment (Fig. 3.4). Additionally, the systematic screening techniques or tools employed must be both accurate and consistent, as they must accommodate the diverse array of breeding lines for selection and seamless integration into the overall breeding process (Fig. 3.4). Utilizing a variety of screening techniques at various phases of the breeding process is crucial for progressing the development of drought-adapted agricultural products.

3.6.1 Utilizing Sensor-Based Approaches for Early-Stage Phenotyping in Large-Scale Breeding Programs

Incorporating sensor-based smart digital technologies, such as the HTP phenotyping platform and UAV-based field phenotyping methods, can significantly accelerate the breeding process during the early stages (F4, F5) of breeding programs. These technologies are particularly valuable for assessing basic traits related to adaptation, including early vigor, canopy index, plant height, and tiller number, and can be applied to both agronomic and adaptation traits.

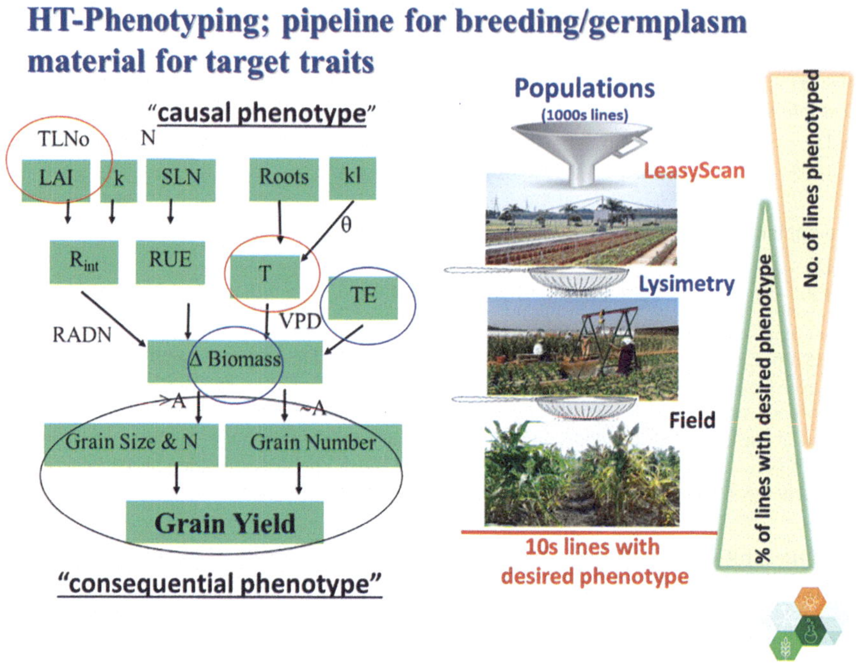

Fig. 3.4 High-throughput phenotyping pipeline for breeding or germplasm material for target traits

3.6.2 High-Throughput Phenotyping Platform (HTP-LeasyScan)

LeasyScan established at ICRISAT, enables the nondestructive assessment of various plant materials during the vegetative stage. Utilizing an optical system called PlantEye, LeasyScan offers accurate measurements of plant canopy traits, encompassing digital biomass, 3D-leaf area, plant height, LAI, leaf angle, leaf inclination, and light penetration depth. Additionally, it can detect diurnal variations in water use parameters through 1500 analytical scales. To illustrate the practical utility of this technology, a group of 181 sorghum RILs, specially developed for enhanced drought adaptation via SG traits, underwent phenotyping using LeasyScan (Chadalavada 2017). The platform assessed a range of canopy development traits (e.g., 3D-leaf area, projected leaf area, plant height, LAI, canopy structure) and conductance-related traits (e.g., evapotranspiration, transpiration, and transpiration rate; Karthika et al. 2019). In summary, the integration of advanced sensor-based technologies offers a nondestructive and comprehensive approach to phenotyping critical plant traits, particularly those pertinent to drought adaptation, in the early stages of breeding. This approach holds the potential to significantly enhance the efficiency and precision of breeding programs.

3.6.3 Unmanned Aerial Vehicles (UAVs)

Unmanned aerial vehicles (UAVs) offer benefits for high-throughput phenotyping, including rapid data collection and adaptability to diverse crops and growth stages. However, they face limitations like reduced payload capacity and susceptibility to environmental factors. Drones are equipped with various sensors, such as RGB cameras and LiDAR sensors, to provide spatial and spectral data for phenotypic trait measurement (Xu et al. 2021). The International Crops Research Institute for the Semi-Arid Tropics (ICRISAT) employs drone-based phenotyping for crop research, contributing to efficient agriculture. Priyanka et al. (2023) explores the use of UAVs with optical sensors for precise crop assessment, addressing the challenge of consistent image quality through image acquisition protocols and quality control procedures. The study offers a web-based application for computing essential image quality indicators and foresees broader application in UAV imaging setups for data interoperability.

3.6.4 HTP-Leasy-Field

Promising lines identified through large-scale screening can undergo detailed characterization specific to drought types relevant to targeted environments, as depicted in Fig. 3.4. This involves artificially inducing water stress and measuring water uptake under controlled conditions like rainout shelters or greenhouses. The utilization of the lysimeter platform, a gravimetric transpiration measurement system offering individual soil profiles for each plant, enables the continuous monitoring of water usage throughout the crop's life cycle (Vadez et al. 2013). This system facilitates critical agronomic assessments of yield and its components. Additionally, it facilitates monitoring of total plant water consumption (the cumulative water usage throughout the crop's lifespan), the fluctuations in water consumption during crop growth, and TE, calculated as the ratio of total biomass produced to total water usage. Lysimeter-based assessments facilitate the investigation of temporal fluctuations in water demands across different genotypes and various crops, including sorghum, pearl millet, finger millet, chickpea, and peanut (Vadez et al. 2011a, b, 2013; Zaman-Allah et al. 2011; Vadez and Ratnakumar 2016; Tharanya et al. 2018; Sivasakthi et al. 2019). This approach has been used to identify genetic variations in TE and grain yield in sorghum (Vadez et al. 2011a, b; Chadalavada 2017; Sivasakthi 2022). Furthermore, it has revealed genotypic variances in water extraction patterns during critical crop growth stages. Drought-adapted genotypes were observed to extract less water before flowering and more water after anthesis, suggesting a potential contribution to terminal drought adaptation (Vadez et al. 2013).

3.6.5 Field Phenotyping

Promising lines identified through HTP-Lysi-Field screening can undergo detailed characterization specific to drought types relevant to targeted environments, as depicted in Fig. 3.4. In field phenotyping for sorghum, a comprehensive evaluation of plant characteristics like growth rate, yield, resilience to drought, resistance to diseases, and others takes place within the authentic context of field conditions. This field-based screening comprises the examination of test materials through trials held in regions prone to drought. Another approach involves planting sorghum during the post-rainy season with an initial water supply, followed by the deliberate withholding of irrigation to induce drought stress at specific growth stages.

3.7 QTLs Are Linked with Canopy Development, Water Use, and Gronomical Traits

Sorghum breeding and genomic research have historically fallen behind other major crops like rice and maize. Nevertheless, significant milestones achieved in the past decade have propelled sorghum research into a promising new era. These milestones include the sequencing of the sorghum reference genome (Paterson et al. 2009), the development of multiple Sorghum Association Panels (SAP) (Casa et al. 2008; Morris et al. 2013; Upadhyaya et al. 2009), the establishment of mutant libraries, the recent completion of sorghum pan-genomes, the availability of a sorghum gene expression atlas, and the creation of online resources like SorghumBase (Addo-Quaye et al. 2018; Jiao et al. 2016; Xin et al. 2008; Tao et al. 2021; Makita et al. 2015; Shakoor et al. 2014). With these extensive resources now accessible to producers and researchers, sorghum is dignified to play a crucial role in addressing global food and energy security in the face of a changing climate (Xin et al. 2021).

Over the past 25 years, numerous linkage mapping studies have been conducted in sorghum to investigate the genetic basis of various traits, including height (Hilley 2017; Lin et al. 1995; Yamaguchi et al. 2016), flowering time (Casto et al. 2019; Guitton et al. 2018; Mace et al. 2013), pigmentation (Wu et al. 2012; Morris et al. 2013), drought tolerance (Hayes et al. 2016; Subudhi et al. 2000; Tuinstra et al. 1996), cold tolerance (Burow et al. 2011; Knoll and Ejeta 2008; Marla et al. 2019), and disease resistance (Upadhyaya et al. 2013).

The primary and extensively utilized resource for sorghum Genome-Wide Association Studies (GWAS) is the Sorghum Association Panel (SAP). SAP's global diversity panel, complete with GBS marker data (Hu et al. 2019; Morris et al. 2013), can be accessed through Germplasm Resources Information Network (GRIN). This panel was meticulously designed with the aim of encompassing the worldwide diversity of sorghum in terms of its physical characteristics, functionalities, and applications (Casa et al. 2008). It predominantly comprises various grain sorghum types, supplemented by a selection of sweet sorghums, forages, and broomcorn varieties (Xin et al. 2021).

The concept of a "mini-core" was introduced, representing 10% of the core collection or 1% of the entire germplasm collection within a sorghum gene bank, as proposed by Upadhyaya and Ortiz (2001). The first sorghum mini-core, comprising 242 accessions, was established using data from 21 morpho-agronomic traits and passport information drawn from a larger pool of 2246 accessions in the core collection, as detailed by Upadhyaya et al. (2009). Both the sorghum core and minicore collections have proven to be valuable resources for identifying accessions possessing various traits of interest, as demonstrated by Upadhyaya et al. (2016). Table 3.3 illustrates a comprehensive list of QTLs associated with component traits related to drought adaptation, as depicted in Fig. 3.4. These QTLs have been previously mapped, and the findings from various studies are compiled and presented in this table.

3.8 Conclusion

In conclusion, the exploration of sorghum's physiological responses, especially concerning environmental stresses like drought, has provided invaluable insights. Physiological characteristics such as regulated transpiration rates, sensitivity to soil drying, and the intricacies of canopy development have been methodically examined, revealing their crucial roles in strengthening sorghum's adaptive resilience under water-limited condition.

The molecular profile of sorghum has been thoroughly investigated, particularly emphasizing AQPs and their active participation in transportation of water. The identification of distinct AQP subfamilies, coupled with their inducible expressions under abiotic stress, underscores the detailed molecular mechanisms governing sorghum's responses. Furthermore, the tissue-specific patterns of sorghum AQPs, particularly within seedling, root, panicle, and seed tissues, have elucidated the molecular intricacies contributing to various developmental stages.

Genetic advancements have significantly advanced sorghum research, as demonstrated by the mapping of QTLs associated with canopy development, water utilization, and agronomic traits. The establishment of Sorghum Association Panels (SAP) combined with genotyping-by-sequencing (GBS) data has empowered Genome-Wide Association Studies (GWAS), leading to the identification of key genetic determinants. Additionally, the exploration of sorghum pan-genomes and gene expression atlases has enriched our understanding of genetic variations and regulatory networks governing sorghum traits.

Novel breeding strategies have been embraced, focusing on physiological traits and genetic markers. The incorporation of the "stay-green" trait, with its profound impact on water utilization and extended grain-filling periods, serves as a testament to the successful integration of genetic advancements into breeding programs.

Technological integration, particularly through sensor-based platforms and UAVs, signifies a transformative leap in sorghum research. These technologies facilitate nondestructive and rapid data collection, enabling precise assessments of critical physiological traits, particularly in the early phases of breeding programs.

Table 3.3 Serves as a comprehensive compilation of previously identified Quantitative Trait Loci (QTLs) and Marker-Trait Associations (MTAs) associated with a diverse set or bi-parental mapping population of agronomic, phenological, plant growth, and physiological traits in sorghum

Sl. No.	Trait list	Population	LG:Start-End (v3.0)	Genes under QTL (v3.0)	Publication	Total No. of QTLs
1	Leaf area	IS2449/IS1488	1:57,244,324–60,329,324	224	Phuong et al. (2013)	6
		HYP/DTP	2:10,604,952–14,181,362	174	Phuong et al. (2013)	
		HYP/DTP	2:62,738,401–65,138,565	312	Phuong et al. (2013)	
		HYP/DTP	3:9,870,542–11,085,377	130	Phuong et al. (2013)	
		HYP/DTP	5:787,751–1,912,781	108	Phuong et al. (2013)	
		IS2449/IS1488	10:10,532,053–50,312,453	687	Phuong et al. (2013)	
2	Leaf numbers	Diversity research set ($n = 107$)	3:1,902,942–1,917,791	1	Sakhi et al. (2013)	5
		Diversity research set ($n = 107$)	3:1,902,942–1,917,791	1	Sakhi et al. (2013)	
		M35-1/B35	1:2,229,144–6,482,923	569	Nagaraja Reddy et al. (2013)	
		M35-1/B35	9:58,706,951–59,398,238	84	Nagaraja Reddy et al. (2013)	
		E36-1/SPV570	10:13,940,494–51,776,001	622	Rajkumar et al. (2013)	
3	Leaf width	Red Kafir/Takakibi (F2:3)	1:49,985,144–66,041,119	1099	Shehzad and Okuno (2015)	5
		BTx623/IS3620C	4:60,890,000–65,072,462	535	McCormick (2017)	
		Tx430/Tx7078	9:2,306,153–2,951,546	72	Kapanigowda et al. (2014)	
		Diversity research set ($n = 107$)	9:4,087,397–4,339,608	24	Sakhi et al. (2013)	
		Red Kafir/Takakibi (F2)	3:58,542,948–74,384,257	2086	Shehzad and Okuno (2015)	

4	Leaf angle	BTx623/IS3620C	7:59,480,000–61,098,149	172	McCormick (2017)	5
		Sorghum Association Panel (SAP)	1:54,369,359–54,929,968	16	Zhao et al. (2008)	
		R07018/R07020	3:14,889,177–51,648,272	518	Truong et al. (2015)	
		Sorghum Association Panel (SAP)	3:4,560,802–4,884,543	44	Mantilla-Perez et al. (2020)	
		Sorghum Association Panel (SAP)	9:1,000,006–1,049,867	11	Zhao et al. (2008)	
5	Stem dry weight	SS79/M71	1:66,062,531–73,867,727	978	Bekele et al. (2014)	5
		SS79/M71	2:59,667,756–63,495,686	416	Bekele et al. (2014)	
		SS79/M71	3:58,289,710–63,625,029	680	Bekele et al. (2014)	
		Shihong137/L-Tian	4:52,931,257–56,139,504	365	Wang et al. (2016)	
		Shihong137/L-Tian	6:57,953,334–61,260,478	461	Wang et al. (2016)	
6	Shoot dry weight	Tx430/Tx7078	1:23,465,053–53,904,060	478	Kapanigowda et al. (2014)	5
		SS79/M71	2:60,376,496–61,756,869	153	Shiringani and Friedt (2011)	
		B923296/ SC170-6-8	5:12,964,702–64,258,310	834	Mace et al. (2012)	
		SS79/M71	7:3,427,571–55,700,505	1020	Bekele et al. (2014)	
		Diverse set	9:54,267,103–59,193,962	689	Zhang et al. (2015)	

(continued)

Table 3.3 (continued)

Sl. No.	Trait list	Population	LG:Start-End (v3.0)	Genes under QTL (v3.0)	Publication	Total No. of QTLs
7	Plant height	NPGS Sudan Sorghum Core collection	1:0–211,157	21	Cuevas and Prom (2020)	5
		BTx623/Hong Ke Zi (CIM)	2:3,803,804–3,948,428	20	Marla et al. (2019)	
		BTx623/IS3620C	3:2,992,535–3,031,013	6	Kong et al. (2018)	
		Nested Association Mapping set US	4:7,130,950–7,299,969	10	Bouchet et al. (2017)	
		CK60/China17	6:39,296,212–43,944,566	209	Gelli et al. (2016)	
8	Tillers	IS8525/ R931945-2-2	4:6,740,670–10,927,759	302	Alam et al. (2014)	6
		B923296/ SC170-6-8	4:8,725,340–12,234,807	192	Alam et al. (2014)	
		B923296/ SC170-6-8	4:8,748,298–12,294,811	192	Alam et al. (2014)	
		IS8525/ R931945-2-2	7:61,938,430–65,460,255	422	Alam et al. (2014)	
		IS8525/ R931945-2-2	8:2,322,131–3,776,419	140	Alam et al. (2014)	
		B923296/ SC170-6-8	10:8,854,460–51,529,292	856	Alam et al. (2014)	

9	Root length	SS79/M71	10:8,636,988–50,276,204	811	Bekele et al. (2014)	5
		Sorghum Association Panel (SAP)	9:8,083,378–8,157,501	3	Chopra et al. (2017)	
		Shihong137/L-Tian	1:71,259,122–74,093,585	365	Wang et al. (2016)	
		HYP/DTP	5:69,995,481–71,830,991	192	Phuong et al. (2013)	
		SS79/M71	8:5,504,935–55,898,690	834	Bekele et al. (2014)	
10	Root volume	E36-1/SPV570	4:56,958,693–61,029,003	491	Rajkumar et al. (2013)	1
11	Root dry weight	HYP/DTP	9:51,942,634–55,721,697	510	Phuong et al. (2013)	5
		B923296/ SC170-6-8	8:55,169,373–58,390,210	244	Mace et al. (2012)	
		Sorghum Association Panel (SAP)	7:2,842,968–2,897,763	7	Hufnagel et al. (2014)	
		Shihong137/L-Tian	6:58,257,246–61,260,478	427	Wang et al. (2016)	
		E36-1/SPV570	4:56,247,781–61,221,600	588	Rajkumar et al. (2013)	
12	Root surface area	Sorghum Association Panel (SAP)	3:6,162,031–7,030,484	104	Hufnagel et al. (2014)	2
		Sorghum Association Panel (SAP)	3:59,582,432–60,273,303	86	Hufnagel et al. (2014)	

(continued)

Table 3.3 (continued)

Sl. No.	Trait list	Population	LG:Start-End (v3.0)	Genes under QTL (v3.0)	Publication	Total No. of QTLs
13	Root biomass	Sorghum Association Panel (SAP)	1:24,700,039–49,720,395	308	Moghimi et al. (2019)	5
		Sorghum Association Panel (SAP)	2:57,175,591–57,369,842	11	Moghimi et al. (2019)	
		Sorghum Association Panel (SAP)	3:3,285,713–3,493,670	16	Moghimi et al. (2019)	
		Sorghum Association Panel (SAP)	4:63,330,527–65,080,596	226	Moghimi et al. (2019)	
		Sorghum Association Panel (SAP)	5:55,940,425–56,351,449	12	Moghimi et al. (2019)	
14	Root angle	Bk7/ Early-Hegari-Sart	3:0–11,227,673	1287	Lopez et al. (2017)	5
		B923296/ SC170-6-8	5:11,160,761–61,009,806	695	Mace et al. (2012)	
		B923296/ SC170-6-8	5:64,894,679–65,069,572	14	Mace et al. (2012)	
		B923296/ SC170-6-8	8:5,070,472–55,930,493	873	Mace et al. (2012)	
		B923296/ SC170-6-8	10:57,102,619–60,429,269	442	Mace et al. (2012)	

15	Nodal root angle	BR23296 X SC170-6-8 F6	5	–	Harris-Shultz et al. (2019)	4
		BR23296 X SC170-6-8 F6	5	–	Harris-Shultz et al. (2019)	
		BR23296 X SC170-6-8 F6	8	–	Harris-Shultz et al. (2019)	
		BR23296 X SC170-6-8 F6	10	–	Harris-Shultz et al. (2019)	
16	Transpiration rate	Tx430/Tx7078	1:11,551,429–17,763,052	548	Kapanigowda et al. (2014)	5
		Sorghum Association Panel (SAP)	2:5,823,924–6,043,857	23	Ortiz et al. (2017)	
		Sorghum Association Panel (SAP)	3:9,360,977–9,478,602	3	Ortiz et al. (2017)	
		Tx430/Tx7078	7:61,509,637–63,163,284	185	Kapanigowda et al. (2014)	
		Sorghum Association Panel (SAP)	10:13,967,670–41,745,111	272	Ortiz et al. (2017)	

(continued)

Table 3.3 (continued)

Sl. No.	Trait list	Population	LG:Start-End (v3.0)	Genes under QTL (v3.0)	Publication	Total No. of QTLs
17	Stomatal conductance	Sorghum Association Panel (SAP)	1:3,989,757–4,161,813	20	Ortiz et al. (2017)	5
		Sorghum Association Panel (SAP)	3:1,945,887–2,235,348	41	Ortiz et al. (2017)	
		Sorghum Association Panel (SAP)	4:9,608,822–10,135,566	41	Ortiz et al. (2017)	
		Sorghum Association Panel (SAP)	5:67,936,235–68,660,797	64	Ortiz et al. (2017)	
		Sorghum Association Panel (SAP)	6:49,500,009–49,647,577	20	Ortiz et al. (2017)	
18	Transpiration ratio (A:E)	Tx430/Tx7078	6:51,324,599–53,491,505	286	Kapanigowda et al. (2014)	4
		Tx430/Tx7078	9:1,355,225–2,272,894	107	Kapanigowda et al. (2014)	
		Tx430/Tx7078	9:2,378,248–2,951,091	63	Kapanigowda et al. (2014)	
		Tx430/Tx7078	10:57,282,934–60,429,689	420	Kapanigowda et al. (2014)	

19	Days to flowering	BTx623/IS3620C	1:22,182,504–22,857,599	39	Kong et al. (2018)	5
		P118	2:67,854,771–72,685,107	658	Guindo et al. (2019)	
		BTx623/Niu Shen Zui (CIM)	3:48,215,495–50,181,334	40	Marla et al. (2019)	
		R931945-2-2*2/*S. bicolor* subsp. verticilliflorum	4:1,051,788–1,295,563	40	Bangbol Sangma (2013)	
		Tx623 x *S. virgatum*	10:8,939,298–50,271,638	780	Liu et al. (2019)	
20	Stay-green	Advanced Yield Test germplasm set	1:67,191,764–67,694,876	59	Wang et al. (2016)	5
		B35/Tx7000	2:60,068,675–61,997,085	207	Xu et al. (2021)	
		QL39/QL41	3:7,461,203–9,984,674	251	Tao et al. (2000)	
		SC56/Tx7000	4:43,661,002–50,999,600	180	Kebede et al. (2001)	
		B35/Tx7000	5:10,923,140–62,813,315	833	Subudhi et al. (2000)	
21	Number of green leaves	M35-1/B35	1:4,155,674–6,629,725	316	Rama Reddy et al. (2014)	5
		M35-1/B35	2:1,212,467–2,545,754	146	Rama Reddy et al. (2014)	
		M35-1/B35	3:16,834,712–53,491,968	538	Rama Reddy et al. (2014)	
		M35-1/B35	4:2,942,349–4,694,129	224	Rama Reddy et al. (2014)	
		M35-1/B35	9:55,813,801–57,991,519	321	Rama Reddy et al. (2014)	

(continued)

Table 3.3 (continued)

Sl. No.	Trait list	Population	LG:Start-End (v3.0)	Genes under QTL (v3.0)	Publication	Total No. of QTLs
22	Grain weight	P114	1:0–11,463,167	1483	Guindo et al. (2019)	5
		Tx623 x *S. virgatum*	2:65,837,295–70,940,526	685	Liu et al. (2019)	
		R931945-2-2*2/*S. bicolor* subsp. verticilliflorum	3:58,432,545–58,756,700	36	Tao et al. (2018)	
		IS8525/ R931945-2-2	4:50,223,794–52,216,695	126	Spagnolli et al. (2016)	
		BTx642/ BTxARG-1	6:44,954,674–46,015,366	98	Boyles et al. (2017)	
23	Grain numbers	BTxARG-1/ PI656056	1:0–2,213,656	285	Boyles et al. (2017)	5
		IS8525/ R931945-2-2	2:6,044,411–6,054,024	1	Spagnolli et al. (2016)	
		Sorghum Association Panel (SAP)	6:54,856,354–54,972,978	17	Zhao et al. (2008)	
		BTx642/ BTxARG-1	9:2,877,552–5,113,141	198	Boyles et al. (2017)	
		Sorghum Association Panel (SAP)	10:51,758,198–52,033,182	19	Zhao et al. (2008)	

24	Grain yield	P114	1:73,031,122–78,904,411	754	Guindo et al. (2019)	5
		Tx436/00MN7645	2:62,279,663–64,238,402	249	Sukumaran et al. (2016)	
		West African Association Panel (WAP)	3:70,932,594–71,325,290	55	Leiser et al. (2014)	
		E-Tian/Ji2731	4:60,965,264–63,347,669	308	Mocoeur et al. (2015)	
		CK60/China17	5:65,122,552–69,628,610	386	Gelli et al. (2016)	

The dataset is sourced from the study conducted by Mace et al. (2019), as presented in the publication titled "The Sorghum QTL Atlas: a potent tool for trait dissection, comparative genomics, and crop enhancement," featured in Theoretical and Applied Genetics with the https://doi.org/10.1007/s00122-018-3212-5

This chapter provides a comprehensive overview of sorghum, emphasizing its significance in arid regions, physiological, molecular insights, genetic advancements, novel breeding strategies, and transformative technological integration for improved adaptation under environmental stresses.

References

Addo-Quaye C, Tuinstra M, Carraro N, Weil C, Dilkes BP (2018) Whole-genome sequence accuracy is improved by replication in a population of mutagenized sorghum. G3 Genes Genomes Genet 8(3):1079–1094

Alam MM, Mace ES, van Oosterom EJ, Cruickshank A, Hunt CH, Hammer GL, Jordan DR (2014) QTL analysis in multiple sorghum populations facilitates the dissection of the genetic and physiological control of tillering. Theor Appl Genet 127:2253–2266

Anonymous. FAOSTAT. Crop Statistics; 2022. Available: http://www.fao.org/faostat/en/#data/QCL (Accessed on 15th February 2023).

Aruna C, Cheruku D (2019) Genetic improvement of grain sorghum. In: Breeding sorghum for diverse end uses. Woodhead Publishing, Sawston, pp 157–173

Baird AS, Taylor SH, Pasquet-Kok J, Vuong C, Zhang Y, Watcharamongkol T, Sack L (2021) Developmental and biophysical determinants of grass leaf size worldwide. Nature 592(7853):242–247

Bangbol Sangma H (2013) Genetic characterization of flowering time in sorghum. PhD Thesis, School of Agriculture and Food Sciences, The University of Queensland

Battisti DS, Naylor RL (2009) Historical warnings of future food insecurity with unprecedented seasonal heat. Science 323(5911):240–244

Bekele WA, Fiedler K, Shiringani A, Schnaubelt D, Windpassinger S, Uptmoor R, Snowdon RJ (2014) Unravelling the genetic complexity of sorghum seedling development under low-temperature conditions. Plant Cell Environ 37(3):707–723

Blum A (2009) Effective use of water (EUW) and not water-use efficiency (WUE) is the target of crop yield improvement under drought stress. Field Crop Res 112(2-3):119–123

Blum A, Golan G, Mayer J, Sinmena B (1997) The effect of dwarfing genes on sorghum grain filling from remobilized stem reserves, under stress. Field Crop Res 52(1-2):43–54

Borrell AK, Mullet JE, George-Jaeggli B, van Oosterom EJ, Hammer GL, Klein PE, Jordan DR (2014) Drought adaptation of stay-green sorghum is associated with canopy development, leaf anatomy, root growth, and water uptake. J Exp Bot 65(21):6251–6263

Bouchet S, Olatoye MO, Marla SR, Perumal R, Tesso T, Yu J, Morris GP (2017) Increased power to dissect adaptive traits in global sorghum diversity using a nested association mapping population. Genetics 206(2):573–585

Boyle A (2018) Climate change, the Paris Agreement and human rights. Int Comp Law Q 67(4):759–777

Boyles RE, Pfeiffer BK, Cooper EA, Zielinski KJ, Myers MT, Rooney WL, Kresovich S (2017) Quantitative trait loci mapping of agronomic and yield traits in two grain sorghum biparental families. Crop Sci 57(5):2443–2456

Boyles RE, Brenton ZW, Kresovich S (2019) Genetic and genomic resources of sorghum to connect genotype with phenotype in contrasting environments. Plant J 97(1):19–39

Burow G, Burke JJ, Xin Z, Franks CD (2011) Genetic dissection of early-season cold tolerance in sorghum (Sorghum bicolor (L.) Moench). Mol Breed 28:391–402

Casa AM, Pressoir G, Brown PJ, Mitchell SE, Rooney WL, Tuinstra MR, Kresovich S (2008) Community resources and strategies for association mapping in sorghum. Crop Sci 48(1):30–40

Casto AL, Mattison AJ, Olson SN, Thakran M, Rooney WL, Mullet JE (2019) Maturity2, a novel regulator of flowering time in Sorghum bicolor, increases expression of SbPRR37 and SbCO in long days delaying flowering. PLoS One 14(4):e0212154

Chadalavada K (2017) Understanding the physiological processes determining sorghum grain nutritional quality (Shodhganga@INFLIBNET: understanding the physiological processes determining sorghum grain nutritional quality)
Chapke RR, Babu S, Subbarayudu B, Tonapi VA (2017) Growing popularity of sorghum in rice fallows: an IIMR Case study. ICAR - Indian Institute of Millets Research, Hyderabad
Chenu K, Van Oosterom EJ, McLean G, Deifel KS, Fletcher A, Geetika G, Hammer GL (2018) Integrating modelling and phenotyping approaches to identify and screen complex traits: transpiration efficiency in cereals. J Exp Bot 69(13):3181–3194
Chopra R, Burow G, Burke JJ, Gladman N, Xin Z (2017) Genome-wide association analysis of seedling traits in diverse Sorghum germplasm under thermal stress. BMC Plant Biol 17:1–15
Choudhary S, Mutava RN, Shekoofa A, Sinclair TR, Prasad PV (2013a) Is the stay-green trait in sorghum a result of transpiration sensitivity to either soil drying or vapor pressure deficit? Crop Sci 53(5):2129–2134
Choudhary S, Sinclair TR, Prasad PV (2013b) Hydraulic conductance of intact plants of two contrasting sorghum lines, SC15 and SC1205. Funct Plant Biol 40(7):730–738
Choudhary S, Guha A, Kholova J, Pandravada A, Messina CD, Cooper M, Vadez V (2020) Maize, sorghum, and pearl millet have highly contrasting species strategies to adapt to water stress and climate change-like conditions. Plant Sci 295:110297
Christopher JT, Christopher MJ, Borrell AK, Fletcher S, Chenu K (2016) Stay-green traits to improve wheat adaptation in well-watered and water-limited environments. J Exp Bot 67(17):5159–5172
Clegg MD, Biggs WW, Eastin JD, Maranville JW, Sullivan CY (1974) Light transmission in field communities of sorghum 1. Agron J 66(4):471–476
Cline WR (2007) Global warming and agriculture: Impact estimates by country. Peterson Institute. Component of water stress tolerance research. Funct Plant Biol 40(12):1310–1322
Cuevas HE, Prom LK (2020) Evaluation of genetic diversity, agronomic traits, and anthracnose resistance in the NPGS Sudan Sorghum Core collection. BMC Genomics 21(1):1–15
Duncan WG (1971) Leaf angles, leaf area, and canopy photosynthesis 1. Crop Sci 11(4):482–485
Duncan WGRS, Loomis R, Williams W, Hanau R (1967) A model for simulating photosynthesis in plant communities. Hilgardia 38(4):181–205
Duresso ME, Lule D, Tirfessa A, Gelmesa D, Tesso T, Menamo T, Serba DD (2023) Genetic diversity in Ethiopian sorghum germplasm for root system architecture and trait association. Rhizosphere 27:100759
Gelli M et al (2016) Mapping QTLs and association of differentially expressed gene transcripts for multiple agronomic traits under different nitrogen levels in sorghum. BMC Plant Biol 16:1–18
George-Jaeggli B, Mortlock MY, Borrell AK (2017) Bigger is not always better: reducing leaf area helps stay-green sorghum use soil water more slowly. Environ Exp Bot 138:119–129
Gholipoor M, Prasad PV, Mutava RN, Sinclair TR (2010) Genetic variability of transpiration response to vapor pressure deficit among sorghum genotypes. Field Crop Res 119(1):85–90
Gholipoor M, Sinclair TR, Prasad PV (2012) Genotypic variation within sorghum for transpiration response to drying soil. Plant Soil 357:35–40
Guindo D, Teme N, Vaksmann M, Doumbia M, Vilmus I, Guitton B, Rami JF (2019) Quantitative trait loci for sorghum grain morphology and quality traits: toward breeding for a traditional food preparation of West-Africa. J Cereal Sci 85:256–272
Guitton B, Thera K, Tékété ML, Pot D, Kouressy M, Teme N, Vaksmann M (2018) Integrating genetic analysis and crop modeling: A major QTL can finely adjust photoperiod-sensitive sorghum flowering. Field Crop Res 221:7–18
Hammer GL, Dong Z, McLean G, Doherty A, Messina C, Schussler J, Cooper M (2009) Can changes in canopy and/or root system architecture explain historical maize yield trends in the US corn belt? Crop Sci 49(1):299–312
Harris-Shultz KR, Hayes CM, Knoll JE (2019) Mapping QTLs and identification of genes associated with drought resistance in sorghum. Sorghum: methods and protocols, 11-40

Hayes CM, Weers BD, Thakran M, Burow G, Xin Z, Emendack Y, Mullet JE (2016) Discovery of a dhurrin QTL in sorghum: co-localization of dhurrin biosynthesis and a novel stay-green QTL. Crop Sci 56(1):104–112

Hilley JL (2017) Taking sorghum to new heights: identification of genes controlling height variation in sorghum (Doctoral dissertation)

Hu Z, Olatoye MO, Marla S, Morris GP (2019) An integrated genotyping-by-sequencing polymorphism map for over 10,000 sorghum genotypes. Plant Genome 12(1):180044

Hufnagel B, de Sousa SM, Assis L, Guimaraes CT, Leiser W, Azevedo GC, Magalhaes JV (2014) Duplicate and conquer: multiple homologs of phosphorus-starvation tolerance1 enhance phosphorus acquisition and sorghum performance on low-phosphorus soils. Plant Physiol 166(2):659–677

Jiao Y, Burke J, Chopra R, Burow G, Chen J, Wang B, Xin Z (2016) A sorghum mutant resource as an efficient platform for gene discovery in grasses. Plant Cell 28(7):1551–1562

Jordan DR, Hunt CH, Cruickshank AW, Borrell AK, Henzell RG (2012) The relationship between the stay-green trait and grain yield in elite sorghum hybrids grown in a range of environments. Crop Sci 52(3):1153–1161

Joshi D, Singh V, van Oosterom E, Mace E, Jordan D, Hammer G (2016) Genetic manipulation of root system architecture to improve drought adaptation in Sorghum. In: The sorghum genome. Springer, Cham, pp 207–226

Joshi DC, Singh V, Hunt C, Mace E, van Oosterom E, Sulman R, Hammer G (2017) Development of a phenotyping platform for high throughput screening of nodal root angle in sorghum. Plant Methods 13(1):1–12

Kaldenhoff R, Bertl A, Otto B, Moshelion M, Uehlein N (2007) Characterization of plant aquaporins. Methods Enzymol 428:505–531

Kamal NM, Gorafi YSA, Tsujimoto H, Ghanim AMA (2018) Stay-green QTLs response in adaptation to post-flowering drought depends on the drought severity. Biomed Res Int 2018:7082095

Kamal NM, Gorafi YSA, Abdelrahman M, Abdellatef E, Tsujimoto H (2019) Stay-green trait: a prospective approach for yield potential, and drought and heat stress adaptation in globally important cereals. Int J Mol Sci 20(23):5837

Kapanigowda MH, Payne WA, Rooney WL, Mullet JE, Balota M (2014) Quantitative trait locus mapping of the transpiration ratio related to preflowering drought tolerance in sorghum (Sorghum bicolor). Funct Plant Biol 41(11):1049–1065

Karthika G, Kholova J, Alimagham S, Ganesan M, Chadalavada K, Kumari R, Vadez V (2019) Measurement of transpiration restriction under high vapor pressure deficit for sorghum mapping population parents. Plant Physiol Rep 24:74–85

Kassahun B, Bidinger FR, Hash CT, Kuruvinashetti MS (2010) Stay-green expression in early generation sorghum [Sorghum bicolor (L.) Moench] QTL introgression lines. Euphytica 172:351–362

Kebede H, Subudhi PK, Rosenow DT, Nguyen HT (2001) Quantitative trait loci influencing drought tolerance in grain sorghum (Sorghum bicolor L. Moench). Theor Appl Genet 103:266–276

Kholová J, Hash CT, Kakkera A, Kočová M, Vadez V (2010a) Constitutive water-conserving mechanisms are correlated with the terminal drought tolerance of pearl millet [Pennisetum glaucum (L.) R. Br.]. J Exp Bot 61(2):369–377

Kholová J, Zindy P, Hash CT, Kočová M, Vadez V (2010b) Genotypes contrasting for terminal drought tolerance also contrast for the developmental pattern of water use in varying environmental conditions

Kholova J, McLean G, Vadez V, Craufurd P, Hammer GL (2013) Drought stress characterization of post-rainy season (rabi) sorghum in India. Field Crop Res 141:38–46

Kholová J, Murugesan T, Kaliamoorthy S, Malayee S, Baddam R, Hammer GL, Vadez V (2014) Modelling the effect of plant water use traits on yield and stay-green expression in sorghum. Funct Plant Biol 41(11):1019–1034

Knoll J, Ejeta G (2008) Marker-assisted selection for early-season cold tolerance in sorghum: QTL validation across populations and environments. Theor Appl Genet 116:541–553

Kong W, Kim C, Zhang D, Guo H, Tan X, Jin H, Paterson AH (2018) Genotyping by sequencing of 393 Sorghum bicolor BTx623× IS3620C recombinant inbred lines improves sensitivity and resolution of QTL detection. G3 Genes Genomes Genet 8(8):2563–2572

Leiser WL, Rattunde HFW, Weltzien E, Cisse N, Abdou M, Diallo A et al (2014) Two in one sweep: aluminum tolerance and grain yield in P-limited soils are associated to the same genomic region in West African sorghum. BMC Plant Biol 14:1–13

Li R, Han Y, Lv P, Du R, Liu G (2014a) Molecular mapping of the brace root traits in sorghum (Sorghum bicolor L. Moench). Breed Sci 64(2):193–198

Li G, Santoni V, Maurel C (2014b) Plant aquaporins: roles in plant physiology. Biochim Biophys Acta 1840(5):1574–1582

Lin YR, Schertz KF, Paterson AH (1995) Comparative analysis of QTLs affecting plant height and maturity across the Poaceae, in reference to an interspecific sorghum population. Genetics 141(1):391–411

Liu H, Liu H, Zhou L, Lin Z (2019) Genetic Architecture of domestication-and improvement-related traits using a population derived from Sorghum virgatum and Sorghum bicolor. Plant Sci 283:135–146

Lobell DB, Burke MB (2008) Why are agricultural impacts of climate change so uncertain? The importance of temperature relative to precipitation. Environ Res Lett 3(3):034007

Lopez JR, Erickson JE, Munoz P, Saballos A, Felderhoff TJ, Vermerris W (2017) QTLs associated with crown root angle, stomatal conductance, and maturity in sorghum. The plant genome 10(2) plantgenome 2016-04

Mace ES, Singh V, Van Oosterom EJ, Hammer GL, Hunt CH, Jordan DR (2012) QTL for nodal root angle in sorghum (Sorghum bicolor L. Moench) co-locate with QTL for traits associated with drought adaptation. Theor Appl Genet 124:97–109

Mace ES, Hunt CH, Jordan DR (2013) Supermodels: sorghum and maize provide mutual insight into the genetics of flowering time. Theor Appl Genet 126:1377–1395

Mace E, Innes D, Hunt C, Wang X, Tao Y, Baxter J et al (2019) The Sorghum QTL Atlas: a powerful tool for trait dissection, comparative genomics and crop improvement. Theor Appl Genet 132:751–766.

Makita Y, Shimada S, Kawashima M, Kondou-Kuriyama T, Toyoda T, Matsui M (2015) MOROKOSHI: transcriptome database in Sorghum bicolor. Plant Cell Physiol 56(1):e6

Mantilla-Perez MB, Salas Fernandez MG (2017) Differential manipulation of leaf angle throughout the canopy: current status and prospects. J Exp Bot 68(21-22):5699–5717

Mantilla-Perez MB, Bao Y, Tang L, Schnable PS, Salas-Fernandez MG (2020) Toward "smart canopy" sorghum: discovery of the genetic control of leaf angle across layers. Plant Physiol 184(4):1927–1940

Marla SR, Burow G, Chopra R, Hayes C, Olatoye MO, Felderhoff T, Morris GP (2019) Genetic architecture of chilling tolerance in sorghum dissected with a nested association mapping population. G3 Genes Genomes Genet 9(12):4045–4057

McCormick RF (2017) High-throughput genotyping analyses and image-based phenotyping in sorghum bicolor (Doctoral dissertation)

McMillen GG, McClendon JH (1978) Leaf angle-adaptive feature of sun and shade leaves. Plant Physiol 61(4):52–52

Mocoeur A, Zhang YM, Liu ZQ, Shen X, Zhang LM, Rasmussen SK, Jing HC (2015) Stability and genetic control of morphological, biomass and biofuel traits under temperate maritime and continental conditions in sweet sorghum (Sorghum bicolour). Theor Appl Genet 128:1685–1701

Moghimi N, Desai JS, Bheemanahalli R, Impa SM, Vennapusa AR, Sebela D et al (2019) New candidate loci and marker genes on chromosome 7 for improved chilling tolerance in sorghum. J Exp Bot 70(12):3357–3371

Morris GP, Rhodes DH, Brenton Z, Ramu P, Thayil VM, Deshpande S, Kresovich S (2013) Dissecting genome-wide association signals for loss-of-function phenotypes in sorghum flavonoid pigmentation traits. G3 Genes Genomes Genet 3(11):2085–2094

Murchie EH, Niyogi KK (2011) Manipulation of photoprotection to improve plant photosynthesis. Plant Physiol 155(1):86–92

Nagaraja Reddy R, Madhusudhana R, Murali Mohan S, Chakravarthi DVN, Mehtre SP, Seetharama N, Patil JV (2013) Mapping QTL for grain yield and other agronomic traits in post-rainy sorghum [Sorghum bicolor (L.) Moench]. Theor Appl Genet 126(8):1921–1939

Nagesh Kumar MV, Ramya V, Govindaraj M, Dandapani A, Maheshwaramma S, Ganapathy KN, Jagadeeshwar R (2022) India's rainfed sorghum improvement: three decades of genetic gain assessment for yield, grain quality, grain mold and shoot fly resistance. Front Plant Sci 13:1056040

Ortiz D, Hu J, Salas Fernandez MG (2017) Genetic architecture of photosynthesis in Sorghum bicolor under non-stress and cold stress conditions. J Exp Bot 68(16):4545–4557

Paterson AH, Bowers JE, Bruggmann R, Dubchak I, Grimwood J, Gundlach H, Rokhsar DS (2009) The Sorghum bicolor genome and the diversification of grasses. Nature 457(7229):551–556

Phuong N, Stützel H, Uptmoor R (2013) Quantitative trait loci associated to agronomic traits and yield components in a Sorghum bicolor L. Moench RIL population cultivated under pre-flowering drought and well-watered conditions. Agric Sci 4(12):781–791

Priyanka G, Choudhary S, Anbazhagan K, Naresh D, Baddam R, Jarolimek J, Kholova J (2023) A step towards inter-operable Unmanned Aerial Vehicles (UAV) based phenotyping; a case study demonstrating a rapid, quantitative approach to standardize image acquisition and check quality of acquired images. ISPRS Open J Photogrammetry Remote Sens 4:100042

Rajkumar, Fakrudin B, Kavil SP, Girma Y, Arun SS, Dadakhalandar D, Kamatar MY (2013) Molecular mapping of genomic regions harbouring QTLs for root and yield traits in sorghum (Sorghum bicolor L. Moench). Physiol Mol Biol Plants 19:409–419

Rama Reddy NR, Ragimasalawada M, Sabbavarapu MM, Nadoor S, Patil JV (2014) Detection and validation of stay-green QTL in post-rainy sorghum involving widely adapted cultivar, M35-1 and a popular stay-green genotype B35. BMC Genomics 15:1–16

Reddy PS, Rao TSRB, Sharma KK, Vadez V (2015) Genome-wide identification and characterization of the aquaporin gene family in Sorghum bicolor (L.). Plant Gene 1:18–28

Richards RA, Passioura JB (1989) A breeding program to reduce the diameter of the major xylem vessel in the seminal roots of wheat and its effect on grain yield in rain-fed environments. Aust J Agric Res 40(5):943–950

Rosenzweig C, Elliott J, Deryng D, Ruane AC, Müller C, Arneth A, Jones JW (2014) Assessing agricultural risks of climate change in the 21st century in a global gridded crop model intercomparison. Proc Natl Acad Sci 111(9):3268–3273

Sakhi S, Shehzad T, Rehman S, Okuno K (2013) Mapping the QTLs underlying drought stress at developmental stage of sorghum (Sorghum bicolor (L.) Moench) by association analysis. Euphytica 193:433–450

Shakoor N, Nair R, Crasta O, Morris G, Feltus A, Kresovich S (2014) A Sorghum bicolor expression atlas reveals dynamic genotype-specific expression profiles for vegetative tissues of grain, sweet and bioenergy sorghums. BMC Plant Biol 14:1–15

Shehzad T, Okuno K (2015) QTL mapping for yield and yield-contributing traits in sorghum (Sorghum bicolor (L.) Moench) with genome-based SSR markers. Euphytica 203:17–31

Shekoofa A, Sinclair TR (2018) Aquaporin activity to improve crop drought tolerance. Cells 7(9):123

Shiringani AL, Friedt W (2011) QTL for fibre-related traits in grain× sweet sorghum as a tool for the enhancement of sorghum as a biomass crop. Theor Appl Genet 123(6):999–1011

Sinclair TR (2012) Is transpiration efficiency a viable plant trait in breeding for crop improvement? Funct Plant Biol 39(5):359–365

Sinclair TR, Rufty TW (2012) Nitrogen and water resources commonly limit crop yield increases, not necessarily plant genetics. Glob Food Sec 1(2):94–98

Sinclair TR, Hammer GL, Van Oosterom EJ (2005) Potential yield and water-use efficiency benefits in sorghum from limited maximum transpiration rate. Funct Plant Biol 32(10):945–952

Singh V, van Oosterom EJ, Jordan DR, Hunt CH, Hammer GL (2011) Genetic variability and control of nodal root angle in sorghum. Crop Sci 51(5):2011–2020

Singh V, van Oosterom EJ, Jordan DR, Hammer GL (2012) Genetic control of nodal root angle in sorghum and its implications on water extraction. Eur J Agron 42:3–10

Sivasakthi K (2022) Tapping large genetic variation in water use, grain yield-related traits from global diversity panel of sorghum and Pearlmillet; new opportunities for enhancing grain yield and drought adaption

Sivasakthi K, Marques E, Kalungwana NA, Carrasquilla-Garcia N, Chang PL, Bergmann EM, Penmetsa RV (2019) Functional dissection of the chickpea (Cicer arietinum L.) stay-green phenotype associated with molecular variation at an ortholog of Mendel's I gene for cotyledon color: implications for crop production and carotenoid biofortification. Int J Mol Sci 20(22):5562

Spagnolli FC, Mace E, Jordan D, Borrás L, Gambin BL (2016) Quantitative trait loci of plant attributes related to sorghum grain number determination. Crop Sci 56(6):3046–3054

Subudhi PK, Rosenow DT, Nguyen HT (2000) Quantitative trait loci for the stay green trait in sorghum (Sorghum bicolor L. Moench): consistency across genetic backgrounds and environments. Theor Appl Genet 101:733–741

Sukumaran S, Li X, Li X, Zhu C, Bai G, Perumal R, Yu J (2016) QTL mapping for grain yield, flowering time, and stay-green traits in sorghum with genotyping-by-sequencing markers. Crop Sci 56(4):1429–1442

Tao Y, Mace E, George-Jaeggli B, Hunt C, Cruickshank A, Henzell R, Jordan D (2018) Novel grain weight loci revealed in a cross between cultivated and wild sorghum. Plant Genome 11(2):170089

Tao YZ, Henzell RG, Jordan DR, Butler DG, Kelly AM, McIntyre CL (2000) Identification of genomic regions associated with stay green in sorghum by testing RILs in multiple environments. Theor Appl Genet 100:1225–1232

Tao Y, Luo H, Xu J, Cruickshank A, Zhao X, Teng F, Mace E (2021) Extensive variation within the pan-genome of cultivated and wild sorghum. Nat Plants 7(6):766–773

Teshome A, Zhang J (2019) Increase of extreme drought over Ethiopia under climate warming. Adv Meteorol 2019:1–18

Tharanya M, Kholova J, Sivasakthi K, Seghal D, Hash CT, Raj B, Vadez V (2018) Quantitative trait loci (QTLs) for water use and crop production traits co-locate with major QTL for tolerance to water deficit in a fine-mapping population of pearl millet (Pennisetum glaucum LR Br.). Theor Appl Genet 131:1509–1529

Truong SK, McCormick RF, Rooney WL, Mullet JE (2015) Harnessing genetic variation in leaf angle to increase productivity of Sorghum bicolor. Genetics 201(3):1229–1238

Tuinstra MR, Grote EM, Goldsbrough PB, Ejeta G (1996) Identification of quantitative trait loci associated with pre-flowering drought tolerance in sorghum. Crop Sci 36(5):1337–1344

Upadhyaya HD, Ortiz R (2001) A mini core subset for capturing diversity and promoting utilization of chickpea genetic resources in crop improvement. Theor Appl Genet 102:1292–1298

Upadhyaya HD, Pundir RPS, Dwivedi SL, Gowda CLL, Reddy VG, Singh S (2009) Developing a mini core collection of sorghum for diversified utilization of germplasm. Crop Sci 49(5):1769–1780

Upadhyaya HD, Wang YH, Sharma R, Sharma S (2013) Identification of genetic markers linked to anthracnose resistance in sorghum using association analysis. Theor Appl Genet 126:1649–1657

Upadhyaya HD, Dwivedi SL, Singh S, Sahrawat KL, Singh SK (2016) Genetic variation and postflowering drought effects on seed iron and zinc in ICRISAT sorghum mini core collection. Crop Sci 56(1):374–383

Upadhyaya HD, Dwivedi SL, Vetriventhan M, Krishnamurthy L, Singh SK (2017) Post-flowering drought tolerance using managed stress trials, adjustment to flowering, and mini core collection in sorghum. Crop Sci 57(1):310–321

Vadez V, Krishnamurthy L, Kashiwagi J, Kholova J, Devi JM, Sharma KK et al (2007) Exploiting the functionality of root systems for dry, saline, and nutrient deficient environments in a changing climate. J SAT Agric Res 4(1):1–61

Vadez V, Ratnakumar P (2016) High transpiration efficiency increases pod yield under intermittent drought in dry and hot atmospheric conditions but less so under wetter and cooler conditions in groundnut (Arachis hypogaea (L.)). Field Crop Res 193:16–23

Vadez V, Krishnamurthy L, Hash CT, Upadhyaya HD, Borrell AK (2011a) Yield, transpiration efficiency, and water-use variations and their interrelationships in the sorghum reference collection. Crop Pasture Sci 62(8):645–655
Vadez V, Deshpande SP, Kholova J, Hammer GL, Borrell AK, Talwar HS, Hash CT (2011b) Stay-green quantitative trait loci's effects on water extraction, transpiration efficiency and seed yield depend on recipient parent background. Funct Plant Biol 38(7):553–566
Vadez V, Kholova J, Zaman-Allah M, Belko N (2013) Water: the most important 'molecular'
Vadez V, Kholova J, Medina S, Kakkera A, Anderberg H (2014) Transpiration efficiency: new insights into an old story. J Exp Bot 65(21):6141–6153
Vadez V, Kholová J, Hummel G, Zhokhavets U, Gupta SK, Hash CT (2015) LeasyScan: a novel concept combining 3D imaging and lysimetry for high-throughput phenotyping of traits controlling plant water budget. J Exp Bot 66(18):5581–5593
Wang WM, Li ZL, Su HB (2007) Comparison of leaf angle distribution functions: effects on extinction coefficient and fraction of sunlit foliage. Agric For Meteorol 143(1-2):106–122
Wang HL, Zhang HW, Du RH, Chen GL, Liu B, Yang YB, Guan YA (2016) Identification and validation of QTLs controlling multiple traits in sorghum. Crop Pasture Sci 67(2):193–203
Wu Y, Li X, Xiang W, Zhu C, Lin Z, Wu Y, Yu J (2012) Presence of tannins in sorghum grains is conditioned by different natural alleles of Tannin1. Proc Natl Acad Sci 109(26):10281–10286
Xin Z, Li Wang M, Barkley NA, Burow G, Franks C, Pederson G, Burke J (2008) Applying genotyping (TILLING) and phenotyping analyses to elucidate gene function in a chemically induced sorghum mutant population. BMC Plant Biol 8:1–14
Xin Z, Wang M, Cuevas HE, Chen J, Harrison M, Pugh NA, Morris G (2021) Sorghum genetic, genomic, and breeding resources. Planta 254(6):114
Xu R, Li C, Bernardes S (2021) Development and testing of a uav-based multi-sensor system for plant phenotyping and precision agriculture. Remote Sens 13(17):3517
Yamaguchi M, Fujimoto H, Hirano K, Araki-Nakamura S, Ohmae-Shinohara K, Fujii A, Sazuka T (2016) Sorghum Dw1, an agronomically important gene for lodging resistance, encodes a novel protein involved in cell proliferation. Sci Rep 6(1):28366
Zaman-Allah M, Jenkinson DM, Vadez V (2011) A conservative pattern of water use, rather than deep or profuse rooting, is critical for the terminal drought tolerance of chickpea. J Exp Bot 62(12):4239–4252
Zhang D, Li J, Compton RO, Robertson J, Goff VH, Epps E et al (2015) Comparative genetics of seed size traits in divergent cereal lineages represented by sorghum (Panicoidae) and rice (Oryzoidae). G3: Genes, Genomes. Genetics 5(6):1117–1128
Zhao CX, Shao HB, Chu LY (2008) Aquaporin structure–function relationships: water flow through plant living cells. Colloids Surf B Biointerfaces 62(2):163–172
Zhi X, Hammer G, Borrell A, Tao Y, Wu A, Hunt C, George-Jaeggli B (2022a) Genetic basis of sorghum leaf width and its potential as a surrogate for transpiration efficiency. Theor Appl Genet 135(9):3057–3071
Zhi X, Tao Y, Jordan D, Borrell A, Hunt C, Cruickshank A, Mace E (2022b) Genetic control of leaf angle in sorghum and its effect on light interception. J Exp Bot 73(3):801–816
Zhu XG, Song Q, Ort DR (2012) Elements of a dynamic systems model of canopy photosynthesis. Curr Opin Plant Biol 15(3):237–244

Advancing Sorghum Improvement Through High-Throughput Phenotyping Technologies

4

Vishal Hivare, Kaliamoorthy Sivasakthi,
and Sunita Choudhary

Abstract

Sorghum, a vital crop in arid regions, confronts the challenges of climate change, prompting the urgent development of resilient varieties to ensure sustained production, food security, and livelihoods. The slow progress of conventional breeding methods, coupled with the complexity of genotype-environment interactions, underscores the urgency for innovative approaches like phenomics. Front-line research is now directed towards enhancing sorghum as a climate-smart crop globally, with phenomics emerging as the future of phenotyping, presenting both opportunities and challenges in the pursuit of standardized, data-driven, and dynamically measured advancements for sustainable agriculture. Phenotyping is essential for crop improvement, but the digitalization gap in research hinders progress. Embracing advanced technologies in India is crucial for future agriculture. This chapter discusses diverse phenomics tools and sensors, illustrating successful cases and ongoing efforts to integrate technology into breeding programs for climate-smart crops at International Crops Research Institute for the Semi-Arid Tropics (ICRISAT), catering to the needs of semiarid tropics. ICRISAT employs advanced high-throughput phenotyping, exemplified by the high-throughput phenotyping (HTP)-LeasyScan facility and Lysimeter facility, to revolutionize sorghum breeding. The HTP-LeasyScan uses 3D laser scanning for rapid plant canopy assessment, aiding climate-smart approaches. The Lysimeter facility pioneers drought tolerance evaluation, bridging field and lab research. Innovative technologies like near-infrared spectroscopy (NIRS), X-ray fluorescence (XRF), and computed tomography (CT) enhance sorghum quality traits, ensuring precise evaluations without grain damage. These strategies align

V. Hivare · K. Sivasakthi · S. Choudhary (✉)
International Crop Research Institute for the Semi-Arid Tropics, Patancheru, Telangana, India
e-mail: Sunita.choudhary@icrisat.org

E. Habyarimana et al. (eds.), *Omics and Biotechnological Approaches for Product Profile-Driven Sorghum Improvement*,
https://doi.org/10.1007/978-981-97-4347-6_4

with climate-smart breeding, promoting resilience in sorghum crops among changing environmental conditions.

Keywords

Sorghum · Phenomics · High throughput phenotyping · Lysimetric system · UAV

4.1 Addressing Technological Gaps in Phenotyping for Sorghum Improvement Programs

Sorghum stands as a crucial lifeline for resource-poor farming communities inhabiting arid and semi-arid tropics. However, the characterization of sorghum's target production environments (TPEs) has raised concerns about the impending impact of climate change on its production (Kholova et al. 2013). Notably, post-rainy sorghum cultivation relies heavily on stored soil moisture, subject to the challenges posed by high year-to-year rainfall variations, typically ranging between 300 and 800 mm. This variability presents a formidable hurdle for breeding programs aiming to develop improved sorghum varieties capable of adapting to the unpredictability of drought episodes (Bänziger and Cooper 2001; Chauhan et al. 2008; Kholova et al. 2013; Chenu et al. 2013; Hajjarpoor et al. 2018).

The breeding of resilient crop varieties that can withstand future climate scenarios, marked by increased drought and heat periods, necessitates the incorporation of multiple traits. Achieving this through conventional breeding methods faces challenges, particularly within low-input agricultural systems, where progress is slower than needed to meet the demands for both human consumption and livestock feed. Compounding this issue is the constraint posed by the lack of high-throughput phenotyping, often referred to as phenomics. Existing tools struggle to capture the intricate interactions between crops and environments ($G \times E$), critical for translating into product development tailored to specific conditions (Doggett 1988; Ludlow and Muchow 1990; Mullet et al. 2002; Sanchez et al. 2002).

Front-line research is now dedicated to positioning sorghum as a climate-smart crop on a global scale. Phenomics, heralded as the future of phenotyping, brings forth both opportunities and challenges. It promises increased throughput, objectivity, precision, dynamic measures, and integration—all crucial for developing climate-smart crops. However, these advancements demand innovative approaches for standardization, data management, and analysis. Bridging the phenotyping technology gap is not merely a scientific necessity but a strategic imperative for ensuring the resilience and sustainability of sorghum cultivation in the face of evolving environmental dynamics.

4.2 Enhancing Crop Improvement Through Phenomics Tools

Phenotyping plays a pivotal role in expediting the characterization of genotypes for more accurate and refined selections in crop improvement programs. Despite the significant strides in technology, the digitalization of this critical aspect in research has been limited, creating a growing bottleneck in crop improvement. This is particularly evident in the measurement of various genotype characteristics in the field or the nondestructive postharvest assessment of nutritional and other traits. Recognizing this gap and aligning with the progress of Digital India, it becomes imperative to leverage advanced technologies to empower agriculture in meeting the demands of a burgeoning population in the future.

To address this technological void, we propose the initiation of our crop improvement and breeding systems, integrating cutting-edge technologies that are now well-characterized and can be seamlessly incorporated into crop improvement programs. A diverse array of phenomics tools and sensors, such as Lidar, Red, Green, Blue (RGB), multispectral, hyperspectral sensors carried by unmanned aerial vehicles (UAVs) or gantry, near-infrared spectroscopy (NIRS), computed tomography (CT), and X-ray fluorescence (XRF), is being employed to enable precise and rapid large-scale phenotyping for various adaptation traits.

However, it is not just about the tools; considerable efforts have been invested in standardizing and automating the data analysis pipeline. This ensures that the wealth of data generated by sensor-based technology does not become a bottleneck in the analysis of target traits. Several cases have demonstrated how high-throughput phenomics has already facilitated the swift development of climate-smart crops. Our ongoing endeavors focus on seamlessly integrating relevant technology into the breeding programs of the International Crops Research Institute for the Semi-Arid Tropics (ICRISAT), providing practical support for the improvement of dry-land cereals and legumes-based cropping systems.

Drawing from 15 years of research in areas of drought and our commitment to serving semiarid tropics, ICRISAT has pioneered innovative digital methods and a high-throughput phenotyping (HTP) platform. This platform not only automates the precise characterization and screening for abiotic stress adaptation and nutrition but also empowers agricultural researchers and partners. It enables the screening of various genetic materials, including elite lines, national checks, advanced breeding lines, and breeding populations, for crop improvement programs geared towards climate adaptation (drought, heat, and salinity adaptation). The integration of these technologies represents a significant leap forward in the journey to transform phenotyping into phenomics, ensuring a more resilient and adaptive agriculture sector.

4.2.1 High-Throughput Highly Automated Phenotyping Platform (HTP-LeasyScan)

In summary, the HTP-LeasyScan Platform (Fig. 4.1) employs advanced nondestructive three-dimensional (3D) point cloud laser scanning technology, specifically the PlantEye system, for precise evaluation of various plant canopy attributes during the vegetative stage. It comprehensively measures traits such as digital biomass, 3D-leaf area, projected leaf area, plant height, leaf area index, leaf angle, leaf inclination, and light penetration depth (Vadez et al. 2015; Sivasakthi et al. 2018, 2019; Tharanya et al. 2018; Kar et al. 2020). Additionally, it has the capability to monitor diurnal variations in water use parameters through 1488 analytical scales (Kar et al. 2020). The HTP-LeasyScan can efficiently evaluate 4800 microplots in just 2 h, with real-time data processing. Environmental sensors closely monitor conditions, and the collected data by PlantEye scanners and scales, then stored in a central database accessible through HortControl software. While handling extensive datasets presents a challenge, the system offers robust tools for data access, filtering, and outlier removal via an R library interface, allowing for the comparison of growth patterns with weather variables (Vadez et al. 2015).

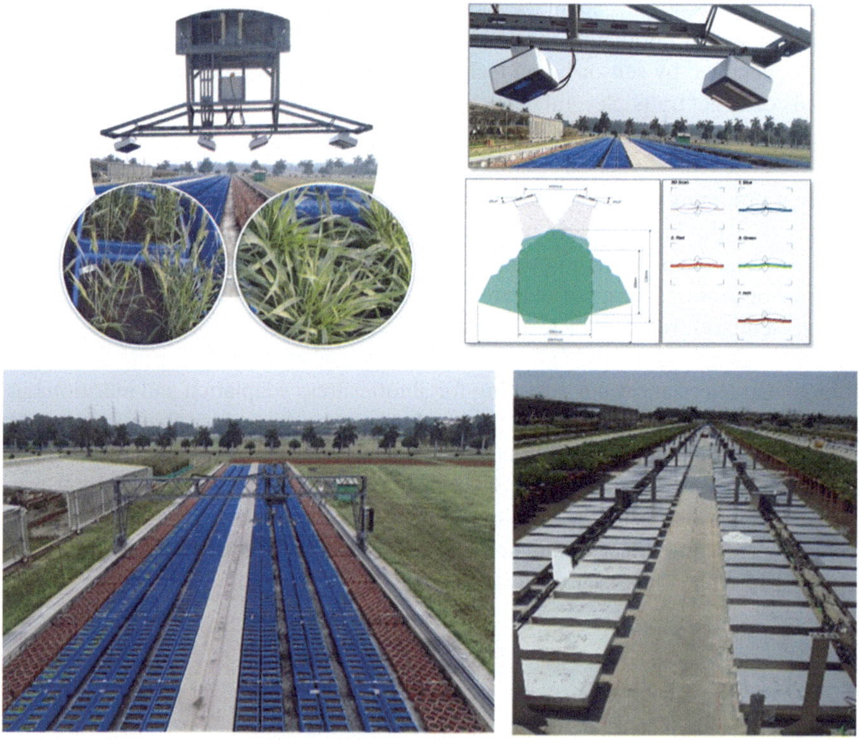

Fig. 4.1 HTP-LeasyScan facility at ICRISAT

LeasyScan, the high-throughput phenotyping platform, analyzes plant canopies in a non-destructive and high-throughput manner. It utilizes PlantEye F600 laser triangulation sensors to generate detailed 3D point clouds of the crop canopy. These point clouds are crucial for extracting plant parameters such as leaf area, height, and leaf angle. The PE F600 scanners are equipped with RGB and NIRS bands based dual sensors, enabling high-resolution images for quantifying additional traits like plant senescence, nutrient deficiency, stay green, foliar diseases (detectable through color distortion), and biomass quality (specifically stover nitrogen), as well as conducting agro-chemical impact studies, among other applications. LeasyScan stands out as the world's largest and most unique platform for phenotyping the component traits of abiotic and biotic stress adaptation.

Data collection is carried out by PlantEye scanners, scales, and environmental sensors, with the information stored in a centralized database. This database can be accessed and visualized through HortControl software, tracking variables such as leaf area over time, identifying outliers, and correlating plant parameters with environmental conditions. Managing extensive datasets is a known challenge for the LeasyScan platform; however, tools have been developed to access data through an R library interface, allowing for queries, filtering, and interpolation needed for outlier removal. The library also extracts weather variables and applies thermal time to facilitate the comparison of growth patterns.

4.2.1.1 HTP-LeasyScan Case Studies on Sorghum

1. **Assessment of crop canopy development in mapping population parents**: A subset of sorghum genotypes from eight established recombinant inbred lines (RILs) mapping populations was tested at LeasyScan platform in a series of experiments. The primary focus was to investigate different aspects of canopy development which play critical role in adaptation to drought and the measurement of canopy in field condition is laborious and time consuming. The study also encompassed an analysis of various other canopy development traits, such as digital biomass, leaf area index, projected leaf area, and plant height, employing the high-throughput phenotyping platform LeasyScan. Interestingly, the study unveiled a wide range of fold differences in these traits: a 2.7-fold variation in digital biomass (ranging from 3,219,018 to 8,749,690 A mm), a twofold difference in leaf area index (from 0.08 to 0.16), a twofold diversity in projected leaf area (ranging from 145 to 299 mm^3), and a 1.45-fold variation in plant height (spanning from 141 to 204 mm). These differences in canopy development parameters among the various genotypes or RILs had a significant impact on the overall development of the plant canopy which further has been exploited in genomics study to assist crop improvement programs (for additional information, refer to Vadez et al. 2015) (Fig. 4.2).
2. **Characterization of evapotranspiration dynamics under natural conditions varying in VPD Using LeasyScan**: To scrutinize the evapotranspiration pattern, FieldScales data were gathered over two consecutive days marked by elevated vapor pressure deficit (VPD) conditions, as documented by Karthika et al. in 2019. Across these 2 days, the maximum VPD ranged from 6.88 to 6.51 kPa.

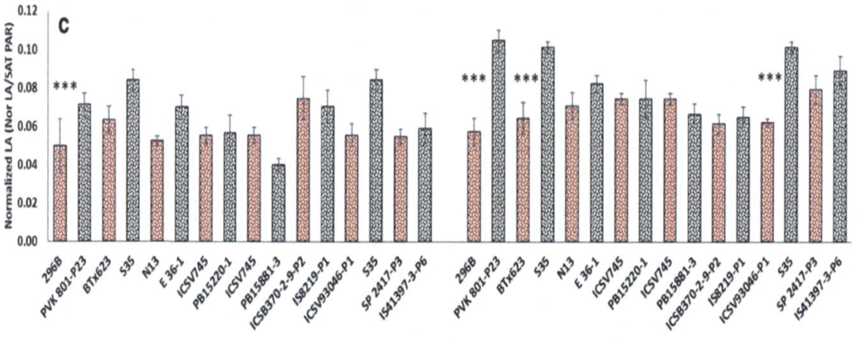

Fig. 4.2 Differences in leaf area normalized to received radiation concerning VPD in sorghum mapping population parents (red color—Parent A; black color—Parent B). (This figure was sourced from Karthika et al. (2019). Plant Physiol. Rep. 24, 74–85. https://doi.org/10.1007/s40502-019-0432-x)

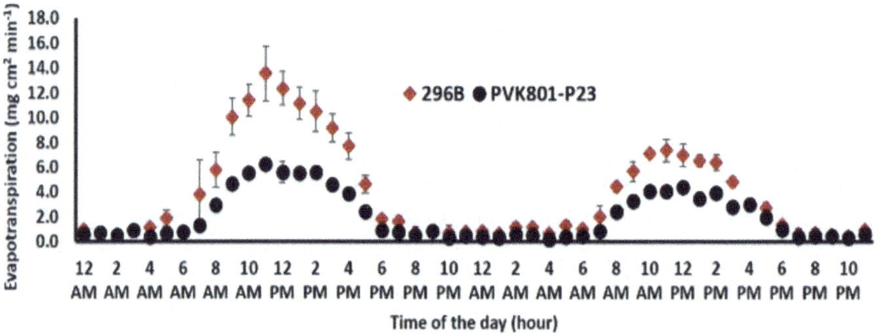

Fig. 4.3 Midday evapotranspiration rate variation for two consecutive days across RIL (296B and PVK801-P23) sorghum mapping population parents (red color—parent A; black color—parent B). (This figure was sourced from Karthika et al. (2019). Plant Physiol. Rep. 24, 74–85. https://doi.org/10.1007/s40502-019-0432-x)

Among the diverse recombinant inbred line (RIL) combinations examined, particularly RIL1 (296B x PVK801), RIL2 (BTx623 x S35), RIL6 (SP2417 x IS413973), and RIL8 (ICSV 93046 x S35), notable variations were discerned between the parental pairs during midday. It is noteworthy that parent 296B exhibited the most substantial increase in evapotranspiration rate during midday compared to all other genotypes under investigation. Furthermore, within the cohort of studied RILs, parent A displayed a higher surge in evapotranspiration rate during midnoon compared to parent B. This suggests that the recurrent parent demonstrated a higher evapotranspiration rate under high VPD conditions than the donor parent. The genotypic variations in the RILs are visually illustrated in Fig. 4.3 (Karthika et al. 2019).

3. **Assessing plant vigor traits in sorghum bi-parental mapping population**: Chadalavada (2017), tested 181 recombinant inbred lines of sorghum of cross

between N13, an Indian durra sorghum renowned for its Striga resistance, and E36-1, an Ethiopian guinea-caudatum hybrid sorghum harboring the valuable stay-green gene generated through the single seed descent method. These lines originated from a cross. It is noteworthy that N13 and E36-1 share the same flowering group and exhibit comparable water use efficiency under irrigated conditions. However, they distinctly differ in grain size and transpiration efficiency.

The study's primary objectivel was to evaluate plant vigor traits, specifically leaf area and growth rate by leveraging the HTP-LeasyScan facility. Phenotyping with HTP-LeasyScan categorized plant vigor data into three groups: early, medium, and late vigor. Remarkably, the genotypic mean values for the 3D-leaf area trait were striking across all three categories: early (38,420 mm^2), medium (75,360 mm^2), and late vigor (96,397 mm^2). Similarly, variations in genotypic mean values for the projected leaf area trait were observed in the early (307 cm^2), medium (592 cm^2), and late vigor (746 cm^2) groups. It is crucial to highlight the substantial twofold difference between the low and high vigor groups concerning plant growth-related traits quantification made possible through such platforms.

4.2.2 High-Throughput Phenotyping Platform (Lysi-Field or Lysimeter Facility)

The Lysi-field (LF) equipped with lysimetric facility and rainout shelter, depicted in Fig. 4.4, serves as a specialized facility designed for evaluating the drought tolerance of crop germplasm and breeding materials under controlled conditions, free from rainfall interference. This facility allows crops to be cultivated in conditions resembling the field while enabling researchers to intentionally induce drought stress. The focus is on studying drought-tolerant traits and understanding the role of roots in supporting the survival of dryland crops.

Key components of the facility include a lysimeter system, comprising PVC tubes filled with soil, specifically designed for measuring plant water use. ICRISAT's lysimeter facility comprises 2800 tubes measuring 1.2 m in length and 20 cm in diameter, along with 1700 tubes measuring 2.0 m in length and 25 cm in diameter. These dimensions closely replicate natural soil depth and aerial spacing conditions. Lysimeters play a vital role in complementing other phenotyping activities, acting as a bridge between field and laboratory research.

The lysimeter system facilitates highly relevant agronomic assessments of yield and yield components. It allows for the measurement of total plant water use (the sum of water use data across the entire life of the crop), dynamics of plant water use (water used during crop development), and transpiration efficiency (TE), which represents the ratio between total biomass produced (pods and vegetative parts) and total water use. This system enables the evaluation of temporal variations in water requirements for individual genotypes and different crops such as sorghum, pearl millet, finger millet, chickpea, and peanut (Vadez et al. 2011a, b, 2013; Zaman-Allah et al. 2011; Vadez and Ratnakumar 2016; Tharanya et al. 2018; Sivasakthi et al. 2019).

Fig. 4.4 Overview of the ICRISAT-lysimeter facility for drought phenotyping, including the activities of growing sorghum in cylinders and weighing the cylinders

ICRISAT's Lysimeter facility for phenotyping stands as the first of its kind globally. Thousands of plants are tested annually at this facility, and its success has led to its replication on a smaller scale by the International Rice Research Institute (IRRI) and the Directorate of Sorghum Research of the Indian Council of Agricultural Research (ICAR), and on a larger scale by CIMMYT. The facility is open for use by other institutions, and a similar one has been established at the ICRISAT station in Niger to address the interaction between drought and low fertility issues (lysimetric facility—Gems [icrisat.org]).

Sorghum Case Studies Examining transpiration efficiency (TE), water extraction, and agronomic traits in both irrigated and drought conditions using a lysimeter facility reveals insights into the drought adaptation of different sorghum materials:

1. **Sorghum reference set (Vadez et al.** 2011a, b): In their 2011 study, Vadez and colleagues assessed yield, transpiration efficiency (TE), and water extraction in 149 germplasm entries from the sorghum reference set, which included three control cultivars. They conducted this research using an outdoor lysimetric system, subjecting the plants to both terminal water stress and full irrigation conditions. The study revealed significant variability in the total water extracted under drought stress among the 149 entries, ranging from 10,600 to 15,200 g per plant. Similarly, under full irrigation, water extraction showed notable vari-

ation, spanning from 10,500 to 42,300 g per plant. Additionally, the research examined the total water uptake under drought stress for different racial groups, highlighting that Durra race entries exhibited the highest water extraction capacity, while the Caudatum-Bicolor and Caudatum-Durra intermediate races displayed lower water extraction capabilities. Regarding transpiration efficiency (TE), the Durra, Caudatum, and Caudatum-Guinea races demonstrated the highest TE, while the Guinea race exhibited the lowest TE.

2. **Sorghum Minicore collection (Sivasakthi** 2022): Sivasakthi (2022) evaluated sorghum minicore collection (n = 242) and farmer preferred post-rainy sorghum cultivars (n = 20) for water use, phenology, and agronomical traits in both irrigated and drought stressed conditions using HTP-Lysi-Field facility. The sorghum minicore collection represents major five races (caudatum 16.1%, durra 12.4%, guinea 12%, kafir 8.7%, bicolor 8.3%) and 10 intermediate races (caudatum-bicolor 12.4%; guinea-caudatum 11.2%; durra-caudatum 7.9%; durra-bicolor and kafir-caudatum each 2.9%; kafir-durra 1.7%; guinea-kafir 1.2%; and guinea-bicolor, guinea-durra, and kafir-bicolor each 0.8% (Upadhyaya et al. 2009). The study uncovered significant variations in key water use and agronomic traits in both well-watered (WW) and water stressed (WS) conditions.
3. **Total water use**: Total water use exhibited a fivefold variation in well-watered (WW) conditions and a 1.5-fold variation in water-stressed (WS) conditions.
4. **Biomass and grain yield**: Large genotypic variations were observed in total biomass (8.5-fold) and grain yield (ninefold) under WW conditions, along with substantial variations in biomass (about threefold) in WS conditions.
5. **Transpiration efficiency (TE)**: TE displayed noteworthy variability, with values ranging from 2.56 to 4.95 under irrigated conditions and from 4.33 to 7.37 under drought-stressed conditions. Few entries in sorghum diversity panels exhibited superior TE and water extraction capabilities under WS conditions, making them promising candidates for breeding programs targeting drought-adapted cultivars in water-limited environments.
6. **Grain yield and TE relationship**: A strong correlation was noted between grain yield and TE (R^2 = 0.60**) (Fig. 4.5).
7. **Races and TE**: Durra (major) and Guinea-Durra (intermediate) races generally demonstrated higher TE values compared to other major and intermediate races, both in WW and WS conditions. This aligns with previous research findings.

   The findings emphasize the significance of genotypic variation in TE, particularly within Durra races, and suggest the potential for harnessing these variations in sorghum breeding programs. The adaptation of Durra races to their origin sites, which feature high temperatures and vapor pressure deficit (VPD), is considered a key factor contributing to their superior TE, and further geographical adaptation may enhance this trait in Durra races.
8. **Sorghum bi-parental mapping population**: Chadalavada and their team (2017) conducted an evaluation of 18 sorghum mapping population parents and six stay-green introgression lines (R16-K349W, R16-K648, S35-6008,

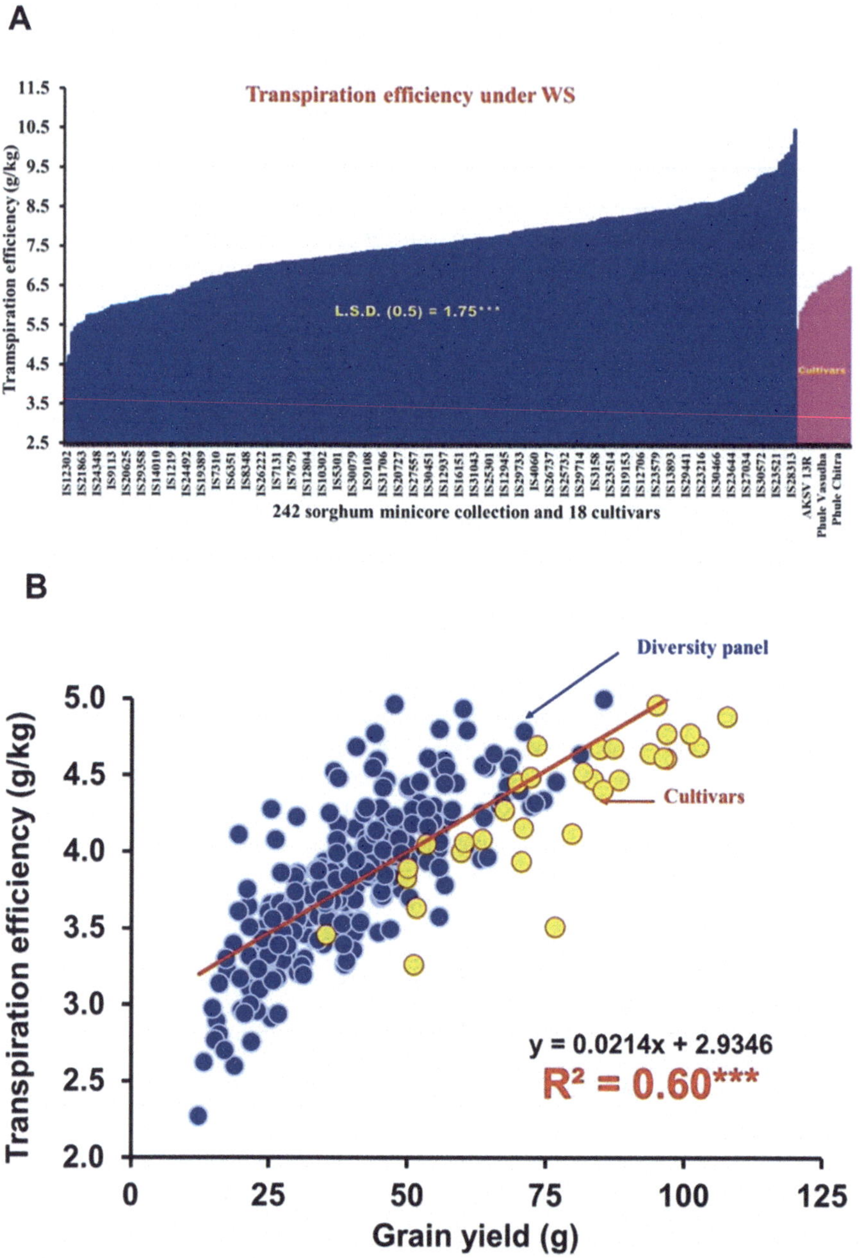

Fig. 4.5 (**a**) Demonstrating significant genetic diversity in transpiration efficiency under water-stressed conditions, encompassing both minicore lines and farmer-preferred cultivars, through the utilization of the HTP-lysimeter phenotyping facility. (**b**) Exploring the correlation between transpiration efficiency and grain yield in well-irrigated conditions. We performed phenotyping on 242 sorghum minicore lines and 18 sorghum cultivars, specifically chosen for their adaptation to post-rainy conditions and favorability among farmers. The assessment of water usage and agronomic

S35-6026, S35-6040, and S35-7001) from R16 and S35 backgrounds. This evaluation was carried out using the HTP-lysimeter facility under both well-irrigated and drought-stressed conditions. The primary research objective was to examine water utilization and agronomic traits to identify genotypes best suited for adaptation to water-limited environments. The findings revealed a broad spectrum of transpiration efficiency (TE) values, ranging from 3.70 to 5.92 g kg^{-1}, and a substantial variability in total water usage, spanning from 7051 to 14,980 g. Moreover, grain yield exhibited a variation from 15.16 to 41.94 g, and the size of grains, measured by 100 seed weight, ranged from 1.15 to 2.92 g. In addition, stover yield displayed notable diversity, falling within the range of 51.19–131.38 g. Remarkably, genotypes with equivalent water extraction levels produced varying amounts of biomass per kilogram of water used, underscoring the significant impact of genetics on adapting to different environmental conditions. Late-duration genotypes typically consumed more water in well-watered settings, while short-duration genotypes extracted more water under water-stress conditions. Stay-green lines like S35 and 6008 exhibited higher transpiration efficiency (TE) by retaining more green leaf area in water-stress situations.

9. **Sorghum advanced breeding material—Stay-green introgression lines (Vadez et al.** 2011a, b): In a 2011 study conducted by Vadez and colleagues, the research focused on assessing variations in water extraction under terminal drought and fully irrigated conditions. This assessment involved 29 introgression lines (IL) designed to target stay-green QTLs, including Stg1, Stg2, Stg3, Stg4, StgA, and StgB, in the S35 background, as well as 16 ILs targeting Stg1, Stg3, Stg4, and StgB in the R16 background. The study's findings revealed that, overall, entries in the R16 background exhibited higher total water extraction (13,900 g kg^{-1}) compared to the S35 background (13,140 g kg^{-1}). In the S35 background, the total water extracted from the soil profile between 30 days after sowing (DAS) and maturity ranged from 10,880 to 14,640 g per plant. Stg1 positively influenced water extraction in the S35 background but had no impact in the R16 background. Furthermore, a negative relationship was observed between pre- and postanthesis water use in both the S35 ($R^2 = 0.51$) and R16 backgrounds ($R^2 = 0.87$). Among the local checks, two durra race entries, Phule Maulee and CRS4, displayed the highest water extraction capacities, measuring 15,180 and 15,260 g per plant, respectively. In contrast, the caudatum race NTJ-2, a popular post-rainy season-adapted sweet and grain plus stover sorghum variety, exhibited lower water extraction capacity.
10. **Sorghum farmer preferred cultivars (Anbazhagan et al.** 2022): In 2022, Anbhazhagan and colleagues conducted research using lysimeter facilities to investigate the influence of drought on various agronomic and quality traits in

Fig. 4.5 (continued) traits was conducted using the HTP-Lysimeter facility. (This image was sourced from Sivasakthi (2022) during the TIGR2EES Summative Workshop and is available at (cam.ac.uk) in the file titled "tapping_large_genetic.jpg" with dimensions 1308 × 925 (tapping_large_genetic.jpg (1308×925) [cam.ac.uk]))

four dual-purpose, open-pollinated sorghum varieties, namely CSV 22, Phule Chitra, M 35-1, and Sevata Jonna. The study evaluated key agronomic and physiological traits in sorghum plants under both waters stressed (WS) and well-watered (WW) conditions. Results revealed that (i) Harvest Index (HI): HI, which represents the percentage of grain yield relative to the total aerial biomass, varied significantly, with a range of 10–34% in WS plants and 24–47% in WW plants. A strong positive correlation between grain yield and HI was observed in WS plants, unlike WW plants; (ii) Leaf Senescence; (iii) Stay-Green Score: Leaf senescence was assessed at two time points. The inverse of leaf senescence was interpreted as the stay-green score. Among the studied genotypes, M35-1 exhibited the highest stay-green score at 80 days after sowing; in vitro organic matter digestibility (IVOMD): IVOMD of stover ranged from 51% to 61% under both WW and WS conditions. There was no significant variation in IVOMD among the tested varieties, and there was no substantial relationship between stover yield and IVOMD, and (iv) grain quality: In WS plants, a negative correlation was observed between protein content and fat content in seeds, whereas in WW plants, a positive correlation between these two components was evident. Overall, the study highlights the variability in these traits and their relationships under different water conditions, providing insights into sorghum adaptation to water-limited environments and the importance of genetic factors in these adaptations.

4.2.3 UAV-Based Field Phenotyping

Currently, the quantification of crop traits in field experiments for breeding programs relies on visual or manual methods, resulting in prolonged programs that demand significant time and resources. To expedite plant selection in breeding pipelines, the integration of state-of-the-art high-throughput phenotyping (HTP) tools is recommended. Numerous sensors and platforms have been developed to capture plant images comprehensively, generating extensive databases. These databases can be processed using advanced image processing techniques, facilitating the extraction of crop features utilized by machine learning or serving as direct inputs for data-hungry deep learning algorithms. This not only automates the phenotyping process but also ensures precise measurements.

Many existing tools fall short of meeting breeders' requirements, often limited to indoor or static settings, necessitating plant transportation to sensors. Unmanned aerial vehicle (UAV)-based high-throughput phenotyping (HTP) technology (Fig. 4.6) addresses this by allowing breeders to characterize plants in the field and extract desired traits. While crop yield remains a primary focus, this article highlights the extraction of closely related yield traits, including plant height (PH), canopy cover (CC), and biomass, using cutting-edge photogrammetry and computer vision techniques like structure from motion. This involves creating a high-density 3D point cloud from UAV-captured RGB imagery, forming a high-resolution orthomosaic. The primary goal is to utilize lightweight, cost-effective RGB sensors for imaging instead of heavier multi and hyperspectral sensors.

Fig. 4.6 Transforming the visual and manual phenotyping into automated field phenotyping using UAV technology

1. **A case study in sorghum illustrates the application of UAV-based HTP for estimating PH, CC, and biomass using RGB imagery**: Advanced photogrammetric techniques are employed to obtain high-resolution orthomosaics from raw drone-captured images across various crop growth stages. Computer vision algorithms are then used to extract spatiotemporal crop information for analysis and selection. The study reveals a significant correlation between estimated and observed PH ($R^2 = 0.90$) and between green CC estimation and manually measured CC ($R^2 = 0.80$). Linear (simple univariate and multivariate) and nonlinear (segmented) regression models are explored using these component traits (PH and CC) and a derived attribute (PH × CC) for predicting biomass. The segmented linear regression model demonstrates the highest prediction potential ($R^2 = 0.86$) (Fig. 4.7) among the three models tested.

   This study demonstrates the potential application of UAV drones and low-cost RGB sensors as scalable, image-based HTP tools for real-time measurement of laborious crop structural traits, including PH, CC, and biomass. This methodology can be extended to various crops and traits, enabling researchers to rapidly screen numerous genotypes with reliable precision. The use of low-cost RGB-based images holds promise as valuable data products in HTP studies.

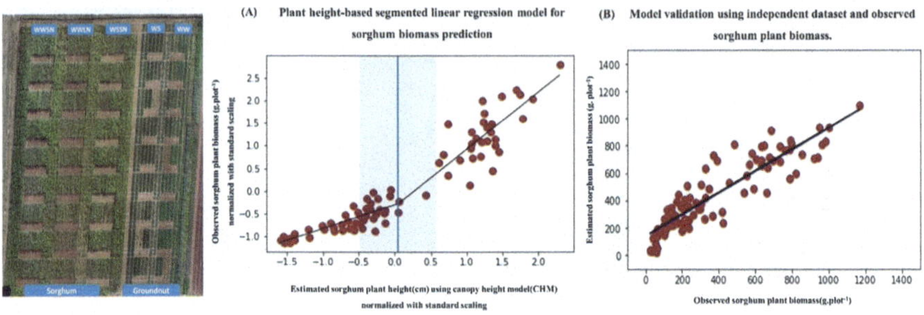

Fig. 4.7 Establishment of a UAV-based field phenotyping pipeline for predictive modeling: conducting precise experiments under water stress, nitrogen stress, and control conditions, contrasting genotypes for plant height, canopy cover, and abiotic stress adaptation

4.2.4 High-Throughput Phenotyping Tools for Sorghum Quality Traits

In semiarid tropics, sorghum improvement programs have primarily concentrated on enhancing grain yield and more recently, yield stability, with limited emphasis on grain quality, despite the potential use of the stover as cattle fodder. Given the projected demand for fodder in India in 2020, estimated at 855 MT of green fodder, 526 MT of dry fodder, and 56 MT of concentrate feed (Dikshit and Birthal 2010), in contrast to the current demand of 666 MT of green fodder and 138 MT of dry crop residues, it is imperative to place greater emphasis on enhancing fodder production. This would enable us to provide more substantial support to the livestock industry in the country.

In semiarid and tropical regions, livestock face productivity challenges due to limited feed availability and land competition with crops. Small-scale farmers, especially during lean periods, rely on livestock for additional income but often feed them low-quality crop residues, negatively impacting livestock and, consequently, farmers' livelihoods. The quality of sorghum fodder is crucial for its value as livestock feed, with in vitro matter digestibility (IVOMD) being a key parameter. Even a slight IVOMD difference of 5% units can lead to a significant price premium of over 20% (Blümmel et al. 2013). To address this, the focus has shifted to developing dual-purpose crop varieties that improve both grain yield and biomass quality, allowing small farmers to run mixed livestock-crop systems while reducing land and water competition (Somegowda et al. 2021, 2022; Anbazhagan et al. 2022).

Sorghum (*Sorghum bicolor* (L.) Moench) holds a significant position as a staple cereal in various Asian and African nations. Consequently, grain quality is a critical attribute for both farmers and consumers, and it is increasingly becoming a significant criterion for breeding programs. The specific requirements for grain quality can differ depending on the intended final product but are fundamentally shaped by the biochemical and physical characteristics of the grain. Amylose (AM), protein (PR), and lipid (LI) contents play a pivotal role in affecting the rheological and sensory properties of certain traditional sorghum dishes.

Therefore, there is a clear imperative to comprehend the natural variability of grain quality parameters such as protein, fat, and starch, along with stover quality traits like nitrogen dry matter (NDM) and in-vitro organic matter digestibility (IVOMD) in sorghum crop. This understanding can significantly enhance the nutritional quality of sorghum through crop improvement initiatives. To achieve this, ICRISAT employs a diverse array of phenotyping tools, encompassing NIRS, XRF, and CT, to assess the quality of both grain and stover. In the subsequent section, we will delve into the utilization of these technologies in the phenotyping of quality traits through a case study.

(a) **Near-infrared spectroscopy (NIRS)**: Near-infrared spectroscopy (NIRS) is a rapid, low-cost, simple, and nondestructive technology to predict organic compounds in grain materials through interactions with electromagnetic waves. It offers a cost-effective way to analyze grain and stover quality (Chadalavada et al. 2022; Ejaz et al. 2021; see Fig. 4.8). An NIR spectroscopic system typically consists of interconnected components: a light source, a spectrometer, and a computer. The light source emits infrared photons, interacting with a sample

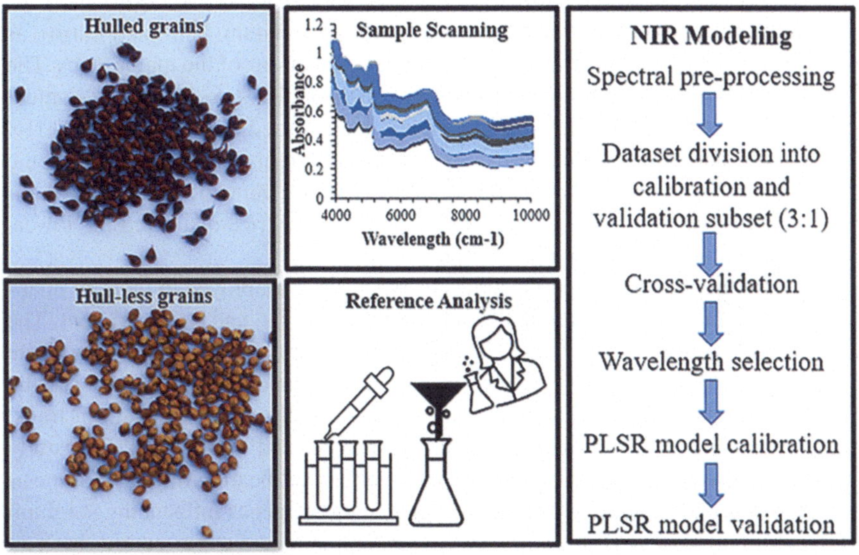

Fig. 4.8 The categorization of sorghum grains based on their morphology, distinguishing between hulled and hull-less grains, which are further utilized as whole grains and flours. This classification is achieved through Fourier transform near-infrared spectroscopy (FT-NIR) spectral data collection, followed by reference analysis. The spectral data are processed through a series of pre-treatments, cross-validation, calculation of principal components (PCs) via principal component analysis (PCA), division into calibration and validation subsets, and finally, the development of partial least square regression (PLSR) models using both FT-NIR spectral and reference data. (The images sourced from Ejas et al.'s work published in Frontiers in Plant Science on September 16, 2021, in the section titled "Technical Advances in Plant Science," Volume 12–2021. https://doi.org/10.3389/fpls.2021.720022)

to create molecular vibrations and stretching, resulting in a spectrum unique to the sample's properties and chemical bonds. Specific spectral characteristics are observed in defined parts of the spectrum, such as chlorophyll vibrations in the 500–750 nm range and O–H bonds in water within the 970–1150 nm range. Spectral data require preprocessing to remove noise and irrelevant information. Effective models can classify samples accurately, but careful system design is essential for precise, timely analysis. Standard error in the laboratory (SEL) impacts prediction accuracy (SEP), emphasizing the need for minimizing errors in quantitative predictions through proper methods (Pandiselvam et al. 2022). Near-infrared spectroscopy (NIRS) technique can be employed to evaluate the quality of sorghum rain and fodder. The International Livestock Research Institute (ILRI-India) has established standardized procedures for implementing NIRS in the assessment of sorghum fodder quality in India. This method has demonstrated both accuracy and reliability, making it a convenient means of efficiently evaluating large quantities of fodder.

(b) **Evaluation of grain quality traits using NIRS technology**:

Case Study 1: Rhodes et al. (2017) conducted an assessment involving 390 sorghum accessions from the Sorghum Association Panel (SAP). This panel encompasses crucial breeding lines from the United States and traditional varieties from all five major races (bicolor, guinea, caudatum, kafir, and durra), as well as 10 intermediate races representing combinations of the major races. The findings indicated that protein content ranged from 8.1% to 18.8%, fat content spanned from 1.0% to 4.3%, and starch content varied from 61.7% to 71.1%. Notably, durra and bicolor-durra sorghum from Ethiopia and India exhibited the highest protein and fat levels but the lowest starch content. In contrast, kafir sorghum from the United States, India, and South Africa displayed the lowest protein content but the highest starch content.

Genome-wide association studies (GWAS) were employed to identify quantitative trait loci (QTL) related to sorghum protein, fat, and starch content. The study also utilized previously published RNAseq data to identify candidate genes within a GWAS QTL region. A potential candidate identified was the alpha-amylase 3 gene, known to be associated with grain composition traits, and it emerged as a strong candidate for protein and fat content. This study identified valuable genetic sources for potential modification of grain composition traits and revealed potential loci and candidate genes influencing sorghum grain composition. The comprehensive examination of grain composition in sorghum germplasm, along with the identification of QTLs linked to protein, fat, and starch, enhances our understanding of the genetic basis underlying the natural variations in sorghum grain nutrition.

Case Study 2: In 2006, Figueiredo and colleagues conducted an analysis using a core collection of 205 accessions, which represented five primary sorghum races: Guinea (60 accessions), Caudatum (43 accessions), Durra (29 accessions), Bicolor (24 accessions), Kafir (18 accessions), and five intermediate races (31 accessions). They employed near-infrared reflectance spectroscopy (NIRS) to evaluate key food grain quality traits, including amylose content, protein content and lipid content. During this process, a total of 278 sorghum samples were scanned in both whole and ground grain forms to create calibration equations. NIRS predictions indicated that amylose content ranged from 15.2% to 23.6%, protein content ranged from 9.2% to 18.0% (refer to Fig. 4.10), and lipid content ranged from 2.89% to 5.61%. Additionally, there was a strong correlation between laboratory data (wet lab) and NIRS-predicted values for various grain quality traits, such as protein content (R^2 = 0.981; n = 131; see Fig. 4.9), lipid content (R^2 = 0.906; n = 101), and amylose content (R^2 = 0.751; n = 236). The performance of the NIRS equations developed in this study for amylose, protein, and lipid content was assessed using SECV (standard error of cross-validation) and RPD (ratio of performance deviation) statistics. Notably, the equation for protein content demonstrated the highest quality and can be employed for quality control in breeding and food technology programs.

Case Study 3: In the study by Chadalawada in 2017, the focus was on developing and validating calibration models for essential qualitative traits in sorghum grains, including moisture, protein, and fat. Samples were collected from two postrainy seasons, 2013–2014 and 2014–2015, with a varying number of genotypes. Prior to chemical analyses, all milled samples were scanned using a FOSS XDS NIR spectrometer, with data stored as the logarithm of reciprocal reflectance. These samples were then divided into calibration and validation sets.

The calibration model showed ranges of 6.96–13.95% for protein, 1.5–5.41% for fat, and 9.07–10.56% for moisture. In the validation set, the sample ranges were 12.45–15.75% for protein, 2.75–4.78% for fat, and 9.22–10.26% for moisture. The calibration and validation statistics for sorghum protein, fat, and moisture were as follows: the protein equation demonstrated an R^2 of 0.94 and SECV (standard error of cross-validation) of 0.39, the fat equation showed an R^2 of 0.94 and SECV of 0.26, and the moisture equation displayed an R^2 of 0.93 and SECV of 0.14. All components in the sorghum calibration model achieved R^2 values exceeding 0.91.

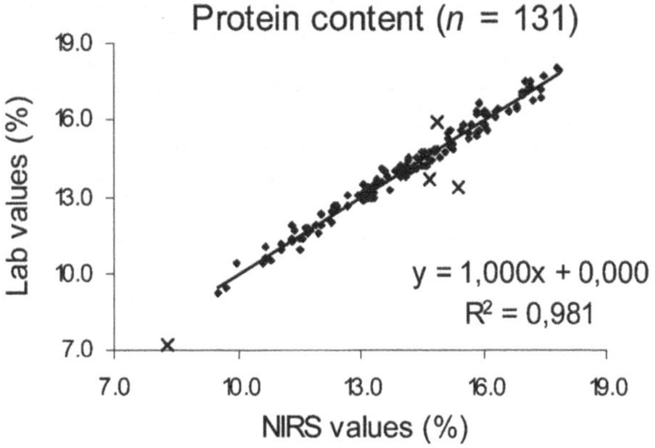

Fig. 4.9 Laboratory data was regressed against NIRS-predicted data using ground grain spectra, with t-outliers excluded, but still marked as (x) on the figure for each trait. (This picture was sourced from Figueiredo et al.'s work in the Journal of Agricultural and Food Chemistry in 2006, Volume 54, page no. 8501–8509)

The study's results suggest that these calibration models have practical applications in various fields and can potentially replace laborious reference analyses. The researchers also highlighted the potential for enhancing these calibrations by including more variability and a wider range of sorghum samples from different environments, ultimately contributing to the improvement of the global grain value chain.

Case Study 4: A total of 3000 samples of sorghum were evaluated for organic composition of grains using near-infrared spectroscopy (NIRS) method. Grains were powdered for 1 mm size and the spectral absorption profile evaluated by NIRS. Sorghum subset was inclusive of germplasm composite collects, exotic collections, and others. Sorghum samples were evaluated for many organic traits including protein. The genetic variation in protein across sub-species and regions has been document. The ranges of variability found in the analyzed samples for protein are presented in Fig. 4.10.

(c) **Assessment of Stover quality traits using NIRS technology**: For assessing fodder quality, ILRI-India has standardized near-infrared spectroscopy (NIRS). Despite involving destructive sampling, this method has been refined using wet lab data for the traits in question. The quantity and quality of biomass are influenced by various factors, including biotic and abiotic stresses, genotype differences, crop harvest stage, and environmental effects during growth. Drought delays flowering and reduces biomass quality and quantity and increases lignin content in plants. Crop harvest stage and environmental factors greatly influence forage quality. Analyzing the relationship between fodder quantity and quality helps breed traits without sacrificing either aspect (Somegowda et al. 2021, 2022; Blümmel et al. 2013).

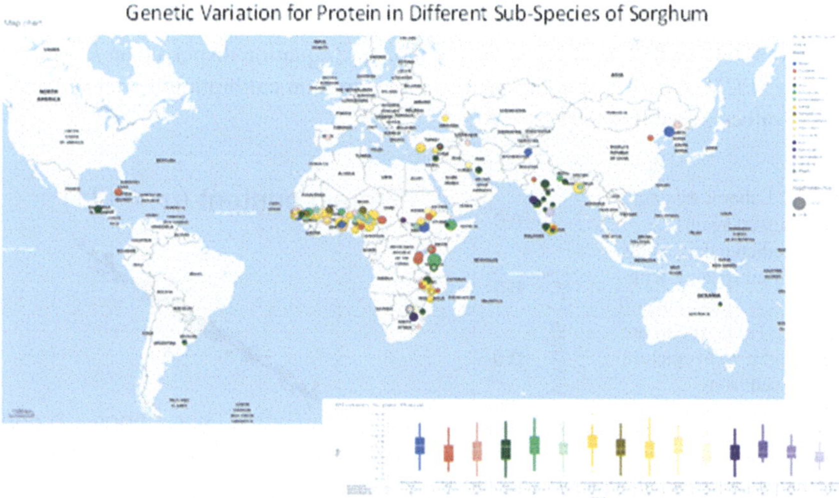

Fig. 4.10 Utilizing near-infrared spectroscopy (NIRS) for crude protein estimation in germplasm lines derived from GenBank, representing various races collected from different Sorghum regions

Case Study 1: Somegowda et al. (2021) conducted a 3-year evaluation of 24 sorghum genotypes under both irrigated and terminal drought conditions to assess the genetic variability concerning fodder yield and quality traits. Near-infrared reflectance spectroscopy (NIRS) was utilized to predict fodder quality traits. The results revealed that drought stress significantly reduced in vitro organic matter digestibility (IVOMD) and increased the levels of neutral detergent fiber and acidic detergent fiber (NDF and ADF). In the irrigated conditions, IVOMD ranged from 46.74 to 51.74 in 2013 and 45.33 to 49.93 in 2015. Similarly, under drought stress, IVOMD varied from 45.79 to 51.31 in 2013 and 43.28 to 49.42 in 2015. The genotypic variance for IVOMD was more pronounced under stress conditions compared to the control conditions. However, the broad-sense heritability for IVOMD was lower under stress (32%) than in the control environment (82%). Near-infrared reflectance spectroscopy (NIRS) predictions demonstrated variations in IVOMD, nitrogen, and fiber fractions from year to year in both control and stress conditions. IVOMD displayed a negative correlation with fiber fractions under both irrigated and drought conditions. These results highlight the potential of using high-throughput NIRS to effectively phenotype large sorghum germplasm sets for fodder quality traits.

Case Study 2: In 2022, Somegowda et al. conducted a study on a population of 319 RILs, focusing on stover quality traits and its associated QTLs regions in both irrigated and drought stress conditions. The RIL population was created via a plant-to-plant cross between two lines: ICSV1, which is considered a less favorable parent for IVOMD, and ICSV700, recognized as a superior parent for IVOMD (Somegowda et al. 2021). ICSV1 is an open-pollinated variety with dual-purpose utility, while ICSV700 is a dual-purpose breeding line. The findings indicated that IVOMD content exhibited a range of 45.58–50.16 in 2016 and 47.91–50.0 in 2017. Likewise, nitrogen dry matter (NDM) content displayed variations from 0.68 to 1.28 in 2016 and 0.41 to 0.58 in 2017. The study found that there was no significant correlation between biomass and digestibility traits in control conditions, but a positive correlation was observed under drought stress. However, there was a negative relationship between digestibility and grain yield in control conditions, with no significant correlation under drought stress. The research involved a population of 319 RILs that were genotyped using the genotyping-by-sequencing (GBS) method, identifying 1141 informative SNP markers. A linkage map was created, revealing a total of 294 quantitative trait loci (QTLs) and 534 epistatic interactions across the studied traits. QTLs for agronomic traits were mapped to linkage group 7, while QTLs for IVOMD were located on LG1, LG2, and LG8. Further, these QTL can be used for stover quality improvement in elite breeding lines.

Case Study 3: In their 2022 study, Anbazhagan and colleagues focused on evaluating stover and grain quality traits, which included stover digestibility and grain protein and fat content. They employed near infrared reflectance spectroscopy (NIRS) for this assessment, utilizing two dual-purpose sorghum cultivars (CSV 22 and Phule Chitra), an improved landrace (MS 35-1), and a local check (Sevata Jonna). These cultivars are favored by farmers in Adilabad district, Telangana state, due to their appealing attributes like taste, grain yield, and stover yield. The study's findings highlighted significant genetic variations in organic matter digestibility (IVOMD). However, no substantial correlation was observed between stover yield and IVOMD, regardless of whether the conditions were well-watered or subjected to water stress, as tested in lysimeters. Additionally, the grain quality analysis revealed an inverse relationship between fat and protein content in water-stressed plants compared to those well-watered.

(d) **X-ray fluorescence (XRF)**:

Sorghum is an important source of energy and micronutrients, but it often lacks sufficient iron and zinc. Populations in Central India and Africa depend on sorghum for their nutritional needs. Insufficient minerals and vitamins can lead to serious health problems, especially for vulnerable groups. To address this issue, there's a pressing need to develop sorghum varieties with higher micronutrient content. Advanced phenotyping methods for assessing iron and zinc in sorghum grains have been developed, ranging from basic staining to complex instruments like AAS and ICP-OES (see Fig. 4.11). However, these methods can be costly and time-consuming. X-ray fluorescence spectrometry (XRF) offers a reliable, cost-effective, and nondestructive alternative, showing a strong correlation with ICP for assessing iron and zinc levels.

X-ray fluorescence spectrometry (XRF) works by bombarding a sample with high-energy X-rays, which excite inner-shell electrons in the sample's atoms. When these electrons return to their lower energy levels, they emit characteristic X-rays, which are detected and used to identify and quantify the elements present in the sample. XRF is a fast, nondestructive method widely used for elemental analysis in various industries, especially in agricultural sector.

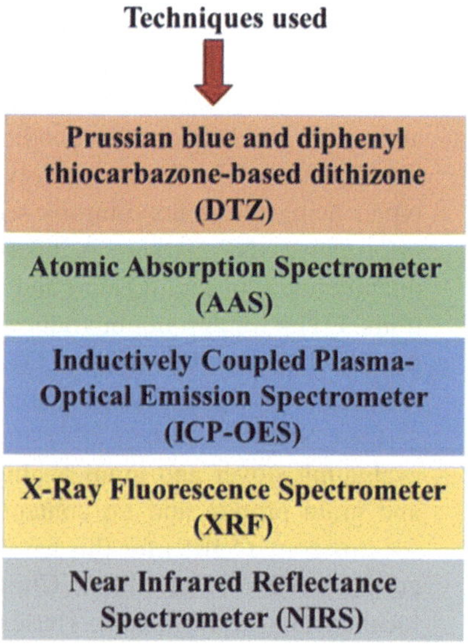

Fig. 4.11 Evaluation different of technologies for grain quality assessment. (This image retrieved from Vadez et al. (2013))

Ashok Kumar et al. (2013) demonstrated the reliability of the X-ray fluorescence spectrometer (XRF) for assessing iron and zinc in sorghum grain, with a strong correlation to inductively coupled plasma-optical emission spectrometer (ICP-OES). Breeding efforts have shown potential for enhancing iron and zinc content in sorghum, as seen in the biofortified variety "Parbhani Shakti." XRF offers a cost-effective and nondestructive method to efficiently measure these minerals in large sample populations, replacing laborious wet lab methods.

In 2013, Ashok Kumar and his research team assessed 43 sorghum commercial cultivars, encompassing 24 hybrids and 19 varieties. This study was carried out over the post-rainy seasons of 2009 and 2010 at ICRISAT-Patancheru, with the primary aim of examining variations in grain mineral content, specifically iron (Fe) and zinc (Zn), using NIRS technology. The investigation unveiled a spectrum of Fe content ranging from 25 to 41 mg/kg and Zn content spanning from 17 to 25 mg/kg within the different sorghum cultivars. Additionally, a highly significant positive correlation was noted between the concentrations of Fe and Zn in the grains ($r = 0.853$; $P < 0.01$).

In 2015, Kumar et al. extended this research by investigating the genetic variability of Fe and Zn content in selected sorghum land races, again utilizing NIRS technology. Their findings indicated a range of Fe variability from 62 to 72 mg/kg and Zn variability from 24 to 39 mg/kg. Notably, one sorghum land race, known as IS 12750, exhibited higher Fe content (72 mg/kg) and Zn content (39 mg/kg) when compared to the commercial sorghum cultivars from the 2013 study conducted by Kumar et al. This suggests the potential of using sorghum land races as donors in the process of sorghum biofortification due to their elevated levels of Fe and Zn content.

These two sets of findings strongly suggest a genetic link between the concentrations of these two micronutrients or the possibility of interconnected physiological mechanisms in the uptake and translocation of Fe and Zn within the grains. These results highlight the promising prospect of simultaneous genetic improvement for both essential micronutrients.

(e) **X-ray computed tomography (CT)**: X-ray CT is a non-destructive imaging technique that creates both 2D and 3D images of an object's internal structure by measuring X-ray attenuation as it passes through the object (Okochi et al. 2007). During a CT scan, the object is rotated while an X-ray source and detector move around it. This technology is extensively used to examine various structural, physical, and physiological characteristics in agricultural products (Domhoefer et al. 2022).

In India, the manual evaluation of raw peanuts from small farms is a slow process that can result in procurement challenges and fungal contamination due to delays and storage limitations. To tackle these issues, the Crop Physiology team at ICRISAT has investigated the potential of X-ray technology for the swift assessment of crucial peanut characteristics. Utilizing a CT system, we generated X-ray images to predict both kernel and shell weights, which are vital factors in determining the price of peanuts. Both X-ray image transformation and a convolutional neural

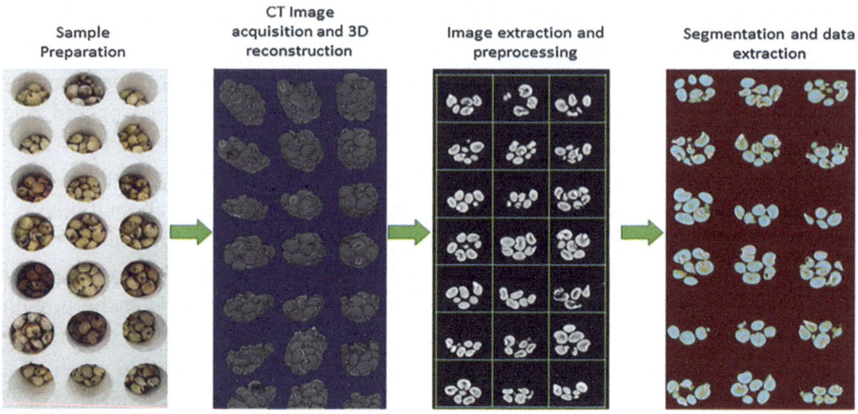

Fig. 4.12 CT imaging of sorghum and data acquisition process. (Adapted from Crozier et al. Plant Methods (2022) 18:3, available at https://doi.org/10.1186/s13007-022-00837-7)

network accurately predicted kernel weight ($R^2 > 0.93$), with the CNN performing better for shell weight ($R^2 = 0.91$). This technology can enhance peanut procurement and crop improvement programs, reducing post-harvest losses and ensuring fair payments to farmers (Domhoefer et al. 2022).

X-Ray CT for Sorghum Grain Analysis In a study conducted by Crozier et al. in 2022, X-ray CT technology was employed to assess the structural quality of grains in 19 sorghum inbred lines with variations in kernel traits, including pericarp color, mesocarp thickness, presence or absence of the testa layer, kernel size, kernel hardness, and endosperm texture. The results revealed substantial variations in seed size (ranging from 11.82 to 29.12 mm^3), pericarp volume (ranging from 1.24 to 2.26 mm^3), embryo volume (ranging from 1.73 to 4.14 mm^3), endosperm volume (ranging from 8.82 to 22.77 mm^3), and a nearly fivefold variation in endosperm texture. These structural characteristics enable the selection of genotypes for specific end-use purposes. For instance, for milling purposes, a genotype like BTx399, with larger kernels (2.02 mm^3), a higher percentage of endosperm (18.79 mm^3), and harder endosperm, is preferred. Therefore, CT imaging provides a unique and non-destructive means of studying sorghum grain (refer to Fig. 4.12).

4.3 Conclusion

Sorghum, a crucial crop for arid regions, faces the challenges of climate change, particularly erratic rainfall patterns, necessitating the development of resilient varieties. High-throughput phenotyping and innovative technologies play a pivotal role in enhancing sorghum's adaptability. This digital transformation accelerates crop improvement programs by utilizing advanced tools like Lidar, RGB, multispectral, hyperspectral sensors, NIRS, CT, and XRF for precise phenotyping. Automated data analysis pipelines ensure efficient handling of large datasets. ICRISAT's success in employing high-throughput phenomics for climate-smart crops and integrating these technologies into breeding programs underscores their significance. This digital revolution promises to enhance food security in arid regions by improving sorghum's resilience and nutritional value.

References

Anbazhagan K, Voorhaar M, Kholová J, Chadalavada K, Choudhary S, Mallayee S et al (2022) Dual-purpose sorghum: a targeted sustainable crop-livestock intervention for the smallholder subsistence farming communities of Adilabad, India. Front Sustain Food Syst 6:742909

Ashok Kumar A, Anuradha K, Ramaiah B (2013) Increasing grain Fe and Zn concentration in sorghum: progress and way forward. J SAT Agric Res 11(12):1–5

Bänziger M, Cooper M (2001) Breeding for low input conditions and consequences for participatory plant breeding examples from tropical maize and wheat. Euphytica 122:503–519

Blümmel M, Homann-Kee Tui S, Valbuena D, Duncan AJ, Herrero MT (2013) Biomass in crop-livestock systems in the context of the livestock revolution, vol 24. Secheresse, p 330

Chadalavada K (2017) Understanding the physiological processes determining sorghum grain nutritional quality (Shodhganga@INFLIBNET: understanding the physiological processes determining sorghum grain nutritional quality)

Chadalavada K, Anbazhagan K, Ndour A, Choudhary S, Palmer W, Flynn JR et al (2022) NIR instruments and prediction methods for rapid access to grain protein content in multiple cereals. Sensors 22(10):3710

Chauhan Y, Wright G, Rachaputi N, McCosker K (2008) Identifying chickpea homoclimes using the APSIM chickpea model. Aust J Agric Res 59(3):260–269

Chenu K, Deihimfard R, Chapman SC (2013) Large-scale characterization of drought pattern: a continent-wide modelling approach applied to the Australian wheatbelt–spatial and temporal trends. New Phytol 198(3):801–820

Crozier D, Riera-Lizarazu O, Rooney WL (2022) Application of X-ray computed tomography to analyze the structure of sorghum grain. Plant Methods 18(1):1–11

De Alencar Figueiredo LF, Davrieux F, Fliedel G, Rami JF, Chantereau J, Deu M et al (2006) Development of NIRS equations for food grain quality traits through exploitation of a core collection of cultivated sorghum. J Agric Food Chem 54(22):8501–8509

Dikshit AK, Birthal PS (2010) India's livestock feed demand: estimates and projections. Agric Econ Res Rev 23(1):15–28

Doggett H (1988) Sorghum, 2nd edn. Longman, London

Domhoefer M, Chakraborty D, Hufnagel E, Claußen J, Wörlein N, Voorhaar M, Anbazhagan K et al (2022) X-ray driven peanut trait estimation: computer vision aided agri-system transformation. Plant Methods 18(1):1–14

Ejaz I, He S, Li W, Hu N, Tang C, Li S et al (2021) Sorghum grains grading for food, feed, and fuel using NIR spectroscopy. Front Plant Sci 12:720022

Hajjarpoor A, Vadez V, Soltani A, Gaur P, Whitbread A, Babu DS et al (2018) Characterization of the main chickpea cropping systems in India using a yield gap analysis approach. Field Crop Res 223:93–104

Kar S, Tanaka R, Iwata H, Kholova J, Durbha SS, Adinarayana J, Vadez V (2020) Multi-scale time series analysis of evapotranspiration for high-throughput phenotyping frequency optimization. In: 2020 IEEE Latin American GRSS & ISPRS remote sensing conference (LAGIRS). IEEE, pp 98–103

Karthika G, Kholova J, Alimagham S, Ganesan M, Chadalavada K, Kumari R, Vadez V (2019) Measurement of transpiration restriction under high vapor pressure deficit for sorghum mapping population parents. Plant Physiol Rep 24:74–85

Kholova J, McLean G, Vadez V, Craufurd P, Hammer GL (2013) Drought stress characterization of post-rainy season (rabi) sorghum in India. Field Crop Res 141:38–46

Kumar AA, Anuradha K, Ramaiah B, Grando S, Frederick H, Rattunde W, Pfeiffer WH (2015) Recent advances in sorghum biofortification research. Plant Breed Rev 39:89–124

Ludlow MM, Muchow RC (1990) A critical evaluation of traits for improving crop yields in water-limited environments. Adv Agron 43:107–153

Mullet JE, Klein RR, Klein PE (2002) Sorghum bicolor—an important species for comparative grass genomics and a source of beneficial genes for agriculture. Curr Opin Plant Biol 5(2):118–121

Okochi T, Hoshino Y, Fujii H, Mitsutani T (2007) Nondestructive tree-ring measurements for Japanese oak and Japanese beech using micro-focus X-ray computed tomography. Dendrochronologia 24(2–3):155–164

Pandiselvam R, Prithviraj V, Manikantan MR, Kothakota A, Rusu AV, Trif M, Mousavi Khaneghah A (2022) Recent advancements in NIR spectroscopy for assessing the quality and safety of horticultural products: a comprehensive review. Front Nutr 9:973457

Rhodes DH, Hoffmann L, Rooney WL, Herald TJ, Bean S, Boyles R, Kresovich S (2017) Genetic architecture of kernel composition in global sorghum germplasm. BMC Genomics 18(1):1–8

Sanchez AC, Subudhi PK, Rosenow DT, Nguyen HT (2002) Mapping QTLs associated with drought resistance in sorghum (Sorghum bicolor L. Moench). Plant Mol Biol 48:713–726

Sivasakthi (2022) Tapping large genetic variation in water use, grain yield-related traits from global diversity panel of sorghum and pearlmillet; new opportunities for enhancing grain yield and drought adaption. *TIGR2EES* Summative Workshop Research Poster tapping_large_genetic.jpg (1308×925) (cam.ac.uk)

Sivasakthi K, Thudi M, Tharanya M, Kale SM, Kholová J, Halime MH et al (2018) Plant vigour QTLs co-map with an earlier reported QTL hotspot for drought tolerance while water saving QTLs map in other regions of the chickpea genome. BMC Plant Biol 18:1–18

Sivasakthi K, Marques E, Kalungwana NA, Carrasquilla-Garcia N, Chang PL, Bergmann EM et al (2019) Functional dissection of the chickpea (Cicer arietinum L.) stay-green phenotype associated with molecular variation at an ortholog of Mendel's I gene for cotyledon color: implications for crop production and carotenoid biofortification. Int J Mol Sci 20(22):5562

Somegowda VK, Prasad KV, Naravula J, Vemula A, Sivasubramani S, Rathore A, Deshpande SP (2021) Fodder quality comparison in two sorghum populations under drought. Tropentag 2021, hybrid conference September 15-17, 2021

Somegowda VK, Prasad KV, Naravula J, Vemula A, Selvanayagam S, Rathore A, Deshpande SP (2022) Genetic dissection and quantitative trait loci mapping of agronomic and fodder quality traits in sorghum under different water regimes. Front Plant Sci 13:810632

Tharanya M, Kholova J, Sivasakthi K, Seghal D, Hash CT, Raj B et al (2018) Quantitative trait loci (QTLs) for water use and crop production traits co-locate with major QTL for tolerance to water deficit in a fine-mapping population of pearl millet (Pennisetum glaucum LR Br.). Theor Appl Genet 131:1509–1529

Upadhyaya HD, Pundir RPS, Dwivedi SL, Gowda CLL, Reddy VG, Singh S (2009) Developing a mini core collection of sorghum for diversified utilization of germplasm. Crop Sci 49(5):1769–1780

Vadez V, Ratnakumar P (2016) High transpiration efficiency increases pod yield under intermittent drought in dry and hot atmospheric conditions but less so under wetter and cooler conditions in groundnut (Arachis hypogaea (L.)). Field Crop Res 193:16–23

Vadez V, Deshpande SP, Kholova J, Hammer GL, Borrell AK, Talwar HS, Hash CT (2011a) Stay-green quantitative trait loci's effects on water extraction, transpiration efficiency and seed yield depend on recipient parent background. Funct Plant Biol 38(7):553–566

Vadez V, Krishnamurthy L, Hash CT, Upadhyaya HD, Borrell AK (2011b) Yield, transpiration efficiency, and water-use variations and their interrelationships in the sorghum reference collection. Crop Pasture Sci 62(8):645–655

Vadez V, Kholova J, Zaman-Allah M, Belko N (2013) Water: the most important 'molecular' component of water stress tolerance research. Funct Plant Biol 40(12):1310–1322

Vadez V, Kholová J, Hummel G, Zhokhavets U, Gupta SK, Hash CT (2015) LeasyScan: a novel concept combining 3D imaging and lysimetry for high-throughput phenotyping of traits controlling plant water budget. J Exp Bot 66(18):5581–5593

Zaman-Allah M, Jenkinson DM, Vadez V (2011) A conservative pattern of water use, rather than deep or profuse rooting, is critical for the terminal drought tolerance of chickpea. J Exp Bot 62(12):4239–4252

Sorghum Potential for Nutrition and Food Security

5

Catarina Vila-Real, Ana Sofia Pimenta-Martins, and Ana Maria Pereia Gomes

Abstract

Approximately 238 million people are currently experiencing acute food and nutrition insecurity across the world, in which the major drivers are conflicts, economic challenges, and climate-induced weather extremes. There is an urgent need to take immediate action at global level, but small-scale initiatives are also crucial. This chapter focuses precisely on the potential of the sorghum crop within the context of food security, addressing its nutrition value, techno-functional properties, climate-resilient attributes, and its role as a food source. Regarding the nutritional profile, sorghum's high content in dietary fiber, balanced protein and fat profiles, presence of bioactive compounds, and gluten-free essence may be highlighted. Combining those with the techno-functional characteristics of this grain, whole grain sorghum shows to be a promising food ingredient to develop nutritionally balanced novel products, with interesting, but still improvable, sensory properties. In this sense, sorghum brings together unique and essential features, rendering it a viable and efficient alternative to the cultivation of other cereals such as wheat or corn. Sorghum is widely produced in Africa; nevertheless America and Asia also have a significant production share. The production and cultivation of this resilient crop, capable of withstanding adverse climate conditions, open doors to the development of nutritionally rich solutions, especially in areas where its use as a food source is currently limited, such as Europe. Popularity of sorghum-based products is on the rise, and some efforts are being done in the scientific research field; however, the exploration

C. Vila-Real · A. S. Pimenta-Martins · A. M. P. Gomes (✉)
Universidade Católica Portuguesa, CBQF—Centro de Biotecnologia e Química Fina—Laboratório Associado, Escola Superior de Biotecnologia, Porto, Portugal
e-mail: amgomes@ucp.pt

E. Habyarimana et al. (eds.), *Omics and Biotechnological Approaches for Product Profile-Driven Sorghum Improvement*,
https://doi.org/10.1007/978-981-97-4347-6_5

and refinement of novel research and processing techniques are still required. Moreover, effective communication among different sectors, including farmers, food manufacturers, researchers, and other stakeholders, is imperative in order to mark the beginning of a new phase for sorghum-based food products, where they can play a significant role in addressing pressing global challenges.

Keywords

Climate resilience · Food security · Nutrition value · Sorghum · Techno-functional properties

5.1 Framework

According to the Global Report on Food Crises 2023 Mid-Year Update, approximately 238 million people are currently facing acute food and nutrition insecurity across 48 countries (FSIN and Global Network Against Food Crises 2023). The food insecurity is primarily driven by conflicts, economic challenges, and climate-induced weather extremes (The World Bank 2023). Major contributing factors are climate change, supply chain disruptions, the COVID-19 pandemic, rising interest rates, and the Ukraine-Russia conflict is the major contributor (FSIN and Global Network Against Food Crises 2023). To address this issue and prevent a worsening of the food and nutrition security crisis, it is imperative for global leaders and decision-makers to take immediate action.

Food security is defined by the Food and Agriculture Organization (FAO) of the United Nations (UN) as "Food security exists when all people, at all times, have physical, social and economic access to sufficient, safe and nutritious food which meets their dietary needs and food preferences for an active and healthy life. Household food security is the application of this concept to the family level, with individuals within households as the focus of concern" (Clay 2002). In addition to global actions, small-scale initiatives remain significant and have the potential to influence the trajectory of this food and nutrition security crisis. In line with the previous definition of food security and the concept of nutrition, this chapter aims to summarize the potential of the sorghum crop within the context of food security, addressing its nutrition value, techno-functional properties, climate-resilient attributes, and its role as a food source. These topics will be framed within the four dimensions of food security: food availability, food access, utilization, and stability.

5.2 Nutrition Value of Sorghum

Sorghum, a North-Eastern African native cereal, belongs to the Poaceae grass family (Dahlberg 2000). Cereal whole grains, a major source of energy, protein, and dietary fiber, are important for human health. They consist of three main anatomical components: the outer pericarp, the endosperm (storage tissue), and the germ (embryo and scutellum). Considering the sorghum particular grain composition, it has been described to be 3–6% pericarp, 84–90% endosperm, and 5–10% germ, depending on variety and environmental growth conditions (Khoddami et al. 2023). Nutrient distribution varies within different parts of the cereal grain: the pericarp (or bran) is rich in non-starch polysaccharides and phenolic compounds; the endosperm contains proteins, starch, water-soluble vitamins, and minerals; and the germ is primarily composed of lipids, fat-soluble vitamins, minerals, some proteins, and a small amount of non-starch polysaccharides (Vila-Real et al. 2017) (see Fig. 5.1.).

There are around 30 species of sorghum; however, the most used for grain cultivation is *Sorghum bicolor* (L.) Moench, and so this chapter will be mainly focused on this specific species, with reference to different varieties. Despite different classifications (Serna-Saldivar and Espinosa-Ramírez 2018; Dykes 2019; Khalid et al. 2022), these varieties may be classified according to grain pericarp color and total phenolic compounds present (mainly tannins), namely, white, red, brown, and black sorghum.

The environmental conditions found in the different cultivation regions will strongly influence the appearance and nutritional composition of the crops. Albeit those differences, carbohydrates are one of the main nutrients of sorghum, varying on average between 56% and 70% of the dry weight (Khoddami et al. 2023). The main constituent is the polysaccharide starch (average range from 56% to 79% of grain weight, but some varieties have much less (Cardoso et al. 2017; Batariuc et al. 2023); nevertheless, the content in nonstarch polysaccharides (5.9–7.2 g/100 g of dry flour) and lignin (1.8–3.9 g/100 g of dry flour) is also considerable (Knudsen et al. 1988). The ratio of amylopectin and amylose of the traditional (nonwaxy) sorghum starch is 75/25; however, the waxy sorghum varieties have little to none amylose (Pontieri et al. 2020; Batariuc et al. 2023). Within carbohydrates, sorghum has an interesting content of fiber, ranging from 6% to 15% of grain, of which

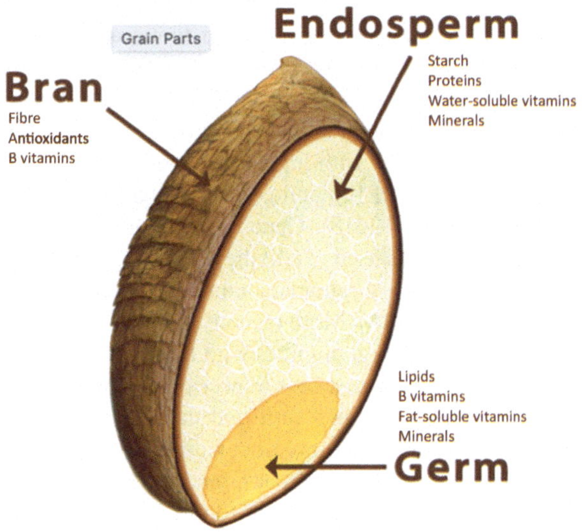

Fig. 5.1 Anatomy of a whole grain, with the nutritional richness of each part: bran, endosperm and germ. (Adapted from: wholegrainscouncil.org)

75–90% is insoluble fiber and 10–25% is soluble fiber (Cardoso et al. 2017). The protein content of cereals is characteristically low, and the essential amino acids are also scarce, in comparison with high-protein foods. Cereal proteins may be categorized according to different classifications. One way of dividing those proteins is based on their solubility (Osborn classification). According to this classification, cereal proteins may be divided into albumins (soluble in water), globulins (soluble in diluted salts solutions), prolamins (soluble in water-alcohol solutions), and glutelins (soluble in diluted acid solutions) (De Mesa-Stonestreet et al. 2010). Albumins have a physiologic role in the plant, while the other three types are storage proteins (Chéreau et al. 2016). Others divide the group of proteins in prolamins and nonprolamins (which include albumins, globulins, and glutelins). Albumins and globulins are predominant in sorghum germ, while prolamins and glutelins are found mainly in the endosperm (Serna-Saldivar and Espinosa-Ramírez 2018). Prolamins are plant storage proteins rich in proline and glutamine, and each cereal has its own specific prolamin, with kafirin being unique to sorghum. There are four main forms of kafirin, namely α-, β-, γ-, and δ-kafirin, differing depending on the molecular weight and the profile of amino acids residues (Belton et al. 2006; Castro-Jácome et al. 2020). The α-kafirin is the major form, which forms agglomerations, interacting with other molecules, reducing in consequence the digestibility (Belton et al. 2006; De Mesa-Stonestreet et al. 2010; Castro-Jácome et al. 2020). A detailed discussion about the hydrophobic structure of kafirins can be found in the work pieces of these authors (Belton et al. 2006; De Mesa-Stonestreet et al. 2010; Castro-Jácome et al. 2020). Generally, the amino acids found in higher quantities are glutamic acid, leucine, alanine, and proline. On the other hand, the most deficient amino acids are lysine, threonine, tryptophan, methionine, cysteine, isoleucine, and valine. On the contrary, the study of Zhao et al. (2022) revealed L-arginine, L-aspartic acid, and L-phenylalanine as the amino acids present in higher quantities; however, they analyzed only two types of grains from China. Interestingly, specialized metabolites derived from two amino acids, namely phenylalanine and tyrosine, are very important since they confer resistance to biotic and abiotic stresses, one of the main advantages of this crop (Simpson et al. 2021). Furthermore, breeding strategies or fortification may overcome essential amino acids deficiency (Galili and Amir 2013). Rebalancing of protein content may be essential in this endeavor, and it has been shown that varieties with higher levels of albumins, globulins, and glutelins and a lower proportion of prolamins have a higher protein quality (Serna-Saldivar and Espinosa-Ramírez 2018). Gene editing has emerged as a promising strategy and CRISPR/Cas9 was used to edit the α-kafirin gene family in sorghum variety Tx430, which resulted in a sorghum containing increased lysine content and protein digestibility (Li et al. 2018). More recently, breeding improvements have produced transgene-free, edited sorghum lines with homozygous edited alleles (19Q4-13). One of these lines was used to assess transcriptional and chromatin changes occurring during seed development in a large and stable genome-edited sorghum variety. Results showed that genome editing of α-kafirin led to changes in chromatin accessibility and to an upregulation of genes encoding lysine enrichment revealing promising potential for breeders in the near future (Hurst et al. 2023).

Kafirins naturally interact with different components in the grain, specifically phenolic compounds, lipids, phytic acid, and cell wall components, forming complexes that hamper the digestibility of those proteins (De Mesa-Stonestreet et al. 2010). During cooking, these interactions are enhanced and consequently protein digestibility is reduced. As sorghum is mostly consumed in cooked preparations, such as porridges, bread, snacks, this aspect must not be undervalued and is one of the limiting factors in terms of nutritional value. Also, sorghum proteins form intramolecular disulfide bonds, resulting in higher resistance to the action of peptidases (Cardoso et al. 2017). Nevertheless, there are several formulation strategies and technological processes that might increase the digestibility of this macronutrient, for instance, the use of fermentation, the conjugation of proteins and polysaccharides, the use of reducing agents, enzymatic hydrolyses, and irradiation (De Mesa-Stonestreet et al. 2010).

Lipids are minor constituents of cereals, including sorghum, with content varying between 1% and 4%. They are primarily found in the sorghum scutellum of the germ and to a lesser extent in the aleurone layer (Cardoso et al. 2017; Serna-Saldivar and Espinosa-Ramírez 2018; Osman et al. 2022). Analytical research has shown that sorghum extractable lipids (hexane-ether) are composed mainly by the nonpolar triglycerides (85%), followed by diglycerides (4%) and free sterols (4%). The predominant fatty acids in sorghum are unsaturated fatty acids (ca. 81% of total fat), primarily polyunsaturated (ca. 57% of unsaturated total fat). The major fatty acids found in sorghum, in descending order of magnitude, are linoleic (43.9% of total fat), oleic (34.3% of total fat), palmitic (13.4% of total fat), linolenic, and stearic acids (each at 2.7% of total fat) (Cardoso et al. 2017; Serna-Saldivar and Espinosa-Ramírez 2018; Pontieri et al. 2020). Even in low amounts, fat plays an important role in what concerns flavor, aroma, and shelf-life (Khoddami et al. 2023).

Sorghum is a good source of minerals, such as zinc (Zn), iron (Fe), potassium (K), and phosphorus (P) (Table 5.1). Reported differences are consequence of different abiotic factors characteristic across the different cultivation regions, with specific climate, temperature, and soil compositions. This fact is particularly evident for the case of iron content in sorghum grown either in Brazil (1.64 mg/100 g) or in Ethiopia (6.1 mg/100 g). Albeit the high content in phosphorus, its bioavailability may be quite low depending on the amount found bound to phytates. In this case, fermentation performed by phytase-producing microorganisms can contribute to the release of these essential minerals thus increasing associated bioavailability and increasing nutritional value.

The sorghum's content in vitamins is not deeply explored within the literature, although the presence of B-vitamins, in particular niacin, riboflavin and thiamine, is well-known. The values presented in Table 5.1 are values reported by food composition tables or from one research work that was focused on a variety of sorghum from Brazil (Cardoso et al. 2017). According to Serna-Saldivar and Espinosa-Ramírez (2018), vitamin B which is found mainly in the aleurone layer and germ are easily lost by decortication leading to loss of nutritional value. On the other hand, several processing technologies, including fermentation, sprouting and alkalinization, can lead to increased bioavailability of these vitamins. Sorghum also harbors fat-soluble

Table 5.1 Nutrient composition of sorghum (per 100 g edible portion) (Adapted from Vila-Real et al. (2017))

Energy (Kcal)	329 (Saleh et al. 2013)—388 (Osman et al. 2022)
Protein (g)	8.7 (Osman et al. 2022)—12.1 (Pontieri et al. 2020)
Fat (g)	1.24 (Cardoso et al. 2017)—4.0 (Osman et al. 2022)
Carbohydrates (g)	61.7 (Mawouma et al. 2022)—71.6 (Pontieri et al. 2020)
Dietary fiber (g)	6.0 (Cardoso et al. 2017)—15.0 (Cardoso et al. 2017) (Crude fiber: 2.0 (Saleh et al. 2013)—2.9 (Osman et al. 2022))
Zn (mg)	1.65 (Cardoso et al. 2017)—2.2 (Osman et al. 2022)
K (mg)	289 (Cardoso et al. 2017)—383 (Osman et al. 2022)
Fe (mg)	1.64 (Cardoso et al. 2017)—6.1 (Cardoso et al. 2017)
P (mg)	256 (Mawouma et al. 2022)—320 (Osman et al. 2022)
Thiamine (mg)	0.38 (Saleh et al. 2013)—0.70 (Pinheiro et al. 2021)
Riboflavin (mg)	0.04 (Pinheiro et al. 2021)—0.15 (Saleh et al. 2013)
Niacin (mg)	4.3 (Saleh et al. 2013)
Total vitamin E (μg)	1950.2 ± 179.2 (Pinheiro et al. 2021)
Total phenolics (micrograms/g)	Phenolic acids: 135.5–479.40 (Cardoso et al. 2017)

Values from Osman et al. (2022) and Mawouma et al. (2022) calculated from quantity per dry matter, based on sorghum water content

vitamins in the germ (Vitamins D and E) and in the yellow endosperm of certain sorghum varieties (Serna-Saldivar and Espinosa-Ramírez 2018). In this case, the high carotenoid germplasm can be used in biofortification efforts to improve the nutritional quality of sorghum in pro-vitamin A, particularly in those countries where sorghum is a staple crop and where high vitamin A deficiency prevails (Cruet-Burgos et al. 2020; Cruet-Burgos and Rhodes 2023).

The profile of bioactive compounds in sorghum deserves special attention, particularly the content of phenolic compounds, such as phenolic acids, proanthocyanidins and flavonoids, that vary according to cultivar and pericarp color. Some of the major phenolic acids found in sorghum are ferulic (105–343 μg/g), protocatechuic (7–141 μg/g), and caffeic acids (26–52 μg/g) (Vanamala et al. 2017). Also the contents in chlorogenic, gallic, and vanillic acids are interesting in particular cultivars, namely yellow-pale and yellow (Mawouma et al. 2022). Awika et al. (2004) showed that black cultivars had three to fourfold higher total contents of the most common sorghum anthocyanins, the 3-deoxyanthocyanidins which include apigeninidin (yellow) and luteolinidin (orange), than red and brown cultivars. The less common nature of these 3-deoxyanthocyanidins, given the absence of a hydroxyl group in position 3 of the C-ring, confers on them a high stability to pH-induced color loss. According to the study of Mawouma et al. (2022), the total contents of phenolic compounds and flavonoids are clearly higher in the red cultivars, supporting that pigmented cultivars are richer in those compounds when compared to non-pigmented ones. The main sorghum flavanones, eriodictyol and naringenin, were found to be the highest in yellow pericarp cultivars and the lowest in white cultivars (Dykes et al. 2011).

The presence of phenolic compounds is directly related to the antioxidant capacity of the grain, and so, higher concentrations are desired when looking for antioxidant capacity. Nevertheless, it is important to remember that sorghum cultivars are also source of protease inhibitors such as tannins and phytates, as previously mentioned. Protease inhibitors such as trypsin and chymotrypsin inhibitors inactivate the gastrointestinal peptidases trypsin and chymotrypsin, respectively, decreasing digestibility (Joye 2019). Tannins are prone not only to associate with minerals, but also with proteins forming complexes that precipitate, reducing the availability for the protein to become hydrolyzed (Joye 2019). Phytates sequester minerals that are essential cofactors for peptidases activity. Dietary fiber also hinders protein digestibility by increasing viscosity in the gastrointestinal tract preventing enzymes to easily access their substrates (Joye 2019). Despite this downside, such richness in naturally bioactive compounds may uphold the use of this whole grain, or derived fractions, in food product development.

Overall, highlighting the high dietary fiber content, the well-balanced protein and fat profiles, the existence of bioactive compounds, and its gluten-free nature, whole grain sorghum can be considered a highly balanced food choice that can be consumed on its own, as a whole grain, or as an ingredient in composite food products. In the Sect. 5.5—Sorghum as a food source—the positive impact of sorghum and sorghum-based products in the management of many disorders will be explored.

5.3 Techno-functional Properties of Sorghum

In addition to its nutritional value, sorghum also has techno-functional properties that make it a desirable ingredient in food products. Given some of the specific functional properties of this crop, it can be used to improve texture, viscosity, or even shelf-life of food products. Bulk density, oil and water absorption capacity, and swelling capacity are important parameters of ingredients that influence the way that the food matrix will behave in terms of gelling, emulsification, solubility, or foaming capacities. The study of Adebo and Kesa (2023) studied different fractions—whole grain, dehulled and bran—of two varieties of sorghum grains, white and red sorghum. Regarding bulk density, the authors concluded that dehulled fractions showed higher density when compared to the other fractions and, on the opposite, the bran fractions recorded the lowest values for this parameter. In consequence, in terms of food processing the denser parts of the grain could be applied for food formulations, while the less dense parts can be used for complementary foods, for which the interest is not to produce high-density foods but to provide additional nutrition and functionality, such as texture improvement, facilitate digestion, or nutritional composition enrichment (Adebo and Kesa 2023).

The water absorption capacity (WAC) of an ingredient is the ability of that ingredient to incorporate water, by capillary, osmotic, chemical, or solvent action (Tiefenbacher 2019). Mathematically, the WAC is the ratio of the weight of water absorbed by the ingredient in saturated state over the ingredient dry weight. Proteins and carbohydrate enhance WAC of an ingredient, since these components provide hydrophilic parts that have the ability to link and hold water molecules (Htet et al. 2022). Food ingredients that are source of amino acids with polar or charged side chains would have higher WAC. In the case of carbohydrates, this enhancement is related to the presence of hydrophilic polysaccharides. In the study of Adebo and Kesa (2023), it was observed that the content of polar amino acids and fiber properties impacted the values of WAC. The bran component exhibited higher WAC values compared to both the whole grain and the dehulled sorghum flour. These results were attributed to its elevated content of not only polar amino acids but also hydrophilic polysaccharides (Adebo and Kesa 2023). The oil absorption capacity of an ingredient is linked to the interaction between the nonpolar side chains of amino acids and the hydrocarbon chains of lipids (Htet et al. 2022). In the food industry, this interaction plays a major role on the mouthfeel and flavor retention (Htet et al. 2022). Sorghum, especially as whole grain rather than polished grain, by having both polar and nonpolar amino acids, and hydrophilic and hydrophobic polysaccharides can exhibit high WAC and OAC (Adebo and Kesa 2023). Nevertheless, the entire proximate composition—protein, carbohydrates, lipids contents—of sorghum contributes to its functional properties. Parameters such as size particles (of flour, for instance) can also dictate how those parameters will be affected. Batariuc et al. (2023) concluded that WAC was positively affected by sorghum flour particle size, while OAC was adversely affected.

The swelling capacity indicates the extent to which an ingredient swells and physically expands when it absorbs a liquid. According to some recent studies, sorghum flour revels low values of swelling capacity or power (Adebo and Kesa 2023; Htet et al. 2022). According to Adebo and Kesa, no significant differences were observed between the different formats of sorghum (wholegrain vs. dehulled grain vs. bran). As the above cited parameters, swelling power is also affected by particle size, processing, and cereal varieties (Adebo and Kesa 2023). Batariuc et al. (2023) observed that the particle size negatively affected the swelling power. Considering the influence of particle size on WAC, it might mean that the increase in surface area increases the ability of absorbing water, but an expansion of the particles is not necessarily observed. One reason behind this lower swelling power might be related to the lower content in amylopectin (compared with other cereals) and the presence of amylose–lipid complexes in the whole sorghum (Batariuc et al. 2023), as suggested by Tester and Morrison, when comparing swelling capacities between different cereal starches (Tester and Morrison 1990). This parameter plays a crucial role when viscosity, texture, or mouthfeel are important for food product development. In gluten-free processing, including sorghum processing, the study of the abovementioned parameters is particularly valuable for mimicking the binding and thickening qualities of gluten (Nikinmaa et al. 2022).

The combination of the nutrition profile of sorghum as well as its techno-functional properties turns this cereal an excellent food option to be utilized as a nutritious source in a well-balanced diet, thus contributing to achieve one out of the four dimensions of food security: utilization.

5.4 Climate Resilient Properties of Sorghum

Cultivated sorghum is considered a C4 plant-type, which has a different and more efficient way to carry out photosynthesis, when compared to C3 plants (for example wheat). Briefly, in C3 photosynthesis, named for the three-carbon compound produced during the initial stages, carbon dioxide (CO_2) enters the plant through stomata and is converted into sugars by the enzyme Ribulose-1,5-bisphosphate carboxylase-oxygenase (RuBisCO) (Meacham-Hensold and Nguyen 2020). However, in this process, the stomata must open to allow CO_2 entry, which results in the loss of water vapor. This can be a disadvantage, especially for plants in hot and arid environments. Additionally, RuBisCO not only fixes CO_2 but also binds with oxygen (O_2), leading to the formation of a toxic two-carbon compound (Meacham-Hensold and Nguyen 2020). This compound needs to be recycled by the plant through a process known as photorespiration. On the other hand, C4 plants like sorghum have a distinct leaf anatomy that keeps CO_2 primarily in contact with the enzyme phosphoenolpyruvate carboxylase (PEPC), which is abundant in the mesophyll cells. PEPC efficiently converts CO_2 into a four-carbon compound (Meacham-Hensold and Nguyen 2020). This enzyme exhibits a much higher affinity for CO_2 than O_2, limiting the production of the toxic compound and preventing the need for photorespiration (Meacham-Hensold and Nguyen 2020). Furthermore, the need to open the stomata is decreased when compared with C3 plants, and so the preservation of water is more efficient. These characteristics make C4 plants more competitive in challenging climate conditions. Based on those features, sorghum reveals to be a drought-tolerant crop that can be cultivated in arid and semiarid regions of sub-Saharan Africa, Asia, the Middle East, Central and North America, as well as Australia. While optimum temperatures for high-yield sorghum varieties typically exceed 25 °C, there are resilient varieties capable of thriving in cooler climates with temperatures around 12–15 °C, and minimum temperature for germination varies from 7 to 10 °C (Mashao and Prinsloo 2021). In terms of soil preferences, sorghum generally thrives on a wide range of soils, but it flourishes particularly well in light to medium-textured soils (Food and Agriculture Organization of the United Nations 2021). Ideally, the soil should be well-aerated and well-drained (Food and Agriculture Organization of the United Nations 2021). Sorghum exhibits a notable tolerance for brief periods of water-logged conditions and can adapt to high-rainfall areas, making it a versatile crop suitable for a broad range of farming conditions. Furthermore, given its extensive root system and salinity tolerance, sorghum contributes to soil conservation and biodiversity. These properties contribute to the cultivation of sorghum under a wide range of climate conditions, which together with its short maturity period adds value to it as a cereal and characterizes it as a stable and resilient crop, thus proving its coherence as an essential food source in the search for food security. In addition, its requirement of less agricultural inputs in comparison to other crops favors those farmers with lower incomes and less resources.

Sorghum is widely harvested in African regions (estimate of 68% of worldwide production—40,925,310 ha, in 2021), with Western Africa being the most representative with 33% of the area harvested globally (FAOSTAT 2021). Then, America (35%) followed by Asia (13%) are the areas where sorghum is mostly harvested (FAOSTAT 2021). Europe represents less than 1% in this context (FAOSTAT 2021). The same players are observed in terms of production, in which Africa covers 42% of global production, followed by America (38%, mainly by Northern and South America), and Asia representing 14% of worldwide production of sorghum, in 2021 (FAOSTAT 2021). Europe with 2% of global production is one of the least producing regions of this cereal (FAOSTAT 2021). In contrast, when looking at wheat production, Asia and Europe emerge as the top leaders in terms of harvested area and overall production (FAOSTAT 2021). Figure 5.2 represents the worldwide production/yield quantities of sorghum in average, from 2016 to 2021, and Fig. 5.3 shows the production share of sorghum by region, considering the same period.

Sorghum serves as a fundamental food crop for small-scale subsistence farmers in sub-Saharan Africa and Southeast Asia, regions with fragile agroecosystems (Pontieri et al. 2020). In contrast, in various regions worldwide, sorghum primarily finds its role as animal feed (Pontieri et al. 2020) or as a sustainable bioenergy source (Tanwar et al. 2023). However, and as it will be explored in the next section, there has been a notable surge in recent years in exploring the versatile applications of sorghum in food products (Pontieri et al. 2020). There is a clear window of opportunity for regions like Europe to join efforts to increase the sorghum production share, and consequently enhance self-sufficiency, ultimately targeting the development of new sorghum-based products. This will also contribute to increasing sorghum's availability and access across the world. Given the inevitable global climate change, notably global warming, sorghum emerges as an excellent food source, contributing to ensure food security. Moreover, given the possibility of using the whole crop sorghum can provide economic opportunities for farmers, including through the production of value-added products from food to feed to industry applications. The latter may include natural antioxidant preservatives or colorants and glucans from sorghum bran or lignocellulosic biomass from sorghum

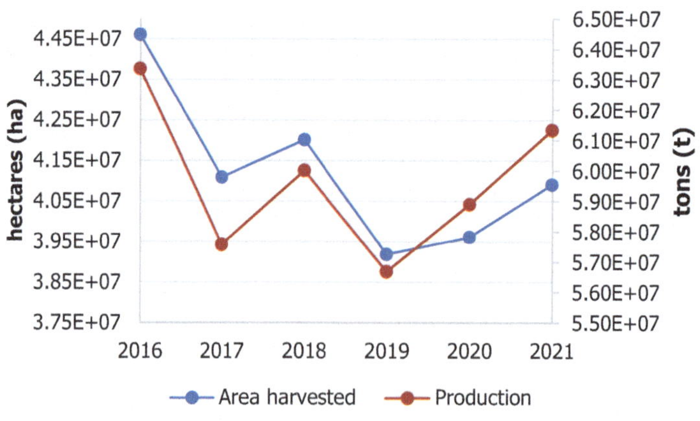

Fig. 5.2 Worldwide production/yield quantities of sorghum, from 2016 to 2021 (FAOSTAT 2021)

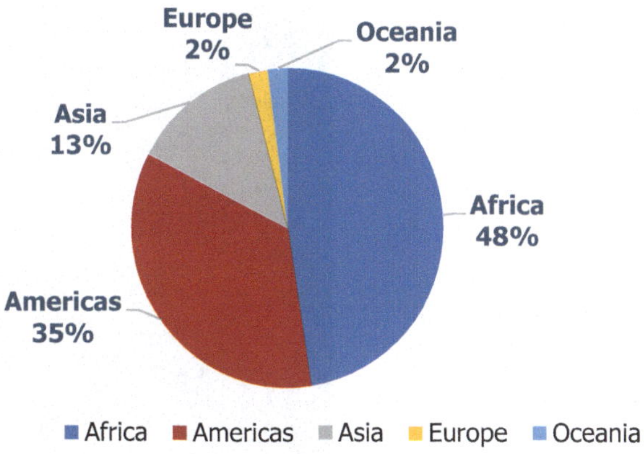

Fig. 5.3 Production share of sorghum by region, from 2016 to 2021 (FAOSTAT 2021)

crop residues or even carbon adsorbents from sorghum stem and roots (Tanwar et al. 2023). Inevitably, sorghum potential should be thoroughly explored across agriculture and industry, with investment opportunities in these sectors strongly considered.

5.5 Sorghum as a Food Source

Based on the above-mentioned evidence there is a triad of reasons that should encourage the food industry to focus on the development of new sorghum-based foods: the excellent nutritional and techno-functional properties of sorghum and the potential positive impact on human health; the crop's unique ability to withstand adverse climate conditions like drought and heat; and last, but not least, its underutilization as a food source in Western diets.

For decades, the most common incorporation of sorghum into food products was fermented porridges and alcoholic and nonalcoholic beverages, using the traditional manufacturing process. Generally, this process is carried out by women in households, and starts with the collection of the cereals, followed by sieving for removal of stones, debris, and foreign materials. The next two phases are cleaning and manually grinding either with a stone or with a wooden pestle and mortar over several cycles until the obtaining a paste (Fig. 5.4 presents the Kenyan traditional production of *Uji*). Afterward, the paste is put in a wood container, and it is let to ferment for some days (depending on the desired type of product). Also, in Mexico and China, sorghum is traditionally used for the production of alcoholic drinks, such as beer and spirit, and vinegar (Khoddami et al. 2023).

In recent years, sorghum has found new applications in the food industry, leading to the emergence of novel formats. These innovative products are being produced using different sorghum varieties sourced from various parts of the world, including India, China, and African countries. The range of products includes beverages, porridges, snacks, confectionery, bakery items, extruded products, and even incorporation into main meals such as pasta, couscous, noodles. Additionally, sorghum is used in the production of fermented foods (Thilakarathna et al. 2022; Khoddami

Fig. 5.4 Manual stone grinding of cereal grain, typically done in Kenyan villages to produce sorghum-based fermented products

et al. 2023). Table 5.2 presents the most recent sorghum-based product developments. The type of sorghum-based foods can be entirely prepared with sorghum, or using cereal blends, normally including wheat flour (Thilakarathna et al. 2022). The ratio of blends is dependent of the sorghum flour characteristics, namely particle size, baking process and sensory properties, such as aroma, flavor and color, and of the target characteristics of the final product (Thilakarathna et al. 2022).

The gluten-free bakery industry is one of the sectors in which sorghum might be applied; however, the formulation of this kind of products is quite challenging, so many of the researchers and food developers choose to use composite flours. In the sector of ready-to-eat foods, there have been few publications, namely on the production of extruded products such as flakes, instant porridge (with cowpea), pasta (with wheat) or bakery products, such as cookies (with cocoyam), and flat-bread (with wheat) (Khoddami et al. 2023). Alavi and collaborators (2019) explored the application of sorghum in extruded foods and concluded that sorghum presents specific functional features that give it a high extrusion cooking capacity. The process of extrusion is an excellent way to increase the shapes and textures portfolio, providing attractive novel food options (Alavi et al. 2019; Thilakarathna et al. 2022). One interesting example of its application is the production of popped sorghum, with sensory profile similar to the traditional popcorns (Thilakarathna et al. 2022). The research work of Mishra et al. (2015) observed that thickness of pericarp or grain size, both dependent of the sorghum variety, impacts popping qualities. One of the challenges associated with developing sorghum-based products is their often less favorable sensory qualities—bitter and astringent taste—especially when using sorghum varieties with high tannin content. One way to reduce these less positive aspects is to decorticate the grains; however, this would have implications in the final content of phenolic compounds, for instance. Also, processes such as fermentation or enzymatic treatment have been applied to enhance the sensory and functional properties of sorghum-based products (Nikinmaa et al. 2022). Therefore, the path to follow when it comes to developing a new product must consider which

Table 5.2 Food product formulations from different sorghum types, from around the world (Adapted from Xiong et al. (2019), Thilakarathna et al. (2022), and Khoddami et al. (2023))

Potential food applications	Food name	Region	References
Part of the main meal	Noodles	China	Liu et al. (2012)
	Pasta	Argentina and Australia	Khan et al. (2015), Palavecino et al. (2019)
	Couscous Annam (Sora) Kaoliang mi fan	India and Sahel India China	Ratnavathi and Patil (2013)
	Nufro	Ethiopia	Worku et al. (2016)
	Okababa	Nigeria	Thilakarathna et al. (2022)
Confectionary or sweet	Rawa laddu	India	Ratnavathi and Patil (2013)
	Kesari	India	Ratnavathi and Patil (2013)
Bakery products	Roti	India	Ratnavathi et al. (2008)
	Bhakri Dosa Kisra Injera	India India Sudan Ethiopia	Ratnavathi and Patil (2013)
	Mantou	China	Khoddami et al. (2023)
	Tortilla	Texas	Dunn et al. (2015)
Porridges	Sankati (Muda or Kali) Kanji (Ambali)	India India	Ali et al. (2003)
	Bogobe	Botswana	Boling and Nancy (1981)
	Ogi	Nigeria	Tunde Obilana (1981)
	Tô Nasha Aceda	West Africa Sudan Sudan	Ali et al. (2003)
	Ugali	Uganda, Rwanda and Tanzania	Mukuru et al. (1981)
	Traditional porridges	Mali and Burkina Faso	Cercamondi et al. (2014)
Beverages	Rabadi	India	Hussain et al. (2014)
	Pito	Nigeria and Ghana	Ali et al. (2003), Djameh et al. (2015)
	Baijiu	China	Khoddami et al. (2023)
	Obiolor	Nigeria and Ghana	Ajiboye et al. (2014)
Breakfast	Upma Idli Noodles	India India China	Ratnavathi and Patil (2013)
Snacks	Popped sorghum	India	Mishra et al. (2015)
	Other extruded products		Ratnavathi and Patil (2013)

characteristics should be prioritized in order to satisfy consumer expectations. It is evident that sorghum has the potential to offer cost-effective and nutritious food choices, especially in regions with limited access to alternative food sources;

nevertheless novel and deep research is needed to reach desirable characteristics. Furthermore, one should consider expanding the European portfolio of sorghum-based foods, recognizing the underappreciated beneficial properties of this crop.

As investments in scientific research on this ancient crop continue to grow, along with its undeniable potential, it becomes crucial to analyze its impact on disease management and its potential contribution to improve human health.

Simnadis et al. (2016) carried out a systematic review on this topic, and highlighted the work of Poquette et al. (2014) which observed that in a sample of 10 healthy men after the consumption of sorghum muffins, glucose and insulin responses decreased by up to 26% and 55%, respectively. One reason for this result is the higher levels of slowly digestible and resistant starch present in sorghum products, compared with other cereals (wheat was used as control). This low digestibility might be mainly justified by the strong association between the starch granules and proteins and tannins (Cardoso et al. 2017). Also, the sorghum richness in dietary fiber may also contribute to this reduction, as it is well-founded in terms of scientific evidence (Reynolds et al. 2020). Moreover, this attenuation of blood glucose and insulin responses was confirmed after the consumption of other sorghum-based products such as flatbread, porridge, *dhokla*, and *roti* (traditional foods), and pasta and biscuits (more common in Western diets) (Simnadis et al. 2016).

The previously mentioned systematic review also focused on the effect of sorghum consumption on different types of cancer, namely gastric, oral, and esophageal cancers. The results observed among those case–control studies were somehow controversial. It seems not only that the crop origin impacts the results (for example, China vs. South Africa), but also does the crop processing (for example, whole grain vs. dehulled grain). While the authors concluded that an inverse relationship appears to exist between the risk of oral and esophageal cancers and the frequency of sorghum consumption, especially among women, the exact role of sorghum in the etiology of these cancer types remains unclear, warranting further research (Simnadis et al. 2016).

The impact of sorghum on oxidative stress was also explored, and it seems that the consumption of red sorghum flour-containing pasta has a significant reduction in oxidative stress (higher than the control, wheat pasta), most likely given the high presence of phenolic compounds. The study of Anunciação et al. (2019) investigated the effect of sorghum consumption on body composition and metabolic variables in overweight men. It is worth noting that the consumption of extruded sorghum (40 g/day at breakfast) for eight consecutive weeks led to a reduction in body fat percentage and an increase in serum glutathione peroxidase concentration when compared to the consumption of extruded wheat (38 g/day at breakfast). However, there were no significant differences observed in enzyme concentration between the two treatments. The authors suggested that the high content of bioactive compounds in sorghum may justify this lack of difference. The researchers propose that the consumption of whole grain extruded sorghum could potentially reduce oxidative stress by enhancing antioxidant enzymes like glutathione peroxidase. This presents a promising alternative for managing obesity in overweight men (Anunciação et al. 2019).

As previously mentioned, sorghum-based products have been included in the gluten-free industry; nevertheless, given its limited use as a food source for humans in current and global society, it is important to guarantee its safety for this population-group, the celiac patients. The review of Simnadis et al. (2016) underlined that the only included study, despite its conclusions were in agreement with the safety of sorghum consumption by celiac patients, was categorized as low-quality study. The researchers suggested that long-term studies should be carried out. Moreover, it would be valuable to have more insights on the identification and characterization of the types of protein present in sorghum. In this context, Pontieri et al. (2013) conducted research aimed at characterizing the protein profiles of storage proteins in seven different sorghum varieties. The researchers confirmed the absence of toxic gliadin-like peptides in all varieties. Immunochemical experiments involving the use of the monoclonal antibody R5, which reacts with gliadin fractions from wheat and corresponding prolamins from rye and barley, provided further support for this result (Pontieri et al. 2013).

In the context of developing countries, the work of Simnadis et al. (2016) also highlights the potential of sorghum for application in the treatment of other diseases, such as diarrhea and HIV (Human Immunodeficiency Virus). For example, sorghum shows promise as an ingredient for inclusion in oral rehydration solutions and appears to be a suitable option for supplementing antiretroviral therapy in HIV patients, potentially enhancing its effectiveness.

Based on the information set out above, individuals facing clinical complications such as diabetes, obesity, celiac disease, lipid disorders, or inflammatory and oxidative conditions should contemplate incorporating sorghum-based products into their regular diets. To achieve this, the food industry should prioritize nutritional awareness and employ strategies for the development and enhancement of these products. Despite the fact that this cereal is mainly grown in Africa and Asia, Europe should consider exploring new product developments that incorporate sorghum whole grains, especially with the increasing occurrence of noncommunicable diseases (NCDs). In parallel, further biological validation of potential NCDs prevention of sorghum with animal models and human trials still need to be performed.

5.6 Conclusion

Based on the scientific evidence outlined in this chapter, sorghum brings together unique and essential features, rendering it a viable and efficient alternative to the cultivation of other cereals such as wheat or corn (Fig. 5.5). The production and cultivation of this resilient crop, capable of withstanding adverse climate conditions, open doors to the development of nutritionally rich solutions. These solutions have the potential to significantly impact human health, particularly in regions where sorghum consumption is well established, while also offering an opportunity to assert itself in areas where its use as a food source is currently limited, such as

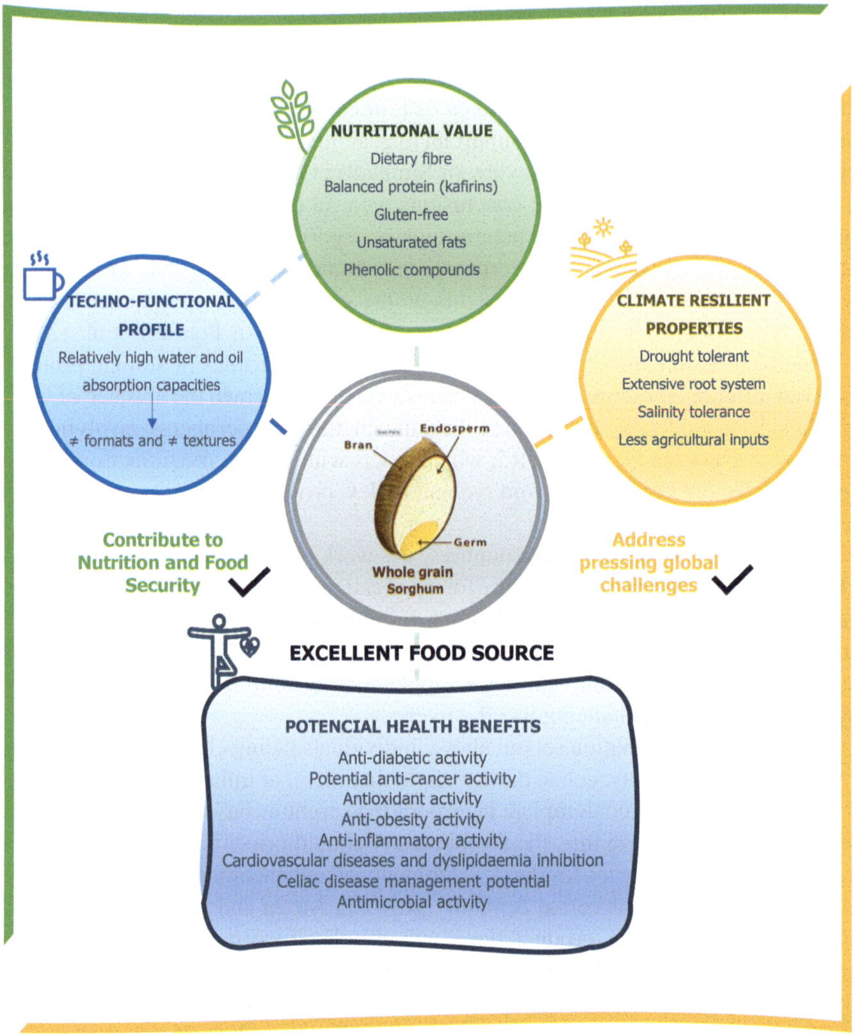

Fig. 5.5 Essential features associated with sorghum that help render it a staple crop for nutrition and security promotion

Europe. Reinforcing sorghum's production makes even more sense when we are facing global climate challenges, including issues such as global warming, water scarcity, and agricultural crops losses. Sorghum's inherent advantages over other commercial crops become increasingly evident in this context. Moreover, genomics and biofortification production strategies can be explored to enhance sorghum nutritional profiles and efficacy contributing to nutritional security. Popularity of sorghum-based products is on the rise, and this presents an opportune moment for farmers and the cereal industry to wager on this ancient and under valorized crop.

Nevertheless, the wide range of sorghum varieties, each with its unique nutritional and functional profiles, needs further characterization. In parallel, and in alignment with United Nations Sustainable Development Goal 15, it is of utmost importance to enhance functionality and sensory characteristics and preserve as much as possible nutritional quality of sorghum-based products. This requires the exploration and refinement of novel research and processing methods and technologies. In this sense, more research is required. Effective communication among different sectors, including farmers, food manufacturers, researchers, and other stakeholders, is imperative in order to mark the beginning of a new phase for sorghum-based food products, where they can play a significant role in addressing pressing global challenges and contribute positively to nutrition and food security.

References

Adebo JA, Kesa H (2023) Evaluation of nutritional and functional properties of anatomical parts of two sorghum (Sorghum bicolor) varieties. Heliyon 9(6):e17296. https://doi.org/10.1016/j.heliyon.2023.e17296

Ajiboye TO et al (2014) Nutritional and antioxidant dispositions of sorghum/millet-based beverages indigenous to Nigeria. Food Sci Nutr 2(5):597–604. https://doi.org/10.1002/fsn3.140

Alavi S et al (2019) Use of grain sorghum in extruded products developed for gluten-free and food aid applications. In: Sorghum: state of the art and future perspectives, pp 425–440. https://doi.org/10.2134/agronmonogr58.2018.0001

Ali SZ, Meera MS, Malleshi NG (2003) Processing of sorghum and pearl millet for promoting wider utilization for food. In: Alternative uses of sorghum and pearl millet in Asia—proceedings of the expert meeting ICRISAT, pp 169–187

Anunciação PC et al (2019) Extruded sorghum consumption associated with a caloric restricted diet reduces body fat in overweight men: a randomized controlled trial. Food Res Int 119:693–700. https://doi.org/10.1016/j.foodres.2018.10.048

Awika JM, Rooney LW, Waniska RD (2004) Properties of 3-deoxyanthocyanins from sorghum. J Agric Food Chem 52(14):4388–4394. https://doi.org/10.1021/jf049653f

Batariuc A, Coţovanu I, Mironeasa S (2023) Sorghum flour features related to dry heat treatment and milling. Foods 12(11). https://doi.org/10.3390/foods12112248

Belton PS et al (2006) Kafirin structure and functionality. J Cereal Sci 44(3):272–286. https://doi.org/10.1016/j.jcs.2006.05.004

Boling MB, Nancy E (1981) Bogobe: sorghum porridge of Botswana. In: Processing of the international symposium on sorghum grain quality—session 2: traditional food preparations and their quality parameters, pp 32–35

Cardoso LDM et al (2017) Sorghum (Sorghum bicolor L.): nutrients, bioactive compounds, and potential impact on human health. Crit Rev Food Sci Nutr 57(2):372–390. https://doi.org/10.1080/10408398.2014.887057

Castro-Jácome TP, Alcántara-Quintana LE, Tovar-Pérez EG (2020) Optimization of sorghum kafirin extraction conditions and identification of potential bioactive peptides. Biores Open Access 9(1):198–208. https://doi.org/10.1089/biores.2020.0013

Cercamondi CI et al (2014) Sodium iron EDTA and ascorbic acid, but not polyphenol oxidase treatment, counteract the strong inhibitory effect of polyphenols from brown sorghum on the absorption of fortification iron in young women. Br J Nutr 111(3):481–489. https://doi.org/10.1017/S0007114513002705

Chéreau D et al (2016) Combination of existing and alternative technologies to promote oilseeds and pulses proteins in food applications. OCL 41(1):D406. https://doi.org/10.1051/ocl/2016020

Clay E (2002) Food security: concepts and measurement. trade reforms and food security: conceptualizing the linkages. Paper for FAO expert consultation on trade and food security: conceptualising the linkages Rome. Published as Chapter 2 of Trade Reforms and Food Security: conceptualising the linkages. FAO, Rome

Cruet-Burgos C, Rhodes DH (2023) Unraveling transcriptomics of sorghum grain carotenoids: a step forward for biofortification. BMC Genomics 24(1):1–19. https://doi.org/10.1186/s12864-023-09323-3

Cruet-Burgos C et al (2020) Advancing provitamin A biofortification in sorghum: genome-wide association studies of grain carotenoids in global germplasm. Plant Genome 13(1):1–14. https://doi.org/10.1002/tpg2.20013

Dahlberg J (2000) Classification and characterization of sorghum. In: Smith CW, Frederiksen RA (eds) Sorghum: origin, history, technology, and production. Wiley, New York

De Mesa-Stonestreet NJ, Alavi S, Bean SR (2010) Sorghum proteins: the concentration, isolation, modification, and food applications of kafirins. J Food Sci 75(5):R90. https://doi.org/10.1111/j.1750-3841.2010.01623.x

Djameh C et al (2015) Optimization of the sorghum malting process for pito production in Ghana. J Inst Brew 121(1):106–112. https://doi.org/10.1002/jib.191

Dunn KL et al (2015) Interaction of sorghum tannins with wheat proteins and effect on in vitro starch and protein digestibility in a baked product matrix. J Agric Food Chem 63(4):1234–1241. https://doi.org/10.1021/jf504112z

Dykes L (2019) Tannin analysis in sorghum grains. Methods Mol Biol 1931:109–120. https://doi.org/10.1007/978-1-4939-9039-9_8

Dykes L et al (2011) Flavonoid composition of lemon-yellow sorghum genotypes. Food Chem 128(1):173–179. https://doi.org/10.1016/j.foodchem.2011.03.020

FAOSTAT (2021) Crops and livestock products—sorghum. https://www.fao.org/faostat/en/#data/QCL. Accessed 10 Oct 2023

Food and Agriculture Organization of the United Nations (2021) Land & water—sorghum, databases & softwares

FSIN and Global Network Against Food Crises (2023) GRFC 2023 mid-year update. Rome

Galili G, Amir R (2013) Fortifying plants with the essential amino acids lysine and methionine to improve nutritional quality. Plant Biotechnol J 11(2):211–222. https://doi.org/10.1111/pbi.12025

Htet MNS et al (2022) Comparative assessment of nutritional and functional properties of different sorghum genotypes for ensuring nutritional security in dryland agro-ecosystem. Front Nutr 9:1–13. https://doi.org/10.3389/fnut.2022.1048789

Hurst JP et al (2023) Large and stable genome edits at the sorghum alpha kafirin locus result in changes in chromatin accessibility and globally increased expression of genes encoding lysine enrichment. Front Plant Sci 14:1–11. https://doi.org/10.3389/fpls.2023.1116886

Hussain SA, Garg FC, Pal D (2014) Effect of different preservative treatments on the shelf-life of sorghum malt based fermented milk beverage. J Food Sci Technol 51(8):1582–1587. https://doi.org/10.1007/s13197-012-0657-4

Joye I (2019) Protein digestibility of cereal products. Foods 8:199. https://doi.org/10.3390/foods8060199

Khalid W et al (2022) Nutrients and bioactive compounds of Sorghum bicolor L. used to prepare functional foods: a review on the efficacy against different chronic disorders. Int J Food Properties 25(1):1045–1062. https://doi.org/10.1080/10942912.2022.2071293

Khan I et al (2015) Acute effect of sorghum flour-containing pasta on plasma total polyphenols, antioxidant capacity and oxidative stress markers in healthy subjects: a randomised controlled trial. Clin Nutr 34(3):415–421. https://doi.org/10.1016/j.clnu.2014.08.005

Khoddami A et al (2023) Sorghum in foods: functionality and potential in innovative products. Crit Rev Food Sci Nutr 63(9):1170–1186. https://doi.org/10.1080/10408398.2021.1960793

Knudsen KEB, Munck L, Eggum BO (1988) Effect of cooking, pH and polyphenol level on carbohydrate composition and nutritional quality of a sorghum (Sorghum bicolor (L.) Moench) food, ugali. Br J Nutr 59(1):31–47. https://doi.org/10.1079/bjn19880007

Li A et al (2018) Editing of an alpha-kafirin gene family increases digestibility and protein quality in sorghum. Plant Physiol 177(4):1425–1438. https://doi.org/10.1104/pp.18.00200

Liu L et al (2012) Characterization of sorghum grain and evaluation of sorghum flour in a Chinese egg noodle system. J Cereal Sci 55(1):31–36. https://doi.org/10.1016/j.jcs.2011.09.007

Mashao J, Prinsloo T (2021) Sorghum production, sorghum production guide. https://www.arc.agric.za/arc-gci/Fact%20Sheets%20Library/Sorghum%20Production.pdf

Mawouma S et al (2022) Chemical composition and antioxidant profile of sorghum (Sorghumbicolor (L.) Moench) and pearl millet (Pennisetumglaucum (L.) R.Br.) grains cultivated in the far-north region of Cameroon. Foods 11:13p. https://doi.org/10.3390/foods11142026

Meacham-Hensold K, Nguyen A (2020) Ripe—realizing increased photosynthetic efficiency for sustainable increases in crop yield. The difference between C3 and C4 plants

Mishra G et al (2015) Varietal influence on the microwave popping characteristics of sorghum. J Cereal Sc 65:19–24. https://doi.org/10.1016/j.jcs.2015.06.001

Mukuru SZ, Mushonga JN, Murty DS (1981) Sorghum ugali. In: Processing of the international symposium on sorghum grain quality—session 2: traditional food preparations and their quality parameters, pp 39–44

Nikinmaa M et al (2022) Effect of bioprocessing on techno-functional properties of climate-resilient African crops, sorghum and cowpea. Foods 11(19):1–19. https://doi.org/10.3390/foods11193049

Osman A et al (2022) Nutrient composition and in vitro fermentation characteristics of sorghum depending on variety and year of cultivation in northern Italy. Foods 11(20):3255. https://doi.org/10.3390/foods11203255

Palavecino PM et al (2019) Gluten-free sorghum pasta: starch digestibility and antioxidant capacity compared with commercial products. J Sci Food Agric 99(3):1351–1357. https://doi.org/10.1002/jsfa.9310

Pinheiro SS et al (2021) Stability of B vitamins, vitamin E, xanthophylls and flavonoids during germination and maceration of sorghum (Sorghum bicolor L.). Food Chem 345:1–8. https://doi.org/10.1016/j.foodchem.2020.128775

Pontieri P et al (2013) Sorghum, a healthy and gluten-free food for celiac patients as demonstrated by genome, biochemical, and immunochemical analyses. J Agric Food Chem 61(10):2565–2571. https://doi.org/10.1021/jf304882k

Pontieri P et al (2020) Nutritional composition of a selected white food-grade waxy sorghum variety grown in Mediterranean environment. Aust J Crop Sci 14(9):1525–1532. https://doi.org/10.21475/ajcs.20.14.09.p2783

Poquette NM, Gu X, Lee SO (2014) Grain sorghum muffin reduces glucose and insulin responses in men. Food Funct 5(5):894–899. https://doi.org/10.1039/c3fo60432b

Ratnavathi CV, Patil JV (2013) Sorghum utilization as food. J Nutr Food Sci 4(1):1–8. https://doi.org/10.4172/2155-9600.1000247

Ratnavathi C, Nadoor S, Chavan UD (2008) Dough and roti making qualities of sorghum genotypes. Jowar Samachar 4:2–4

Reynolds AN, Akerman AP, Mann J (2020) Dietary fibre and whole grains in diabetes management: systematic review and meta-analyses. PLoS Med 17(3):e1003053. https://doi.org/10.1371/journal.pmed.1003053

Saleh ASM et al (2013) Millet grains: nutritional quality, processing, and potential health benefits. Compr Rev Food Sci Food Saf 12:281–295. https://doi.org/10.1111/1541-4337.12012

Serna-Saldivar SO, Espinosa-Ramírez J (2018) Grain structure and grain chemical composition. In: Sorghum and millets: chemistry, technology, and nutritional attributes. https://doi.org/10.1016/B978-0-12-811527-5.00005-8

Simnadis TG, Tapsell LC, Beck EJ (2016) Effect of sorghum consumption on health outcomes: a systematic review. Nutr Rev 74(11):690–707. https://doi.org/10.1093/nutrit/nuw036

Simpson JP et al (2021) Identification of the tyrosine- and phenylalanine-derived soluble metabolomes of sorghum. Front Plant Sci 12:1–17. https://doi.org/10.3389/fpls.2021.714164

Tanwar, R. et al. (2023) 'Nutritional, phytochemical and functional potential of sorghum: a review', Food Chem Adv, 3(4) p. 100501. doi: https://doi.org/10.1016/j.focha.2023.100501

Tester RF, Morrison WR (1990) Swelling and gelatinization of cereal starches. I. Effects of amylopectin, amylose, and lipids. Cereal Chem 67(6):551–557. https://www.cerealsgrains.org/publications/cc/backissues/1990/documents/67_551.pdf

The World Bank (2023) Joint statement by the Heads of the Food and Agriculture Organization, International Monetary Fund, World Bank Group, World Food Programme and World Trade Organization on the Global Food and Nutrition Security Crisis. https://www.worldbank.org/en/news/statement/2023/02/08/joint-statement-on-the-global-food-and-nutrition-security-crisis. Accessed 10 Oct 2023

Thilakarathna RCN, Madhusankha GDMP, Navaratne SB (2022) Potential food applications of sorghum (Sorghum bicolor) and rapid screening methods of nutritional traits by spectroscopic platforms. J Food Sci 87(1):36–51. https://doi.org/10.1111/1750-3841.16008

Tiefenbacher KF (2019) Glossary of terms in wafers, waffles and adjuncts. In: The technology of wafers and waffles II—recipes, product development and know-how, pp 325–411. https://doi.org/10.1016/b978-0-12-809437-2.00010-1

Tunde Obilana A (1981) Traditional sorghum foods in Nigeria: their preparation and quality parameters. In: Processing of the international symposium on sorghum grain quality—session 2: traditional food preparations and their quality parameters, pp 45–54

Vanamala JKP et al (2017) Grain and sweet sorghum (Sorghum bicolor L. Moench) serves as a novel source of bioactive compounds for human health. Crit Rev Food Sci Nutr 58(17):2867–2881. https://doi.org/10.1080/10408398.2017.1344186

Vila-Real C et al (2017) Nutritional value of African indigenous whole grain cereals millet and sorghum. Nutr Food Sci Int J 4(1). https://doi.org/10.19080/NFSIJ.2017.03.555628

Worku BB, Woldegiorgis AZ, Gemeda HF (2016) Indigenous processing methods of Cheka: a traditional fermented beverage in southwestern Ethiopia. J Food Process Technol 07(01):1–7. https://doi.org/10.4172/2157-7110.1000540

Xiong Y et al (2019) Sorghum grain: from genotype, nutrition, and phenolic profile to its health benefits and food applications. Compr Rev Food Sci Food Saf 18(6):2025–2046. https://doi.org/10.1111/1541-4337.12506

Zhao Y et al (2022) Metabolomics reveals nutritional diversity among six coarse cereals and antioxidant activity analysis of grain sorghum and sweet sorghum. Antioxidants 11(10):1984. https://doi.org/10.3390/antiox11101984

Crop Wild Relatives of Sorghum: A Novel Source of Genetic Variation for Crop Improvement

6

Mani Vetriventhan, K. Jagadeesh, S. Baraskar, Naresh Reddy Dagunapur, Raja Srinivas, Kanchanapally Anilkumar, and Kuldeep Singh

Abstract

Sorghum is one of the most important dryland crops with greater prominence in terms of food and fodder, contribution to global food security and nutrition. Sorghum production is constrained by several biotic and abiotic stresses; on the other hand, the repeated use of identical sources in breeding program has resulted in narrow genetic base of crop cultivars. This enhanced the need to broaden the genetic base of crop cultivars by using diverse landraces and crop wild relatives. Currently, 259,595 accessions of sorghum spp. including 2465 wild and weedy relatives are being conserved ex situ in the genebanks across the globe. The exploitation of wild species harboring several desirable alleles for various biotic and abiotic stress tolerance, productivity, and nutritional traits can aid in diversifying the trait sources. Gap analysis of the sorghum wild relatives indicated significant taxonomical and geographical gaps in the global collection, requiring immediate efforts to enrich the gene pool. Using novel breeding strategies including the application of genomics approaches like genome sequencing, trait mapping and pangenome development provides more comprehensive insights

M. Vetriventhan (✉) · N. R. Dagunapur · R. Srinivas · K. Anilkumar · K. Singh
Genebank, International Crops Research Institute for the Semi-Arid Tropics (ICRISAT), Hyderabad, India
e-mail: Vetriventhan.Mani@icrisat.org

K. Jagadeesh · S. Baraskar
Genebank, International Crops Research Institute for the Semi-Arid Tropics (ICRISAT), Hyderabad, India

Professor Jayashankar Telangana State Agricultural University (PJTSAU), Rajendranagar, Hyderabad, India

E. Habyarimana et al. (eds.), *Omics and Biotechnological Approaches for Product Profile-Driven Sorghum Improvement*,
https://doi.org/10.1007/978-981-97-4347-6_6

into trait-specific sources. This can aid in identifying and transferring novel traits from wild relatives and can lead to sustainable production levels in sorghum.

Keywords

Crop wild relatives (CWRs) · Genebank · Sorghum germplasm · Sorghum classification · Sorghum utilization

6.1 Introduction

Sorghum [*Sorghum bicolor* (L.) Moench] is one of the most important C_4 dryland crops cultivated in arid and semi-arid climates and serves as an important source of energy in developing countries (Adebo 2020). Globally sorghum occupies an area of 40.9 million ha, yielding 61.36 million tons (FAO 2021). The United States, Nigeria, and India are the three major producers of sorghum globally (FAO 2021). Climate change accelerates, posing challenges to crop growth, while the escalating population demands increased production. Beyond drought, various biotic and abiotic stressors compound the situation, necessitating the development of resilient sorghum varieties. However, repeated use of identical sources in breeding for target traits has led to a narrowing of genetic diversity among cultivars.

Incorporating wild relatives of sorghum, which offer diverse traits, becomes essential to address this. These wild relatives, closely related to cultivated crops, possess immense untapped potential, as demonstrated in various crops. According to FAO, 75% of the genetic diversity among the agricultural varieties had been lost due to the replacement of diverse landraces with genetically uniform crop cultivars (www.cropwildrelatives.org/cwr/threats/). This emphasizes the need to integrate novel diversity from landraces and crop wild relatives (CWRs) into breeding programs. Overcoming challenges like crossability barriers and linkage drag requires modern plant breeding methods. This allows easy access to novel alleles from CWRs, enriching the domesticated gene pool. Ultimately, the conservation, characterization, and effective utilization of CWRs as genetic sources pave the way for sustainable food production, steering us away from hunger. This chapter provides a comprehensive insight into the importance of sorghum wild relatives and their contribution towards achieving sustainable production levels in sorghum.

6.2 Origin and Taxonomy

Sorghum was domesticated in the Ethiopia-Sudan region of Northeast Africa from its wild progenitor *S. bicolor* subsp. *verticilliflorum*, while secondary centers of origin are found in India and China. The genus *Sorghum* belongs to the family Poaceae (grass family), tribe Andropogoneae, subtribe Sorghastrae, and is subdivided into five subgenera or sections: Chaetosorghum, Heterosorghum, Parasorghum, Stiposorghum, and Eu-Sorghum (Table 6.2) (Garber 1950). The Section Eu-Sorghum includes two wild perennials [*Sorghum halepense* (L.) Pers. and *S. propinquum*

(Kunth) Hitch.] and an annual *S. bicolor* (L.) Moench (de Wet 1978). The *S. bicolor* has three subspecies namely (i) subsp. *bicolor* (cultivated sorghum); (ii) subsp. *drummondii* (Steud.) de Wet comb. nov, derivatives of hybridization among cultivated sorghums and their closest wild relatives; and (iii) subsp. *verticilliflorum* (Steud.) the wild progenitors of cultivated sorghums.

Based on crossability and fertility of progeny, subspecies of *S. bicolor* and *S. propinquum* are placed in the crops' primary gene pool. These two species are fully interfertile and show high level of fertility (Baker 1972; Doggett and Prasada Rao 1995; Doggett and Majisu 1968). *S. propinquum* is a diploid robust wild perennial species with rhizomes which is distributed across Sri Lanka, southern India, and part of southeast Asia (de Wet 1978). It shows close relativeness towards *S. halepense* but differs in ploidy level. It shows great degree of crossability with *S. bicolor* without any barriers which underlines the reason behind the existence of wide range of intermediate types in the nature (Ejeta and Grenier 2005; Venkateswaran et al. 2019). The Subsp. *bicolor* is divided into the five basic interfertile races Bicolor, Kafir, Caudatum, Durra, and Guinea on the basis of spikelet morphology and ten intermediate races (Lazarides et al. 1991). Subsp. *verticilliflorum* was initially trifurcated into *verticilliflorum, arundinaceum,* and *aethiopicum* (de Wet and Huckabay 1967), but the fourth race called *virgatum* was added later (Harlan and de Wet 1971). The third subspecies called *drummondii* is also known as Sudan grass which is especially utilized as forage crop. Sudan grass is believed to be the progeny of the cross between cultivated sorghum and *S. bicolor subsp. verticilliflorum* (Andersson and De Vicente 2010; Ejeta and Grenier 2005).

The secondary gene pool consists of *S. halepense,* commonly known as Johnson grass and *S. almum,* commonly known as Columbus grass. *S. halepense* is distributed in Indian subcontinent as well as in Southern Eurasia. It is tetraploid in nature and believed to be originated from the cross between *S. propinquum* and *S. bicolor* subsp. *verticilliflorum* which was further underwent chromosome doubling (de Wet 1978). It has outcrossing rate less than 10% and it is self-compatible in nature (Black 1983; Burke et al. 2007). The cross with diploid cultivated sorghum is reported to be successful in natural as well as controlled environments despite of marked differences in their ploidy levels leading to generation of triploids which are sterile in nature or partial fertile tetraploids (Morrell et al. 2005). *S. almum* Parodi is another tetraploid member of secondary gene pool which is thought to be originated from the hybridization between *S. bicolor* and *S. halepense* (Parodi 1943). Endrizzi (1957) reported the progenies obtained from the cross *S. almum* and *S. bicolor* exhibited chromosome number $2n = 30$ and $2n = 40$ having intermediate characteristics and shown heterosis for vegetative growth, while the cross between *S. almum* and *S. halepense* generated hybrids carrying partial vigor with chromosome number $2n = 40$.

The tertiary gene pool includes all the species belonging to subgenera other than *Eusorghum*. Due to the presence of various fertilization barriers, the gene transfer from this pool is relatively difficult as compared to first two gene pools. It consists of a total 19 species which include one species from the subgenera *Heterosorghum* (*S. laxiflorum*) and one from the subgenera *Chaetosorghum* (*S. macrospermum*), seven species from the subgenera *Parasorghum* and ten species from the subgenera *Stiposorghum*. Sun et al. (1991) carried out crossing between *S. bicolor* and *S. versicolor* and reported that pollen tube inhibition occurred mostly at the stigma and very few pollen tube managed to grow up to the ovary. The major reason behind the unsuccessful crossing events between domesticated sorghum and species belonging to tertiary gene pool lies in the pollen-pistil interaction, where the pollen tubes of wild sorghum species exhibited abnormal growth in pistils of *S. bicolor* (Hodnett et al. 2005). Crossing was carried out between *S. bicolor* and *S. macrospermum* by Kuhlman et al. (2008). These wild species are rich in allelic diversity and consist of highly underutilized source of beneficial genes which can be utilized for cultivated sorghum improvement. The primary reason behind such underutilization is the restriction of these wild species in particular geographical areas (Kole 2011).

Liu et al. (2014) suggested a new subgeneric classification of *Sorghum* Moench into three distinct subgenera, (i) Chaetosorghum E.D Garber with two sections (section Chaetosorghum (E.D. Garber) Ivanjuk. & Doronina and section Heterosorghum (E.D. Garber) Ivanjuk. & Doronina), (ii) Parasorghum (Snowden) E.D. Garber, and (iii) Sorghum. The species *S. macrospermum* E.D. Garber comes under section Chaetosorghum, while *S. laxiflorum* F.M. Bailey belongs to section Heterosorghum. The Parasorghum includes 17 species such as *S. amplum* Lazarides, *S. angustum* S.T. Blake, *S. bulbosum* Lazarides, *S. brachypodium* Lazarides, *S. ecarinatum* Lazarides, *S. exstans* Lazarides, *S. grande* Lazarides, *S. interjectum* Lazarides, *S. intrans* F. Muell. ex Benth., *S. leiocladum* (Hack.) C.E. Hubb., *S. matarankense* E.D. Garber & L.A. Snyder, *S. nitidum* (Vahl) Pers., *S. plumosum* (R.Br.) P. Beauv., S. *purpureosericeum* (Hochst. ex A. Rich.) Asch. & Schweinf., *S. stipoideum* (Ewart & Jean White) C.A. Gardner & C.E. Hubb., *S. timorense* (Kunth) Bűse, *S. versicolor* Andersson), and subgenera *Sorghum* includes nine species such as *S. alum* Parodi, *S. arundinaceum* (Desv.) Stapf, *S. bicolor* (L.) Moench, *S. x drummondii* (Nees ex Steud.) Millsp. & Chase, *S. halepense* (L.) Pers., *S. miliaceum* (Roxb.) Swoden, *S. propinquum* (Kunth) Hitchc., *S. sudanense* (Piper) Stapf, and *S. virgatum* (Hack.) Stapf). But according to recent classification USDA-ARS, the species belonging to the two sections Parasorghum and Stiposorghum are classified under a single section Parasorghum. The name Stiposorghum is considered as synonym of Parasorghum (Search Taxonomy GRIN-Global 2023). The species *S. laxiflorum* F.M. Bailey and *Sorghum macrospermum* E. D. Garber are the only species fallen under the sections Heterosorghum and Chaetosorghum respectively (Species GRIN-Global 2023). To overcome the ambiguity in several classifications, there is need for more molecular based evidences.

6.3 Current Status, Conservation, and Utility of Sorghum Wild Species

Globally 259,595 accessions of sorghum spp. including 2465 wild and weedy relatives are conserved in the 135 genebanks located across the globe (Bramel et al. 2022). The wild and weedy species of sorghum accessions held by various genebanks in the world are presented in Table 6.1. The International Crops Research Institute for the Semi-Arid Tropics (ICRISAT) genebank conserves 42,869 accessions originating (collection site) from 93 countries, and it is the largest collection of *Sorghum* spp. germplasm in the world. This collection includes 42,416 cultivated

Table 6.1 *Sorghum* wild and weedy relatives conserved in the genebanks globally

Sl. no	Name of institute	No. of wild species	Total no of wild accessions
1	Agriculture Technology Transfer Center Fush Kruja	1	44
2	Banco Base de Germoplasma, Instituto de Recursos Biológicos, Instituto Nacional de Tecnología Agropecuaria (BBC-INTA)	5	405
3	Australian Tropical Crops & Forages Genetic Resources Centre (ATCFA)	24	346
4	Empresa Pernambucana de Pesquisa Agropecuária (IPA)	1	21
5	Grassland Research Institute, Chinese Academy of Agricultural Sciences	1	15
6	International Livestock Research Institute (ILRI-Ethiopia)	4	30
7	Laboratoire des Ressources Génétiques et Amélioration des Plantes Tropicales, ORSTOM (ORSTOM-MONTP)	1	27
8	Collection Nationale Céréales à Paille, Unité expérimentale du Magneraud, Groupe d'Étude et de contrôle des Variétés et des Semences (GEVES)	2	102
9	National Bureau of Plant Genetic Resources (NBPGR), New Delhi (NBPGR)	3	72
10	International Crop Research Institute for the Semi-Arid Tropics (ICRISAT)	16	453
11	National Plant Genebank of Iran, Seed, and Plant Improvement Institute (NPGBI-SPII)	1	25
12	Department of Genetic Resources I, National Institute of Agrobiological Sciences (NIAS)	3	264
13	National Genebank of Kenya, Crop Plant Genetic Resources Centre—Muguga (KARI-NGBK)	6	101
14	Plant Genetic Resources Program (PGRP)	1	16
15	Research Institute for Cereals and Technical Plants Fundulea (ICCPT Fundul)	1	49
16	Division of Plant and Liquor Control, Department of Agriculture, Economics and Marketing	8	26
17	Millennium Seed Bank Project, Seed Conservation Department, Royal Botanic Gardens, Kew, Wakehurst Place (RBG)	10	30
18	Plant Genetic Resources Conservation Unit, Southern Regional Plant Introduction Station, University of Georgia, USDA-ARS	14	242
19	SADC Plant Genetic Resources Centre (SRGB)	3	35
Total			**2292**

Source: http://www.fao.org/wiews-archive/germplasm_query.html

and 453 wild species accessions, distributed over 0.5 million seed samples of sorghum to 112 countries including wild species accessions following the standard material transfer agreement (SMTA) of the International Treaty on Plant Genetic Resources for Food and Agriculture (ITPGRFA) of the FAO. Among the 2292 wild accessions being conserved at various global genebanks, around 60% of accessions are from primary gene pool (1441) followed by tertiary gene pool with 383 accessions and 261 accessions from secondary gene pool (Table 6.2).

6.4 Gap Analysis in Sorghum Germplasm Collection

Gap analysis provides us the information regarding the current status of biodiversity conservation of a specified area and the requirement of additional measures for conservation of the existing genetic diversity (Maxted et al. 2008). Several studies were conducted regarding the existing gaps in sorghum germplasm collection from various regions. Among the sorghum germplasm accessions at ICRISAT, collection from east Africa countries holds 12,750 accessions including 11,672 landraces, 877 breeding materials, 6 improved cultivars, and 195 wild accessions. The geographical gaps identified includes 153 districts located in 50 provinces of 10 countries. The species collected from the east African countries include *S. bicolor*, *S. halepense*, *S. lanceolatum*, *S. macrochaeta*, *S. purpureosericeum,* and *S. versicolor*. There is a need for exploration of remaining species which can bridge the existing gaps (Upadhyaya et al. 2017b). Similarly, sorghum collection at ICRISAT from west and central African countries was assessed for geographical gaps and existing diversity and identified 386 districts located in 11 West and Central African countries as major gaps with Burkina Faso and Nigeria holding 140 and 118 districts. The 43 accessions of wild species collected from this region belong to only three species of sorghum, i.e., *S. bicolor* ssp. *drummondii* and ssp. *verticilliflorum*, *S. hevisonii*, and *S. macrochaeta* that envisaged the need for exploration and enriching the collection with other species (Upadhyaya et al. 2017c). Based on the information from 5340 georeferenced accessions of which 5322 were characterized at ICRISAT, gap analysis was carried out for sorghum germplasm collected from south Asian countries (Upadhyaya et al. 2017a). The gaps identified include 110 districts in 20 provinces of India, 13 districts in three provinces of Pakistan, 3 districts in Bangladesh, and 5 districts in four provinces of Sri Lanka. The species needed to be considered further for collection includes *Sorghum bicolor subsp. verticilliflorum*, *S. halepense,* and *S. propinquum*. The exploration in above identified districts can aid collecting the species underrepresented in the collection from south Asian countries (Upadhyaya et al. 2017a). Among the wild relatives of sorghum, 88% of them are considered high priority for collecting. The gene pool is concentrated in tropical northern Australia and southern Africa, and the most significant area identified as in need of collecting is northern Australia (Crop Wild Relatives | Gap Analysis Results—Crop Wild Relatives 2023). The conservation status and threat assessment for 23 taxa of wild sorghum across the world, i.e., Australia, New Guinea, Asia, Africa and Central America at both in-situ and ex-situ identified 3 taxa are categorized as high priority,

Table 6.2 Species wise classification of *Sorghum* wild and weedy relatives conserved globally

Wild species	Section	Current classification	Life form	Chromosome no. (2*n*)	Gene Pool	No. of accessions available
Sorghum drummondii	Eu-Sorghum	*Sorghum bicolor nothosubsp. drummondii (Steud.) de Wet ex Davidse*	Annual	20	Primary	276
S. japonicum	Eu-Sorghum	*S. bicolor subsp. bicolor*	Annual	20	Primary	28
S. saccharatum	Eu-Sorghum	*S. bicolor subsp. bicolor*	Annual	20	Primary	237
S. sudanense	Eu-Sorghum	*S. bicolor nothosubsp. drummondii (Steud.) de Wet ex Davidse*	Annual	20	Primary	406
S. technicum	Eu-Sorghum	*S. bicolor subsp. bicolor*	Annual	20	Primary	107
S. arundinaceum	Eu-Sorghum	*S. bicolor subsp. verticilliflorum (Steud.) de Wet ex Wiersema & J. Dahlb.*	Annual	20	Primary	76
S. bicolor ssp. verticilliflorum	Eu-Sorghum	*S. bicolor subsp. verticilliflorum (Steud.) de Wet ex Wiersema & J. Dahlb.*	Annual	20	Primary	215
S. halepense	Eu-Sorghum	*S. halepense (L.) Pers.*	Perennial	40	Secondary	208
S. nigricans	Eu-Sorghum	*S. bicolor subsp. bicolor*	Annual	20	Primary	2
S. propinquum	Eu-Sorghum	*S. propinquum (Kunth) Hitchc.*	Perennial	20	Primary	5
S. virgatum	Eu-Sorghum	*Sorghum bicolor subsp. verticilliflorum (Steud.) de Wet ex Wiersema & J. Dahlb.*	Annual	20	Primary	1
S. × almum	Eu-Sorghum	*S. ×almum Parodi*	Perennial	40	Secondary	53
Sorghum ×drummondii	Eu-Sorghum	*S. bicolor nothosubsp. drummondii (Steud.) de Wet ex Davidse*	Annual	20	Primary	39
S. angustum	Stiposorghum	*S. angustum S. T. Blake*	Annual	10	Tertiary	11
S. bulbosum	Stiposorghum	*Sorghum bulbosum Lazarides*	Annual	10	Tertiary	22
S. ecarinatum	Stiposorghum	*Sorghum ecarinatum Lazarides*	Annual	10	Tertiary	12
S. interjectum	Stiposorghum	*S. interjectum Lazarides*	Annual/ perennial	30	Tertiary	20
S. intrans	Stiposorghum	*S. intrans F. Muell. ex Benth*	Annual	10	Tertiary	35

(continued)

Table 6.2 (continued)

Wild species	Section	Current classification	Life form	Chromosome no. (2*n*)	Gene Pool	No. of accessions available
S. leiocladum	Parasorghum	*S. leiocladum (Hack.) C. E. Hubb.*	Perennial	10, 20	Tertiary	29
S. matarankense	Parasorghum	*S. matarankense E. D. Garber & Snyder*	Perennial	10	Tertiary	11
S. nitidum	Parasorghum	*S. nitidum (Vahl) Pers.*	Perennial	10, 20	Tertiary	12
S. plumosum	Stiposorghum	*S. plumosum (R. Br.) P. Beauv.*	Annual	10, 20, 30, 40	Tertiary	62
S. purpureosericeum	Parasorghum	*S. purpureosericeum (Hochst. ex A. Rich.) Schweinf. & Asch.*	Annual	10	Tertiary	16
S. stipoideum	Stiposorghum	*S. stipoideum (Ewart & Jean White) C. A. Gardner & C. E. Hubb.*	Annual	10	Tertiary	49
S. timorense	Parasorghum	*S. timorense (Kunth) Büse*	Perennial	10, 20	Tertiary	43
S. versicolor	Parasorghum	*S. versicolor Andersson*	Annual	10, 20	Tertiary	19
S. amplum	Stiposorghum	*S. amplum Lazarides*	Annual	10, 30	Tertiary	2
S. australiense	Parasorghum	*Sorghum timorense (Kunth) Büse*	Perennial	10, 20	Tertiary	3
S. brachypodium	Stiposorghum	*S. brachypodium Lazarides*	Annual	10	Tertiary	3
S. exstans	Stiposorghum	*S. exstans Lazarides*	Annual	10	Tertiary	4
S. grande	Parasorghum	*S. grande Lazarides*	Perennial	30, 40	Tertiary	1
S. laxiflorum	Heterosorghum	*S. laxiflorum F. M. Bailey*	Annual	40	Tertiary	27
S. macrospermum	Chaetosorghum	*S. macrospermum E. D. Garber*	Annual	40	Tertiary	4

Note: Around 211 accessions remained as unclassified species and 5 accessions of hybrids from wilds; Source: Reviewed by Ananda et al. 2020; http://www.fao.org/wiews-archive/germplasm_query.html

19 as medium priority, and 1 as low priority. The preliminary assessment also revealed that 12 taxa are endangered, four vulnerable, and four are nearly threatened (Myrans et al. 2020).

6.5 Wild Relatives as Trait-Specific Sources

The screening of wild and weedy relatives of sorghum identified trait-specific sources for various biotic and abiotic stresses. But still huge potential exists in the available wild accessions that need to be explored further through screening for other biotic and abiotic stresses. Drought is an abiotic stress that adversely impacts yield traits of sorghum such as grains per panicle and seed size and physiological functions (Abreha et al. 2022). The higher photosystem II functionality, photosynthetic rate, and relative water content of wild species, i.e., *S. macrospermum and S. brachypodium* resulted in greater drought tolerance than cultivated species, i.e., *S. bicolor* (Cowan et al. 2020). The greater expression of the stay-green trait can confer drought tolerance by modifying the canopy development and water uptake pattern (Borrell et al. 2014). The accessions of wild species *S. bicolor* subsp *arundinaceum (Desv.) Stapf* exhibited drought tolerance in terms of relative chlorophyll content, the number of green leaves at maturity, and a combination of relative chlorophyll content and yield (Ochieng et al. 2021). The superiority of morphological traits, i.e., leaf rolling and drying, root length, plant height, and flag leaf width in the wild species, resulted in preflowering drought tolerance (Techale et al. 2014). Apart from drought tolerance, the above-ground dhurrin content was negligible in *S. macrospermum and S. brachypodium* species which can be exploited for forage purposes (Cowan et al. 2020).

Among various biotic stresses observed in sorghum, downy mildew is a major disease that has been reported in around 44 countries of tropical and subtropical regions (Perumal et al. 2008). The screening of 308 wild and weedy species of sorghum for downy mildew incidence through inoculation of seedlings with conidia and scoring means for downy mildew incidence identified 29 accessions with resistance among which 21 were completely free from downy mildew and 8 accessions with <10% of mean incidence. Among the 29 species screened, no accession belonging to 12 species was free from downy mildew and all the resistant accessions were from 13 different species and every accession of Parasorghum showed downy mildew resistance (Karunakar et al. 1994). Similarly, the wild sorghum germplasm of 103 wild and weedy sorghum accessions and six cultivated sorghums belonging to five sections, i.e., Parasorghum, Stiposorghum, Chaetosorghum, Heterosorghum and and Eusorghum, representing 17 species, originating from Asia, Australia, Africa and the USA, were screened in glasshouse for downy mildew resistance for a period of 2 years at ICRISAT. Forty-five accessions belonging to 15 species from four sections exhibited immunity to downy mildew, while cultivated types and wild races of the section Eusorghum showed greater susceptibility. Among the screened accessions, 40% of accessions with downy mildew resistance originated from Australia and in terms of sections, Parasorghum holds the maximum accessions (23) comprising 8 different species and 5 accessions of unclassified species followed by Stiposorghum

(11 accessions) of 6 different species which are exclusively from Australia, 7 from Heterosorghum and 1 from Chaetosorghum (Kamala et al. 2002).

The major pests of sorghum include stem borer and shoot fly causing yield loss of about 18–27% and 4–20%, respectively (Reddy 1988). The identification of genetic resistance through screening can be a potential solution. Apart from identification of trait specific sources, understanding the mechanisms responsible for stem borer resistance can enhance the utility (Kishore Kumar et al. 2006). Fifty-five wild species accessions of sorghum belonging to 17 different species were screened for resistance to stem borer, C. partellus damage under artificial infestation in the field condition, resulted in identification of 19 accessions belonging to sections Heterosorghum (1), Parasorghum (10) and Stiposorghum (8) which were immune to stem borer which was reflected in terms of plant damage (0%), leaf damage score (1) and dead heart formation (0%) and 14 other accessions showed high resistance but with a plant damage (<10%), leaf damage score (1) and with dead heart formation around 1–2%. The accessions of section Heterosorghum, Parasorghum, and Stiposorghum exhibited greater resistance to stem borer damage when compared to Chaetosorghum and Eusorghum (Kamala et al. 2012). Under multi- and no-choice tests in the greenhouse conditions, no oviposition were observed on the species of Stiposorghum (*S. extans*) and Parasorghum (*S. versicolor and S. purpureosericeum*) indicating high oviposition non-preference and even several other accessions were less preferred for oviposition which led to stem borer resistance. The complete absence of plant damage in all the accessions of Stiposorghum and one accession of Parasorghum under no choice condition might be due to the existence of antibiosis or antixenosis as mechanism conferring resistance (Kamala et al. 2012). Kamala et al. (2009) identified 32 accessions belonging to the sections Parasorghum (22), Heterosorghum (2), and Stiposorghum (8) were free from shoot fly damage. All the Stiposorghum and Heterosorghum accessions exhibiting shoot fly resistance were of Australian origin and Parasorghum accession includes accession from India and Africa. The wild sources identified for green bug resistance being limited and the screening of cultivated species revealed that bloom less nature of sorghum conferred tolerance to green bugs through antibiosis mechanism. The traits bloomless was recessive in nature and its consideration during screening for green bug resistance in wilds can result in identification of resistant sources (Peiretti et al. 1980).

In addition to pests and diseases, striga commonly known as witchweed, parasitize monocot crops such as sorghum, maize, millet, which hampers crop growth and yield. The mechanism of striga resistance can be pre-attachment stage, i.e., reducing seed germination rate striga, toxic compound secretion in the host plants' roots and reducing the haustoria initiation; or post attachment, i.e., hypersensitive responses or cell wall modification (Jhu et al. 2023). Muchira et al. (2021) reported that *S. purpureosericeum (Hochst. ex A. Rich.) Schweinf. & Asch.* of tertiary gene pool and section Parasorghum exhibited striga resistance in terms of area under striga number progress curve, maximum above ground striga and number of striga forming capsules. Rich et al. (2004) reported several wild accessions of *S. bicolor* (6) and *S. almum* (1) with lowest mean germination distance (<10 mm) and mean haustorial distance (<2 mm) which confers strigs resistance. The low haustorial initiation trait was not identified among striga resistance sources of cultivated gene pool and has limited its exploitation. Apart from above wild species, *aethiopicum*, *drummondii*, and *arundinaceum* races also exhibited greater resistance for striga. The number of striga attached to plant and striga biomass was very low. The hindrance for haustoria introgression secondary and metabolite accumulation at host parasite interphase might be responsible for striga resistance in these species (Mbuvi et al. 2017). The transfer of this traits from wild to cultivated species can enhance striga resistance potential of sorghum cultivars. CWRs as a sources of genetic variation for different traits listed in the Table 6.3.

6.6 Novel Breeding Strategies for Enhancing the Utilization of CWR

First, most research work targeted towards cultivated gene pool, while the wild and weedy relatives are largely neglected, in terms of collection, conservation, and their utilization. Therefore, considering the taxonomy and geographical gaps, enriching the sorghum wild species in the genebank collection is foremost important. The advancement in technology has resulted in the development of several novel phenotyping approaches that can assist in characterization and evaluation of CWRs of sorghum as in case of cultivated sorghum (Wang et al. 2018; Salas Fernandez et al. 2017; Young et al. 2019; Malambo et al. 2019; Pugh et al. 2018). Apart from the phenotypic characterization data, the availability of whole genome sequence information of wild species can provide the basis for understanding the genetic variation existing in wilds and can support to improve several traits in sorghum. The whole-genome resequencing of a sorghum panel of 172 lines including 19 *S. bicolor* × *S. halpense* advanced lines and 153 S. bicolor lines identified 567,046,841 SNPs, 91,825,474 indels, 1,532,171 SVs, and 4,973,961 CNVs. The variants and mutations were more in *S. bicolor* × *S. halpense* with powerful effects on genetic differentiation. Among the 5548 genes private to *S. bicolor* × *S. halpense* mapped to biological process gene ontogeny enrichment terms; 34 of these genes mapped to root system development and the two of these root-specific genes, i.e., *ROOT PRIMORDIUM DEFECTIVE* 1 (RPD1) and *RETARDED ROOT GROWTH* (RRG) were found to exert direct effect on root growth and development (Habyarimana et al. 2022). Mitochondria and chloroplasts are organelles that regulate key functions such as cellular respiration, cytonuclear incompatibility and disease resistance. The genomic information of these organelles provides the genetic basis essential for the improvement of these functions and associated traits (Zhang et al. 2023). The chloroplast genome of *S. sudanense* and *S. propinquum* ranged from 140,629 to 140,755 bp and comparison of these with *S. bicolor* and *S. timorense* identified that gene contents, gene orders, and GC contents were like those for other Poaceae species but were slightly different in the number of SSRs. The comparative analyses among the four chloroplast genomes revealed 651 variable sites, 137 indels, and 9 small inversions. The highly divergent DNA regions *rps16-trnQ*, *trnG-trnM*, *rbcL-psaI*, and *rps15-ndhF* can be valuable resource which can be used as molecular markers for phylogeny and species identification in *Sorghum* (Song et al. 2019). Several protein-coding genes (81), tRNA (38), and rRNA genes (4) were

Table 6.3 Wild species accessions of sorghum a sources of genetic variation for different traits

Trait	Germplasm accessions	References
Fodder (Low dhurrin)	Low dhurrin# *Sorghum macrospermum*—JC 2253, *S. brachypodium*—JC 2125	Cowan et al. (2020)
Drought	*Sorghum arundinaceum (Desv.) Stapf*—GBK 016109 and GBK 048156	Ochieng et al. (2021), Cowan et al. (2020)
	Sorghum arundinaceum (Desv.) Stapf—GBK 047293, GBK 048917	
	Sorghum macrospermum—JC 2253 *S. brachypodium*—JC 2125	
Downy mildew	*S. macrospermum*—PQ24	Kamala et al. (2002), Karunakar et al. (1994)
	S. laxiflorum—PQ 155, PQ 156, PQ 157, PQ 158, PQ 159, PQ 162 and IS 18958	
	S. australiense—IS 18954, IS 18955 and IS 18956	
	S. brevicallosum—PQ2/1, RN320 and IS 18957	
	S. matarankense—PQ 25, RN 341 and IS 18952	
	S. purpureosericeum subsp. *Deccanense*—RN 285, IS 18943, IS 18947 and IS 22191	
	S. purpureosericeum subsp dimidiatum—IS 18944 and IS 18945	
	S. timorense—PQ 14-1 and PQ 26-1	
	S. versicolor—IS 18926, IS 14262, IS 14275, IS 18940 and IS 18941	
	Sorghum sp.—IS 18929, IS 23147, IS 23159, IS 23175, and IS 18942,	
	S. angustum—PQ 10, PQ 2, PQ 3, PQ 5, PQ 7, and PQ 9	
	S. ecarinatum—PQ 19	
	S. extans—PQ 35	
	S. intrans—PQ 30	
	S. interjectum—PQ 47	
	S. stiposorghum—PQ 100	
	Species of para sorghum—IS 14262, IS 14275, IS 18926, IS 18941, IS 18942, IS 18946, IS 23159, IS 23177	
	Sorghum purpureosericeum—IS 18947, IS 22191 and IS 18939	
	S. nitidum—PQ 22–1, IS 18958	
	S. stipodeum—IS 18963, IS 18965	
	Sorgastrum—IS 23176	

(continued)

Table 6.3 (continued)

Trait	Germplasm accessions	References
Stem borer	*S. purpureosericeum*—RN 285, IS 18947, IS 18951, IS 18943, IS 18944, IS 18945	Kamala et al. (2012)
	S. australiense—IS 18954, IS 18955, IS 18956	
	S. brevicallosum—TRC 243491, IS 18957	
	S. matarankense—TRC 243576, RN 341	
	S. laxiflorum—TRC 243492, IS 18958	
	S. nitidum—TRC 243514	
	S. timorense—TRC 243437, TRC 243498	
	S. versicolor—IS 14262, IS 14275, IS 18926, IS 18940, IS 18941, IS 23177	
	S. angustum—TRC 243598, TRC 243499	
	S. ecarinatum—TRC 243574	
	S. exstans—TRC 243601	
	S. interjectum—TRC 243461	
	S. intrans—TRC 243571, TRC 243602	
	S. stipoideum—TRC 243399	
Shoot fly resistance	*S. laxiflorum*—TRC 243492, IS 18958	Kamala et al. (2009)
	S. australiense—IS 18954, IS 18955, IS 18956	
	S. brevicallosum—TRC 243491, IS 18957	
	S. matarankense—TRC 243576, RN 341	
	S. nitidum—TRC 243514	
	S. purpureosericeum—RN 285, IS 18947, IS 18951, IS 18943, IS 18944, IS 18945	
	S. timorense—TRC 243437, TRC 243498	
	S. versicolor—IS 14262, IS 14275, IS 18926, IS 18940, IS 18941, IS 23177	
	S. angustum—TRC 243598, TRC 243499	
	S. ecarinatum—TRC 243574	
	S. exstans—TRC 243601	
	S. interjectum—TRC 243461	
	S. intrans—TRC 243571, TRC 243602	
	S. stipoideum—TRC 243399	
Striga resistance	WSA-1, WSE-1, and WSA-2	Mbuvi et al. (2017)
	S. purpureosericeum (Hochst. ex A. Rich.) Schweinf. & Asch.—GBK045827	Muchira et al. (2021)
	S. bicolor sub. sp drummodii—PQ-434	Rich et al. (2004)
	S. bicolor sub. sp verticilliflorum—IS 14313, IS 18803, IS 14569, IS 14301, IS 14232,	
	S. almum—IS 18852	

identified through exploring the genome assembly of chloroplast and 99 common single-copy concatenated nuclear genes of 15 sorghum accessions representing five subgenera. The phylogenetic study based on chloroplast and nuclear genome classified the species of three subgenera, i.e., *Eusorghum*, *Chaetosorghum*, and *Heterosorghum* under single clade and species of other two subgenera *Parasorghum* and *Stiposorghum* classified under another clade. This better understanding of phylogeny of wild species can improve the utility of wilds species in crop improvement (Ananda et al. 2021). The genome sequence of Sudan grass was around 715.95 Mb with 35,243 protein-coding genes (Li et al. 2023). Sudangrass accessions contained significantly lower dhurrin at seedling stage as measured by hydrocyanic acid potential than the cultivated sorghum accessions. Genome-wide association study identified a QTL most tightly associated with HCN-p and the linked SNPs were located in the 3' UTR of Sobic.001G012300 which encodes CYP79A1, the enzyme that catalyzes the first step in dhurrin biosynthesis. The abundance of copia/gypsy long terminal repeat retrotransposons in cultivated than in wild sorghum as in other grassess such as maize and rice led to a conclusion that domestication in these crops is accompanied by increased copia/gypsy long terminal repeat retrotransposon insertions in the genomes (Li et al. 2023). Rather than studying the genetic diversity at individual genome level, the concept of pangenome provides an opportunity to view the entire genomic diversity existing within the species (Jayakodi et al. 2021). The development of sorghum pan-genome considering the wild species can aid in exploration of existing genetic variability among cultivated and wild species present under various gene pools. The pangenome was developed by considering the information of 13 genomes including both cultivated and wild species along and integrated them with with 3 other published genomes. The exploration of the pangenome identified 44,079 gene families with 222.6 Mb of new sequence. The pan-genome displays substantial gene-content variation, with 64% of gene families showing presence/absence variation among genomes (Tao et al. 2021).

Several approaches like linkage mapping using biparental mapping populations, association mapping based on linkage dis-equilibrium and multiparent parent populations like MAGIC (Multiparent advanced generation intercross), NAM (Nested association mapping) and BC-NAM (Back cross-Nested association mapping) are employed to carry out trait mapping (Guden et al. 2023; Kumar et al. 2023; Olatoye et al. 2020; Perumal et al. 2021; Winans et al. 2023). Development of BCNAM populations by including sorghum wild species accessions can integrate the novel diversity from those wild species accessions into the cultivated gene pool. The genotyping of the nested association mapping population of sorghum developed using donor parents which came from the *S. bicolor (L.) Moench subsp. verticilliflorum (Steud.) de Wet ex Wiersema & J. Dahlb., S. bicolor (L.) Moench subsp. drummondii (Steud.) de Wet ex Davidse, and S. bicolor (L.) Moench subsp. bicolor margaritiferum* taxa with diversity array technology (DArT) markers and sampling in a range of environments across Africa. identified 42,372 unique single nucleotide polymorphism (SNP) markers covering the genome (Mace et al. 2021). The integration of phenotypic data with the results of Omics approaches, i.e., transcriptomics, proteomics, and metabolomics, can provide in-depth understanding regarding trait functionality and aid plant breeders in improving target traits (Fig. 6.1). The RNA-Seq analysis for differential expression of cyanogenesis-related genes in *S. bicolor* and the Australian native wild species *S. macrospermum* revealed that genes regulating dhurrin synthesis, i.e., *CYP79A1, CYP71E1 and UGT85B1* and genes *DHR* and *HNL* encoding the dhurrinase and α-hydroxynitrilase involved in bioactivation of dhurrin are also highly expressed in *S. bicolor* resulting in high dhurrin content

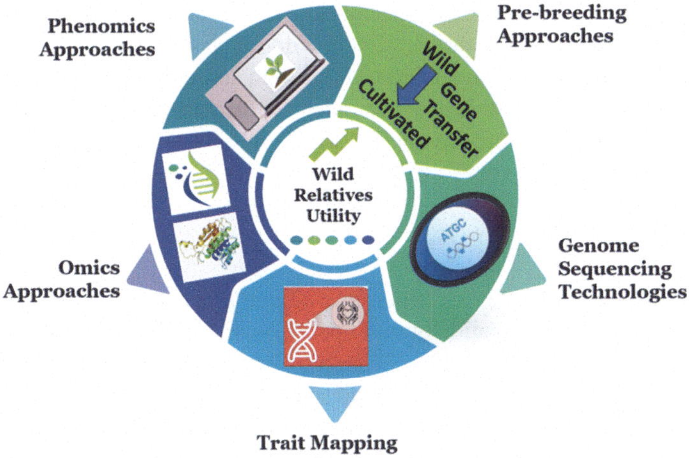

Fig. 6.1 Novel approaches enhancing contribution of wild relatives in sorghum breeding

compared to wild sorghum (Ananda et al. 2022). The tissue-specific gene expression profile of *S. propinquum* using whole rice genome oligonucleotide microarray identified 548 tissue-enriched genes including 31 and 114 unique genes that were expressed predominantly in the rhizome tips (RT) and internodes (RI), respectively. The prominence of several phytohormones, i.e., abscisic acid, gibberellic acid, and salicylic acid in regulating rhizome development through varied levels of gene expression is identified through comparative analysis of rhizome-enriched genes and rhizome-specific genes in *Oryza longistaminata* and *S. propinquum* (Zhang et al. 2013).

Though several trait-specific sources have been identified in wild species and genomic information is available, the factors like poor adaptability, crossability barriers, linkage drag, and many others are hindering the effective usage of novel genetic resources existing in the genebanks (Sharma et al. 2013). Prebreeding is an effective strategy that can overcome these barriers and can enhance the utility of plant genetic resources. The transfer of novel genes from species of Chaetosorghum, Parasorghum, and Stiposorghum to Eusorghum is being restricted due to pollen–pistil incompatibilities and abortion of seeds before maturity due to endosperm break down (Hodnett et al. 2005). The strategies like floral treatments using hormones like synthetic auxin, 2,4-dichlorophenoxyacetic acid, and the in vitro plant regeneration from immature embryos developed can achieve the target of gene transfer (Kurella 2020; Oldach et al. 2001; Price et al. 2005).

The introgression of complementary traits like biotic and abiotic stress and quality and quantity of biomass has been attempted between sugarcane and sorghum. The factors like breeding and selection of sugarcane pollen parents can enhance success rate of hybridization and the avoidance of vivipary through premature seed harvest and seed coat removal can increase the seedling recovery rates from 1.5% to 33% (Hodnett et al. 2010). As reviewed by Ananda et al. (2020), several hybrids were produced by crossing the cultivated species with the available wilds such as *S. almum*, *S. angustum*, *S. halepense*, and several other species and this has further led to successful introgression of desirable alleles for yield, various biotic and abiotic stress from wild species into cultivated gene pool. The effective utilization of available trait-specific sources and information made available through several strategies can enhance the contribution of wild relatives towards sorghum crop improvement.

6.7 Conclusion

Several trait-specific sources identified prove huge potential of wilds species that can be explored in diversification of trait-specific sources. The geographical and the taxonomic gaps identified through gap analysis provide the key insights on further measures required for conservation of wild species. The application of modern phenomics approaches can aid in further characterization of wild species. The available genome sequence of sorghum wild relatives and pangenome developed including the wild relatives, the genomic information available through trait mapping, and omics approaches in wild species can provide a basis for further study of the wild and weedy relatives of sorghum. Several successful attempts of crossing cultivated and wild species and introgression of several traits paved the way for enhancing the utilization of wilds and broadening the genetic base of trait-specific sources.

References

Abreha KB, Enyew M, Carlsson AS, Vetukuri RR, Feyissa T, Motlhaodi T, Ng'uni D, Geleta M (2022) Sorghum in dryland: morphological, physiological, and molecular responses of sorghum under drought stress. Planta 255(1):1–23. https://doi.org/10.1007/s00425-021-03799-7

Adebo OA (2020) African sorghum-based fermented foods: past, current and future prospects. Nutrients 12(4):1111

Ananda GKS, Myrans H, Norton SL, Gleadow R, Furtado A, Henry RJ (2020) Wild sorghum as a promising resource for crop improvement. Front Plant Sci 11:1–14. https://doi.org/10.3389/fpls.2020.01108

Ananda G, Norton S, Blomstedt C, Furtado A, Møller B, Gleadow R, Henry R (2021) Phylogenetic relationships in the Sorghum genus based on sequencing of the chloroplast and nuclear genes. Plant Genome 14(3):1–16. https://doi.org/10.1002/tpg2.20123

Ananda GKS, Norton SL, Blomstedt C, Furtado A, Møller BL, Gleadow R, Henry RJ (2022) Transcript profiles of wild and domesticated sorghum under water-stressed conditions and the differential impact on dhurrin metabolism. Planta 255(2):1–19. https://doi.org/10.1007/s00425-022-03831-4

Andersson MS, De Vicente MC (2010) Gene flow between crops and their wild relatives. JHU Press, Baltimore, MD

Baker HG (1972) Human influences on plant evolution. Econ Bot 26(1):32–43

Black LD (1983) D. 1983. Can J Plant Sci 63:997–114

Borrell AK, Mullet JE, George-Jaeggli B, van Oosterom EJ, Hammer GL, Klein PE & Jordan DR (2014). Drought adaptation of stay-green sorghum is associated with canopy development, leaf anatomy, root growth, and water uptake. Journal of Experimental Botany, 65(21), 6251–6263. https://doi.org/10.1093/jxb/eru232

Bramel P, Kresovich S, Giovannini P (2022) Global strategy for the conservation and use of sorghum (Sorghum bicolor (L.) Moench) genetic resources. Global Crop Diversity Trust, Germany. https://doi.org/10.5281/zenodo.8192869

Burke IC, Holland JB, Burton JD, York AC, Wilcut JW (2007) Johnsongrass (Sorghum halepense) pollen expresses ACCase target-site resistance. Weed Technol 21(2):384–388. https://doi.org/10.1614/wt-06-061.1

Cowan MF, Blomstedt CK, Norton SL, Henry RJ, Møller BL, Gleadow R (2020) Crop wild relatives as a genetic resource for generating low-cyanide, drought-tolerant Sorghum. Environ Exp Bot 169:103884

Crop Wild Relatives | Gap Analysis Results—Crop Wild Relatives (2023). https://www.cwrdiversity.org/gap-analysis-results/

de Wet JMJ (1978) Systematics and evolution of sorghum sect. Sorghum (Gramineae). Am J Bot 65(4):477. https://doi.org/10.2307/2442706

de Wet JMJ, Huckabay JP (1967) The origin of Sorghum bicolor. II. Distribution and dometication. Evolution 21(4):787–802

Doggett HJ, Majisu B (1968) Disruptive selection in crop development. Heredity 23, 1–23. https://doi.org/10.1038/hdy.1968.1

Doggett H, Prasada Rao KE (1995) Sorghum. In: Smartt J, Simmonds NW (eds) Evolution of crop plants. Longman, Harlow

Ejeta G, Grenier C (2005) Sorghum and its weedy hybrids. In: Crop ferality and volunteerism, pp 123–135

Endrizzi JE (1957) Cytological studies of some species and hybrids in the Eu-sorghums. Bot Gaz 119(1):1–10

FAO (2021) FAOSTAT. https://www.fao.org/faostat/en/#data/QCL

Garber ED (1950) Cytotaxonomic studies in the genus Sorghum. Univ Calif Publ Bot 23:283–362

Guden B, Yol E, Erdurmus C, Lucas SJ, Uzun B (2023) Construction of a high-density genetic linkage map and QTL mapping for bioenergy-related traits in sweet sorghum [Sorghum bicolor (L.) Moench]. Front Plant Sci 14:1–13. https://doi.org/10.3389/fpls.2023.1081931

Habyarimana E, Gorthy S, Baloch FS, Ercisli S, Chung G (2022) Whole-genome resequencing of Sorghum bicolor and S. bicolor × S. halepense lines provides new insights for improving plant agroecological characteristics. Sci Rep 12(1):1–14. https://doi.org/10.1038/s41598-022-09433-0

Harlan JR, de Wet JMJ (1971) Toward a rational classification of cultivated plants. Taxon 20(4):509–517

Hodnett GL, Burson BL, Rooney WL, Dillon SL, Price HJ (2005) Pollen-pistil interactions result in reproductive isolation between Sorghum bicolor and divergent Sorghum species. Crop Sci 45(4):1403–1409. https://doi.org/10.2135/cropsci2004.0429

Hodnett GL, Hale AL, Packer DJ, Stelly DM, da Silva J, Rooney WL (2010) Elimination of a reproductive barrier facilitates intergeneric hybridization of Sorghum bicolor and Saccharum. Crop Sci 50(4):1188–1195. https://doi.org/10.2135/cropsci2009.09.0486

Jayakodi M, Schreiber M, Stein N, Mascher M (2021) Building pan-genome infrastructures for crop plants and their use in association genetics. DNA Res 28(1):1–9. https://doi.org/10.1093/dnares/dsaa030

Jhu M-Y, Kawa D, Brady SM (2023) The genetic basis of plants' battle against witchweeds: linking immune responses to distinct resistance mechanisms. J Exp Bot 74(17):4903–4909. https://doi.org/10.1093/jxb/erad305

Kamala V, Singh SD, Bramel PJ, Manohar Rao D (2002) Sources of resistance to downy mildew in wild and weedy sorghums. Crop Sci 42(4):1357–1360. https://doi.org/10.2135/cropsci2002.1357

Kamala V, Sharma HC, Manohar Rao D, Varaprasad KS, Bramel PJ (2009) Wild relatives of sorghum as sources of resistance to sorghum shoot fly, Atherigona soccata. Plant Breed 128(2):137–142. https://doi.org/10.1111/j.1439-0523.2008.01585.x

Kamala V, Sharma HC, Manohar Rao D, Varaprasad KS, Bramel PJ, Chandra S (2012) Interactions of spotted stem borer Chilo partellus with wild relatives of sorghum. Plant Breed 131(4):511–521. https://doi.org/10.1111/j.1439-0523.2012.01966.x
Karunakar RI, Narayana YD, Pande S, Mughogho L, Singh S (1994) Evaluation of wild and weedy sorghums for downy mildew resistance. ISMN 35:104–106
Kishore Kumar V, Sharma HC, Dharma Reddy K (2006) Antibiosis mechanism of resistance to spotted stem borer, Chilo partellus in sorghum, Sorghum bicolor. Crop Prot 25(1):66–72. https://doi.org/10.1016/j.cropro.2005.04.001
Kole C (2011) Wild crop relatives: genomic and breeding resources. Springer, Berlin
Kuhlman LC, Burson BL, Klein PE, Klein RR, Stelly DM, Price HJ, Rooney WL (2008) Genetic recombination in Sorghum bicolor x S. macrospermum interspecific hybrids. Genome 51(9):749–756. https://doi.org/10.1139/G08-061
Kumar N, Boatwright JL, Brenton ZW, Sapkota S, Ballén-Taborda C, Myers MT, Cox WA, Jordan KE, Kresovich S, Boyles RE (2023) Development and characterization of a sorghum multiparent advanced generation intercross (MAGIC) population for capturing diversity among seed parent gene pool. G3 13(4):1–14. https://doi.org/10.1093/g3journal/jkad037
Kurella BRSV (2020) Overcoming interspecific barriers in Sorghum and molecular characterization of the derivatives. Plant Breed 139(5):892–905. https://doi.org/10.1111/pbr.12831
Lazarides M, Hacker JB, Andrew MH (1991) Taxonomy, cytology and ecology of indigenous Australian sorghums (Sorghum Moench: Andropogoneae: Poaceae). Aust Syst Bot 4(4):591–635. https://doi.org/10.1071/SB9910591
Li J, Wang L, Bible PW, Tu W, Zheng J, Jin P, Liu Y, Du J, Zheng J, Wang YH, Zhan Q (2023) A chromosome-scale genome sequence of Sudan grass (Sorghum sudanense) highlights the genome evolution and regulation of dhurrin biosynthesis. Theor Appl Genet 136(3):1–21. https://doi.org/10.1007/s00122-023-04262-9
Liu Q, Liu H, Wen J, Peterson PM (2014) Infrageneric phylogeny and temporal divergence of Sorghum (Andropogoneae, Poaceae) based on low-copy nuclear and plastid sequences. PLoS One 9(8):e104933. https://doi.org/10.1371/journal.pone.0104933
Mace ES, Cruickshank AW, Tao Y, Hunt CH, Jordan DR (2021) A global resource for exploring and exploiting genetic variation in sorghum crop wild relatives. Crop Sci 61(1):150–162. https://doi.org/10.1002/csc2.20332
Malambo L, Popescu SC, Horne DW, Pugh NA, Rooney WL (2019) Automated detection and measurement of individual sorghum panicles using density-based clustering of terrestrial lidar data. ISPRS J Photogramm Remote Sens 149:1–13. https://doi.org/10.1016/j.isprsjprs.2018.12.015
Maxted N, Dulloo E, Ford-Lloyd BV, Iriondo JM, Jarvis A (2008) Gap analysis: a tool for complementary genetic conservation assessment. Divers Distrib 14(6):1018–1030. https://doi.org/10.1111/j.1472-4642.2008.00512.x
Mbuvi DA, Masiga CW, Kuria E, Masanga J, Wamalwa M, Mohamed A, Odeny DA, Hamza N, Timko MP, Runo S (2017) Novel sources of witchweed (Striga) resistance from wild sorghum accessions. Front Plant Sci 8:116. https://doi.org/10.3389/fpls.2017.00116
Morrell PL, Williams-Coplin TD, Lattu AL, Bowers JE, Chandler JM, Paterson AH (2005) Crop-to-weed introgression has impacted allelic composition of johnsongrass populations with and without recent exposure to cultivated sorghum. Mol Ecol 14(7):2143–2154. https://doi.org/10.1111/j.1365-294X.2005.02579.x
Muchira N, Ngugi K, Wamalwa LN, Avosa M, Chepkorir W, Manyasa E, Nyamongo D, Odeny DA (2021) Genotypic variation in cultivated and wild sorghum genotypes in response to Striga hermonthica infestation. Front Plant Sci 12:1–16. https://doi.org/10.3389/fpls.2021.671984
Myrans H, Diaz MV, Khoury CK, Henry RJ, Gleadow R, Carver D (2020) Modelled distributions and conservation priorities of wild sorghums (Sorghum Moench). pp 1727–1740. https://doi.org/10.1111/ddi.13166
Ochieng G, Ngugi K, Wamalwa LN, Manyasa E, Muchira N, Nyamongo D, Odeny DA (2021) Novel sources of drought tolerance from landraces and wild sorghum relatives. Crop Sci 61(1):104–118. https://doi.org/10.1002/csc2.20300

Olatoye MO, Marla SR, Hu Z, Bouchet S, Perumal R, Morris GP (2020) Dissecting adaptive traits with nested association mapping: genetic architecture of inflorescence morphology in sorghum. G3 10(5):1785–1796. https://doi.org/10.1534/g3.119.400658

Oldach KH, Morgenstern A, Rother S, Girgi M, O'Kennedy M, Lörz H (2001) Efficient in vitro plant regeneration from immature zygotic embryos of pearl millet [Pennisetum glaucum (L.) R. Br.] and Sorghum bicolor (L.) Moench. Plant Cell Rep 20(5):416–421. https://doi.org/10.1007/s002990100335

Parodi LR (1943) Una nueve especie de Sorghum cultivada en la Argentina. Rev Argent Agron 10:361–372

Peiretti RA, Amini I, Weibel DE, Starks KJ, McNew RW (1980) Relationship of "bloomless" (bm bm) sorghum to greenbug resistance 1. Crop Sci 20(2):173–176

Perumal R, Nimmakayala P, Erattaimuthu SR, No EG, Reddy UK, Prom LK, Odvody GN, Luster DG, Magill CW (2008) Simple sequence repeat markers useful for sorghum downy mildew (Peronosclerospora sorghi) and related species. BMC Genet 9:1–14. https://doi.org/10.1186/1471-2156-9-77

Perumal R, Tesso TT, Morris GP, Jagadish SVK, Little CR, Bean SR, Yu J, Prasad PVV, Tuinstra MR (2021) Registration of the sorghum nested association mapping (NAM) population in RTx430 background. J Plant Regist 15(2):395–402. https://doi.org/10.1002/plr2.20110

Price HJ, Hodnett GL, Burson BL, Dillon SL, Rooney WL (2005) A Sorghum bicolor x S. macrospermum hybrid recovered by embryo rescue and culture. Aust J Bot 53(6):579–582. https://doi.org/10.1071/BT04213

Pugh NA, Han X, Collins SD, Thomasson JA, Cope D, Chang A, Jung J, Isakeit TS, Prom LK, Carvalho G, Gates IT, Vree A, Bagnall GC, Rooney WL (2018) Estimation of plant health in a sorghum field infected with anthracnose using a fixed-wing unmanned aerial system. J Crop Improv 32(6):861–877. https://doi.org/10.1080/15427528.2018.1535462

Reddy KVS (1988) Assessment of on-farm yield losses in sorghum due to insect pests. Int J Trop Insect Sci 9(6):679–685

Rich PJ, Grenier C, Ejeta G (2004) Striga resistance in the wild relatives of sorghum. Plant Genet Resources 44(6):2221–2229

Salas Fernandez MG, Bao Y, Tang L, Schnable PS (2017) A high-throughput, field-based phenotyping technology for tall biomass crops. Plant Physiol 174(4):2008–2022. https://doi.org/10.1104/pp.17.00707

Search Taxonomy GRIN-Global (2023). https://npgsweb.ars-grin.gov/gringlobal/taxon/taxonomysearch

Sharma S, Upadhyaya HD, Varshney RK, Gowda CLL (2013) Pre-breeding for diversification of primary gene pool and genetic enhancement of grain legumes. Front Plant Sci 4:309. https://doi.org/10.3389/fpls.2013.00309

Song Y, Chen Y, Lv J, Xu J, Zhu S, Li M (2019) Comparative chloroplast genomes of Sorghum species: sequence divergence and phylogenetic relationships. Biomed Res Int 2019:5046958. https://doi.org/10.1155/2019/5046958

Species GRIN-Global (2023). https://npgsweb.ars-grin.gov/gringlobal/taxon/taxonomyspecieslist?id=27889&type=section

Sun Y, Suksayretrup K, Kirkham MB, Liang GH (1991) Pollen tube growth in reciprocal interspecific pollinations of Sorghum bicolor and S. versicolor. Plant Breed 107(3):197–202

Tao Y, Luo H, Xu J, Cruickshank A, Zhao X, Teng F, Hathorn A, Wu X, Liu Y, Shatte T, Jordan D, Jing H, Mace E (2021) Extensive variation within the pan-genome of cultivated and wild sorghum. Nat Plants 7(6):766–773. https://doi.org/10.1038/s41477-021-00925-x

Techale B, Tamene Y, Ferede T (2014) Response of wild type sorghum (Sorghum bicolor (L.) Moench) accessions to pre-flowering drought stress. Afr J Agric Res 9(41):3077–3081. https://doi.org/10.5897/ajar2014.9064

Upadhyaya HD, Reddy KN, Vetriventhan M, Gumma MK, Ahmed MI, Reddy MT, Singh SK (2017a) Status, genetic diversity and gaps in sorghum germplasm from South Asia conserved at ICRISAT genebank. Plant Genet Resour 15(6):527–538. https://doi.org/10.1017/S147926211600023X

Upadhyaya HD, Reddy NK, Murali KG, Ahmed IM, Manyasa E, Reddy TM, Singh SK (2017b) Geographical distribution, diversity and gap analysis of East African sorghum collection conserved at the ICRISAT genebank. Aust J Crop Sci 11(4):424–437

Upadhyaya HD, Narsimha Reddy K, Vetriventhan M, Irshad Ahmed M, Murali Krishna G, Thimma Reddy M, Singh SK (2017c) Sorghum germplasm from West and Central Africa maintained in the ICRISAT genebank: status, gaps, and diversity. Crop J 5(6):518–532. https://doi.org/10.1016/j.cj.2017.07.002

Venkateswaran K, Elangovan M, Sivaraj N (2019) Origin, domestication and diffusion of Sorghum bicolor. In: Breeding sorghum for diverse end uses. Elsevier, Amsterdam, pp 15–31

Wang X, Singh D, Marla S, Morris G, Poland J (2018) Field-based high-throughput phenotyping of plant height in sorghum using different sensing technologies. Plant Methods 14(1):1–16. https://doi.org/10.1186/s13007-018-0324-5

Winans ND, Klein RR, Fonseca JMO, Klein PE, Rooney WL (2023) Evaluating introgression sorghum germplasm selected at the population level while exploring genomic resources as a screening method. Plants 12(3):444. https://doi.org/10.3390/plants12030444

Young SN, Kayacan E, Peschel JM (2019) Design and field evaluation of a ground robot for high-throughput phenotyping of energy sorghum. Precis Agric 20(4):697–722. https://doi.org/10.1007/s11119-018-9601-6

Zhang T, Zhao X, Huang L, Liu X, Zong Y, Zhu L, Yang D, Fu B (2013) Tissue-specific transcriptomic profiling of Sorghum propinquum using a rice genome array. PLoS One 8(3):1–13. https://doi.org/10.1371/journal.pone.0060202

Zhang S, Wang J, He W, Kan S, Liao X, Jordan DR, Mace ES, Tao Y, Cruickshank AW, Klein R, Yuan D, Tembrock LR, Wu Z (2023) Variation in mitogenome structural conformation in wild and cultivated lineages of sorghum corresponds with domestication history and plastome evolution. BMC Plant Biol 23(1):1–17. https://doi.org/10.1186/s12870-023-04104-2

Hybrid Sorghum Production: Considerations According to Breeder and End-User

7

Sonal Chavan, Krishna Kasanaboina, Niranjan Thakur, Awio Bruno, Y. Vinod Kumar Reddy, Pramod Sargar, Uttej Karla, and Ephrem Habyarimana

Abstract

The development of high-yielding, climate-resilient and product profile-driven cultivars preferred by the end users is the only way out in the context of changing climatic scenario and growing world population. The hybrids of the versatile crop, sorghum, are grown across the globe; however, the genetic gains have been limited as compared to other major crops. This chapter provides an overview of the trend of sorghum breeding from the time before the advent of heterosis till

S. Chavan · K. Kasanaboina · A. Bruno
Professor Jayashankar Telangana State Agricultural University, College of Agriculture, Rajendranagar, Telangana, India

International Crops Research Institute for the Semi-Arid Tropics, Patancheru, Telangana, India

N. Thakur · P. Sargar
International Crops Research Institute for the Semi-Arid Tropics, Patancheru, Telangana, India

Vasantrao Naik Marathwada Agriculture University, Parbhani, Maharashtra, India

Y. V. K. Reddy
International Crops Research Institute for the Semi-Arid Tropics, Patancheru, Telangana, India

Sam Higginbottom University of Agriculture Technology and Sciences (SHUATS), Prayagraj (Allahabad), Uttar Pradesh, India

U. Karla
Maharana Pratap University of Agriculture and Technology, Udaipur, Rajasthan, India

E. Habyarimana (✉)
International Crops Research Institute for the Semi-Arid Tropics, Hyderabad, Telangana, India
e-mail: Ephrem.Habyarimana@icrisat.org

E. Habyarimana et al. (eds.), *Omics and Biotechnological Approaches for Product Profile-Driven Sorghum Improvement*,
https://doi.org/10.1007/978-981-97-4347-6_7

date where heterosis is accompanied by biotechnological and 'omic-' based approaches to study the genetic basis of traits and the concept behind it. An adequate strategy to apply different techniques in the selection of parental lines will aid in achieving the desired genetic gains specific to targeted environments.

Keywords

Accelerated line breeding · Biotechnological tools · Heterosis · Heterotic grouping · Heterosis maximisation · Hybrid development · Omic-approaches · Parental development

7.1 Introduction

The world's food security is under unprecedented pressure and challenges as a result of the growing global population and increasingly frequent natural disasters brought on by climate change (Hao et al. 2021). A key factor in achieving food security is the rise in yield per unit area (Liu et al. 2023). Genetic enhancement through hybridisation has been instrumental in raising agricultural productivity since the days of 'green revolution' up until the present for several crops including sorghum. The versatility of the crop as food, feed, fodder, fuel (4Fs) and its ability to thrive under harsh environments makes sorghum a prominent choice to take up as a crop for cultivation in semiarid tropics of the world. However, the global average yield of sorghum in 2021/2022 was only ~1.51 MT/ha, far lower than maize (5.88 MT/ha), rice (4.64 MT/ha) and wheat (3.52 MT/ha) (USDA report). The current worldwide sorghum yield picture highlights the urgent need for the development of superior hybrids adapted for varying climatic conditions while utilising modern breeding principles and technologies to accommodate a range of end uses.

7.1.1 Concept of Heterosis

The phenomenon of heterosis (or hybrid vigour) describes the morphological and functional superiority of traits in F1 (obtained by crossing genetically diverse individuals) over its parents. The concept of heterosis was born with its initial observation in tobacco hybrids by Koelreuter (1733–1806). Charles Darwin first described it in the vegetable kingdom in 1876, and it has since been scientifically characterised in a wide range of crop species (Srivastava et al. 2020). However, the idea of using heterosis for crop improvement was established only after the rediscovery of Mendel's laws in 1900, which demonstrated that hybrids were more heterozygous than their inbred parents and the subsequent elaborate work on maize by George H Shull and Edward M East in documenting the heterosis in 1908 (Timberlake 2013).

Although heterosis has been effectively used in crop breeding, the enigma underlying the molecular mechanism remains a mystery. Three basic genetic hypotheses, namely dominance, over-dominance, and epistasis, have been postulated as the driving forces behind heterosis as a result of a number of significant studies that have been undertaken to understand the genetic basis of heterosis (Shull 1911; Jones 1917; Schnell and Cockerham 1992; Yu et al. 1997; Birchler et al. 2003, 2006; Hochholdinger and Hoecker 2007). Although these classical models remain valid, they are insufficient to explain the molecular mechanism of heterosis. As a genome-wide phenomenon, heterosis is widely acknowledged to be caused by changes in gene expression, while epigenetic modifications (such as DNA methylation, histone modifications and sRNA regulations) have recently been proposed as a potential factor in the improvement of hybrid performance (Zhang et al. 2023). With the advent of molecular markers and next-generation sequencing (NGS) technologies, a substantial number of 'omic' studies have been performed to dissect the molecular architecture of heterosis. Omics is a term used to describe the comprehensive study of the functions, interactions and roles of various types of molecules in biological cells (Xiang et al. 2022).

The massive increase in knowledge of genomics (study of the genome), transcriptomics (study of gene expression), proteomics (study of proteins), metabolomics (study of metabolites) and other '-omics' fields consistently provide new insights to understand the causes for the expression of heterosis. The significance of dominant complementation in heterosis has been uncovered by research on the transcriptome and metabolome of hybrid potatoes to better understand the molecular mechanisms underlying heterosis (Li et al. 2022). Using a multi-omics approach, Zhang et al. (2023) found that non-additively expressed small RNAs played significant roles in heterosis for plant height in maize. Omic-approaches have also been employed for the prediction of hybrid performance. In order to predict hybrid performance in canola, Knoch et al. (2021) employed metabolic and transcriptomic data, along with genetic markers. Their findings indicate that the accuracy of hybrid prediction can be enhanced by incorporating transcriptomic data in place of or in addition to genetic markers. Similarly, combining mRNA and genomic data as predictors resulted in high predictive abilities in a study on maize (Schrag et al. 2018). In rice, a study on metabolome-based predictive models has revealed accurate predictions of yield heterosis when accompanied by correctly dissected population structure and screened predictive variables (Dan et al. 2020).

7.2 Sorghum Breeding in Pre-hybrid Era

Sorghum breeding was shown to have distinct status across the globe, based on the research emphasis and priorities of policy makers in the respective nations. Historically, the consideration of this crop as a coarse grain with a subsistence level of production has failed to draw attention from the scientific community, especially in Africa and Asia (Reddy and Reddy 2019).

Africa With its origin in Ethiopia-Sudan, the sorghum genus has a wide range of variations within Africa (OECD 2016). Sorghum improvement research began in the late 1930s in eastern Africa. Formal scientific research efforts were launched in the late 1950s with the establishment of Jimma Agricultural Technical School

(JATS) involving the collection and evaluation of local germplasm for identification and release of 'improved local selections' (Obilana 2004; Yali and Begna 2022). Popular local choices include L 28 from Uganda and Dobbs from western Kenya (Doggett 1988). The current composition of highly adapted and diversified sorghum landraces in Ethiopia is a result of genetic interactions between cultivated and wild sorghum as well as recurring gene flow that is influenced by selection (Ejeta and Grenier 2005).

Asia In Asia, sorghum has been grown since ages in nations such as China, India, Pakistan, Thailand and Korea (Mundia et al. 2019). The research on this crop was of less prominence before the 1940s, which has comparatively improved from the 1940s to the 1960s. Selections were an important breeding method during this time. The two most common methods for genetic improvement of sorghum were recurrent selection and pedigree breeding. The recurrent selection was primarily intended for variety development, enabling the breeding programmes to handle the germplasm effectively (Rai et al. 1999).

Selections from local landraces were made in India during the pre-hybrid era (before the mid-1960s), leading to the release of outstanding cultivars like M 35-1 and Maldandi in the 1930s (Patil et al. 2014). These cultivars with high stable yields and tolerance to biotic and abiotic stresses still dominate as famous cultivars among farmers (Sanjana Reddy et al. 2009). Landraces such as Moli, irungu cholam and bidri are famous for their grain and quality traits (Venkateswaran et al. 2019). The breeding material developed during this era has laid a foundation for hybrid development. In China, breeding efforts began in the 1920s, with collection, evaluation, classification and pedigree selection among local varieties being the most commonly used breeding strategies. With the establishment of the People's Republic of China in the 1950s, sorghum research began in earnest (Qingshan and Dahlberg 2001).

Europe In Europe, very limited breeding efforts have been undertaken due to its minor importance as a food crop. However, the crop has gained industrial importance with its applications in making brooms, paper and ethanol. The area remained stagnant due to a lack of efforts in the improvement of the crop before the advent of heterosis breeding (Berenji and Dahlberg 2004).

Australia Australia's success in the grain industry has served as a model for the rest of the world, demonstrating the importance of long-term investment in breeding and agronomic research. In 1958, the sorghum-breeding programme was started by the Queensland Government at Hermitage Research Station and it began delivering to Australian grain growers through the private seed industry in the mid-1960s. When compared to other public pre-breeding programmes around the world, the programme has about 2400 licenced lines, placing it at the top in terms of genetic material utilisation (source: https://qaafi.uq.edu.au/article/2016/10/how-sorghum-became-queenslands-top-crop).

North America With its introduction in the mid-nineteenth century from West Africa, sorghum has been grown for use as animal feed in North America. United States and Mexico are the major sorghum-growing countries in the continent

(Mundia et al. 2019). Efforts to improve sorghum started immediately after introduction by a selection of dwarf and early maturing lines. However, with the discovery of the cytoplasm male sterility system in 1954, breeding for improved sorghum underwent an evolutionary shift paving a path for the development of hybrids (Zhao and Dahlberg 2019).

7.3 Hybrids as a Means to Achieve Yield Gains: An Ultimate Target of Every Breeder

The goal of any plant breeder is to achieve genetic gain in the form of a greater economic yield along with resistance to biotic and abiotic stresses, and improvement of quality traits. Among different objectives, high yield is the ultimate target in any crop including sorghum. The yield advantage can be achieved either by selections from transgressive segregants or by taking advantage of heterosis in hybrids (Rakshit and Bellundagi 2019). The use of hybrids has resulted in a yield quantum jump of 20% to more than 50%, accounting for more than half of the total crop production of major crops and formed the foundation of a multibillion-dollar global agri-business (Virmani et al. 2004; Tester and Langridge 2010). In addition to higher yield, hybrids outperformed the parental lines in terms of disease resistance and performance in different environments (Begna 2022). The success of hybrid maize has sparked interest in developing hybrid sorghum varieties. The advantage of heterosis in terms of yield gain in sorghum was recognised as early as in maize (Conner and Karper 1927; Windpassinger 2016).

7.3.1 Types of Hybrids

The conventional hybrids involving inbred parents can be classified into three types (Fig. 7.1) based on the number of inbred lines involved, such as

1. Single cross hybrid: Two unrelated inbred parents are crossed to produce a single cross hybrid. These hybrids are characterised by high heterosis and uniformity. When produced, these hybrids will produce the best grain yield and are considered as the most standardised hybrids.
2. Three-way cross hybrid: The single cross F_1 when crossed to an unrelated inbred line (third parent) produces a three-way cross hybrid. It is next after single cross hybrids in terms of heterosis.
3. Double cross hybrids: The F_1s of two single crosses are used as parents and are crossed to produce a double cross hybrid. It follows three-way cross hybrids with respect to heterosis.

The distinction between these hybrids is that instead of using inbred lines as in the generation of single cross hybrids; three-way and double cross hybrids use F_1s as one of the parents or as both parents, respectively. High-yielding F_1s are used to

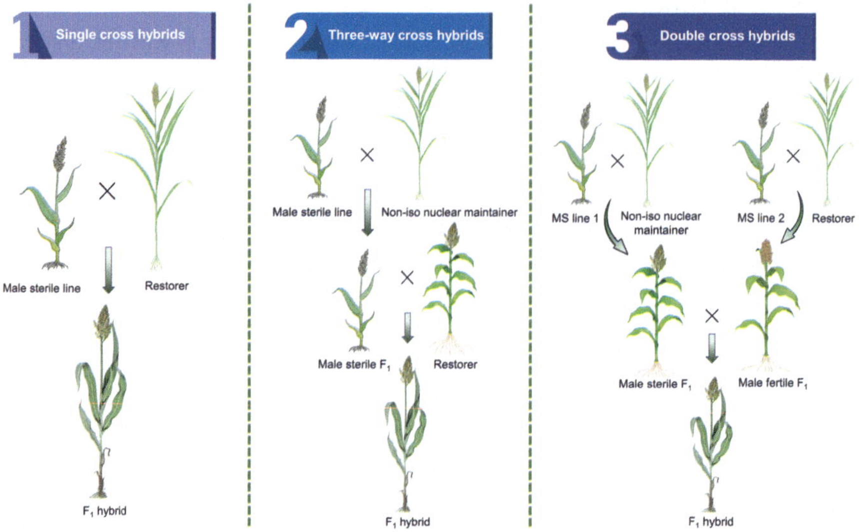

Fig. 7.1 Illustration of breeding scheme used to develop single, three-way, and double cross hybrids

boost male parent pollen output and seed parent seed yield. Limited research support has been extended for hybrid sorghum seed systems compared to maize, rice and vegetable crops. It is considered as a secondary crop in the portfolio of seed companies due to the low margin of sorghum seeds as compared to maize and vegetable seeds (Hambloch et al. 2021). Three-way and double-cross hybrids will help in the large-scale adoption of hybrids in crops with minor emphasis on hybrid production. One of the reasons involved in large-scale adoption is the growth of a highly developed seed industry and economical seed production (Yadav et al. 2015). The ratio of seeds planted to harvested is critical for both grain farmers and commercial seed producers (Chase 1971).

Apart from high seed yield, three-way and double-cross hybrids provide more variability and a broad genetic base as they allow the involvement of three and four diverse parents as compared to only two inbred parents involved in the case of single-cross hybrids (Lakshmi pathy et al. 2018). These three-way and double-crosses can be more stable in performance across different environment conditions as compared to single-cross hybrids (Yıldırım and Çakmak 2018).

7.3.2 Development of First Sorghum Hybrid

In sorghum, hybrid technology has the potential to increase productivity while retaining high biomass and large grain size. Heterosis was first realised in sorghum as early as in 1920s. Male sterile plants were observed in the segregating generation of milo × kafir crosses by Conner and Karper in 1921 (Axtell et al. 1999). However, Stephens must be credited for the isolation and use of male sterility to create sorghum hybrids in 1937 (Miller and Kebede 1984). The Day Milo variety's genetic male sterility was employed in the late 1940s to create hybrids by using male sterility sources in sorghum (Windpassinger 2016). The hybrids had a yield advantage of over 40% compared to other cultivars under cultivation at that time (Klein et al. 2008). However, the Day Milo genetic male sterility was soon replaced with the discovery of the cytoplasmic-genetic male-sterility (CGMS) system by Stephens and Holland (1954). The sterile Combine Kafir-60 plants were 100% male sterile and have managed to grab the attention of public and private industry to begin sorghum hybrid breeding programmes. Commercial production of hybrid seeds based on CMS began in the United States in 1956, and its cultivation has doubled grain yield as compared to the pre-hybrid era (Smith et al. 2010). In addition, the hybrids have also exhibited increased yield stability over inbred lines, particularly in stressful environmental conditions (Reddy et al. 2006).

Followed by the United States, hybrid breeding started in China with the introduction of CMS line Tx 3197A and its maintainer Tx 3197B in 1956. Yiza and Yuanza series of hybrids were released in 1958 with a yield advantage of 20–60% (Reddy and Reddy 2019). However, the Jinza 5 hybrid released in the 1970s was widely accepted by farmers due to its short stature unlike the previous hybrids (Qingshan and Dahlberg 2001). Meanwhile in India, CSH-1 hybrid (a cross combination of CK60A × IS84) was released in 1964. Sorghum became the second crop after maize to develop hybrids using the CGMS system of male sterility. While in Sub-Saharan Africa, the programme for hybrid breeding was started in the 1970s but the first sorghum hybrid 'Hageen-Dura 1 (HD-1)' was released in 1983 (Axtell et al. 1999).

7.3.3 Production Trend of Hybrids

Sorghum grain yields in the United States ranged between 1.2 and 1.4 t/ha from 1874 to 1955 (Murty 1999). With the advent of hybrids in the year 1956 on a commercial scale, yield increased by about 139% (1.6 t/ha in 1956 to 3.8 t/ha in 1997) (Assefa 2010). When observed in the past 60 years of the hybrid era where different hybrids were released, it resulted in a 58% increase in productivity (2.74 t/ha in 1961 to 4.33 t/ha in 2021/2022). In Europe, the yield advantage was reported to be about 396% (1.09 t/ha in 1961 to 5.41 t/ha in 2021/2022) (USDA Foreign Agriculture Service/United States Department of Agriculture 2023). While in Asia, although with an increase of 167%, the productivity is still low with an average of 1.55 t/ha in 2021. An increase of 81% was observed for productivity in Australia from 1.58 t/ha in 1961 to 2.86 t/ha in 2021. Compared to other continents, Africa had an overall lower percentage of productivity gain of 16% with a yield advantage of only about 0.13 t/ha where it was about 0.80 t/ha in 1961 and 0.93 t/ha in 2021 (FAOSTAT 2023) (Fig. 7.2). The adoption of hybrids in Asia and India has attributed to the higher increase in productivity as compared to Africa (Reddy et al. 2004). However, the yield gains are not attributed exclusively due to genetic gains but have also been due to the changes in agronomic practices.

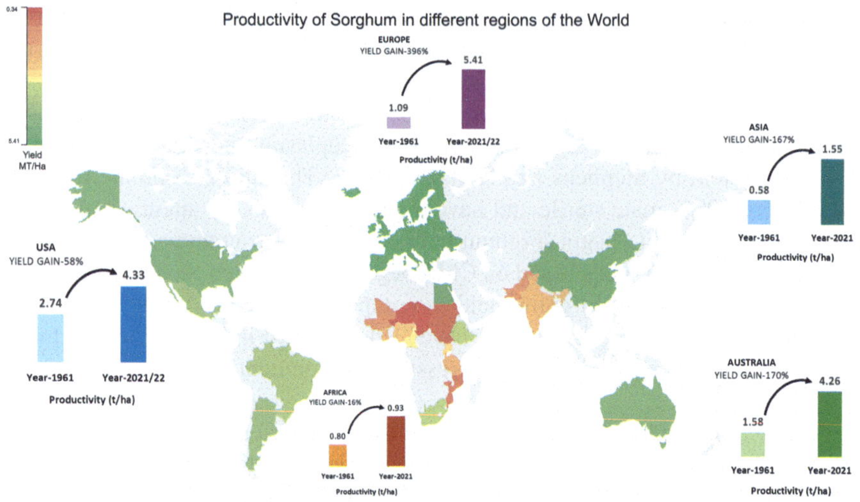

Fig. 7.2 Productivity of sorghum in different regions of the world and productivity gain over a period of six decades

7.4 Hybrid Product Profile Preferred by the Consumers

The product preference of sorghum varies with the target environments and nations based on the end uses. In nations such as India, where sorghum grain is used as food and feed, hybrids with bold white seeds and medium height are preferred as dual-type cultivars. Similarly, in nations like the United States, Australia and South America, grain-type hybrids with small to medium height are chosen for cultivation owing to their suitability for mechanical harvesting. Hybrids with tall heights are favoured in Africa and red grain types are especially preferred by brewing industries (Ashok Kumar 2019). In forage hybrids, it is important to concentrate on quality traits such as digestibility and crude protein apart from high biomass. In sweet sorghum, traits such as tall height, thick stem diameter with high biomass, juice yield and brix form an important criterion in the hybrid product profile chosen by the end users (Rooney 2014). In addition to grain yield, stover yield and quality traits, the traits such as photo and thermo-period insensitivity and resistance to biotic and abiotic stresses specific to different environments form an important selection criterion by end users while accepting the new hybrids. The breeder needs to be precautious while selecting the hybrid parents keeping in view the desired traits to be included in the new hybrids along with the desired genetic gains in terms of yield. Below is the current list (Table 7.1) of sorghum product profiles at the International Crops Research Institute for the Semi-Arid Tropics.

Table 7.1 Compiled product concepts of ICRISAT crops in Asia (as on 15 March 2021)

S. no.	Product concept	Target and spill over agro-ecologies	Product development goals
1.	**Post-rainy season sorghum** for food (parents with high GCA and yield)	**Target**: Indian sub-continent, predominantly Maharashtra, Karnataka, Telangana, Andhra Pradesh and Madhya Pradesh states in India **Spill over**: Parts of Sudan, Chad (Lake Chad areas)	**Must have traits (3–5 years)** 1. White, bold, globular, lustrous grains (100 grain weight >3.0 g) with 5% increase in grain yield and/or 10% increase in stover yield (with In-vitro organic matter digestibility IVOMD >48%) over the best checks (M 35-1; Parbhani Moti, Phule Revati, CSV 29R)) 2. Maturity: 120–130 days (for shallow soils 90–100 days) 3. Plant height: 2.0–2.2 m 4. Shoot fly resistance (≤40% shoot fly dead hearts, SFDH) 5. Charcoal rot resistance (≤10 cm infection length and 0% soft rot) 6. Post-flowering drought tolerance (Least yield penalty under post-flowering drought conditions vis-à-vis normal conditions—calculated based on drought tolerance index) **Long-term goals (>5 years)** 1. Aphid tolerance (≤4 incidence at 0–9 scale) 2. High grain Zn (>32 ppm) and Fe (>45 ppm) 3. Nitrogen use efficiency, NUE (same yield at 30% less N) 4. Water use efficiency (WUE) 5. Reproductive stage cold tolerance (maintaining >80% seed set)
2.	**Rainy season sorghum** for food, feed and industrial uses (brewing) (parents with high GCA and yield)	**Target**: Maharashtra, Rajasthan, Madhya Pradesh, Karnataka, Telangana, Andhra Pradesh, Tamil Nadu and Gujarat states of India **Spill over**: Ethiopia, Eritrea, Kenya, Uganda, Tanzania, Sudan, Zimbabwe, Malawi, Zambia, Nigeria, China.	**Must have traits (3–5 years)** 1. 5% increase in grain yield (100 seed weight >2.8 g) and 10% increase in stover yield over best check (CSH 16; CSH 30), white, bold grains for food and feed use 2. For industrial use: lines with higher starch (>68%) and medium protein content (8–10%) and higher protein digestibility 3. Plant height: 1.8–2.0 m 4. Maturity: 110–120 days (90–100 days specific to India) 5. Shoot fly (≤40% SFDH), stem borer (≤5 on 0–9 scale), grain mould resistance (≤3 for colour and ≤ 5 for white grains on 0–9 scale; grain hardness) 6. Coloured grain sorghums (red and yellow) **Long-term goals (>5 years)** 1. Aphid tolerance (≤4 at 0–9 scale) 2. High grain Zn (>32 ppm) and Fe (>45 ppm) 3. NUE (same yield at 30% less N); project with Damaris

(continued)

Table 7.1 (continued)

S. no.	Product concept	Target and spill over agro-ecologies	Product development goals
3.	Sorghum for **forage** (parents with high GCA and yield)	**Target**: Punjab, Haryana, Uttarakhand, Uttar Pradesh, Bihar, Madhya Pradesh, Tamil Nadu and Gujarat states of India **Spill over**: Ethiopia, Eritrea, Kenya, Uganda, Tanzania, Sudan, Zimbabwe, Malawi, Zambia, China	**Must have traits (3–5 years)** 1. 10% increase in forage yield over best check (Forage SSG 59-3; CSH 24MF) Tan plant, fast growth, high tillering ability; higher in vitro organic matter digestibility 2. Plant height 2.2–2.5 m with single cut/multi-cut types 3. Shoot fly (≤40% SFDH), stem borer (≤5 on 0–9 scale), resistance to anthracnose, leaf blight, rust, aphids (sugarcane aphids) <3–4 4. High seed yield in hybrid seed production 5. IVOMD >55% 6. Protein content (8–10%) 7. Low HCN content (<100 ppm) 8. Brix% (>12%) **Long-term goals (>5 years)** 1. Low-lignin content 2. BMR6&12
4.	Sorghum for **1G bioethanol** (parents with high GCA and Theoretical ethanol yield (TEY))	**Target**: All India (because sweet sorghum is an industrial crop) **Spill over**: Africa, China	**Must have traits (3–5 years)** 1. 10% increase in 1G TEY yield over best check (sweet sorghum RVICSH 28 (ICSSH 28) and CSH 22SS, Tan plant, fast growth, monoculm; higher in vitro organic matter digestibility 2. Brix >16 cfr Keller variety 3. Plant height 3–4 (5) m 4. Shoot fly (≤40% SFDH), stem borer (≤5 on 0–9 scale), resistance to anthracnose, leaf blight, rust 5. High seed yield in hybrid seed production 6. IVOMD >55% **Long-term goals (>5 years)** 1. Low-lignin content 2. BMR6&12

7.5 Biotechnological Interventions in Sorghum Heterosis Breeding

A global effort to anticipate heterosis has been sparked by the availability of genomic tools including DNA markers and gene expression platforms. Genomic-assisted breeding can be used to choose parents for crossing, validate the genetic purity of F1 offspring, map features for introgression and select enhanced types based on the molecular profiles of breeding populations (Baloch et al. 2023). Different markers have been employed in sorghum to study the genetic diversity (Table 7.2). Efforts have been taken to map the heterotic trait loci (HTL) (Ben-Israel

Table 7.2 A sample of studies using molecular markers in sorghum heterosis breeding

S. no.	Kind of study	Markers used/ stage of analysis	References
1.	Genetic diversity study	SSR and RAPD markers	Agrama and Tuinstra (2003)
		RFLP markers	Jordan et al. (2003)
		AFLP and SSR markers	Menz et al. (2004)
		SSR markers	Ktavii et al. (2014)
		RNA	Stokes et al. (2010)
		SSR markers	Wang et al. (2013)
		SSR markers	Amelework et al. (2017)
2.	Heterotic trait locus (HTL) mapping	SSR markers	Ben-Israel et al. (2012)
3.	Prediction of hybrid performance by multi-environment models using marker-based kinship information	SNP markers	Hunt et al. (2020)
4.	Genetic differentiation analysis to identify complementary parental pools for hybrid breeding	SNP markers	Mindaye et al. (2015)
5.	Genetic diversity and heterotic grouping	SNP markers	Silva et al. (2020)
6.	Genomic prediction	SNP markers	Velazco et al. (2019)
7.	GBS-based genetic similarity for prediction of heterosis	SNP markers	Crozier et al. (2020)
8.	Genotyping-by-sequencing (GBS) -for modelling used to predict heterosis	SNP markers	Fonseca et al. (2021)
9.	Genomic prediction to screen Inbred lines in hybrid test crosses	SNP markers	Kent et al. (2023)
10.	Transcriptomics to predict hybrid performance	RNA	Stokes et al. (2010)
11.	Proteomic analysis of heterosis	Proteins	Han et al. (2016)
12.	Remote sensing and genetic markers for prediction of hybrid yields	Not mentioned	Wang et al. (2023b)

et al. 2012) and to develop heterotic groups (Silva et al. 2020) in sorghum using molecular markers. The development of straightforward and accurate strategies for heterosis prediction is greatly facilitated by the use of cutting-edge genomic techniques and mathematical modelling (Rajendrakumar 2015). With the advent of next-generation and third-generation sequencing technologies, sorghum breeding's scope and toolkits have been transformed by genome-wide association studies (GWAS), genome editing (GE), molecular modules (MMs), genome selection (GS) and non-invasive high throughput phenomics (Hao et al. 2021). Using a range of training populations, including natural populations, testcross hybrids and mixed populations, GS mostly concentrated on model training in sorghum (Yu et al. 2016; Dos Santos et al. 2020; Habyarimana et al. 2020; Fonseca et al. 2021). Once the DNA markers for heterosis prediction have been established, DNA marker-based analysis can be used to forecast potential parental combinations, speeding up the generation of hybrids without extensive field testing. Transcriptomic and proteomic approaches have also been used to predict the hybrid performance and heterosis in sorghum (Stokes et al. 2010; Han et al. 2016), aiding in the identification of genes or metabolic networks implicated in heterosis.

7.6 Floral Biology

Sorghum is a short-day plant, and panicle initiation marks the start of the reproductive stage. The photoperiod must drop below a particular day length for the floral initiation to occur. It requires short days and long nights, with 12 or more hours of darkness in traditional varieties. However, photoperiod sensitivity varies between cultivars. Panicle initiation occurs 30–40 days after germination, marking the end of the vegetative stage. With floral initiation, inflorescence development begins 30–100 days after germination, depending on the genotype. The developing inflorescence inside the leaf sheath of the flag leaf will appear as a boot-shaped structure about 6–10 days before flowering. The inflorescence of sorghum is a raceme and about 21 days following the start of the reproductive phase, the many racemes and spikelets that make up the panicle begin to form. Flowers begin to open 2 days after the inflorescence emerges from the boot (Quinby and Karper 1947; Folliard et al. 2004; Aruna and Audilakshmi 2008; OECD 2017; Venkateswaran et al. 2019).

The panicle consisting of many primary branches, varies in size and density based on the genotype. The length of the panicle varies according to the length of rachis, the number of nodes and the length of internodes. The average length of panicle for cultivated sorghum varies from 5 to 18 in. Each primary branch bears secondary branches, which in turn bear spikelets. Each spikelet occurs in pedicelled and sessile pairs, except the terminal sessile spikelet is accompanied by two pedicelled spikelets (Vinall et al. 1936; Aruna and Audilakshmi 2008). The pedicelled spikelets are staminate and sessile spikelets are fertile. Two florets are encased by the glumes of the sessile spikelet. On average, a panicle bears approximately 6000 florets (Karper and Quinby 1947). The lower floret of the sessile spikelet is sterile and has a lemma, while the upper floret is perfect and it comprises a lemma, palea, two lodicules, three stamens and an ovule with two pistils protruding into two feathery stigmas (Doggett 1988).

Anthesis starts first with the blooming of the uppermost flower and thereafter progresses from tip to downwards of the panicle. Blooming of sessile spikelets is followed by the pedicelled spikelets 2–4 days later. Duration of flowering ranges from 2 to 15 days based on the genotype, size of the panicle and temperature (Stephens and Quinby 1934; Quinby 1974). The time of flowering usually occurs between 10 pm and 8 am. The swelling of lodicules helps the flower to open, allowing the emergence of stigmas and anthers. Pollen may shed through the apical pore immediately from the anthers or after the dehiscence of the dry anthers (Doggett 1988; OECD 2017; Venkateswaran et al. 2019). The amount of pollen shed is highly variable and in the case of cultivated sorghum panicles can produce up to 24 million pollen grains. The pollen remains highly functional for a few minutes, but may remain viable only for up to 3–6 h. Receptivity of the stigma is for 2 days before anthesis but may remain receptive for 5–16 days based on the cultivar (Ayyangar and Rao 1931; Stephens and Quinby 1934; Karper and Quinby 1947; Ross 1957; Maunder and Sharp 1963; Doggett 1988). Upon reaching the receptive stigma, pollen germinates immediately and fertilises the egg cell within 2 h to develop into an embryo (OECD 2017).

Crop yield is influenced by various plant architecture factors, of which inflorescence architecture and its component traits have a significant effect (Wang et al. 2018). Knowledge of the genetic basis of sorghum inflorescence architecture and its components can complement the breeder's efforts to improve sorghum (Witt Hmon et al. 2014). Different studies involving biotechnological tools have been used in sorghum to study inflorescence ranging from the identification of genes to the modern omic-based approaches (Table 7.3), providing valuable insights about inflorescence morphology and its impact on seed development.

Table 7.3 Molecular studies undertaken in sorghum by different researchers across the globe to understand inflorescence in sorghum

S. no.	Kind of study	Major findings	References
1.	Inheritance study for inflorescence architecture	Two candidate genes—*Dw3* and *Sb-ra2* were identified to co-localise with QTL of large effects	Brown et al. (2006)
2.	Mutation study for multi-seeded trait	Both sessile and pedicellate spikelets were fertile in the multiseeded (msd1) mutant	Burow et al. (2014)
3.	GWAS for Inflorescence architecture	Found 16 loci related to panicle traits – 4 QTLs-length based traits, 3 QTLs-Dimensional traits, 4 QTLs-panicle type and shape, 1 QTL-grain weight, 2 QTLs-total node number and branch number	Witt Hmon et al. (2014)
4.	GWAS for Panicle architecture	35 unique SNps were associated with panicle architecture	Zhou et al. (2019)

(continued)

Table 7.3 (continued)

S. no.	Kind of study	Major findings	References
5.	Dissection of genetic architecture for Inflorescence morphology	Identified new loci for inflorescence morphology	Olatoye et al. (2020)
6.	Comparative expression profiling of maize and sorghum for Inflorescence	Revealed a large number of novel isoforms in both species – In maize-genes in B genome and A genome were highly expressed in pollen and endosperm respectively – Considerable expression and splicing differences in both species	Wang et al. (2018)
7.	Genome-wide identification and characterisation of *miR396 gene* family	Regulation of floral organ and seed development by *miR396d/e–GRF1/5/8* modules	Wang et al. (2023a)

7.7 Techniques for Selfing and Crossing

Selfing and crossing are the most crucial procedures in sorghum breeding. Due to the fact that sorghum is predominantly a self-pollinated crop, creating new 'varieties' is a natural way to enhance the crop. However, depending on the genotype, humidity, and wind direction, sorghum exhibits 5–15% outcrossing (House 1985), making it suitable for use in population improvement and hybrid development to take advantage of heterosis (Reddy et al. 2008).

The selfing and crossing techniques differ with the species and depend on the structure of flower and mode of pollination. In the perfect flowered plant like sorghum, it needs accurate emasculation of female parent to make efficient crossings. Emasculation is the removal of functional anthers from a bisexual flower. Breeders use paper or cloth bags to enclose a single flower or whole inflorescence to perform selfing and to prevent pollen contamination in case of manual crossing. Bagging is done before the opening of flowers and is also required after emasculation.

Hand emasculation, plastic bag method and hot water treatment are the methods used to achieve emasculated females in sorghum. Pollen collected from a male parent is dusted on the emasculated female.

7.7.1 Emasculation Techniques

1. Hand emasculation: This technique is a widely preferred method that involves the use of sharp forceps to emasculate the fertile florets. The panicle of the appropriate plant is selected when it is just about to flower. The upper and lower florets are clipped off with a secateur or scissors. The sterile (pedicellate) florets of the remaining 200–300 florets of the central portion are removed, retaining the fertile (sessile) florets. The floret to be emasculated is held between the thumb and forefinger and the anthers are surgically removed with the help of forceps. Care is taken to avoid damage of stigma. The emasculated

Fig. 7.3 Steps in emasculation and pollination: (**a**) selection of panicle to emasculate (**b**) clipping off of upper and lower florets and pedicellate spikelets (**c**) removal of anthers (emasculation) (**d**) bagging of emasculated panicle to avoid cross-pollination (**e** and **f**) collection of pollen from desired pollen parent (**g**) dusting of pollen of emasculated panicle (**h**) bagging of dusted panicle to avoid further cross-pollination (**i**) seed set on the panicle

panicle is covered with a butter paper bag to avoid unwanted pollination (Fig. 7.3).

2. Plastic bag method of emasculation: The panicles to be emasculated are covered by plastic bags, creating high humidity within the panicle. High humidity exposes the anthers out but will not dehisce to shed pollen.
3. Hot water treatment of emasculation: This method focuses on making the pollen non-functional by dipping the panicle with florets, which are about to flower the next day in hot water for 10 min at 42 °C. The florets which have already opened and the lower florets which will flower later are removed prior to the treatment. Floret opening is promoted with the treatment and in response to the high temperature of the water, the anthers will turn non-functional (Rakshit and Bellundagi 2019; Reddy and Kumar 2008).

7.8 Male Sterility Systems

7.8.1 Need for Male Sterility

Emasculation in hybrid production by manual methods was time, energy and cost-inefficient, necessitating the importance for the establishment of male-sterile female lines. Male sterility (MS) in plants refers to the inability of a plant to produce fertile (or functional) pollens, anthers or male gametes without affecting the female fertility of the plant (Kaul 1988). Male sterile plants are incapable of self-pollination but can be fertilised by male fertile plants to produce hybrids (Farinati et al. 2023). The sterility is generally caused by the hindrance of normal biological processes of pollen production by some bio-chemical events (Saxena et al. 2010). Male sterility systems facilitate the exploitation of heterosis in both auto and allogamous species. It also helps in ensuring the seed purity and improved seed production efficiency along with reduced costs and labour requirements.

7.8.2 Different MS Sources (CMS Sources, Chemical Sources)

Male sterility is divided into three types based on the type of origin namely: (1) Genetic male sterility (GMS), also commonly referred to as Nuclear/Genic/Mendelian male sterility, (2) Cytoplasmic male sterility and (3) Chemically induced (or non-genetic) male sterility (Lasa and Bosemark 1993).

Genetic male sterility (or two-lines breeding system) is generally caused by spontaneous mutation in one or more nuclear genes. Most of the time, it is a recessive mutation of a single nuclear gene (ms) that affects a large number of functions such as meiosis, metabolism of hormones, biosynthesis of lipid molecules and secondary metabolites (Budar and Pelletier 2001; Glover et al. 1998; Moffatt and Sommerville 1988; Aarts 1995; Van der Meer et al. 1992; Farinati et al. 2023). Nine GMS lines, ms1 to ms9 and msal (antherless) lines have been reported so far in sorghum (Chen et al. 2019; Andrews and Webster 1971; Xin et al. 2017).

Cytoplasmic male sterility is maternally inherited male sterility caused due to any mitochondrial mutation that impairs the normal functionality of a specific mitochondrial gene affecting the formation of viable pollen and leading to male sterility (Budar and Pelletier 2001). This sterility caused by cytoplasm can be overcome by the use of nuclear genes called as a restorer of fertility (Rf) that are functional in the dominant state and can downregulate the CMS genes and revert the male sterility (Farinati et al. 2023). This system of reversion of CMS by nuclear genes is referred to as cytoplasmic genetic male sterility (or three-line breeding system).

In comparison with CMS, it is difficult to efficiently maintain the male sterility in GMS, making it unsuitable for hybrid seed production. However, the discovery of environmentally sensitive GMS (EGMS) has enabled its use in hybrid crop breeding (Chen et al. 2019). The two-line breeding system dramatically increases the possibilities for making hybrids between accessions that cannot be generated with the current three-line breeding system (Dong et al. 2000; Zhang et al. 2010, 2013) and it also simplifies the hybrid seed production process. The use of GMS has reported an increase of 5–10% grain yield over hybrids produced using the CGMS system, reflecting its potential to increase future agricultural production (Chang et al. 2016; Kim and Zhang 2018; Zhou et al. 2014; Chen et al. 2019).

Chemically induced male sterility is induced in plants by the use of chemicals commonly referred to as gametocides or chemical hybridising agents (CHA). This kind of male sterility is used when there is a lack of suitable CMS or GMS systems. The CHAs are sprayed on plants at a specific stage before flowering leading to male sterility in plants (Saxena et al. 2015). The CHA that causes no or least damage to the female fertility is preferred. Various CHAs have been used for hybrid seed production in sorghum such as trifluoromethanesulphonamide (TFMSA) (Hodnett and Rooney 2018; Boerman et al. 2019), ethrel, trifluoromethanesulphonamide [TFMSA], ethyl 4-fluorooxanilate [E4FO] (Yahaya et al. 2021; Amelework et al. 2016).

7.9 Parental Line Development in Sorghum

The success of hybrid breeding majorly depends on the performance of the parental lines. In the case of sorghum, cytoplasmic genetic male sterility (CGMS) is being explored for hybrid seed production for which the parental lines are to be developed, that is, A-line (male sterile), B-line (maintainer) and R-line (restorer). In order to improve the performance of parental lines, there is a need for the development of new parental lines and for the introgression of traits of interest into the existing ones. The A-line is a male sterile line that is used as female and the B-line is fertile but unable to restore the fertility in the progeny, which is used as the maintainer line in hybrid breeding. Restorer lines (R lines) are employed in hybrid production plots to restore fertility in hybrids after crossing with male sterile A-lines. Hybrids with improved yield, biomass or other features have been extensively cultivated with the use of CMS restorer systems. The mitochondrial chimeric gene hypothesis is the most basic explanation for CMS. Fertility in CMS plants may often be restored with the help of a nuclear-encoded fertility restoration gene (Rf), which may show a range of fertility levels. Fertility restorer (Rf) genes inhibit or correct for particular mitochondrial dysfunctions observed during pollen formation. Such lines were identified from the existing breeding material, released varieties and several inbreds developed for desirable traits by carrying out pollen and fertility restoration studies (Eckardt 2006; Chen and Liu 2014).

7.9.1 Development of Maintainer and Restorer Lines

In most sorghum breeding programmes, inbred line development involves some form of pedigree selection. The process of parental line development begins with the crossing of the elite maintainer (B-lines) and restorer (R-lines) with other elite B-lines and R-lines, respectively. The elite pollen parent collection could consist of a global collection of germplasm, cultivars or advanced generations of genetic stocks. The selection of elite lines can be sometimes done for a specific trait(s) improvement by selecting the line possessing the trait of interest(s). The F_1s obtained are planted in F_2 nurseries at sites representative of target geographies. In the following generation, F_3 nurseries are planted using selections made from segregating F_2 populations. The handling of the B-line and R-line populations is similar from F_1 to F_3 generations, while the selections from F_4 generation are handled differently. The selected B-lines and R-lines from the respective F_4 populations are test-crossed with elite seed parents (A-lines) for identification of new B-lines and R-lines. It is recommended that new selections be tested in early segregating generations such as F_3 or F_4 (Andrews et al. 1996; Murty et al. 1994).

7.9.2 Identification of B and R-Lines

The material from which the parental lines, that is, B-lines and R-lines are to be identified are used as pollen sources and crossed with male sterile lines and progeny are evaluated for the seed set (Fig. 7.4). The progeny plants are bagged to prevent outcrossing and observed for the restoration of fertility. The absence of seed formation indicated the male sterility of the hybrid resulting from the failure of fertility restoration. Thus, the pollen source is confirmed as the maintainer line (B-line). In some, there may be partial restorers resulting in a partial seed set and some lines segregating for fertility restoration genes resulting in seed set in some panicles and failure in others. Among the three cases, only the initial one is of importance and can be directly used in the breeding programmes. The lines segregating for fertility restoration locus can be stabilised and further evaluation of these lines for fertility restoration might be beneficial (Reddy et al. 2008). The lines for which the testcrosses had exhibited a full seed set are identified as restorer (R-lines). The selected lines are also observed for general combining ability, per se performance and genetic diversity. In case of B-lines, other traits such as shorter height (1.25–1.75 m) and high-yielding lines are preferred, while taller (1.75–2.50 m) lines are preferred as restorer lines.

7.9.3 Maintenance of B and R-Lines

The B and R-lines are maintained indefinitely by selfing.

7.9.3.1 Diversification of R-Lines

In spite of their prominent efficacy, cytoplasmic male sterility (CMS) methods have a few major drawbacks. The limitation of germplasms possessing Rf genes makes it difficult to develop better hybrids. Breeders must select for the qualities they desire

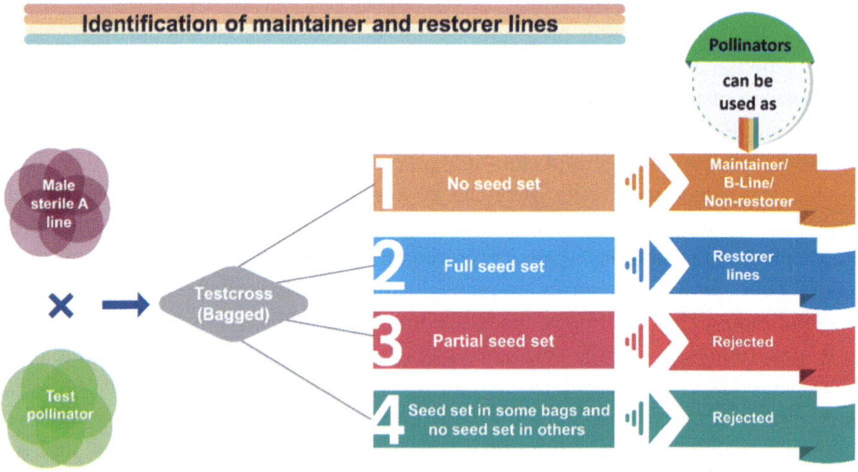

Fig. 7.4 Breeding scheme for identification of new maintainer and restorer lines in sorghum

(plant growth, resilience to stress etc.), but they must also maintain the stability of the Rf genes and combine abilities throughout time in order for the CMS genes to be expressed. Although many hybrid sorghums have been developed, only a small number of genotypes have shown to be reliable CMS restorers (Awad-Allah et al. 2022). Collecting, evaluating, improving and conserving parental lines are crucial to developing hybrid sorghum for these larger purposes. Scientists developed a novel approach to establishing new restorer lines by first identifying restorer lines specific to cytoplasmic backgrounds and then selecting for its restoration ability on diverse cytoplasmic bases from elite sorghum germplasms. This strategy allows for the diversification of not only the cytoplasmic background but also the nuclear background of parental lines, thereby excluding the uniformity of nuclear and cytoplasmic basis of parental lines. Different Rf genes having restoration ability in different cytoplasmic back grounds were identified and mapped on different chromosomes in sorghum (Table 7.4).

7.9.4 Development of Male Sterile A-Lines

It is a male sterile line which is the near-isogenic line of B-line for the fertility locus. Once after identification of maintainer lines, these maintainer lines are crossed with existing male sterile lines. The hybrids are again backcrossed with maintainer lines until the male sterile lines are derived with maintainer line morphology. The maintainer lines and male sterile lines are to be sown in alternate rows and backcrossing must be carried out on plant-to-plant basis which is morphologically much identical to male sterile plants (Reddy et al. 2008). During the several generations of backcrosses, the percentage of male sterility inheritance has been identified to be in the range of 68–95% which may vary depending upon the crosses.

7.9.5 Maintenance of A-Lines

Even after the development of parental lines, there is a need for regular monitoring for the stable expression of parental lines, because some of the A-lines with A1-male sterile cytoplasm witnessed fluctuation in their sterility due to several factors. The

Table 7.4 List of fertility restoration genes having restoration ability in different cytoplasmic back grounds

Rf gene	Restores fertility on	Marker type	Location of gene	References
Rf1	A1	AFLP	SBI-08	Klein et al. (2001)
Rf2	A1	SSR	SBI-02	Jordan et al. (2010)
Rf3	A3	SSR	SBI-01	Praveen et al. (2015)
Rf4	A3	AFLP derived KASP	SBI-01	Wen et al. (2002)
Rf5	A1&A2		SBI-05	Jordan et al. (2011)
Rf6	A1&A2	SSR	SBI-04	Praveen et al. (2015)

temperature at flower initiation and the physiological condition between the early vacuolated and engorged pollen stages have a profound impact on the stability of sterility and may lead to fertility. In order to prevent the breakdown of sterility in parental lines, there is a need to consider the above factors during parental line development.

7.9.6 CMS Sources in Sorghum and Its Diversification

In sorghum, several CMS sources such as A1, A2, A3, A4, A5, A6 and 9E systems were discovered (Table 7.5), among which A1 (milo source), A2 (IS 12662C or TAM 428), A3 (IS 1112C), A4 (Guntur, VZM and Maldandi) and 9E (a selection made in 9E) from Ghana are commonly available and milo (A1) cytoplasm being the source for most of the commercial hybrids worldwide (Reddy et al. 2008). Because of the A1 cytoplasmic homogeneity, sorghum hybrids may be more susceptible to diseases leading to epidemics as in the case of CMS-T cytoplasm in maize hybrids in the 1970s.

One of the challenges to exploit other than A1 sources is the difficulty in fertility restoration on these sources. Studies are being undertaken in sorghum to test fertility restoration by different restorers on these cytoplasmic sources (Sandeep and Biradar 2020; Praveen et al. 2015). The development of markers for the identification of cytoplasmic male sterility is carried out in sorghum to facilitate hybrid development (Choe et al. 2023). Transcriptomic analysis was performed by Dhaka et al. (2020a) on developing anthers from three stages corresponding to meiotic, microspore and mature pollen stages. High concordance between the transcriptional patterns of differentially expressed genes in rice and sorghum and their cis-regulatory components supported the existence of conserved developmental mechanisms governing interspecies development. It provided candidates for engineering male sterility in sorghum (Dhaka et al. 2020a). These studies on restoration of fertility, identification of genes and molecular markers will aid in diversification of CMS sources in sorghum.

Table 7.5 List of different CMS sources and its origin

S. no.	CMS group	Source of CMS group	Origin of CMS source	References
1.	A1	Milo source	– (Crosses between Dwarf Yellow Sooner Milo and Texas Black Hull Kafir)	Stephens and Holland (1954)
2.	A2	IS 12662C	IS 12662C—Origin of Ethiopia	Schertz and Ritchey (1978)
3.	A3	IS 1112C	IS 112C—Origin of India	Quinby (1981)
4.	A4	IS 7920C	IS 7920—Origin of Nigeria	Worstell et al. (1984)

Source for Information on Origin: Worstell et al. (1984)

7.10 Testcrossing of New Parental Lines

The final step in the development of parental lines is to test cross the newly developed lines for assessment of combining ability. Good general combining ability is required in the first testcross, while subsequent testing determines the commercial potential of the specific combinations. The evaluation of inbred lines themselves is not a good use of resources to predict the hybrid performance. The new seed parents are test-crossed with two or three elite pollen parents while the new pollen parents are test-crossed with two or three elite seed parents. The best practice to test cross newly uncovered maintainers is to first cross with a seed parent with a good GCA, and then carry out a polycross on the obtained hybrid F1s. A knowledge of genetic diversity and the nature of parents will aid in the selection of parents. Different studies (Table 7.6) have been taken up to understand the diversity and differential expression of genes at the parent level.

Table 7.6 Studies applying biotechnological tools to understand parental diversity in sorghum

S. no.	Type of study	Markers used/ stage of analysis	Major findings	References
1.	Development and mapping of markers linked to fertility restorer (rf) gene	AFLP	– LW7 and LW8 markers were mapped to rf4 – LW9 marker was mapped on the flanking side of rf4	Wen et al. (2002)
2.	Application of molecular markers to identify progenies with male sterility gene	LW7 (AFLP marker)	Successful use of marker assisted selection (MAS) for male sterility	Bunphan et al. (2013)
3.	Estimation of allele-specific diversity for fertility restoration gene	5 SSR markers	–	Chauhan et al. (2015)
4.	GWAS for important traits in sorghum	SNPs	The bHLH transcription factor-ABORTED MICROSPORES was identified as strong candidate gene for male fertility	Girma et al. (2019)
5.	Small RNA profiling and Annotation of microRNAs (miRNAs)	Meiotic and post-meiotic anthers in sorghum	Identified 262 miRNAs and 5622 modules −38 target genes were identified as candidates for genetic engineering of male fertility	Dhaka et al. (2020b)
6.	Transcriptomic analysis	Meiotic, microspore and mature pollen stages	5286 genes-differential regulation – Among which 890-exhibited anther-preferential expression	Dhaka et al. (2020a)

7.11 Hybrid Seed Evaluation

Single replication trials are used to evaluate the testcrosses at a minimum of three to four locations along with commercial check hybrids. The lines that performed well on multiple testers are identified as good general combiners (GCA). These lines are taken a step ahead in a hybrid breeding programme where they are crossed with additional elite lines and are taken for advanced hybrid testing.

7.11.1 Seed Production Technology

Hybrid seed production of sorghum involves crossing of A × R-line according to the prescribed standards of production and processing in terms of isolation distance, genetic purity and seed quality. Seed is multiplied in four stages in India viz., (1) Nucleus seed, (2) Breeder seed, (3) Foundation seed and (4) Certified seed. Seed certification and seed law enforcement agencies have an important role to play in certified hybrid seed production and distribution as the crop performance is dependent on the quality of the certified seed used. The male sterile (A) line and restorer (R) line are sown in the field following a planting ratio of 4A:2R, surrounded by four to five rows of restorer (R) line on all four sides of the hybrid seed production field to ensure adequate supply of pollen.

7.11.2 Isolation Distance

For hybrid seed production of sorghum, it is advised to follow an isolation distance of 300–400 m for foundation seeds and 200–400 m for certified seeds. In the presence of Johnsongrass and other forage and grassy relatives, a minimum distance of 400 m must be maintained (Kannababu et al. 2020).

7.11.3 Synchronisation of Flowering

A typical issue is the synchronisation of flowering between the A and R lines. Despite taking steps like staggered sowings of the pollen parent, synchronisation might occasionally be an issue. The parent that is lagging should be given additional watering or should be treated with 1% urea solution 2–3 times at intervals of 2–3 days (Kannababu et al. 2020). Instead of completely relying on the natural winds, it is desirable to aid the pollen dispersal by blowing air through an empty duster over the male panicles or to tap the male plants for pollen dispersal. A spray of 2% borax will also enhance the pollen production and dispersal (Kannababu 2008).

7.11.4 Roguing

It is an important operation to maintain the genetic purity by removal of off-types, volunteers and pollen shedders from the field. Roguing is carried out at three stages of the crop growth viz., one before flowering, the other at the time of flowering and the last at the pre-harvest stage. Roguing before flowering helps in the identification and removal of off-types, volunteers and pollen shedders. The proper removal of these unwanted pollen shedders prevents contamination before pollination. The pollen shedders that remain undetected until the time of flowering can be observed at the flowering stage by observing the fertile anthers which are rich yellow in colour. These fertile plants are rouged regularly at the flowering stage. As the last sort, another roguing is done at the pre-harvest stage to remove off-type heads by observing the true plant and seed characters and those infected with diseases such as kernel smut, grain smut and head smut (Kannababu 2008; Kannababu et al. 2020).

7.11.5 Harvesting and Seed Processing

To avoid seed contamination at the time of harvesting, the R-line rows are harvested prior to A-line rows. Care is taken while harvesting A-line rows such as to inspect any off-types and admixtures during the hybrid seed harvest. The harvested seed is then checked and cleaned to maintain physical purity to make sure it is free from weed seeds, other seeds, or any kind of undesirable, damaged and pest and disease-infested seed. The seeds are dried to a moisture level of 9% for safe storage (Kannababu 2008).

7.12 Heterosis Maximisation

In comparison to other major field crops, sorghum hybrid improvement programmes had limited success in terms of yield gains (Crozier et al. 2020). This necessitates the need to enhance the rate of genetic gain by further maximisation of heterosis to levels marked in other crops.

Among the several concepts to increase heterosis, identification of best parental combinations and prediction of hybrid performance are the critical ones to increase the rate of genetic gain (Hussain et al. 2022; Singh and Gupta 2019).

7.12.1 Heterotic Grouping

Sorghum breeding programmes depend primarily on homozygous inbred lines with relatively narrow genetic bases for the development of hybrids (Silva et al. 2020). The parental lines in sorghum are assessed for genetic relatedness by various genetic diversity studies based on morphological and molecular characteristics (Crozier et al. 2020; Singh and Gupta 2019). However, the superior heterosis combinations are not always due to superior parents (Yingheng et al. 2018). It is extremely useful to have information on genetic diversity and heterotic groups for the development of inbred lines and the identification of complementary lines to use as parents in hybrids so as to maximise the heterosis (Silva et al. 2020).

A heterotic group consists of individuals or populations, which show similar heterotic responses when crossed with complementary and genetically distinct groups (Melchinger and Gumber 1998; Vogt et al. 2020). The maximisation of heterosis requires the establishment of divergent heterotic groups in the available germplasm and the assignment of germplasm into these heterotic groups (Beukert et al. 2017; Gurung et al. 2010). Heterotic groups can be developed based on the evaluation of testcross performance among available germplasm in the field trials or based on molecular marker data or by usage of genomic tools.

A heterotic pattern reflects the degree of heterosis exploited in a specific combination of two heterotic groups (Vogt et al. 2020). Identification of heterotic patterns is crucial in any hybrid breeding programme as it aids to enhance the heterosis by prediction of best cross combinations and thereby reducing the number of test-crosses (Pembleton et al. 2015; Zhao et al. 2015).

The development of inbred lines to be used as hybrid parents in sorghum has been a continual endeavour by several public and private organisations. The establishment of distinct groups for sorghum seed (A-/B-lines) and pollen (R-lines) parents will help in the efficient utilisation of these hybrid parents to boost the heterosis. As compared to maize, less divergence was observed among sorghum heterotic groups (Niebur et al. 2004; Duvick 2005). This suggests the opportunities for further selection and breeding of more divergent complementary groups (Crozier et al. 2020).

7.12.2 Accelerated Line Breeding

The commercial hybrid seed production in sorghum is based on the cytoplasmic-nuclear male sterility (CGMS) system and involves A, B and R lines. Development of these parental inbreds through conventional breeding strategies like pedigree and bulk methods is a challenging and time-consuming process. It takes seven to eight generations to obtain complete homozygous lines from heterozygous founder stocks. To speed up this process, accelerated line breeding approaches such as doubled-haploid (DH) production and rapid generation advancement (RGA) can be applied in sorghum for increased productivity of inbred lines per year (Nepolean et al. 2018; Yan et al. 2021).

The technique of doubled-haploids (DHs) enables the breeder to develop homozygous genotypes in a single generation from the F_1 or in two generations from the F_2 of heterozygous genotypes (Rutkoski et al. 2022). By creating entirely homozygous lines in just two generations, it is credited with the drastic reduction of the breeding cycle and streamlining of logistics (Geiger and Gordillo 2009). The huge success of this technology in both the public and private sectors of maize demonstrates the need for its application in sorghum (De La Fuente et al. 2013).

Rapid generation advancement (RGA), commonly referred to as 'Speed breeding', is a novel method that speeds up the breeding pipeline by shortening the breeding cycle and allowing the breeder to complete more generations per year. This technique can also be applied to rapid generation of inbreds (Nepolean et al. 2018).

It has been attempted in several crops and needs to be standardised in sorghum for its effective utilisation in rapid inbred line development.

The primary challenge in sorghum hybrid breeding is selection of parents for crosses. Also, the phenotypic evaluation of numerous crosses resulting from a hybrid breeding programme is a difficult task that necessitates extensive resources (Robertsen et al. 2019). Considering the fact that the screening of lines largely relies on genotypes, genomic selection (GS) has paved the way to overcome this limitation of phenotypic evaluation (Yan et al. 2021). GS uses the genome-wide DNA markers and phenotypic data for target traits in a training population to predict the genomic estimated breeding values (GEBVs) of candidates in a test population that has been genotyped but not phenotyped (Crossa et al. 2017). Due to the enormous progress made in increasing the genetic gain of animal breeding, GS has been used in crop breeding in a variety of ways, including inbred performance prediction, parental selection, and hybrid prediction. In crop hybrid breeding programmes, it provides an opportunity to use phenotypically untested parents based on GEBVs (Robertsen et al. 2019). After superior performance is confirmed, predicted superior hybrids are field evaluated and released as new hybrid cultivars (Cui et al. 2020).

7.13 Role of Omics in Maximisation of Heterosis

Recent advancements in the field of sequencing and phenotyping technologies have enabled the extraction of unique genetic variants from a wide range of germplasm for use in plant breeding. Genomic studies constitute a powerful tool in plant breeding by facilitating the identification of numerous genes responsible for the improvement of several traits. Because trait development during breeding is partially a result of how superior and inferior alleles are balanced by artificial selection, genetic interactions between these alleles are of particular interest (Li et al. 2015). The functions of superior and inferior alleles and how they have been modified in contemporary breeding germplasms to support heterosis performance and environmental adaptation, respectively, call for a thorough analysis (Jiang et al. 2020).

To understand the role of gene interactions in heterosis, it will be necessary to analyse the genes and their underlying molecular networks using multi-omics data, such as transcriptomics, epigenomics, proteomics, metabolomics and phenomics data (Jiang et al. 2020). Transcriptomic studies aid to unravel the mechanism of heterosis by understanding the complex changes in genome and detection of allelic interactions, including the up-regulation of one allele and silencing of another allele (Das et al. 2021). The findings of on transcriptome profiles have revealed that in hybrid broccoli, most genes displayed additive expression patterns due to hybridisation. Below is the table (Table 7.7) that provides a summary of diverse studies on various omics techniques in different crops that have helped to comprehend the complex mechanism of heterosis in order to further maximise heterosis.

Table 7.7 Omic approaches used in sorghum for different traits to unravel the complex mechanism of heterosis

Kind of Omic study	Crop	Trait studied	No. of differentially expressed genes / proteins/ metabolites	Expression state	References
Genomic analysis	Rice	Yield and agronomic traits	130 associated loci	Accumulation of numerous superior alleles in hybrids compared to parents	Huang et al. (2015)
Transcriptomic analysis	Arabidopsis	Biomass	1053 differentially expressed genes	724—up-regulated genes 329—down-regulated genes for photosynthetic capacity	Fujimoto et al. (2012)
Transcriptome and DNA methylome analysis	Broccoli	Curd yield	53 differentially expressed genes	Enhanced expression of DEGs in hybrids	Li et al. (2018)
Proteomic analysis	Wheat	Root development	25 differentially expressed protein spots	2—down regulated in hybrid 2—up regulated in hybrid 16—dominant expression of uniparental genes in hybrids 3—high-dominant in hybrid 2—low-dominant in hybrid	Song et al. (2007)
Proteomic analysis	Maize	Embryo development	141 proteins	Upregulation in hybrids	Marcon et al. (2010)
Proteomic study	Rice	Yield	297 differentially expressed proteins corresponding to 222 proteins	Upregulated in hybrids- especially at flowering and grain filling stages	Zhang et al. (2012)
Proteomic analysis	Maize	Radicle emergence	ZmACT2 and 155 differentially expressed protein spots	Upregulated in hybrids	Guo et al. (2013)
Combined transcriptome and metabolome analysis	Maize	Effect of light quantity on heterosis	Several differentially expressed genes (DEGs) and differentially accumulated Metabolites (DAMs) – Five genes and seven metabolites were identified as potential contributors	Nonadditive effects of DEGs and DAMs in hybrids	Zhan et al. (2023)
Multi-Omics study	Maize	Plant height	59 (18.61%) microRNAs (miRNAs) and 64,534 (54.00%) 24-nt small interfering RNAs (siRNAs) clusters	Non-additive expression of Small RNAs in hybrids	Zhang et al. (2023)

7.14 Conclusion

A scientific understanding of different traits and concepts underlying heterosis coupled with its use in the development, maintenance and selection of parental lines for hybrid development will aid in enhancing the genetic gains. Understanding the product profile preferred by end users and developing such hybrids for the targeted environments are essential to the success of hybrid delivery.

References

Aarts MGM (1995) Molecular characterisation of the CER1 gene of *Arabidopsis* involved in epicutilar wax biosynthesis and pollen fertility. Plant Cell 7:2115–2127

Agrama HA, Tuinstra MR (2003) Phylogenetic diversity and relationships among sorghum accessions using SSRs and RAPDs. Afr J Biotechnol 2(10):334–340

Amelework A, Laing M, Shimelis H (2016) Evaluation of effective gametocides for selective induction of male sterility in sorghum. Czech J Genet Plant Breed 52(4):163–170

Amelework B, Shimelis H, Laing M (2017) Genetic variation in sorghum as revealed by phenotypic and SSR markers: implications for combining ability and heterosis for grain yield. Plant Genet Resources 15(4):335–347

Andrews DJ, Webster OJ (1971) A new factor for genetic male sterility in *Sorghum bicolor* (L.) Moench. Crop Sci 11:308–309. https://doi.org/10.2135/cropsci1971.0011183X001100020046x

Andrews DJ, Ejeta G, Gilbert M, Goswami P, Anand Kumar K, Maunder AB, Porter K, Rai KN, Rajewski JF, Reddy BVS, Stegmeier W (1996) Breeding hybrid parents. In: Proceedings of an international conference on the genetic improvement of sorghum and pearl millet, held at Lubbock, Texas, pp 22–27

Aruna C, Audilakshmi S (2008) Reproductive biology and breeding behaviour of sorghum. In: Sorghum improvement in the new millennium. International Crops Research Institute for the Semi-Arid Tropics, Patancheru, pp 7–12

Ashok Kumar A (2019) Sorghum hybrids development for important traits: progress and way forward. In: Sorghum: a state of the art and future perspectives, vol 58, pp 97–117

Assefa Y (2010) Grain sorghum in the hybrid-era, 1957–2008: yield with hybrid advancement and improved agronomic practices. Kansas State University, Manhattan

Awad-Allah MMA, Elekhtyar NM, El-Abd MA, Abdelkader MFM, Mahmoud MH, Mohamed AH, El-Diasty MZ, Said MM, Shamseldin SAM, Abdein MA (2022) Development of new restorer lines carrying some restoring fertility genes with flowering, yield and grains quality characteristics in rice (Oryza sativa L.). Genes (Basel) 13(3):458. https://doi.org/10.3390/genes13030458. PMID: 35328012; PMCID: PMC8952259

Axtell J, Kapran I, Ibrahim Y, Ejeta G, Andrews DJ (1999) Heterosis in sorghum and pearl millet. In: Genetics and exploitation of heterosis in crops, pp 375–386

Ayyangar GR, Rao VP (1931) Studies in sorghum: I, anthesis and pollination. Indian J Agric Sci 1:445–454

Baloch FS, Altaf MT, Liaqat W, Bedir M, Nadeem MA, Cömertpay G, Çoban N, Habyarimana E, Barutçular C, Cerit I, Ludidi N (2023) Recent advancements in the breeding of sorghum crop: current status and future strategies for marker-assisted breeding. Front Genet 14:1150616

Begna T (2022) Hybrid sorghum development mechanisms to enhance production and productivity. Int J Res Stud Agric Sci 8(2):20–31. ISSN No. (Online) 2454–6224. https://doi.org/10.20431/2454-6224.0802003

Ben-Israel I, Kilian B, Nida H, Fridman E (2012) Heterotic trait locus (HTL) mapping identifies intra-locus interactions that underlie reproductive hybrid vigor in Sorghum bicolor. PLoS One 7(6):e38993

Berenji J, Dahlberg J (2004) Perspectives of sorghum in Europe. J Agron Crop Sci 190(5):332–338

Beukert U, Li Z, Liu G, Zhao Y, Ramachandra N, Mirdita V, Pita F, Pillen K, Reif JC (2017) Genome-based identification of heterotic patterns in rice. Rice 10(1):1–10
Birchler JA, Auger DL, Riddle NC (2003) In search of the molecular basis of heterosis. Plant Cell 15(10):2236–2239
Birchler JA, Yao H, Chudalayandi S (2006) Unraveling the genetic basis of hybrid vigor. Proc Natl Acad Sci U S A 103(35):12957–12958
Boerman NA, Hlavinka KB, Zhu W, Dabney AR, Hodnett GL, Rooney WL (2019) Efficacy of the chemical trifluoromethanesulfonamide as a male gametocide in field-grown sorghum [Sorghum bicolor (L.) Moench]. Euphytica 215(5):1–10
Brown PJ, Klein PE, Bortiri E, Acharya CB, Rooney WL, Kresovich S (2006) Inheritance of inflorescence architecture in sorghum. Theor Appl Genet 113:931–942
Budar F, Pelletier G (2001) Male sterility in plants: occurrence, determinism, significance and use. C R Acad Sci Ser III Sci Vie 324(6):543–550
Bunphan D, Jaisil P, Sanitchon J (2013) Application of LW7 marker for identification of progenies with male sterility gene in sweet sorghum population. Crop Breed Appl Biotechnol 13:59–66
Burow G, Xin Z, Hayes C, Burke J (2014) Characterization of a multiseeded (msd1) mutant of sorghum for increasing grain yield. Crop Sci 54(5):2030–2037
Chang Z, Chen Z, Wang N, Xie G, Lu J, Yan W, Zhou J, Tang X, Deng XW (2016) Construction of a male sterility system for hybrid rice breeding and seed production using a nuclear male sterility gene. Proc Natl Acad Sci U S A 113(49):14145–14150
Chase SS (1971) Estimates of commercial profitability in hybrid seed production of the small grains. Econ Bot 25(3):330–334
Chauhan P, Pandey PK, Shrotria PK, Singh NK, Pandey G, Singh S (2015) Allele specific diversity using fertility restoration related SSR markers in sorghum [Sorghum bicolor (L.) Moench]. Forage Res 40(4):222–227
Chen L, Liu YG (2014) Male sterility and fertility restoration in crops. Annu Rev Plant Biol 65:579–606
Chen J, Jiao Y, Laza H, Payton P, Ware D, Xin Z (2019) Identification of the first nuclear male sterility gene (Male-sterile 9) in sorghum. Plant Genome 12(3):190020
Choe M-E, Kim J-Y, Syed Nabi RB, Han S-I, Cho K-S (2023) Development of InDels markers for the identification of cytoplasmic male sterility in Sorghum by complete chloroplast genome sequences analysis. Front Plant Sci 14:1188149. https://doi.org/10.3389/fpls.2023.1188149
Conner AB, Karper RE (1927) Hybrid vigor in sorghum. Texas Agricultural Experimental Station Bulletin 359. Texas A & M University, College Station
Crossa J, Pérez-Rodríguez P, Cuevas J, Montesinos-López O, Jarquín D, De Los Campos G, Burgueño J, González-Camacho JM, Pérez-Elizalde S, Beyene Y, Dreisigacker S (2017) Genomic selection in plant breeding: methods, models, and perspectives. Trends Plant Sci 22(11):961–975
Crozier D, Hoffmann L Jr, Klein PE, Klein RR, Rooney WL (2020) Predicting heterosis in grain sorghum hybrids using sequence-based genetic similarity estimates. J Crop Improv 34(5):600–617
Cui Y, Li R, Li G, Zhang F, Zhu T, Zhang Q, Ali J, Li Z, Xu S (2020) Hybrid breeding of rice via genomic selection. Plant Biotechnol J 18(1):57–67
Dan Z, Chen Y, Zhao W, Wang Q, Huang W (2020) Metabolome-based prediction of yield heterosis contributes to the breeding of elite rice. Life Sci Alliance 3(1):e201900551
Das AK, Choudhary M, Kumar P, Karjagi CG, Yathish KR, Kumar R, Singh A, Kumar S, Rakshit S (2021) Heterosis in genomic era: advances in the molecular understanding and techniques for rapid exploitation. Crit Rev Plant Sci 40(3):218–242
De La Fuente GN, Frei UK, Lübberstedt T (2013) Accelerating plant breeding. Trends Plant Sci 18(12):667–672. https://doi.org/10.1016/j.tplants.2013.09.001. Epub 2013 Sep 27. PMID: 24080381
Dhaka N, Krishnan K, Kandpal M, Vashisht I, Pal M, Sharma MK, Sharma R (2020a) Transcriptional trajectories of anther development provide candidates for engineering male fertility in sorghum. Sci Rep 10(1):897

Dhaka N, Sharma S, Vashisht I, Kandpal M, Sharma MK, Sharma R (2020b) Small RNA profiling from meiotic and post-meiotic anthers reveals prospective miRNA-target modules for engineering male fertility in sorghum. Genomics 112(2):1598–1610

Doggett H (1988) Sorghum. Tropical agriculture series, 2nd edn. Longman Scientific and Technical Publishers, UK, IDRC, Canada. 512 pp

Dong NV, Subudhi PK, Luong PN, Quang VD, Quy TD, Zheng HG et al (2000) Molecular mapping of a rice gene conditioning thermosensitive genic male sterility using AFLP, RFLP and SSR techniques. Theor Appl Genet 100:727–734. https://doi.org/10.1007/s001220051345

Dos Santos JPR, Fernandes SB, McCoy S, Lozano R, Brown PJ, Leakey ADB, Buckler ES, Garcia AAF, Gore MA (2020) Novel Bayesian networks for genomic prediction of developmental traits in biomass sorghum. G3 (Bethesda) 10:769–781

Duvick DN (2005) The contribution of breeding to yield advances in maize (Zea mays L.). Adv Agron 86:83–145. https://doi.org/10.1016/S0065-2113(05)86002-X

Eckardt NA (2006) Cytoplasmic male sterility and fertility restoration. Plant Cell 18(3):515–517. PMCID: PMC1383628. https://doi.org/10.1105/tpc.106.041830

Ejeta G, Grenier C (2005) Sorghum and its weedy hybrids. In: Crop ferality and volunteerism, pp 123–135

FAOSTAT (2023). http://www.fao.org/faostat/en/#data/QCL

Farinati S, Draga S, Betto A, Palumbo F, Vannozzi A, Lucchin M, Barcaccia G (2023) Current insights and advances into plant male sterility: new precision breeding technology based on genome editing applications. Front Plant Sci 14:1223861. https://doi.org/10.3389/fpls.2023.1223861

Folliard A, Traoré PCS, Vaksmann M, Kouressy M (2004) Modeling of sorghum response to photoperiod: a threshold–hyperbolic approach. Field Crop Res 89(1):59–70

Fonseca JM, Klein PE, Crossa J, Pacheco A, Perez-Rodriguez P, Ramasamy P, Klein R, Rooney WL (2021) Assessing combining abilities, genomic data, and genotype× environment interactions to predict hybrid grain sorghum performance. Plant Genome 14(3):e20127

Fujimoto R, Taylor JM, Shirasawa S, Peacock WJ, Dennis ES (2012) Heterosis of Arabidopsis hybrids between C24 and Col is associated with increased photosynthesis capacity. Proc Natl Acad Sci U S A 109:7109–7114

Geiger HH, Gordillo GA (2009) Doubled haploids in hybrid maize breeding. Maydica 54(4):485

Girma G, Nida H, Seyoum A, Mekonen M, Nega A, Lule D, Dessalegn K, Bekele A, Gebreyohannes A, Adeyanju A, Tirfessa A (2019) A large-scale genome-wide association analyses of Ethiopian sorghum landrace collection reveal loci associated with important traits. Front Plant Sci 10:691

Glover J, Grelon M, Craig S, Chaudury A, Dennis L (1998) Cloning and characterisation of *MS5* from *Arabidopsis*: a gene critical in male meiosis. Plant J 15:345–356

Guo B, Chen Y, Zhang G, Xing J, Hu Z, Feng W, Yao Y, Peng H, Du J, Zhang Y et al (2013) Comparative proteomic analysis of embryos between a maize hybrid and its parental lines during early stages of seed germination. PLoS One 8:e65867

Gurung DB, George MLC, Delacruz QD (2010) Determination of heterotic groups in Nepalese yellow maize populations. Nepal J Sci Technol 10:1–8. https://doi.org/10.3126/njst.v10i0.2802

Habyarimana E, Lopez-Cruz M, Baloch FS (2020) Genomic selection for optimum index with dry biomass yield, dry mass fraction of fresh material, and plant height in biomass sorghum. Genes (Basel) 11:61–76

Hambloch C, Kahwai J, Mugonya J (2021) Contextualizing private sector-based seed system development: the case of sorghum in Eastern Africa. Outlook Agric 50(4):378–391

Han P, Lu X, Mi F, Dong J, Xue C, Li J, Han B, Zhang X (2016) Proteomic analysis of heterosis in the leaves of sorghum–sudangrass hybrids. Acta Biochim Biophys Sin 48(2):161–173

Hao H, Li Z, Leng C, Lu C, Luo H, Liu Y, Wu X, Liu Z, Shang L, Jing HC (2021) Sorghum breeding in the genomic era: opportunities and challenges. Theor Appl Genet 134:1899–1924

Hochholdinger F, Hoecker N (2007) Towards the molecular basis of heterosis. Trends Plant Sci 12(9):427–432

Hodnett GL, Rooney WL (2018) Male sterility induction of sorghum using chemical hybridizing agent TFMSA, trifluoromethanesulfonamide. Can J Plant Sci 98(5):1102–1108

House LR (1985) A guide to sorghum breeding, 2nd edn. International Crops Research Institute for the Semi-Arid Tropics, Patancheru

https://qaafi.uq.edu.au/article/2016/10/how-sorghum-became-queenslands-top-crop

Huang X, Yang S, Gong J, Zhao Y, Feng Q, Gong H et al (2015) Genomic analysis of hybrid rice varieties reveals numerous superior alleles that contribute to heterosis. Nat Commun 6:6258

Hunt CH, Hayes BJ, Van Eeuwijk FA, Mace ES, Jordan DR (2020) Multi-environment analysis of sorghum breeding trials using additive and dominance genomic relationships. Theor Appl Genet 133:1009–1018

Hussain I, Ali S, Liu W, Awais M, Li J, Liao Y, Zhu M, Fu C, Liu D, Wang F (2022) Identification of heterotic groups and patterns based on genotypic and phenotypic characteristics among rice accessions of diverse origins. Front Genet 13:811124. https://doi.org/10.3389/fgene.2022.811124

Jiang S, Cheng Q, Yan J, Fu R, Wang X (2020) Genome optimization for improvement of maize breeding. Theor Appl Genet 133:1491–1502

Jones DF (1917) Dominance of linked factors as a means of accounting for heterosis. Genetics 2:466–479

Jordan D, Tao Y, Godwin I, Henzell R, Cooper M, McIntyre C (2003) Prediction of hybrid performance in grain sorghum using RFLP markers. Theor Appl Genet 106:559–567

Jordan DR, Mace ES, Henzell RG, Klein PE, Klein RR (2010) Molecular mapping and candidate gene identification of the Rf2 gene for pollen fertility restoration in sorghum [Sorghum bicolor (L.) Moench]. Theor Appl Genet 120:1279–1287

Jordan DR, Klein RR, Sakrewski KG, Henzell RG, Klein PE, Mace ES (2011) Mapping and characterization of Rf 5: a new gene conditioning pollen fertility restoration in A1 and A2 cytoplasm in sorghum (Sorghum bicolor (L.) Moench). Theor Appl Genet 123:383–396

Kannababu N (2008) Principles of seed production and seed quality maintenance in sorghum. In: Sorghum improvement in the new millennium. International Crops Research Institute for the Semi-Arid Tropics, Patancheru. 340 pp, 48, p 208

Kannababu N, Dinni S, Talwar HS, Tonapi VA (2020) Sorghum hybrid seed production and quality management: important considerations. In: Sorghum in the 21st century: food–fodder–feed–fuel for a rapidly changing world, pp 295–313

Karper RE, Quinby JR (1947) Sorghum—its production, utilization and breeding. Econ Bot 1(4):355–371

Kaul MLH (1988) Male sterility in higher plants. Springer, Berlin. https://doi.org/10.1007/978-3-642-83139-3

Kent MA, Fonseca JM, Klein PE, Klein RR, Hayes CM, Rooney WL (2023) Use of genomic prediction to screen sorghum B-lines in hybrid testcrosses. Plant Genome 16:e20369

Kim YJ, Zhang D (2018) Molecular control of male fertility for crop hybrid breeding. Trends Plant Sci 23(1):53–65

Klein RR, Klein PE, Chhabra AK, Dong JIANMIN, Pammi SUJATA, Childs KL, Mullet JE, Rooney WL, Schertz KF (2001) Molecular mapping of the rf1 gene for pollen fertility restoration in sorghum (Sorghum bicolor L.). Theor Appl Genet 102:1206–1212

Klein RR, Mullet JE, Jordan DR, Miller FR, Rooney WL, Menz MA, Franks CD, Klein PE (2008) The effect of tropical sorghum conversion and inbred development on genome diversity as revealed by high-resolution genotyping. Crop Sci 48:S-12

Knoch D, Werner CR, Meyer RC, Riewe D, Abbadi A, Lücke S, Snowdon RJ, Altmann T (2021) Multi-omics-based prediction of hybrid performance in canola. Theor Appl Genet 134(4):1147–1165

Ktavii MN, Kiambi DK, Haussman B, Semagn K, Muluvi G, Kairichi M, Machuka J (2014) Assessment of the genetic diversity and pattern of relationship of West African sorghum accessions using microsatellite markers. Afr J Biotechnol 13(14):1503

Lakshmi Pathy T, Mohan Rao A, Ramesh S (2018) Assessing breeding potential of three-way cross and double-cross hybrids in chilli (Capsicum annuum). Agric Res 7:129–134

Lasa JM, Bosemark NO (1993) Male sterility. In: Plant breeding: principles and prospects. Springer, Dordrecht, pp 213–228

Li X, Li X, Fridman E, Tesso TT, Yu J (2015) Dissecting repulsion linkage in the dwarfing gene Dw3 region for sorghum plant height provides insights into heterosis. Proc Natl Acad Sci U S A 112(38):11823–11828

Li H, Yuan J, Wu M, Han Z, Li L, Jiang H, Jia Y, Han X, Liu M, Sun D, Chen C (2018) Transcriptome and DNA methylome reveal insights into yield heterosis in the curds of broccoli (Brassica oleracea L var. italic). BMC Plant Biol 18(1):1–17

Li D, Lu X, Zhu Y, Pan J, Zhou S, Zhang X, Zhu G, Shang Y, Huang S, Zhang C (2022) The multi-omics basis of potato heterosis. J Integr Plant Biol 64(3):671–687

Liu X, Gu S, Wen W, Lu X, Jin Y, Zhang Y, Guo X (2023) Disentangling the heterosis in biomass production and radiation use efficiency in maize: a phytomer-based 3D modelling approach. Plants 12(6):1229

Marcon C, Schützenmeister A, Schuütz W, Madlung J, Piepho H-P, Hochholdinger F (2010) Nonadditive protein accumulation patterns in maize (Zea mays L.) hybrids during embryo development. J Proteome Res 9:6511–6522

Maunder AB, Sharp GI (1963) Localization of outcrosses within the panicle of fertile sorghum 1. Crop Sci 3(5):449–449

Melchinger AE, Gumber RK (1998) Overview of heterosis and heterotic groups in agronomic crops. In: Concepts and breeding of heterosis in crop plants, vol 25, pp 29–44

Menz MA, Klein RR, Unruh NC, Rooney WL, Klein PE, Mullet JE (2004) Genetic diversity of public inbreds of sorghum determined by mapped AFLP and SSR markers. Crop Sci 44(4):1236–1244

Miller FR, Kebede Y (1984) Genetic contributions to yield gains in sorghum, 1950 to 1980. In: Genetic contributions to yield gains of five major crop plants, vol 7, pp 1–14

Mindaye TT, Mace ES, Godwin ID, Jordan DR (2015) Genetic differentiation analysis for the identification of complementary parental pools for sorghum hybrid breeding in Ethiopia. Theor Appl Genet 128:1765–1775

Moffatt BA, Sommerville CR (1988) Positive selection for male-sterile mutants of *Arabidopsis thaliana* lacking adenine phosphoribosyltransferase activity. Plant Physiol 86:1150–1154

Mundia CW, Secchi S, Akamani K, Wang G (2019) A regional comparison of factors affecting global sorghum production: the case of North America, Asia and Africa's Sahel. Sustainability 11(7):2135

Murty UR (1999) Hybrid seed production in sorghum. In: Basra AS (ed) Heterosis and hybrid seed production in agronomic crops. Haworth Press, Binghamton, pp 119–145

Murty DS, Tabo R, Ajayi O (1994) Sorghum hybrid seed production and management. Information Bulletin no. 41, International Crops Research Institute for the Semi-Arid Tropics, Patancheru

Nepolean T, Kaul J, Mukri G, Mittal S (2018) Genomics-enabled next-generation breeding approaches for developing system-specific drought tolerant hybrids in maize. Front Plant Sci 9:361

Niebur WS, Rafalski JA, Smith OS, Cooper M (2004) Applications of genomics technologies to enhance rate of genetic progress for yield of maize within a commercial breeding program. In: New directions for a diverse planet: proceedings for the 4th international crop science congress. Regional Institute, Gosford, Australia. www.cropscience.org.au/icsc2004

Obilana AB (2004) Sorghum breeding research in Africa. In: Sorghum genetic enhancement: research process, dissemination and impacts. International Crops Research Institute for the Semi-Arid Tropics, Patancheru. 320 pp, 33, p 105

OECD (2016) Consensus document on the biology of sorghum (Sorghum bicolor L. Moench), 62nd edn. Organization for Economic Co-operation and Development, Paris

OECD (2017) Sorghum (*Sorghum bicolor*). In: Safety assessment of transgenic organisms in the environment. Chapter 1, vol 7: OECD consensus documents, harmonisation of regulatory oversight in biotechnology. OECD Publishing, Paris, pp 29–65. https://doi.org/10.1787/9789264279728-en

Olatoye MO, Marla SR, Hu Z, Bouchet S, Perumal R, Morris GP (2020) Dissecting adaptive traits with nested association mapping: genetic architecture of inflorescence morphology in sorghum. G3 10(5):1785–1796

Patil JV, Reddy PS, Umakanth AV, Sunil G, Ganapathy KN (2014) History of post-rainy season sorghum research in India and strategies for breaking the yield plateau. Indian J Genet Plant Breed 74(3):271–285

Pembleton LW, Shinozuka H, Wang J, Spangenberg GC, Forster JW, Cogan NO (2015) Design of an F1 hybrid breeding strategy for ryegrasses based on selection of self-incompatibility locus-specific alleles. Front Plant Sci 6:764

Praveen M, Anurag UG, Madhusudhana R (2015) Inheritance of fertility restoration in sorghum cytoplasmic male sterility systems. Int J Sci Res 4:438–439

Qingshan L, Dahlberg JA (2001) Chinese sorghum genetic resources. Econ Bot 55(3):401–425

Quinby JR (1974) The genetic control of flowering and growth in sorghum. Adv Agron 25:125–162

Quinby JR (1981) Interaction of genes and cytoplasms in male sterility in sorghum. In: Proceedings of the... annual corn and sorghum industry research conference American Seed Trade Association, Corn and Sorghum Division, Corn and Sorghum Research Conference, vol 35, pp 175–184

Quinby JR, Karper RE (1947) The effect of short photoperiod on sorghum. J Agric Res 75(11/12):295–300

Rai KN, Murty DS, Andrews DJ, Bramel-Cox PJ (1999) Genetic enhancement of pearl millet and sorghum for the semi-arid tropics of Asia and Africa. Genome 42(4):617–628

Rajendrakumar P (2015) Heterosis prediction using DNA markers. In: Sorghum molecular breeding, pp 101–114

Rakshit S, Bellundagi A (2019) Conventional breeding techniques in sorghum. In: Breeding sorghum for diverse end uses. Woodhead Publishing, Cambridge, pp 77–91

Reddy PS, Kumar AA (2008) Emasculation and selfing techniques in sorghum. In: Sorghum improvement in the new millennium. International Crops Research Institute for the Semi-Arid Tropics, Patancheru. 340 pp, 48, p 13

Reddy PS, Reddy BV (2019) History of sorghum improvement. In: Breeding sorghum for diverse end uses. Woodhead Publishing, Cambridge, pp 61–75

Reddy BVS, Ramesh S, Reddy PS (2004) Sorghum breeding research at ICRISAT-goals, strategies, methods and accomplishments. Int Sorghum Millets Newsl 45:5–12

Reddy BV, Sharma HC, Thakur RP, Ramesh S, Rattunde F, Mgonja M (2006) Sorghum hybrid parents research at ICRISAT-strategies, status and impacts. J SAT Agric Res 2(1):1–24

Reddy BVS, Ashok Kumar A, Sanjana Reddy P, Ramaiah B (2008) Development of male-sterile lines in sorghum. In: Development of male-sterile lines in sorghum, pp 72–78

Robertsen CD, Hjortshøj RL, Janss LL (2019) Genomic selection in cereal breeding. Agronomy 9(2):95

Rooney WL (2014) Sorghum. In: Karlen DL (ed) Cellulosic energy cropping systems. Wiley, Chichester, pp 109–129. https://doi.org/10.1002/9781118676332.ch7

Ross WM (1957) Stigma receptivity in cytoplasmic male-sterile sorghum 1. Agron J 49(4):219–220

Rutkoski JE, Krause MR, Sorrells ME (2022) Breeding methods: line development. In: Wheat improvement. Springer, Cham, pp 69–82

Sandeep N, Biradar B (2020) Identification of stable restorers for diverse cytoplasmic source in Sorghum [Sorghum bicolor (L.) Moench.]. Indian J Genet Plant Breed 80(02):218–220

Sanjana Reddy P, Reddy BV, Ashok Kumar A (2009) M 35-1 derived sorghum varieties for cultivation during the postrainy season. J SAT Agric Res 7:1–4

Saxena KB, Sultana R, Mallikarjuna N, Saxena RK, Kumar RV, Sawargaonkar SL, Varshney RK (2010) Male-sterility systems in pigeonpea and their role in enhancing yield. Plant Breed 129(2):125–134

Saxena KB, Choudhary AK, Hingane AJ (2015) Considerations for breeding, maintenance, and utilization of TGMS lines for a two-parent hybrid system in pigeonpea. Int J Sci Res 4(3):344–348

Schertz KF, Ritchey JM (1978) Cytoplasmic-genic male-sterility systems in sorghum 1. Crop Sci 18(5):890–893

Schnell FW, Cockerham CC (1992) Multiplicative vs. arbitrary gene action in heterosis. Genetics 131:461–469

Schrag TA, Westhues M, Schipprack W, Seifert F, Thiemann A, Scholten S, Melchinger AE (2018) Beyond genomic prediction: combining different types of omics data can improve prediction of hybrid performance in maize. Genetics 208(4):1373–1385

Shull GH (1911) The genotypes of maize. Am Nat 45(532):234–252

Silva KJD, Pastina MM, Guimarães CT, Magalhães JV, Pimentel LD, Schaffert RE, Pinto MDO, Souza VFD, Bernardino KDC, Silva MJD, Borém A (2020) Genetic diversity and heterotic grouping of sorghum lines using SNP markers. Sci Agric 78

Singh S, Gupta SK (2019) Formation of heterotic pools and understanding relationship between molecular divergence and heterosis in pearl millet [Pennisetum glaucum (L.) R. Br.]. PLoS One 14(5):e0207463. https://doi.org/10.1371/journal.pone.0207463

Smith S, Primomo V, Monk R, Nelson B, Jones E, Porter K (2010) Genetic diversity of widely used US sorghum hybrids 1980–2008. Crop Sci 50(5):1664–1673

Song X, Ni Z, Yao Y, Xie C, Li Z, Wu H, Zhang Y, Sun Q (2007) Wheat (Triticum aestivum L.) root proteome and differentially expressed root proteins between hybrid and parents. Proteomics 7:3538–3557

Srivastava KR, Bollam S, Pujarula V, Pusuluri M, Singh RB, Potupureddi G, Gupta R (2020) Exploitation of heterosis in pearl millet: a review. Plan Theory 9(7):807

Stephens JC, Holland RF (1954) Cytoplasmic male sterility for hybrid sorghum seed production. Agron J 46:20–23

Stephens JC, Quinby JR (1934) Anthesis, pollination, and fertilization in sorghum. J Agric Res 49(2):123–136

Stokes D, Fraser F, Morgan C, O'Neill CM, Dreos R, Magusin A, Szalma S, Bancroft I (2010) An association transcriptomics approach to the prediction of hybrid performance. Mol Breed 26:91–106

Tester M, Langridge P (2010) Breeding technologies to increase crop production in a changing world. Science 327(5967):818–822

Timberlake WE (2013) Heterosis in Brenner's encyclopedia of genetics, 2nd edn. Academic Press, San Diego, pp 451–453. ISBN 9780080961569. https://doi.org/10.1016/B978-0-12-374984-0.00705-1

USDA Foreign Agriculture Service/United States Department of Agriculture (2023) World Agricultural Production; circular series WAP 9–23. USDA, Washington, DC

Van der Meer IM, Stam ME, Van Tunen AJ, Mol JN, Stuitje AR (1992) Antisense inhibition of flavonoid biosynthesis in *Petunia* anthers results in male sterility. Plant Cell 4:253–262

Velazco JG, Malosetti M, Hunt CH, Mace ES, Jordan DR, Van Eeuwijk FA (2019) Combining pedigree and genomic information to improve prediction quality: an example in sorghum. Theor Appl Genet 132:2055–2067

Venkateswaran K, Sivaraj N, Pandravada SR, Reddy MT, Babu BS (2019) Classification, distribution and biology. In: Breeding sorghum for diverse end uses. Woodhead Publishing, Cambridge, pp 33–60

Vinall HN, Martin JH, Stephens JC (1936) Identification, history, and distribution of common sorghum varieties (No. 506). US Department of Agriculture, Washington, DC

Virmani SS, Pandey MP, Singh IS, Xu WJ (2004) Classical and molecular concepts of heterosis. In: Plant breeding. Springer, Dordrecht, pp 407–418

Vogt M, Yates S, Sykes T, Luesink W, Koch M, Studer B (2020) Developing heterotic groups for successful hybrid breeding in perennial ryegrass. Agronomy 10(9):1410

Wang L, Jiao S, Jiang Y, Yan H, Su D, Sun G, Yan X, Sun L (2013) Genetic diversity in parent lines of sweet sorghum based on agronomical traits and SSR markers. Field Crop Res 149:11–19

Wang B, Regulski M, Tseng E, Olson A, Goodwin S, McCombie WR, Ware D (2018) A comparative transcriptional landscape of maize and sorghum obtained by single-molecule sequencing. Genome Res 28(6):921–932

Wang H, Zhang Y, Liang D, Zhang X, Fan X, Guo Q, Wang L, Wang J, Liu Q (2023a) Genome-wide identification and characterization of miR396 family members and their target genes GRF in sorghum (Sorghum bicolor (L.) Moench). PLoS One 18(5):e0285494

Wang T, Crawford MM, Tuinstra MR (2023b) A novel transfer learning framework for sorghum biomass prediction using UAV-based remote sensing data and genetic markers. Front Plant Sci 14:1138479

Wen L, Tang HV, Chen W, Chang R, Pring DR, Klein PE, Childs KL, Klein RR (2002) Development and mapping of AFLP markers linked to the sorghum fertility restorer gene rf4. Theor Appl Genet 104:577–585

Windpassinger SM (2016) Breeding strategies for the adaptation of sorghum (Sorghum bicolor L. Moench) as a novel crop for temperate Europe. Verlag nicht ermittelbar

Witt Hmon KP, Shehzad T, Okuno K (2014) QTLs underlying inflorescence architecture in sorghum (Sorghum bicolor (L.) Moench) as detected by association analysis. Genet Resour Crop Evol 61:1545–1564

Worstell JV, Kidd HJ, Schertz KF (1984) Relationships among male-sterility inducing cytoplasms of sorghum 1. Crop Sci 24(1):186–189

Xiang Z, Li J, Lu D, Wei X, Xu X (2022) Advances in multi-omics research on viral hepatitis. Front Microbiol 13:987324. https://doi.org/10.3389/fmicb.2022.987324

Xin Z, Huang J, Smith AR, Chen J, Burke J, Sattler SE, Zhao D (2017) Morphological characterization of a new and easily recognizable nuclear male sterile mutant of sorghum (Sorghum bicolor). PLoS One 12(1):e0165195

Yadav OP, Hossain F, Karjagi CG, Kumar B, Zaidi PH, Jat SL, Chawla JS, Kaul J, Hooda KS, Kumar P, Yadava P (2015) Genetic improvement of maize in India: retrospect and prospects. Agric Res 4:325–338

Yahaya MA, Shimelis H, Laing M, Mathew I (2021) Comparative analysis of the efficacy of selected gametocide agents in sorghum ('Sorghum bicolor' [L.] Moench). Aust J Crop Sci 15(6):816–820

Yali W, Begna T (2022) Sorghum breeding in Ethiopia: progress, achievements and challenges. Int J Agric Sci Food Technol 8(1):045–051

Yan J, Xu Y, Cheng Q, Jiang S, Wang Q, Xiao Y, Ma C, Yan J, Wang X (2021) LightGBM: accelerated genomically designed crop breeding through ensemble learning. Genome Biol 22(1):1–24

Yıldırım M, Çakmak M (2018) Evaluation of hybrid seeds of three-way and single cross for grain number and weight in bread and durum wheat. Turk J Agric Food Sci Technol 6(1):22–27

Yingheng W, Qiuhua CAI, Hongguang X, Fangxi W, Ling L, Wei H, Liping C, Hua'an XIE, Jianfu Z (2018) Determination of heterotic groups and heterosis analysis of yield performance in indica rice. Rice Sci 25(5):261–269

Yu SB, Li JX, Xu CG, Tan YF, Gao YJ, Li XH, Zhang Q, Maroof MS (1997) Importance of epistasis as the genetic basis of heterosis in an elite rice hybrid. Proc Natl Acad Sci U S A 94(17):9226–9231

Yu XQ, Li XR, Guo TT, Zhu CS, Wu YY, Mitchell SE, Roozeboom KL, Wang DH, Wang ML, Pederson GA, Tesso TT, Schnable PS, Bernardo R, Yu JM (2016) Genomic prediction contributing to a promising global strategy to turbocharge gene banks. Nat Plants 2:16150

Zhan W, Guo G, Cui L, Rashid MAR, Jiang L, Sun G, Yang J, Zhang Y (2023) Combined transcriptome and metabolome analysis reveals the effects of light quality on maize hybrids. BMC Plant Biol 23(1):1–16

Zhang H, Liang W, Yang X, Luo X, Jiang N, Ma H et al (2010) Carbon starved anther encodes a MYB domain protein that regulates sugar partitioning required for rice pollen development. Plant Cell 22:672–689. https://doi.org/10.1105/tpc.109.073668

Zhang C, Yin Y, Zhang A, Lu Q, Wen X, Zhu Z, Zhang L, Lu C (2012) Comparative proteomic study reveals dynamic proteome changes between superhybrid rice LYP9 and its parents at different developmental stages. J Plant Physiol 169:387–398

Zhang H, Xu C, He Y, Zong J, Yang X, Si H et al (2013) Mutation in CSA creates a new photoperiod-sensitive genic male sterile line applicable for hybrid rice seed production. Proc Natl Acad Sci U S A 110:76–81. https://doi.org/10.1073/pnas.1213041110

Zhang J, Xie Y, Zhang H, He C, Wang X, Cui Y, Heng Y, Lin Y, Gu R, Wang J, Fu J (2023) Integrated multi-omics reveals significant roles of non-additively expressed small RNAs in heterosis for maize plant height. Int J Mol Sci 24(11):9150

Zhao ZY, Dahlberg J (eds) (2019) Sorghum: methods and protocols. Springer, New York
Zhao Y, Li Z, Liu G, Jiang Y, Maurer HP, Würschum T, Mock HP, Matros A, Ebmeyer E, Schachschneider R, Kazman E (2015) Genome-based establishment of a high-yielding heterotic pattern for hybrid wheat breeding. Proc Natl Acad Sci U S A 112(51):15624–15629
Zhou H, Zhou M, Yang Y, Li J, Zhu L, Jiang D et al (2014) RNase Z^{S1} processes *UbL40* mRNAs and controls thermosensitive genic male sterility in rice. Nat Commun 5:4884. https://doi.org/10.1038/ncomms5884
Zhou Y, Srinivasan S, Mirnezami SV, Kusmec A, Fu Q, Attigala L, Salas Fernandez MG, Ganapathysubramanian B, Schnable PS (2019) Semiautomated feature extraction from RGB images for sorghum panicle architecture GWAS. Plant Physiol 179(1):24–37

Breeding Techniques for the Major Sorghum Market Segments: Grain, Forage, Biomass, and Soluble Sugars

8

Wilfred Vermerris, B. Amasiddha, and A. V. Umakanth

Abstract

Sorghum (*Sorghum bicolor* (L.) Moench) is a versatile crop that can be cultivated for the production of food, feed, fodder, and as a feedstock for the production of renewable chemicals and fuels. Given these diverse uses of sorghum, four distinct types of cultivated sorghum have been developed to meet specific needs: Grain sorghums, forage sorghums, sweet sorghums, and biomass sorghums. Plant breeding has as its main goal the release of improved germplasm that meets the needs of the end user(s) in a particular geographic area. These needs will change over time as a function of the preferences of consumers, new industrial processes, and the identification of novel traits, some of which may result in new uses for the crop. This chapter starts with a brief summary of the different breeding methods for sorghum cultivars and hybrids, and then provides a general overview of the uses of the different types of sorghum, the target traits in breeding programs, and the available genetic variation that can be exploited to achieve those targets. The focus is on traits with global relevance, with the understanding that cultivation in specific geographic areas or for specific industrial purposes will require additional breeding targets.

W. Vermerris (✉)
Department of Microbiology and Cell Science and UF Genetics Institute, University of Florida, Gainesville, FL, USA
e-mail: wev@ufl.edu

B. Amasiddha · A. V. Umakanth
ICAR-Indian Institute of Millets Research (IIMR), Hyderabad, Telangana, India

E. Habyarimana et al. (eds.), *Omics and Biotechnological Approaches for Product Profile-Driven Sorghum Improvement*,
https://doi.org/10.1007/978-981-97-4347-6_8

Keywords

A-line · B-line · Cultivar · Cytoplasmic male sterility · Ethanol · Hybrid · Male sterile · Pedigree · R-line · Starch

8.1 Introduction

Sorghum is a versatile crop that can be cultivated for the production of food, feed, fodder, and as a feedstock for the production of renewable chemicals and fuels. Given these diverse uses of sorghum, four distinct types of cultivated sorghum have been developed to meet the specific needs: Grain sorghums, forage sorghums, sweet sorghums, and biomass sorghums. Grain sorghums are typically short-statured (<1.5 m), combine-compatible plants whose large panicles are used as food feed, and a source of industrial starch that can, for example, be hydrolyzed to glucose and subsequently fermented to ethanol. The other sorghum types tend to be tall (>2.5 m) and are used to feed animals or to generate fermentable sugars for industrial processes, including the production of biofuels and monomers for bio-degradable polymers. The main goal of plant breeding is the release of improved germplasm that meets the needs of the end user(s) in a particular geographic area. These needs will change over time as a function of the preferences of consumers, new industrial processes, and the identification of novel traits, some of which may result in new uses for the crop. The geographic area where the crop is to be cultivated is an important consideration, because location affects temperature, day length, UV-radiation, precipitation and availability of irrigation, soil properties, and which insects and microbial pathogens present a potential threat to the crop.

This chapter starts with a brief summary of the different breeding methods for cultivars and hybrids, and then provides a general overview of the uses of the different types of sorghum and the target traits in breeding programs. The focus is on traits with global relevance, with the understanding that cultivation in specific geographic areas or for specific industrial purposes will require additional breeding targets that go beyond the scope of this chapter.

8.2 A Brief Review of Cultivar Breeding Techniques for Sorghum

The choice of the breeding techniques applied in any crop depends on the crop's mode of reproduction, genetic constitution (ploidy level), and target traits set for improvement. The mode of reproduction determines the genetic architecture of the crop. Specifically, whether a crop plant is propagated via obligate cross pollination, self-pollination, vegetative propagation, or apomixis impacts how feasible it is to attain homozygosity and how feasible it is to exploit heterosis (hybrid vigor) in breeding programs.

Sorghum is a hermaphrodite plant with the male and female reproductive structures in a single organ, the panicle, which is formed at the top of the stem. Sorghum is a mostly self-pollinated crop with rates of cross pollination averaging 6%, with a range of 5–15% (Poehlman 1987). It is possible to make crosses between different genotypes following manual emasculation (removal of anthers) with tweezers, or via heat-induced anther dehiscence, either with a bowl of warm water or by placing a plastic bag over the panicle for several days, when the panicle has emerged but is not yet flowering (Schertz and Clark 1967). The plastic bag creates a hot, humid environment that causes the pollen to abort, while leaving the ovule unaffected. The resulting male-sterile plant can be pollinated with pollen from the selected male parent. The plastic-bag method is the most common way to emasculate plants in breeding programs in warm climates.

Cultivar breeding involves making a cross between two parents with complementary traits, followed by inbreeding of selected progenies, typically for six generations. At that point, the plants are mostly homozygous and can be propagated by self-pollination. The goal is to generate germplasm in which the desirable traits from both parental lines are represented. This means that the parents need to be selected carefully, from the breeder's existing pool of germplasm, or from one of the global germplasm collections (India, United States, and China). The use of core collections and diversity panels can help identify germplasm with desirable traits or can be used for allele mining, that is, the identification of novel alleles associated with a trait of interest to augment the genetic variation in breeding populations. Useful resources for allele mining include the sorghum association panel (SAP; Casa et al. 2008), the sorghum ICRISAT minicore (Upadhyaya et al. 2009), the bioenergy association panel (BAP; Brenton et al. 2016), and the sweet sorghum diversity panel (SWDP; Cuevas et al. 2023).

The selection of progenies with a combination of desirable traits can occur at different stages in the inbreeding process, as outlined in the next sections. The selected inbred progenies (F5 or F6) need to be subjected to multi-location and/or multi-year yield trials to determine the performance relative to the best available (commercial) germplasm. From the perspective of the breeder and seed company, sorghum cultivars have the drawback that the producer can harvest some of the seed for next season's crop without having to purchase new seed. The use of hybrids circumvents this concern because hybrid seed is generated by crossing two inbred lines, resulting in progenies that are heterozygous at many loci across the genome. As a consequence, the progeny from hybrid plants will segregate and is phenotypically not homogeneous. Hybrid production is discussed in Sect. 8.3.

Below follows a brief summary of common cultivar breeding strategies used for the genetic improvement of sorghum. Some of these breeding strategies can also be applied for the development of the parent lines for hybrid production. The information summarized below is based on two classic plant breeding textbooks, Fehr (1991) and Poehlman (1987). Other useful resources include Doggett (1988), Smith and Frederiksen (2000) and Prabhakar et al. (2022).

8.2.1 Pedigree Method

The pedigree method is a widely used breeding strategy in sorghum (and many other crops), in which two parents with complementary traits are crossed to generate an F_1 population. The F_1 is self-pollinated to generate a segregating F_2 population of at least 1000 plants from which the best 25–40 individual plants are selected and self-pollinated. During the subsequent season, the resulting F_3 seed from each plant is planted in rows, and after selecting the best F_3 families, the best individual plants within those families are self-pollinated to generate F_4 seed. This process continues until the F_6 generation, at which point seed from all plants in a row is bulked and used in replicated yield trials. The best performing F_6 lines will be released as improved cultivars. The drawback of the pedigree method is that the selection of the best families, lines, and individual plants needs to happen in the environment where the improved germplasm is intended to be cultivated in the future. This means that progress is relatively slow (typically one generation per year). In addition, keeping track of the different families and selection entails more record keeping relative to the other methods.

8.2.2 Single-Seed Descent Method

The single-seed descent (SSD) method separates the inbreeding process from the selection of superior genotypes. This approach enables the use of an off-season nursery or greenhouse, so that it takes less time to generate new inbred lines. Just like with the pedigree method, the SSD method relies on generating a segregating F_2 population. In the case of the SSD method, the F_2 plants are self-pollinated and a single $F_{2:3}$ seed is collected from all F_2 plants that produced seed (in reality, a few backup seeds are also collected). This seed is sowed, the resulting plants are self-pollinated, and a single $F_{3:4}$ seed is collected from each plant. The process continues until the desired level of inbreeding has been achieved (typically F_5 or F_6), at which point all the seed from each individual plant is harvested. This seed is sowed in rows in the target environment and this is the moment when the selections are made. Seed from selected rows is bulked and subjected to replicated yield trials. Even though the genetic variability in the populations tends to be high, not every F_2 plant in the population is necessarily represented if seed set on certain plants is poor or seed viability is low. In addition, genotypes are retained without natural selection, resulting in the propagation of seed from which no viable progenies will be selected.

8.2.3 Mass Selection

As with the previous two methods, mass selection starts with a segregating F_2 population. Selections are made based on one or more traits that manifest themselves in the target environment (e.g., disease resistance, plant height, and grain yield), and the seed from the F_2 individuals whose phenotypes were above the threshold value is bulked and (a sample of the selected seed is) sowed in the next season. The process is repeated until the desired level of inbreeding has been achieved (typically F_5 or F_6). The drawback of this method is that the phenotype needs to be expressed and needs to have a sufficiently high heritability in order for the selection to be effective. This means that the use of off-season nurseries and greenhouses is generally not feasible with this method.

8.2.4 Marker-Assisted Introgression

Introgression is a strategy to introduce one or a few desirable traits that are controlled by a limited number of loci in a successful existing cultivar. The strategy involves making an initial cross between the existing cultivar (as the female) and the donor of the trait(s) of interest. The cultivar subsequently serves as the recurrent female parent, first for the F_1 progeny, and then for the resulting backcross (BC) progenies. Each time the progenies are crossed with the recurrent parent, the proportion of the genome of the donor parent is reduced by half, until after four generations on average only 3% of the donor genome is left. By selecting backcross progenies that display the trait of interest, the breeder can ensure the relevant donor alleles are being selected. This becomes complicated for traits controlled by recessive alleles, in which case self-pollinations need to be made after every backcross to identify which progenies harbor the alleles of interest. The use of molecular markers can help streamline the introgression scheme, especially for recessive alleles and for traits that manifest themselves after seed set. The marker-assisted selection (MAS) strategy requires that (1) there is a single donor genotype from which the trait(s) can be obtained, (2) the trait(s) of interest have a known genetic basis that is not complex (i.e., controlled by just one or a few loci), and (3) that there are molecular markers available that are closely linked to the loci that impact the trait of interest. MAS relies on DNA extraction from leaf tissue of individual progeny plants for genotyping with molecular markers. Based on the results, the progenies with the desired alleles are identified and selected for backcrossing to the recurrent parent, typically in combination with visual selection based on similarity to the recurrent parent (e.g., leaf color, leaf angle, plant height, and flowering time).

8.2.5 Genomic Selection

Genomic selection is based on statistical models in which molecular marker data are used to predict the phenotype of the plant (Meuwissen et al. 2001; Heffner et al. 2009). This approach is discussed in detail in Chapters 14, 15, 16, and 17 of this book.

8.3 Sorghum Hybrids

8.3.1 Hybrid Development in Sorghum

Heterosis or hybrid vigor refers to the superior performance of progenies derived from crossing two homozygous parent lines that have contrasting alleles at many loci across the genome. Heterosis is the result of a major contribution of non-additive gene effects (Labroo et al. 2021). Compared to cultivar breeding (Sect. 8.2), the breeding of sorghum hybrids is resource intensive and time consuming because it typically involves the development of three sets of lines: (1) a cytoplasmic male-sterile line (A-line) that serves as the female parent of the hybrid progeny; (2) a maintainer line (B-line) that is genetically identical to the A-line but is male-fertile and used as male to generate seed of the A-line; (3) a fertility restorer line (R-line) that is used to pollinate the A-line to generate hybrid seed. The progenies derived from A × R crosses are fertile, whereas the progenies derived from A × B crosses are male-sterile. In terms of the breeding program, the B- and R-lines are developed through recurrent selection for target traits using one of the methods described for cultivar breeding. The advanced-generation B- and R-lines can be used to assess their suitability as hybrid parents. This will typically involve crosses between an emasculated B-line and an R-line. If the progeny of such a cross appears promising, the B-line needs to be converted to an A-line. This is accomplished by crossing the B-line with an existing A-line that serves as the donor of the male sterility. The F_1 progeny of that cross will be male sterile and is subjected to recurrent backcrossing with the B-line as pollen donor until near-isogenic lines have been obtained that look phenotypically identical but differ in terms of their mitochondrial genomes. Larger breeding programs will generate A- and R-lines as part of routine operations and will generate and evaluate many hybrid combinations in parallel. The subsequent sections describe the A-, B-, and R-lines in greater detail and also discuss the prospects for using nuclear encoded male sterility.

8.3.2 A-Lines

The A-line is the male-sterile line that contains nonfunctional pollen grains. Conner and Karper (1927) first observed this phenomenon of male sterility in sorghum. Seed set happens only after cross pollination with functional pollen grains from a donor parent. Male sterility occurs because of changes in the mitochondrial genome (Sivaramakrishnan et al. 1997). Stephens and Holland (1954) developed the first cytoplasmic male sterility (CMS) system, which they referred to as *milo* cytoplasm and is now commonly referred to as A1 cytoplasm. Alternative sources of CMS were subsequently discovered, including A2, A3, A4, 9E, A5, and A6, and efforts are in progress to utilize them in sorghum improvement programs (Ross and Hackerott 1972; Webster and Singh 1964; Quinby 1981; Schertz and Ritchey 1978; Worstell et al. 1984). A2 (IS12662C or TAM 428) and A3 (IS 1112C) cytoplasm were identified in the United States, A4 cytoplasm (Guntur, VZM, and Maldandi) originated from India, and 9E cytoplasm (a selection made in 9E) originated from Ghana (Reddy et al. 2008). The A2 V4 A cytoplasm-based sorghum hybrid Zinza No. 12 was developed in China and has been released for commercial cultivation (Shan et al. 2000). In India, alternative male-sterile lines developed are based on the genetic background of traditional cultivars such as M35-1A, M31-2A (Maldandi), G1A (Guntur), and VZM2A (Vizianagaram) lines (Aruna and Reddy 2013; Rao et al. 1984). A2 cytoplasm was utilized in multi-cut forage sorghum breeding to develop and release CSH 43MF, the first forage hybrid in India with non-A1 cytoplasm. Alternative cytoplasms are also being used to develop hybrid grain and sweet sorghum. This is prudent in light of the risk of crop failure associated with the widespread use of a single source of CMS, as was evident from the severe yield losses in maize in 1970–1971 due to Southern corn leaf blight (Bruns 2017). Moran and Rooney (2003) developed iso-cytoplasmic A-lines in different genetic backgrounds and reported that the alternative cytoplasms have differential effects on grain yield production in hybrids. The hybrids based on A3 cytoplasm tended to result in lower grain yield, whereas hybrids based on A2 cytoplasm had grain yields comparable to the hybrids based on A1 cytoplasm and could be used as suitable alternative in hybrid seed production in sorghum.

An alternative to the use of three-line system is the use of nuclear male sterility (NMS), which is due to mutations in genes in the nuclear genome. Eight *male sterile* mutants, *ms1* through *ms8*, and the antherless mutant *msal* have been reported in sorghum (Andrews and Webster 1971; Pedersen and Toy 2001; Xin et al. 2018). The seed of five *ms* lines (*ms1, ms2, ms3, ms7*, and *ms8*) and the *msal* mutant are available for use in hybrid development program, whereas the mutants *ms4, ms5*, and *ms6* were lost (Chen et al. 2023). Chen et al. (2023) recently reported two newly identified alleles *ms9-1* and *ms9-3* in the gene *Sobic.002G221000*, which encodes a plant homeodomain transcription factor.

If a plant is heterozygous at an *Ms* locus, the progeny derived from self-pollination will display a 3:1 ratio of segregation for male-fertile:male-sterile plants. An A-line that is male sterile because of homozygosity for one of the *ms* mutations can be maintained indefinitely by crossing it with a heterozygous B-line (*Ms ms*) as maintainer line. That cross generates A- and B-lines in a 1:1 ratio (Fig. 8.1). While this is a more favorable ratio, the segregation for male-sterility still complicates the large-scale commercial production of hybrid seed in which male-sterile lines are used as females.

8.3.3 B-Lines

The B-line is called a maintainer line and used to maintain the sterility of the male-sterile A-line (line). The A- and B-lines are genetically highly similar except for the genes controlling the male sterility, which can either be a cytoplasmic gene or a nuclear gene. In the case of CMS, the progeny of the A × B cross will be 100% A-line because the progeny inherits the mitochondrion from the female parent. In the case of NMS, the B-line will be heterozygous (*Ms ms*) and after crossing with

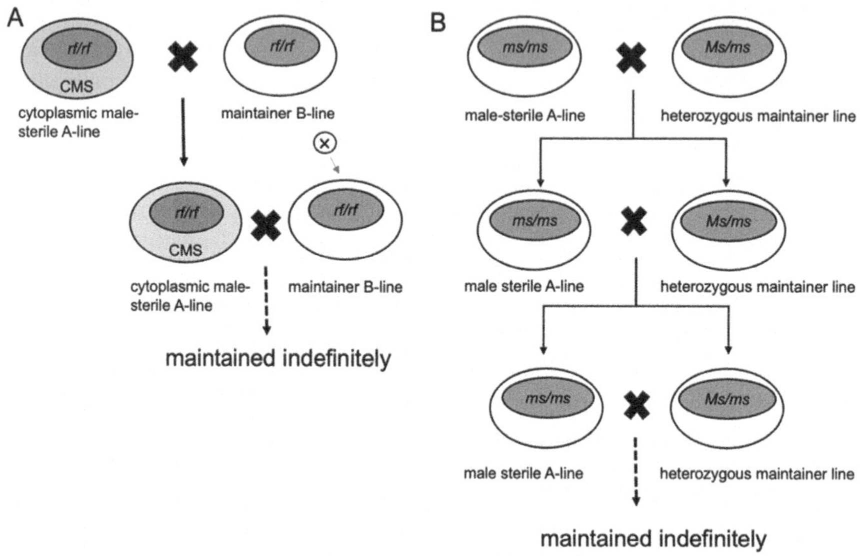

Fig. 8.1 Propagation of male-sterile A-lines. (**a**) A cytoplasmic male-sterile A-line is pollinated by an isogenic maintainer B-line with "fertile" cytoplasm. The nucleus is indicated by the dark gray oval, CMS cytoplasm by the light gray color. The B-line is maintained via self-pollination. Both the A- and B-lines lack a functional copy of a fertility restoring nuclear *Rf* gene. (**b**) A nuclear male-sterile A-line, which harbors two *male sterile* (*ms*) alleles at one of several *ms* loci in the nuclear genome, is maintained by pollinating it with a near-isogenic heterozygous *Ms/ms* maintainer line. The resulting progeny segregates 1:1 for male-sterile (A-line) and male-fertile (B-line) plants

the A-line, the progeny will segregate 1:1 for male fertility and male sterility (Fig. 8.1). Homozygous *ms ms* progenies can be used in hybrid seed production plot with a fertile male plant after removing fertile female plants prior to pollination.

8.3.4 R-Lines

The restorer R-line is used to restore the male fertility in the hybrid progeny derived from the A × R cross. In other words, without the restoration of fertility, the hybrid progenies would be male-sterile and not be able to produce seed via self-pollination. An R-line contains a fertility-restoring nuclear genetic factor (*Rf*) (Fig. 8.2) that compensates for the changes in the mitochondrial genome that underly the CMS. The molecular mechanisms of *Rf* genes in restoring fertility have been dissected and the loci have been mapped on different chromosomes of sorghum. The *Rf* genes encode pentatricopeptide repeat (PPR) proteins that alter the effect of the male sterility genes post-transcriptionally (Klein et al. 2005). The *Rf1* gene is located on chromosome 8 (Klein et al. 2005), *Rf2* on chromosome 2 (Jordan et al. 2010), *Rf5* on chromosome 5 (Jordan et al. 2011), and *Rf6* on chromosome 4 (Praveen et al. 2015). *Rf3* and *Rf4* display complementary gene action leading to gametophytic mode of

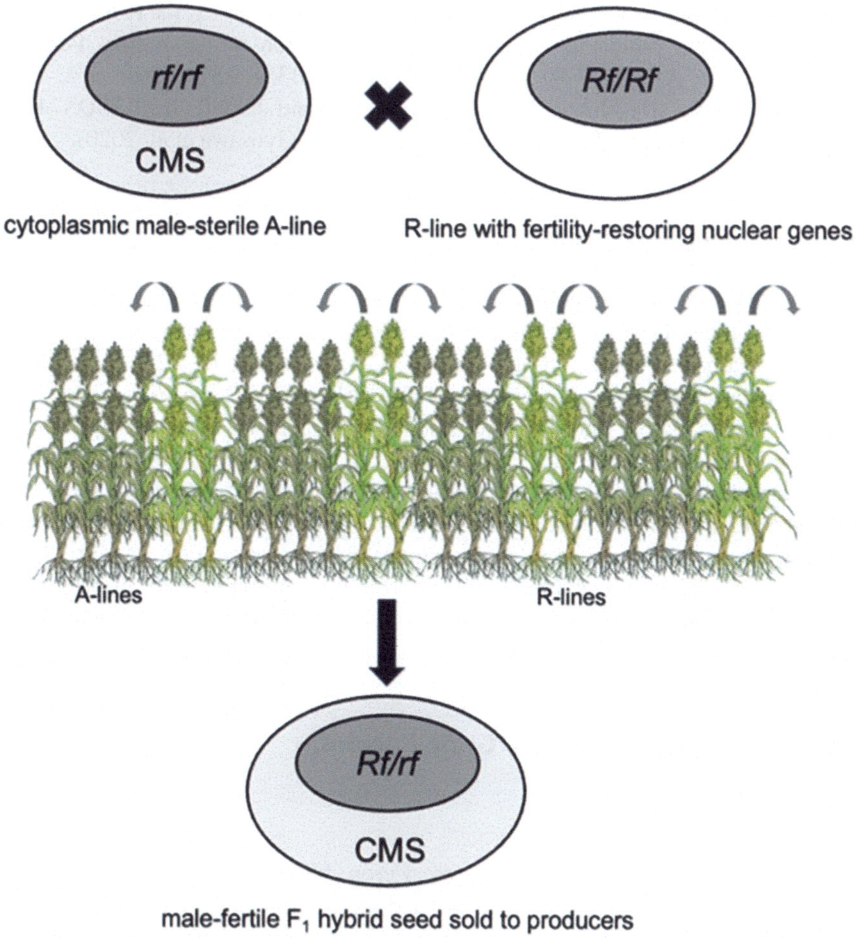

Fig. 8.2 Commercial production of hybrid seed. Alternating rows of cytoplasmic male-sterile A-lines and fertility-restoring R-lines are planted in the field in a ratio of 4:2. The pollen flows from the R-lines to the A-lines (indicated by the arrows), resulting in the production of hybrid seed that is male fertile due to the fertility restoring *Rf* allele in the nucleus. The nucleus is indicated by the dark gray oval, CMS cytoplasm by the light gray color

fertility restoration, specifically when using A3 cytoplasm (Tang et al. 1998). Kiyosawa et al. (2020) identified quantitative trait loci governing fertility restoration in sorghum. The QTL *qRf8* corresponds to the *Rf1* locus on chromosome 8; *qRf5* corresponds to the *Rf5* locus, *qRf2* corresponds to the *Rf2* locus, and *qRf3* corresponds to the *Rf3* locus. Among these *Rf* genes/QTL *Rf1*, *Rf2*, and *Rf5* are most commonly used in hybrid seed production system.

The general combining ability of the R-line should be excellent, so that competitive hybrids can be produced with a host of A-lines. In addition, R-lines should be adaptable to different geographic locations and climatic conditions so that they can be widely used as pollen donors in hybrid seed production. Lastly, the R-line should produce abundant pollen to ensure efficient pollination of the A-line. Popular R-lines include AKR150, RS585, C43, CB11, CB33, RS29, Indore 2, and AKR354 in India (Rakshit and Amasiddha 2019), and JN290, JN43, SDS7444, Chohin237.Daikoukaku, JN503, and JN358 in Japan (Kiyosawa et al. 2020).

8.3.5 Hybrid Seed Production

The F_1 seed produced from pollinating an A-line with an R-line is called hybrid seed (Fig. 8.2). Pollen dispersal from the R-line affects seed set on the A-line, and it is typical to intersperse the A- and R-lines in the field in a ratio of 4:2, meaning four rows of A-lines separated by two rows on R-lines (Fig. 8.2). Frequent monitoring of seed production plots is necessary to maintain the genetic purity and quality of hybrid seed. Removal of off-types ("rogueing") and monitoring of farm practices by a team representing the seed producing entity are necessary to ensure seed yield and quality. Synchrony in flowering between the A- and R-lines is another important factor affecting the seed set on female parent. If lack of synchrony is a problem, spraying with 2% urea or withholding the irrigation on the late flowering parent promotes the parent to enter the reproductive stage and can facilitate synchronous flowering.

8.4 Breeding Targets for Grain Sorghum

8.4.1 Food and Feed

Sorghum is the fifth most cultivated cereal crop, behind wheat, maize, rice, and barley. It is primarily cultivated in the arid and semi-arid regions of the world due to its greater tolerance to heat, salinity, drought, and (periodic) flooding, relative to these other cereals. Sorghum is cultivated in 105 countries representing 41.1 million ha and a total global production of 58.6 million tons of grain, with an average productivity of 1.60 tons/ha (summarized by Prabhakar et al. (2022)). Grain sorghum is grown for subsistence in Asia and Africa while in the Americas, especially the United States, it is largely grown for livestock feed and feedstock for biofuel production. The nutritional quality of sorghum grain, notably flavonoids, is contributing to a greater interest in sorghum as a health food, especially in industrialized nations. The yield potential of sorghum is comparable to those of rice, wheat, and maize, but given the greater propensity of sorghum being cultivated under harsher conditions, the realized yields depend more heavily on the exact growing conditions, especially the availability of water.

Improving grain yield of sorghum is a top priority to ensure a steady supply of food to the increasing world population while the consequences of global climate change pose greater challenges to crop production. Specifically, higher temperatures during the reproductive stages of sorghum are expected to become a major limitation in realizing higher yields. Large opportunities exist to exploit the wealth of genetic diversity in the sorghum world collection to increase productivity. However, the area cultivated with sorghum is on the decline in many parts of the world due to biotic and abiotic stresses including drought, shoot pests, grain mold, and more often, competition from other more lucrative crops including cotton, corn, and soybean, despite the fact these crops have a greater risk of failing and the fact that high-yielding hybrids of sorghum have been developed and replaced up to 90% of the traditional varieties.

As in other cereal crops, grain yield is the most important trait in grain sorghum breeding programs, although stover yield is also important in countries like India. Sorghum grain yield is a complex quantitative trait, influenced by many significant factors: grain weight, the number of grains per panicle, and tiller number (Hao et al. 2021). Mace et al. (2019) highlighted that about 340 quantitative trait loci (QTL) associated with grain yield in sorghum have already been identified. In general, the ideotype of grain sorghum is a plant with high grain yield, a high harvest index, a large number of grains per panicle, high test weight, a large number of leaves, and a large leaf area duration (the time that the combined leaf area of the plant is photosynthetically active). Some of these traits are discussed in greater detail in the sections below.

8.4.1.1 Harvest Index and the Impact of *Dwarfing* Alleles

The harvest index (HI; the proportion of the grain relative to the rest of the plant) of modern cultivars of intensively cultivated grain crops such as wheat, rice, maize, sorghum fall within the range of 0.4–0.6. Hammer and Broad (2003) reported a harvest index between 0.47 and 0.57 for grain sorghum.

The HI of sorghum was impacted significantly by the exploitation of mutations that reduced plant height. The *Dwarfing* genes determine plant height by affecting internode and apex elongation. Four unlinked *Dw* genes have been identified in sorghum based on the mutants *dw1* through *dw4* (Quinby 1974, 1975). Plants homozygous for *dw* alleles at one of the loci will be shorter, and the combination of mutations at multiple loci will display additive effects. This means that the height can be reduced from over 3 m for a plant harboring at least one dominant *Dw* allele at each of the four *Dw* loci to just 60 cm when a plant harbors homozygous recessive *dw* alleles at all four dwarfing loci (Quinby 1967). The *Dwarfing* genes impact the levels of phytohormones in the stem.

The Dw1 protein, encoded by *Sobic.009G228900*, acts as a positive modulator of brassinosteroid signaling by inhibiting the negative regulator Brassinosteroid Insensitive2 (BIN2) (Hirano et al. 2017; Thurber et al. 2013). *Dw2* has been mapped to chromosome 6 (Quinby 1974, 1975; Lin et al. 1995). The locus *Sobic.006G067700* encodes a protein kinase similar to a member of the protein kinase family in Arabidopsis named KIPK (Hilley et al. 2017). Besides plant height, *Dw2* also influences panicle length, seed weight, and leaf area (Graham and Lessman 1966; Pereira and Lee 1995). *Dw3* was the first cloned *dwarfing* gene in sorghum and determined to encode the auxin transporter PGP1/PGP19. *Dw3* is the ortholog of maize *Brachytic2* and *Arabidopsis PGP1* (Multani et al. 2003) and is located on chromosome 7 (Sobic.007G163800) (Brown et al. 2008); Grant et al. 2023). Mutations in the *dw3-reference* allele are unstable because the retrotransposon responsible for disrupting the gene excises spontaneously at a frequency of 0.5–1%, which causes allele to revert to its wild-type allele (Multani et al. 2003). Large fields of sorghum that contain the *dw3-ref* mutation will typically contain a small but noticeable number of taller plants that are otherwise phenotypically identical to the short plants. *Dw4* has not been cloned yet, but it is known to be unlinked to the other three dwarfing loci (Quinby and Karper 1954). Morris et al. (2013) identified a potential location of the *dw4* locus at ~6.6 Mbp on chromosome 6 based on the location of the next most significant peak in genome-wide association study on height and heterozygosity scan. Grant et al. (2023) provided a detailed overview of the different mutant alleles at the three *Dw* loci whose identity has been established.

The source-sink relationship in sorghum also greatly affects the harvest index. Source efficiency is affected during severe stresses such as drought, heat, disease, and insect attack. In cases of sink limitation, genetic enhancement programs for increased grain size, number of grains per unit area, and individual grain weight are needed. The HI could be raised further by reducing the stem and leaf sheath dry matter from current levels, and reallocation of the equivalent amount of fixed carbon to the panicle. Factors that affect the HI include the protein and carbohydrate contents of seeds, and extreme temperatures during crop reproductive development. Crop management can also influence HI, especially delayed sowing, which shortens the length of the vegetative phase and increases HI (Unkovich et al. 2010).

8.4.1.2 Starch and Protein Quality

Starch and protein are the two main components in sorghum grain that account for ~70 and 12% of the grain by dry weight, respectively (Bean et al. 2019). The remainder includes oils, phenolic compounds (flavonoids and tannins), and minerals. Starch in sorghum is comprised of two macromolecules, amylose and amylopectin, in a ratio of about 30:70 (Audilakshmi and Swarnalatha 2018). The starch is deposited in starch granules. The bulk of the protein in sorghum seed is present in the form of seed storage proteins named kafirins, which form protein bodies in the endosperm. From the perspective of human nutrition, the protein quality of sorghum seed is generally considered low due to the low levels of the essential amino acid lysine (Taylor and Schüssler 1986), and the fact that the location of the proteins in protein bodies reduces digestibility, although variation in the structure of protein bodies exist that improves digestibility (Aboubacar et al. 2001). In addition, genome editing approaches targeting *Kafirin* genes show promise in improving protein quality in sorghum (Li et al. 2018).

The overall composition of the grain and the digestibility of starch and protein and can vary widely among different grain sorghums (Rooney and Pflugfelder 1986; Sang et al. 2008; Shewayrga et al. 2012; Sukumaran et al. 2012), which gives promise to advance sorghum biofortification and breeding for specific end-use products. The relative abundance and structure of the macromolecules determine the nutritional properties of the grain and the physico-chemical properties of grain-derived products (Hao et al. 2021). For example, low-amylose sorghums, known as waxy sorghums, have been shown to have increased feed conversion efficiency compared to normal, non-waxy genotypes, and their amylopectin-rich starch is easier to convert by industrial amylases to generate glucose for ethanol production (Yan et al. 2011). Even though selection for nutritive value was not necessarily a major focus of sorghum breeding programs, the average feeding value of grain sorghum is 96–98% that of corn (Hancock 2000).

8.4.2 Production of Ethanol from Sorghum Starch

The bioethanol industry in the United States has developed into a mature technology using first-generation production processes that produced 68.2 billion liters in 2021, with corn starch as the primary substrate (Renewable Fuels Association 2022). The process involves soaking the starch in water, the use of α-amylase and heat (110 °C) to partially hydrolyze the starch to dextrins (liquefaction), followed by complete hydrolysis to glucose with the enzyme glucoamylase (Nichols and Bothast 2008). The glucose is then fermented to ethanol by yeast. The two main industrial processes are dry grinding and wet milling. They differ in the way the grain is processed: The dry grind process is more common and generates distillers dry grain as a co-product that can be used as a feed. Wet mills first separate the starch from other components in the grain and generate a number of (more) valuable co-products (gluten meal, gluten feed, and oil) in addition to the ethanol (Nichols and Bothast 2008).

Sorghum is as suitable as maize for the production of ethanol (Wang et al. 2008), with only a marginally lower yield of ethanol compared to maize (372 vs. 387 L/ton grain). The quantity and composition of the starch and the size of the starch granules affect the viscosity of the gelatized starch and the speed of liquefaction, which contribute to the overall bioconversion efficiency to ethanol. A large amount of genetic variation exists in the sorghum world collection for the above-mentioned traits. The total starch ranges from 50% to 75% for non-waxy sorghum (Yan et al. 2011), while 65–76% has been reported for waxy varieties (Li et al. 2022). Small-scale fermentations indicated that sorghum substrates are good candidates for approaches that deploy consolidated bioprocessing (CBP), whereby the fermenting organism also generates the amylolytic enzymes, with no need for nitrogen supplementation (Cripwell et al. 2020; Malherbe et al. 2023).

Even though corn starch remains the principal bioethanol feedstock in the United States, sorghum may serve as a suitable alternative for new fuel ethanol industries in developing countries, as long as the sorghum production does not compete with the production of food and feed. Regardless, sorghum that has sprouted or that is damaged by insects or mold, can be an alternative feedstock for biorefineries (Chuck-Hernández et al. 2012).

In Africa and India, sorghum grain has traditionally played a major role in the brewing industry. Sorghum malt, which has good proportions of both α- and β-amylases (Malleshi 1987), can be used for various local brews. Bantu beer, a traditional drink in central and southern Africa, is prepared from processed sorghum malt. Clear Beer in Nigeria is produced on a commercial scale through malting and brewing (Olori et al. 1996). Red sorghum malt is used in Burkina Faso for preparation of *dolo*, an opaque beer. Its characteristic pinkish-brown color to the beer is imparted from brown-grained sorghum with soft endosperm.

8.4.3 Priorities for Grain Sorghum Breeding Programs

The major breeding targets for grain sorghum are:

- **Yield Improvement**: The focus should be on developing genotypes with a high grain yield resulting from a combination of a high harvest index, a large number of grains per panicle, high 1000-grain mass (large seeds), producing a large number of leaves, and a greater leaf area duration. In addition, the plant height should be around 1.5–2 m with a maturity of 105–115 days and amenable to mechanical harvesting.
- **Diversification**: The grain yield can be improved beyond the current plateau by broadening the genetic base through the use of diverse germplasm, including diverse sources of cytoplasm for CMS. There should be more emphasis on development of diversified hybrid parents utilizing the racial diversity for exploiting heterosis. The *caudatum* race has been widely exploited in breeding programs, and studies have shown that the *guinea* race can contribute significantly (after *caudatum*) to heterosis and higher grain yield (Aruna and Cheruku 2018).
- **Food**: The important physical grain quality traits that are genetically controlled and influence the milling properties and end-use quality of sorghum are grain size, shape and hardness, grain density, endosperm texture, pericarp thickness, and seed color (Rooney and Miller 1982; Aruna and Audilakshmi 2004; Aruna et al. 2012). Endosperm texture is the property that affects its quality most consistently (Aruna and Cheruku 2018). Other traits to be pursued while developing food-grade sorghum varieties are white pericarp color, thin mesocarp, normal (non-waxy) endosperm, and low tannin content (Tuinstra 2008). The breeding programs have to be formulated based on the genetics underlying these traits, which include starch granule size and shape and amylose/amylopectin ratio (Audilakshmi and Swarnalatha 2018). Protein quality and digestibility can be improved by altering the structure of protein bodies in the endosperm (Aboubacar et al. 2001), with genome editing offering great promise (Li et al. 2018).
- **Feed**: Development of high-yielding sorghums with enhanced levels of lysine and tryptophan would greatly enhance its value for both humans and animals (Venkateswarlu et al. 2018). Emphasis should also be on development of genotypes with large seeds, high in dietary protein and energy, and high digestibility.
- **Biofuel**: The traits associated with high saccharification efficiency, and the traits inhibiting glucose and ethanol production need to be further investigated so that these traits can be incorporated in the breeding program. Known traits include low amylose, higher starch and protein digestibility, soft endosperm, and low tannin content. Both waxy and heterowaxy (intermediate between normal and

waxy) sorghum germplasm needs to be screened for larger germ size and quality and quantity of lipids.

- **Engineering resistance**: Breeding for terminal drought tolerance through development of early maturing sorghum genotypes and incorporation of stay-green and other drought QTL/genes needs attention. Resistance to shoot pests and grain mold needs to be incorporated following molecular dissection of the resistance mechanisms, so that resistance alleles can be incorporated in improved germplasm.
- **Value addition**: Development of cultivars and hybrids with colored grain are of growing interest for use in food, feed, and brewing, and as dietary supplements with health benefits.
- **Mechanical harvesting**: Given the limited availability and the high cost of manual labor in Asia and Africa, and a relatively high incidence of crop lodging, development of locally adapted, short-statured genotypes suitable for mechanical harvesting are needed for these regions, like they are available in the United States, Europe, and Australia.

8.5 Breeding Targets for Forage Sorghums

Unlike with the grain sorghums discussed in Sect. 8.4, the primary focus with forage sorghums is on the vegetative biomass. Forage sorghums serve as fodder for animals, mostly cattle. The biomass these animals consume consists of 40–45% cellulose, 30–35% hemicellulosic polysaccharides, 15–20% lignin, 5% starch, with the remaining ~5% consisting of soluble sugars, proteins, minerals, and epicuticular waxes. The variation in composition is a function of the genotype, cultivation conditions and developmental stage of the plants at the time of harvest or consumption.

High-quality fodder is offered to maximize milk and meat production, which can be accomplished by ensuring high intake, and once the animals have consumed the forage, efficient conversion. This means that forage needs to be palatable (appealing to eat) and easily digestible, so that the animals will want to eat more. Since forage sorghums are cultivated to generate biomass, biomass yield is the third major breeding objective. Taken together, the ideotype of a forage sorghum is a tall plant with a relatively thin stem, multiple tillers, stems containing a high concentration of cell wall polysaccharides and a low concentration of lignin. The softer texture of cell walls with reduced lignin concentration and the presence of soluble sugars in the stems will enhance palatability. Given that sweet sorghums (Sect. 8.7) contain high concentrations of soluble sugars in their stems, there is some overlap among forage sorghums (defined by their end use) and sweet sorghums (defined by their physiology).

Resistance to insect pests and fungal diseases are considered important traits for all sorghums, to limit yield losses and ensure quality. In the case of forage sorghums, there is an inherent risk associated with the presence of disease that can result in the formation of toxic compounds. Of specific concern is ergot, an infection of the panicle with the fungal pathogen *Claviceps africana*, which produces toxic alkaloids that constrict blood vessels in cattle, cause lameness, gangrene, negatively impact milk production, and reduce fertility (Klotz 2015).

Sorghum produces the cyanogenic glucoside dhurrin [(*S*)-(β-D-glucopyranosyloxy)-(4-hydroxyphenyl)acetonitrile] as a defense compound. As insects or fungi attack the plant, dhurrin comes in contact with hydrolytic enzymes that catalyze the formation of the highly toxic hydrogen cyanide (HCN), also known as prussic acid. While this is an effective defense mechanism to protect the plant, there is an inherent risk to animals that consume forage high in dhurrin because HCN will prevent the utilization of oxygen by cells, which can lead to death by asphyxiation. The risk of prussic acid in sorghum was reported by Slade (1903) and remains a concern. Dhurrin levels will vary among sorghum genotypes, tend to be higher in young plants, especially the new leaves, and frost damage, drought, CO_2 levels, and the fertilizer regimen can influence the risk of exposure to prussic acid (Gleadow et al. 2016; Gruss et al. 2022).

8.5.1 Digestibility

The *brown midrib* (*bmr*) mutations have been used effectively to enhance digestibility, by reducing the concentration of lignin in the biomass. The *bmr* mutants were first reported by Porter et al. (1978) following chemical mutagenesis of sorghum with EMS. Additional *bmr* mutants were identified in a TILLiNG population generated by the USDA-ARS in Lubbock, Texas (Xin et al. 2008). The mutants, *bmr6* and *bmr12*, have been widely used to enhance forage quality (Oliver et al. 2005). *Bmr12* was the first *Bmr* gene that was cloned. It was shown to encode the enzyme caffeic acid *O*-methyl transferase (or more correctly, 5-hydroxyconiferyl alcohol/aldehyde *O*-methyltransferase), an enzyme involved in the biosynthesis of monolignols (Bout and Vermerris 2003). Monolignols are hydroxcinnamyl alcohols and related compounds that are exported to the cell wall where they undergo oxidative coupling to form lignin (Vanholme et al. 2010). The *Bmr6* gene was subsequently cloned (Saballos et al. 2009; Sattler et al. 2009) and shown to encode cinnamyl alcohol dehydrogenase2 (CAD2), the enzyme involved in the final step of monolignol biosynthesis. As a result of the reduced enzyme activities, the lignin subunit composition of both the *bmr6* and *bmr12* mutant is altered, with a lower proportion of syringyl residues, derived from sinapyl alcohol, relative to guaiacyl residues, derived from coniferyl alcohol.

The naming of the *bmr* mutants has been cause for confusion because the initial mutants identified by Porter et al. (1978) were given consecutive numbers without knowing their allelic relationships. Based on the allelism studies by Saballos et al. (2008) and Sattler et al. (2014) a total of eight independent *bmr* loci have been identified. Each locus is represented by multiple alleles (Saballos et al. 2008; Sattler et al. 2012, 2014; Scully et al. 2016; Adeyanju et al. 2021). In order to identify the allelic relationships while maintaining a connection to the original mutants reported by Porter et al. (1978), Saballos et al. (2008) proposed a new nomenclature in which the numbers associated with the most widely used mutants were retained and turned into reference alleles (e.g., *bmr6-ref and bmr12-ref*). The mutant *bmr3* was shown to be allelic with *bmr6* and was renamed *bmr6-3*. Similarly, the *bmr7* mutant was renamed to *bmr12-7*. Table 8.1 lists the currently known *bmr* loci, their function (when known), and what is known about the impact of the mutation on lignin, forage quality and the efficiency of enzymatic saccharification of dried sorghum biomass.

8.5.2 Biomass Yield and Palatability

Biomass yield can be enhanced with the use of photoperiod-sensitive sorghums. These sorghums continue to grow vegetatively until the daylength is shorter than approximately 12 h (fall equinox), at which point they start to transition to the reproductive phase. The photoperiod sensitivity in sorghum is controlled by the *maturity* (*ma*) loci, of which six have been identified. This is further discussed in Sect. 8.6.1.1.

Palatability appears to be enhanced by the *brown midrib* mutations, with the reduction in lignin concentration resulting in softer biomass that is easier to chew. This feature is evident when cutting through the midrib of a mature leaf with a scalpel, which is much easier with leaves from *bmr* mutants compared to wild-type leaves. A second approach to improving palatability is the use of sweet sorghums, as discussed in further detail in Sect. 8.7.

8.6 Breeding Targets for High-Biomass Sorghums

High-biomass sorghums are cultivated to generate biomass intended to be processed at biorefineries that produce renewable chemicals and fuels (Ragauskas et al. 2006; Rooney et al. 2007). The most common way of generating these products is via fermentation of sugars obtained from hydrolysis of cell wall polysaccharides. This first requires the biomass to be ground and subjected to a thermo-chemical pretreatment. Several types of pretreatment exist in which heat and chemical agents, sometimes combined with high pressure are applied to disrupt the ordered structure of the plant cell wall (reviewed by Constant et al. (2016), Hu and Ragauskas (2012), Pu et al. (2015)). Following pretreatment, the biomass is subjected to enzymatic saccharification with cellulolytic enzymes obtained from fungi. This results in the production of D-glucose, which is subsequently fermented by one of several microbial species (e.g., *Saccharomyces cerevisiae*, *Escherichia coli*, *Pichia stipitis*, *Bacillus coagulans*, and *Clostridium thermobutyricum*) (Huang et al. 2009; Jang et al. 2012; Karimi et al. 2006; Ou et al. 2015; Yu et al. 2007). The hydrolysis of hemicellulosic polysaccharides present in the cell walls of grasses, including sorghum, results in the formation of the pentoses xylose and arabinose, which can also be fermented, although in the case of *S. cerevisiae*, it requires the use of recombinant strains that have been engineered to be able to convert arabinose and/or xylose (Ho et al. 1998). The fermentation products (e.g., ethanol, butanol, lactic acid, and butyric acid) need

Table 8.1 The sorghum *brown midrib* mutations and their impact on lignin composition, digestibility, and enzymatic saccharification of biomass

Gene	Enzyme	Source	Impact on lignin	Impact on digestibility	Impact on saccharification
Bmr2	4-coumarate-CoA ligase	Saballos et al. (2012)	Reduction in lignin concentration (Saballos et al. 2008)	Greater digestibility (Porter et al. 1978)	More efficient saccharification (Saballos et al. 2008)
Bmr6	Cinnamyl alcohol dehydrogenase	Saballos et al. (2009) Sattler et al. (2009)	Reduction in lignin concentration; (Saballos et al. 2008) S/G reduced (Pillonel et al. 1991)	Greater digestibility (Porter et al. 1978; Oliver et al. 2005)	More efficient saccharification (Vermerris et al. 2007)
Bmr12	Caffeic acid *O*-methyltransferase	Bout and Vermerris (2003)	Reduction in lignin concentration; S/G reduced; (Bout and Vermerris 2003)	Greater digestibility (Porter et al. 1978; Oliver et al. 2005)	More efficient saccharification (Vermerris et al. 2007)
Bmr19	Folylpolyglutamate synthase	Adeyanju et al. (2021)	Reduction in lignin concentration; (Saballos et al. 2008)	Greater digestibility (Porter et al. 1978)	Less efficient saccharification (Saballos et al. 2008)

Bmr29	Unknown	Sattler et al. (2014)	Reduction in lignin concentration; (Sattler et al. 2014)	Unknown	No significant impact (Sattler et al. 2014)
Bmr30	Chalcone isomerase	Tetreault et al. (2021)	Reduction in lignin concentration; (Tetreault et al. 2021)	Unknown	No significant impact (Sattler et al. 2014)
Bmr31	Unknown	Sattler et al. (2014)	No significant reduction in lignin concentration; (Sattler et al. 2014)	Unknown	No significant impact (Sattler et al. 2014)
Bmr32	Unknown	Sattler et al. (2014)	No significant reduction in lignin concentration; (Sattler et al. 2014)	Unknown	No significant impact (Sattler et al. 2014)

to be removed from the fermentation broth via, for example, distillation, before their use as fuel or building block for polymers. Lignin present in the biomass ends up as a residue that can be used as boiler fuel to generate heat for the fermentation, to generate steam for the generation of electricity, or for the production of a variety of chemicals and polymers (Stewart 2008; Grossman and Vermerris 2019).

In the context of the above-described biorefinery process, the target of breeding high-biomass sorghums is to maximize the yield of fermentable sugars on a per-area basis. Furthermore, given the focus on sustainable production of renewable fuels and chemicals, high-biomass sorghums are ideally able to be cultivated on low-fertility land with minimal inputs of water, fertilizer, and chemicals (fungicides and insecticides). Consequently, the ideotype of a high-biomass sorghum is a plant taller than 3.5 m with thick stems containing a high concentration of cell wall polysaccharides and a relatively low concentration of lignin, with resistance against pests and diseases common in the area of production, and tolerance to suboptimal abiotic conditions (periodic flooding, periods of drought, and high temperatures) (Silva and Vermerris 2020). Under optimal conditions (genotype, environment, and management), dry biomass yields as high as 60 Mg/ha have been reported (Olson et al. 2012; Snider et al. 2012), with 25–35 Mg/ha being achievable under less optimal growing conditions typically envisioned for high-biomass sorghums.

8.6.1 Biomass Yield

Given that panicles are not considered important for the use in biorefineries and that leaves make a small contribution to overall biomass yield, especially at the end of the season, stem volume is the key parameter determining biomass yield. Stem volume can be maximized by increasing plant height and stem diameter. Plant height is controlled by maturity (photoperiod sensing) and the activity of plant growth regulators, some of which are controlled by the *Dwarf* genes. A benefit of increasing the stem diameter is that it reduces the proportion of the sclerenchyma fibers in the rind tissue (the perimeter of the stem), which are highly lignified and contribute to biomass recalcitrance in the biorefinery, relative to the parenchyma cells that make up the bulk of the center of the stem.

8.6.1.1 *Maturity* Genes

Sorghum is naturally a short-day plant, and the trigger to transition to the reproductive phase occurs when daylength is shorter than 12 h and 16 min. The genes that control floral initiation in sorghum are called Maturity (*Ma*) genes. Plants with recessive alleles in the maturity genes are photoperiod insensitive and behave like a long-day plant. The use of these recessive alleles enabled the spread of sorghum to temperate regions, such as the United States. The first maturity loci identified were *Ma1*, *Ma2*, *Ma3*, and *Ma4* (Quinby 1967), followed by the identification of *Ma*5 and *Ma6* (Rooney and Aydin 1999). Grant et al. (2023) summarized the genomic location of the *Ma* genes and documented genetic variation at these loci based on analysis of diverse sorghum germplasm (Table 8.2).

Table 8.2 *Maturity* genes, gene identifiers and function, based on Grant et al. (2023)

Gene	Gene identifier	Protein	Source
Ma1	*Sobic.006G057866*	Pseudoresponse regulator protein 37	Murphy et al. (2011)
Ma2	*Sobic.002G302700*	Protein containing a SET (Suppressor of variegation, Enhancer of Zeste, Trithorax) and MYND (Myeloid-Nervy-DEAF1) domain	Casto et al. (2019)
Ma3	*Sobic.001G394400*	Apoprotein of phytochrome B	Childs et al. (1997)
Ma4	Unknown	Unknown	
Ma5	*Sobic.001G087100*	Apoprotein of phytochrome C	Yang et al. (2014)
Ma6	*Sobic.006G004400*	Grain Number, Plant Height and Heading Date7 (Gdh7)	Murphy et al. (2014)

When a sorghum plant contains dominant *Ma5* and *Ma6* alleles, the floral initiation is inhibited, independently of day length (Childs et al. 1997). The observation on the flowering behavior of plants carrying a dominant *Ma5* or *Ma6* allele combined with the notion that there is complementary dominant epistasis between *Ma5* and *Ma6* led Rooney and Aydin (1999) to propose a method for producing photoperiod-sensitive hybrids from photoperiod-*in*sensitive inbred parent lines. Specifically, the genotype of one of the parents is *Ma5/Ma5*; *ma6/ma6* and the genotype of the other parent is *ma5/ma5*; *Ma6/Ma6*. The resulting progeny is heterozygous at both loci and photoperiod-sensitive. Implementation of this approach enables the production of hybrid seed in temperate climates, and avoids the reliance on a winter nursery to produce seed.

8.6.2 Biomass Composition

Since the *bmr* mutations have an impact on cell wall composition, they have been of interest for use in high-biomass sorghums. The enzymatic saccharification of cell wall polysaccharides in a biorefinery relies on the use of cellulolytic enzymes. The presence of lignin in biomass reduces the efficacy of these enzymes: Not only does the lignin shield the cellulose, the enzymes will also irreversibly adsorb on the lignin and become inactive (Zeng et al. 2014). Hence, biomass of *brown midrib* mutants with a reduced concentration of lignin in the cell wall has the potential to yield more fermentable sugars derived from cell wall polysaccharides compared to biomass from wild-type plants, with the same enzyme loading and same amount of hydrolysis time. Vermerris et al. (2007) reported this to be the case for biomass from both *bmr6* and *bmr12* mutants. This was subsequently confirmed by Saballos et al. (2008) and Dien et al. (2009). The latter authors included a *bmr6-bmr12* double mutant in their study, which showed additive effects.

A concern with the use of *bmr* mutants is the reduction in biomass yield. In a 2-year study by Cotton et al. (2013), the *bmr12* forage sorghum cultivar "PaceSetter" was compared with the non-*bmr* forage cultivar "SorghumPartners 1990" under both rainfed conditions and with supplemental irrigation in western Texas, United States. Biomass yield, biomass composition, and the theoretic ethanol yield were calculated and comparisons between treatments and cultivars were made. The two cultivars were not near-isogenic lines, so the impact of the *bmr12* mutation could not be isolated from other genetic differences between the cultivars. Irrigation tended to boost biomass yield, and the non-*bmr* cultivar produced consistently more biomass. Consistent with the observation by Vermerris et al. (2007), the biomass from the *bmr12* cultivar was more amenable to enzymatic saccharification, resulting in greater yields of glucose per gram biomass. Given that the theoretic ethanol yield is a function of both cellulose yield and the efficiency of cellulose-to-glucose conversion, the theoretic ethanol yield of the *bmr12* cultivar was similar to the non-*bmr* cultivar under most conditions. However, when water was not as limited because of higher-than-average rainfall, the theoretic ethanol yield of the non-*bmr* cultivar was higher, indicating the enhanced cellulose conversion efficiency due to the *bmr12* mutation was not able to compensate for the lower biomass yield.

A challenge with the use of the *bmr* mutants as biomass feedstock for the production of fermentable sugars is that the effects of lignin concentration and lignin subunit composition are confounded. An initial indication that a reduced lignin concentration per se is not necessarily predictive of improved yields of fermentable sugars following enzymatic saccharification came from the observation that the glucose yield from *bmr19* stover was lower than that of the wild-type control (Saballos et al. 2008). Sattler et al. (2012) further examined the role of lignin concentration and lignin subunit composition by comparing the properties of near-isogenic mutants harboring different *bmr12* alleles whose impact on lignin concentration and lignin subunit composition varied. The study included the wild-type BTx623, the *bmr12-ref* mutant, which is a null mutant (Bout and Vermerris 2003) with a reduced proportion of syringyl (S) residues in the lignin, and three EMS mutants from the collection of Xin et al. (2008), *bmr12-30*, *bmr12-34*, and *bmr12-35*, whose impact on cell wall composition was intermediate. Based on their data, the S/G ratio, an indicator of lignin subunit composition, was more predictive of the glucose yield than the lignin concentration. This means that it may be possible to generate sorghums that offer superior processing characteristics without compromising agronomic properties, by modifying lignin subunit composition while maintaining lignin concentration at wild-type levels. Table 8.1 lists the impact of the other mutations on the efficiency of biomass conversion to fermentable sugars.

8.7 Breeding Targets for Sweet Sorghums

Sweet sorghums are defined as sorghums that contain juicy stems with a high concentration of soluble sugars. It is common to determine the latter with a refractometer, which determines the concentration of soluble solids in °Brix. The definition of 1°Brix is 1 g soluble solids per 100 g liquid. Sorghums are considered sweet sorghums when the stem juice has a refractometer value >12°Brix, although that is considered low from a commercial perspective. Typical values are 15–18°Brix, with maxima around 24°Brix. The actual values at harvest are a function of genotype, management, developmental age of the plants, and the irrigation status. It is common for sugar concentrations to decrease following heavy rain because the uptake of the water diluted the soluble sugars in the stem. Sweet sorghums were originally used as a source of sweeteners in the form of syrup for human consumption. The presence of glucose in the sorghum juice prevented the formation of crystalline sucrose during the concentration of the juice by boiling. The widespread cultivation of sugarcane and sugarbeet as sources of crystalline sucrose has made sorghum syrup a niche product. Sweet sorghum is currently cultivated as a forage and as a source of fermentable sugars for the production of renewable fuels and chemicals. The exploration of sweet sorghum for the production of fuel ethanol received considerable attention following the oil crisis in 1973, and the desire to become less reliant on imported oil in 2007 (Vermerris 2011). In the United States, the

U.S. Department of Agriculture (USDA) supported a large sweet sorghum breeding program located in Meridian, Mississippi. The station is now closed, but much of the germplasm that was generated there is available via the plant germplasm collection maintained by the USDA (https://www.ars-grin.gov). A subset of the sweet sorghums developed in Meridian is included in the sweet sorghum diversity panel (Cuevas et al. 2023).

8.7.1 Sweet Sorghums as Forage

Due to its height and high concentration of soluble sugars, sweet sorghum exhibits excellent biomass yield and forage quality, and due to its superior adaptability to growing conditions on marginal land (Tang et al. 2017), it could replace maize as a livestock feed (Zerbini and Thomas 2003; Mishra et al. 2017). The presence of the soluble sugars enhances the palatabaility of sweet sorghum, which is highly preferred by livestock (Inal et al. 2021). Good fiber quality combined with high protein content makes sweet sorghum an excellent forage crop, especially as silage for ruminants. In this case, the biomass is chopped and stored under low oxygen conditions, which will result in the fermentation of some of the soluble sugars by microorganisms that are naturally present in the biomass, or that can be added to the chopped biomass as an inoculum. The highest protein content in sweet sorghum is reportedly in the leaves, and can be as high as 22% (Ferraris and Charles-Edwards 1986). Based on a meta-analysis of feeding experiments, cows fed with sweet sorghum silage produced, on average, more milk (1.64 kg/day/cow) with better quality compared to cows fed with maize silage (Sanchez-Duarte et al. 2019).

In sweet sorghum, substantial genotypic variation has been reported for both biomass yield and quality traits. Forage yield is a function of plant height, stem volume, moisture content, tiller number, the number of leaves, leaf area, the ratio of leaf length and width, juice volume, and sugar concentration. In addition, the composition of the cell wall can vary considerably, especially in terms of the lignin concentration due to the incorporation of *brown midrib* mutations, as discussed in Sect. 8.5.1. Heritability is high for plant height, flowering time, and test weight. Farmers benefited from ratoon cropping of sweet sorghum as a convenient double cropping option. A study by Rolz et al. (2014) reported that with two harvests per year, the fresh stalk biomass productivity was the same for the first harvest and the first ratoon crop.

8.7.2 Sweet Sorghum as a Dual-Use Bioenergy Crop: Sugars and Biomass

Sweet sorghum yields both directly fermentable sugars present in the juice and large amounts of bagasse that can be processed as lignocellulosic feedstock to generate fermentable sugars derived from cell wall polysaccharides. The sugars from both sugar streams can be fermented to a variety of products, including monomers for biodegradable plastics and fuel ethanol. As such, sweet sorghum can be cultivated as an alternative to or complement to sugarcane. Under water-limited conditions, sweet sorghum is reliably more productive than sugarcane. Given the desire to avoid competition with food and feed production, sweet sorghum's ability to be cultivated on marginal land that is not suitable for the production of food crops, make it an attractive feedstock for renewable fuels and chemicals (Vermerris and Saballos 2012; Umakanth et al. 2018). One of the concerns is that the sugars in the sweet sorghum juice are prone to undergo microbial degradation at considerably greater rates than is observed for sugarcane. In practical terms, this means that cut stems should not be kept in storage for prolonged periods of time, and that the time between harvest and processing needs to be minimized. The addition of sodium benzoate and sorbic acid at 1000 ppm can slow down the degradation of sugars (Ganesh Kumar et al. 2015). As a consequence, from the perspective of commercial production, it is preferred to have sweet sorghums with a sizeable window during which the yield of fermentable sugars is close to the maximum, so that the harvest can occur in stages over a period of 1–3 weeks, enabling immediate extraction of sugars from the freshly harvested stalks at the sugar mill or biorefinery.

Sweet sorghum can be genetically improved through traditional breeding as well as genomic approaches. There is considerable natural intra-species variation for sugar content in sweet sorghum; it can be useful in identifying genes linked with sugar content and reduced cellulose. The sorghum world collection possesses wide genetic variation for key biofuel traits like total soluble sugars, fresh stalk yield, juice volume, and biomass yield, which helps in development of parental lines and cultivars adapted to different situations (Saballos 2008).

Sweet sorghum bagasse can be considered as a lignocellulosic feedstock for the production of ethanol, similar to high-biomass sorghums (Sect. 8.6). The bagasse needs to be pretreated and subjected to enzymatic saccharification in order to generate fermentable sugars from cellulose and hemicellulosic polysaccharides, which will include pentoses that require the use of suitable fermenting organisms. Often, the pretreatment results in compounds that are toxic to the fermenting microbes. Introgression of the *bmr* mutations (Sect. 8.5.1) into sweet sorghum will enhance the utility of sweet sorghum as a dual-purpose bioenergy crop that can generate fermentable sugars from both juice and bagasse with higher biomass conversion efficiency.

In most of the sweet sorghum breeding programs, emphasis is laid on development of sweet sorghum as a novel dual-purpose bioenergy feedstock that can be utilized for both sugar-based and cellulosic biofuels. The ideal sweet sorghum genotypes produce high yields of soluble sugars and high biomass.

8.7.3 Priorities for Sweet Sorghum Breeding Programs

The ideal characters for sweet sorghum used as forage are as follows (Mishra et al. 2017):

- Tall plants with a high leaf-to-stem ratio to maximize protein content
- Succulent stalks that are highly palatable
- Amenability for ratooning to maximize the fodder yield per growing season
- Photoperiod insensitivity for year-round cultivation when climate permits
- Early flowering to enhance compatibility with other crops
- Resistance to leaf diseases and pests to minimize the intake of harmful compounds
- Adaptability to various agro-climatic conditions and different soils

  The ideal characters for sweet sorghum used as dual-use bioenergy crop are:
- High biomass yield and high sugar content in the stalks
- Juice of high quality (low concentrations of starch and inhibitory compounds) and stability
- Tolerance to biotic stresses (shoot pests and leaf diseases) and abiotic stresses (heat, cold, and drought) to minimize loss of fermentable sugars and negative effects on biomass yield

The availability of germplasm with variable maturities and photo- and thermo-insensitivity will extend the harvest window and facilitate the cultivation of sweet sorghum in different cropping systems. The above-mentioned targets can be met through extensive germplasm screening coupled with informed breeding efforts, genetic and genomic resources, optimization of plant transformation and engineering strategies, cross utilization of information from other closely related species, and a well-defined strategy.

References

Aboubacar A, Axtell JD, Huang CP, Hamaker BR (2001) A rapid protein digestibility assay for identifying highly digestible sorghum lines. Cereal Chem 78:160–165

Adeyanju AO, Sattler SE, Rich PJ, Rivera-Burgos LA, Xu X et al (2021) Sorghum *brown midrib19* (*Bmr19*) gene links lignin biosynthesis to folate metabolism. Genes (Basel) 12:660. https://doi.org/10.3390/genes12050660

Andrews DJ, Webster OJ (1971) A new factor for genetic male sterility in *Sorghum bicolor* (L.) Moench. Crop Sci 11(2):308–309. https://doi.org/10.2135/cropsci1971.0011183x001100020046x

Aruna C, Audilakshmi S (2004) Genetic architecture of grain hardness—a durable resistance mechanism for grain molds in sorghum (*Sorghum bicolor* (L.) Moench). Indian J Genet Plant Breed 64:35–38

Aruna C, Cheruku D (2018) Genetic improvement of grain sorghum. In: Aruna C, Visarada KBRS, Venkatesh Bhat B, Tonapi VA (eds) Breeding sorghum for diverse end uses. Woodhead Publishing, Duxford, pp 157–173. ISBN 9780081018804.

Aruna C, Reddy SP (2013) Cytoplasmic diversification in sorghum hybrid breeding. In: Rakshit S, Patil JV (eds) Basics of sorghum breeding and AICSIP data management. Directorate of Sorghum Research, Hyderabad, p 28e34

Aruna C, Audilakshmi S, Ratnavathi CV, Patil JV (2012) Grain quality improvement of rainy season sorghums. Qual Assur Saf Crop Foods 4(3):149. https://doi.org/10.1111/j.1757-837x.2012.00157.x. Special issue: first ICC India grains conference

Audilakshmi S, Swarnalatha M (2018) Sorghum for starch and grain ethanol. In: Aruna C, Visarada KBRS, Venkatesh Bhat B, Tonapi VA (eds) Breeding sorghum for diverse end uses. Woodhead publishing series in food science, technology and nutrition. Woodhead Publishing, Duxford, pp 239–254. ISBN 9780081018804

Bean SR, Zhu L, Smith BM, Wilson JD, Ioerger BP, Tilley M (2019) Starch and protein chemistry and functional properties. In: Taylor JRN, Duodu KG (eds), Sorghum and Millets, AACC International Press, 2nd edition, pp 131–170

Bout S, Vermerris W (2003) A candidate-gene approach to clone the sorghum *Brown midrib* gene encoding caffeic acid *O*-methyltransferase. Mol Genet Genomics 269:205–214. https://doi.org/10.1007/s00438-003-0824-4

Brenton ZW, Cooper EA, Myers MT, Boyles RE, Shakoor N et al (2016) A genomic resource for the development, improvement, and exploitation of sorghum for bioenergy. Genetics 204:21–33

Brown PJ, Rooney WL, Franks C, Kresovich S (2008) Efficient mapping of plant height quantitative trait loci in a sorghum association population with introgressed *dwarfing* genes. Genetics 637:629–637

Bruns HA (2017) Southern corn leaf blight: a story worth retelling. Agron. J. 109:1218–1224. https://doi.org/10.2134/csa2017.62.0806

Casa AM, Pressoir G, Brown PJ, Mitchell SE, Rooney WL et al (2008) Community resources and strategies for association mapping in sorghum. Crop Sci 48:30–40

Casto AL, Mattison AJ, Olson SN, Thakran M, Rooney WL et al (2019) *Maturity2*, a novel regulator of flowering time in *Sorghum bicolor*, increases expression of SbPRR37 and SbCO in long days delaying flowering. PLoS One 14:e0212154

Chen J, Haydee L, Gloria B, Chand H, Burke JJ, Yves Y, Zhangou X (2023) Registration of two novel grain sorghum nuclear male sterile mutants: BTx623ms9-1 and BTx623ms9-3. J Plant Registr 17:209–213

Childs KL, Miller FR, Cordonnier-Pratt M-M, Pratt LH, Morgan PW et al (1997) The Sorghum photoperiod sensitivity gene, *Ma3*, encodes a phytochrome B. Plant Physiol 113:611–619

Chuck-Hernández C, García-Lara S, Serna-Saldívar SO (2012) Conversion into bioethanol of insect (*Sitophilus zeamais* Motschulsky), mold (*Aspergillus flavus* Link) and sprout-damaged maize (*Zea mays* L.) and sorghum (*Sorghum bicolor* L. Moench). J Cereal Sci 55:285–292

Conner AB, Karper RE (1927) Hybrid vigour in sorghum. Texas Agril Expt Sta Bull 359:21–61

Constant S, Wienk HLJ, Frissen AE, De Peinder P, Boelens R et al (2016) New insights into the structure and composition of technical lignins: a comparative characterisation study. Green Chem 18:2651–2665

Cotton J, Burow G, Acosta-Martinez V, Moore-Kucera J (2013) Biomass and cellulosic ethanol production of forage sorghum under limited water conditions. Bioenergy Res 6:711–718

Cripwell RA, Favaro L, Viljoen-Bloom M, Van Zyl WH (2020) Consolidated bioprocessing of raw starch to ethanol by Saccharomyces cerevisiae: achievements and challenges. Biotechnol Adv 42(107):579

Cuevas HE, Knoll JE, Prom LK, Stutts LR, Vermerris W (2023) Genetic diversity, population structure and anthracnose resistance response in a novel sweet sorghum diversity panel. Front Plant Sci 14:1249555. https://doi.org/10.3389/fpls.2023.1249555

Dien BS, Sarath G, Pedersen JF, Sattler SE, Chen H et al (2009) Improved sugar conversion and ethanol yield for forage sorghum (*Sorghum bicolor* (L.) Moench) lines with reduced lignin contents. Bioenergy Res 2:153–164

Doggett H (1988) Sorghum. Wiley, New York

Fehr WR (1991) Principles of cultivar development. Macmillan Publishing Company, New York. 536 pp

Ferraris R, Charles-Edwards DA (1986) A comparative analysis of the growth of sweet and forage sorghum crops. II. Accumulation of soluble carbohydrates and nitrogen. Aust J Agric Res 37:513–522

Ganesh Kumar C, Srinivasa Rao P, Gupta S, Malapaka J, Kamal A (2015) Chemical preservatives-based storage studies and ethanol production from juice of sweet sorghum cultivar, ICSV 93046. Sugar Tech 17:404–411

Gleadow RM, Ottman MJ, Kimball BA, Wall GW, Pinter PJ et al (2016) Drought-induced changes in nitrogen partitioning between cyanide and nitrate in leaves and stems of sorghum grown at elevated CO_2 are age dependent. Field Crops Res 185:97–102

Graham D, Lessman KJ (1966) Effect of height on yield and yield components of two isogenic lines of *Sorghum vulgare* Pers. Crop Sci 6:372–374

Grant NP, Toy JJ, Funnell-Harris DL, Sattler SE (2023) Deleterious mutations predicted in the sorghum (*Sorghum bicolor*) *Maturity* (*Ma*) and *Dwarf* (*Dw*) genes from whole-genome resequencing. Sci Rep 13(13):16638. https://doi.org/10.1038/s41598-023-42306-8

Grossman A, Vermerris W (2019) Lignin-based polymers and nanomaterials. Curr Opin Biotechnol 56:112–120

Gruss SM, Johnson KD, Johnson SK, Holman JD, Obour A et al (2022) Managing the prussic acid hazard in sorghum. Purdue University Extension. https://www.extension.purdue.edu/extmedia/AY/AY-378-W.pdf

Hammer GL, Broad IJ (2003) Genotype and environment effects on dynamics of harvest index during grain filling in sorghum. Agron J 95:199–206

Hancock JD (2000) Forages and fodder. In: Smith CW, Frederiksen RA (eds) Sorghum-origin, history, technology and production. Wiley, New York, pp 731–749

Hao H, Li Z, Leng C, Lu C, Luo H, Liu Y, Wu X, Liu Z, Shang L, Jing HC (2021) Sorghum breeding in the genomic era: opportunities and challenges. Theor Appl Genet 134(7):1899–1924. https://doi.org/10.1007/s00122-021-03789-z

Heffner EL, Sorrells ME, Jannink JL (2009) Genomic selection for crop improvement. Crop Sci 49:1–12

Hilley JL, Weers BD, Truong SK, McCormick RF, Mattison AJ et al (2017) Sorghum *Dw2* encodes a protein kinase regulator of stem internode length. Sci Rep 7:1–13

Hirano K, Kawamura M, Araki-Nakamura S, Fujimoto H, Ohmae K (2017) Sorghum *DW1* positively regulates brassinosteroid signaling by inhibiting the nuclear localization of BRASSINOSTEROID INSENSITIVE 2. Sci Rep 7(126):1–10. https://doi.org/10.1038/s41598-017-00096-w

Ho NWY, Chen Z, Brainard AP (1998) Genetically engineered *Saccharomyces* yeast capable of effective co-fermentation of glucose and xylose. Appl Environ Microbiol 64:1852–1859

Hu F, Ragauskas A (2012) Pretreatment and lignocellulosic chemistry. Bioenergy Res 5:1043–1066

Huang C-F, Lin T-H, Guo G-L, Hwang W-S (2009) Enhanced ethanol production by fermentation of rice straw hydrolysate without detoxification using a newly adapted strain of *Pichia stipitis*. Bioresour Technol 100:3914–3920

Inal I, Yucel C, Yucel D, Hatipoglu R (2021) Nutritive value and fodder potential of different sweet sorghum genotypes under Mediterranean conditions. Turk J Field Crops 26(1):1–7. https://doi.org/10.17557/tjfc.943445

Jang Y-S, Malaviya A, Cho C, Lee J, Lee SY (2012) Butanol production from renewable biomass by clostridia. Bioresour Technol 123:653–663

Jordan DR, Mace ES, Henzell RG, Klein PE, Klein RR (2010) Molecular mapping and candidate gene identification of the *Rf2* gene for pollen fertility restoration in sorghum [*Sorghum bicolor* (L.) Moench]. Theor Appl Genet 120(7):1279–1287. https://doi.org/10.1007/s00122-009-1255-3

Jordan DR, Klein RR, Sakrewski KG, Henzell RG, Klein PE, Mace ES (2011) Mapping and characterization of *Rf5*: a new gene conditioning pollen fertility restoration in A1 and A2 cytoplasm in sorghum (*Sorghum bicolor* (L.) Moench). Theor Appl Genet 123:383–396

Karimi K, Emtiazi G, Taherzadeh MJ (2006) Ethanol production from dilute-acid pretreated rice straw by simultaneous saccharification and fermentation with *Mucor indicus*, *Rhizopus oryzae*, and *Saccharomyces cerevisiae*. Enzyme Microb Technol 40:138–144

Kiyosawa A, Yonemaru JI, Kawahigashi H, Goto K (2020) Analysis of quantitative trait loci for fertility restoration in seven F2 populations derived from sorghum F1 hybrids bred in Japan. Breed Sci 70(3):379–386. https://doi.org/10.1270/jsbbs.19144

Klein RR, Klein PE, Mullet JE, Minx P, Rooney WL, Schertz KF (2005) Fertility restorer locus *Rf1* of sorghum (*Sorghum bicolor* L.) encodes pentatricopeptide repeat protein not present in the colinear region of rice chromosome 12. Theor Appl Genet 102:994–1012

Klotz JL (2015) Activities and effects of ergot alkaloids on livestock physiology and production. Toxins (Basel) 7:2801–2821

Labroo MR, Studer AJ, Rutkoski JE (2021) Heterosis and hybrid crop breeding: a multidisciplinary review. Front Genet 12:643761. https://doi.org/10.3389/fgene.2021.64376

Li A, Jia S, Yobi A, Ge Z, Sato SJ et al (2018) Editing of an alpha-kafirin gene family increases digestibility and protein quality in sorghum. Plant Physiol 177:1425–1438

Li J, Zhao R, Xu Y, Wu X, Bean SR, Wang D (2022) Fuel ethanol production from starchy grain and other crops: an overview on feedstocks, affecting factors, and technical advances. Renew Energy 188:223–239

Lin Y-R, Schertzt KF, Paterson AH (1995) Comparative analysis of QTLs affecting plant height and maturity across the Poaceae, in reference to an interspecific sorghum population. Genetics 141:391–411

Mace E, Innes D, Hunt C, Wang XM, Tao YF, Baxter J, Hassall M, Hathorn A, Jordan D (2019) The sorghum QTL Atlas: a powerful tool for trait dissection, comparative genomics and crop improvement. Theor Appl Genet 132:751–766

Malherbe SJM, Cripwell RA, Favaro L, Van Zyl WH, Viljoen-Bloom M (2023) Triticale and sorghum as feedstock for bioethanol production via consolidated bioprocessing. Renew Energy 206:498–505

Malleshi NG (1987) Malting of sorghum for food uses. In: Ingle UM, Kulkarni DN, Thorat SS (eds) Proceedings of the national seminar on technology and applications for alternative uses of sorghum. Parbhani, pp 144–152

Meuwissen THE, Hayes BJ, Goddard ME (2001) Prediction of total genetic value using genome-wide dense marker maps. Genetics 157:1819–1829

Mishra JS, Kumar R, Rao SS (2017) Performance of sweet sorghum (*Sorghum bicolor* L. Moench) cultivars as a source of green fodder under varying levels of nitrogen in semi-arid tropical India. Sugar Tech 19:1–7

Moran JL, Rooney WL (2003) Effect of cytoplasm on the agronomic performance of grain sorghum hybrids. Crop Sci 43(3):777–781. https://doi.org/10.2135/cropsci2003.0777

Morris GP, Ramu P, Deshpande SP, Hash CT, Shah T et al (2013) Population genomic and genome-wide association studies of agroclimatic traits in sorghum. Proc Natl Acad Sci U S A 110:453–458

Multani DS, Briggs SP, Chamberlin MA, Blakeslee JJ, Murphy AS et al (2003) Loss of an MDR transporter in compact stalks of maize *br2* and sorghum *dw3* mutants. Science 302:81–84

Murphy RL, Klein RR, Morishige DT, Brady JA, Rooney WL et al (2011) Coincident light and clock regulation of pseudoresponse regulator protein 37 (PRR37) controls photoperiodic flowering in sorghum. Proc Natl Acad Sci U S A 108:16469–16474

Murphy RL, Morishige DT, Brady JA, Rooney WL, Yang S et al (2014) *Ghd7* (*Ma6*) represses sorghum flowering in long days: Ghd7 alleles enhance biomass accumulation and grain production. Plant Genome 7:1–10. https://doi.org/10.3835/plantgenome2013.11.0040

Nichols N, Bothast RJ (2008) Production of ethanol from grain. In: Vermerris W (ed) Genetic improvement of bioenergy crops. Springer, New York, pp 75–88

Oliver AL, Pedersen JF, Grant RJ, Klopfenstein TJ (2005) Comparative effects of the sorghum *bmr-6* and *bmr-12* genes: I. Forage sorghum yield and quality. Crop Sci 45:2234–2239

Olori MO, Makinwa EO, Irefin IA (1996) Indigenous technological capability development in the brewing industry in Nigeria: an engineering economic assessment and policy implications. Food Rev Int 12:511–523

Olson SN, Ritter K, Rooney W, Kemanian A, McCarl BA et al (2012) High biomass yield energy sorghum: developing a genetic model for C4 grass bioenergy crops. Biofuels Bioprod Bioref 6:640–655

Ou MS, Awasthi D, Nieves I, Wang L, Erickson J, Vermerris W, Ingram LO, Shanmugam KT (2015) Sweet sorghum juice and bagasse as feedstocks for the production of optically pure lactic acid by native and engineered *Bacillus coagulans* strains. BioEnergy Research 9:123–131

Pedersen JF, Toy JJ (2001) Registration of N316-N320 sorghum nuclear male-sterility genetic stock. Crop Sci 41(2):607–607. https://doi.org/10.2135/cropsci2001.412607x

Pereira MG, Lee M (1995) Identification of genomic regions affecting plant height in sorghum and maize. Theor Appl Genet 90:380–388

Pillonel C, Mulder MM, Boon JJ, Forster B, Binder A (1991) Involvement of cinnamyl-alcohol dehydrogenase in the control of lignin formation *in Sorghum bicolor* (L.) Moench. Planta 185:538–544

Poehlman JM (1987) Breeding field crops, 3rd edn. Van Nostrand Reinhold, New York. https://doi.org/10.1007/978-94-015-7271-2

Porter KS, Axtell JD, Lechtenberg VL, Colenbrander VF (1978) Phenotype, fiber composition, and in vitro dry matter disappearance of chemically induced *brown midrib* (*bmr*) mutants of sorghum. Crop Sci 18:205–208

Prabhakar, Madhusudhana R, Aruna C (2022) Sorghum breeding. In: Yadava DK, Dikshit HK, Mishra GP, Tripathi S (eds) Fundamentals of field crop breeding. Springer, Singapore. https://doi.org/10.1007/978-981-16-9257-4_7

Praveen M, Anurag Uttam G, Suneeta N, Umakant AV, Patil JV (2015) Inheritance and molecular mapping of *Rf6* locus with pollen fertility restoration on A1 and A2 cytoplasms in sorghum. Plant Sci 238:73–80

Pu Y, Hu F, Huang F, Ragauskas AJ (2015) Lignin structural alterations in thermochemical pretreatments with limited delignification. Bioenerg Res 8:992–1003

Quinby JR (1967) The maturity genes of sorghum. Adv Agron 19:267–305

Quinby J (1974) Sorghum improvement and the genetics of growth. Texas A&M University Press, College Station, TX

Quinby JR (1975) The genetics of sorghum improvement. J Hered 66:56–62

Quinby J (1981) Interaction of genes and cytoplasms in male sterility in sorghum. In: Proceedings of the annual corn and sorghum industry research conference, American Seed Trade Association, Corn and Sorghum Division, Corn and Sorghum Research Conference

Quinby JR, Karper RE (1954) Inheritance of height in sorghum. Agron J 46:211–216

Ragauskas AJ, Williams CK, Davison BH, Britovsek G, Cairney J et al (2006) The path forward for biofuels and biomaterials. Science 311:484–489

Rakshit S, Amasiddha B (2019) Conventional breeding techniques in sorghum. In: Aruna C, Visarada KBRS, Bhat BV, Tonapi VA (eds) Breeding sorghum for diverse end uses. Woodhead Publishing, Duxford, pp 77–91., ISBN 9780081018798,. https://doi.org/10.1016/B978-0-08-101879-8.00005-X

Rao NGP, Tripathi DP, Rana BS (1984) Genetic analysis of cytoplasmic systems of sorghum. Indian J Genet 44:480–496

Reddy BVS, Ashok Kumar A, Kaul SL (2008) Alternative cytoplasmic male sterility systems in sorghum and their utilization. In: Sorghum improvement in the new millennium. International Crops Research Institute for the Semi-Arid Tropics, Patancheru, pp 132–144. ISBN 978-92-9066-512-0

Renewable Fuels Association (2022) Annual industry outlook. https://ethanolrfa.org/resources/annual-industry-outlook

Rolz C, León RD, Montenegro ANM, Cifuentes R (2014) Ethanol from sweet sorghum in a year-round production cycle. Biomass Conv Bioref 4:341–350

Rooney WL, Aydin S (1999) Genetic control of a photoperiod-sensitive response in *Sorghum bicolor* (L.) Moench. Crop Sci 39:397–400

Rooney LW, Miller FR (1982) Variation in the structure and kernel characteristics of sorghum. In: Rooney LW, Murty DS (eds) International symposium on sorghum grain quality. ICRISAT, Patancheru, pp 143–162

Rooney LW, Pflugfelder RL (1986) Factors affecting starch digestibility with special emphasis on sorghum and corn. J Anim Sci 63:1607–1623

Rooney WL, Bean JB, Mullet JE (2007) Designing sorghum as a dedicated bioenergy feedstock. Biofuels Bioprod Bioref 1:147–157

Ross WM, Hackerott HL (1972) Registration of seven isocytoplasmic sorghum germplasm lines. Crop Sci 12:720–721

Saballos A (2008) Development and utilization of sorghum as a bioenergy crop. In: Vermerris W (ed) Genetic improvement of bioenergy crops. Springer, New York, pp 209–248

Saballos A, Vermerris W, Rivera L, Ejeta G (2008) Allelic Association, chemical characterization and saccharification properties of *brown midrib* mutants of sorghum (*Sorghum bicolor* (L.) Moench). Bioenerg Res 1:193–204

Saballos A, Ejeta G, Sanchez E, Kang C, Vermerris W (2009) A genomewide analysis of the *cinnamyl alcohol dehydrogenase* family in sorghum [*Sorghum bicolor* (L.) Moench] identifies *SbCAD2* as the *brown midrib6* gene. Genetics 181:783–795

Saballos A, Sanchez E, Foster TP, Xin Z, Kang CH, Pedersen J, Vermerris W (2012). *Brown midrib2* encodes the major 4-coumarate:CoA ligase involved in lignin biosynthesis in sorghum (S*orghum bicolor* (L.) Moench) The Plant Journal 5:818–830. https://doi.org/10.1111/j.1365-313X.2012.04933.x

Sanchez-Duarte JI, Kalscheur KF, Garcia AD, Contreras Govea FE (2019) Short communication: meta-analysis of dairy cows fed conventional sorghum or corn silages compared with brown midrib sorghum silage. J Dairy Sci 102:419–425

Sang Y, Bean S, Seib PA, Pedersen J, Shi Y-C (2008) Structure and functional properties of sorghum starches differing in amylose content. J Agric Food Chem 56:6680–6685

Sattler SE, Saathoff AJ, Haas EJ, Palmer NA, Funnell-Harris DL et al (2009) A nonsense mutation in a *cinnamyl alcohol dehydrogenase* gene Is responsible for the sorghum *brown midrib6* phenotype. Plant Physiol 150:584–595

Sattler SE, Palmer NA, Saballos A, Greene AM, Xin Z et al (2012) Identification and characterization of four missense mutations in *Brown midrib 12* (*Bmr12*), the caffeic O-methyltranferase (COMT) of sorghum. Bioenergy Res 5:855–865. https://doi.org/10.1007/s12155-012-9197-z

Sattler SE, Saballos A, Xin Z, Funnell-Harris DL, Vermerris W et al (2014) Characterization of novel sorghum *brown midrib* mutants from an EMS-mutagenized population. G3 4:2115–2124

Schertz KF, Clark LE (1967) Controlling dehiscence with plastic bags for hand crosses in sorghum. Crop Sci 7:540–542

Schertz KF, Ritchey JM (1978) Cytoplasmic genic male sterility systems in sorghum. Crop Sci 18:890–893

Scully ED, Gries T, Funnell-Harris DL, Xin Z, Kovacs FA et al (2016) Characterization of novel *brown midrib 6* mutations affecting lignin biosynthesis in sorghum. J Integr Plant Biol 58:136–149

Shan LQ, Jun Ai P, Tuan Yin L, Yao ZF (2000) New grain sorghum cytoplasmic male sterile line A2 V4 A and F1 hybrid Jinza No. 12 for Northwest China. Int Sorghum Millets Newsl 41:31–32

Shewayrga H, Sopade PA, Jordan DR, Godwin ID (2012) Characterization of grain quality in diverse sorghum germplasm using a Rapid Visco-Analyzer and near infrared reflectance spectroscopy. J Sci Food Agric 92:1402–1410

Silva TN, Vermerris W (2020) High-biomass sorghums as a feedstock for renewable fuels and chemicals. In: Aruna C (ed) Sorghum in the 21st century: food–fodder–feed–fuel for a rapidly changing world. Springer, Singapore, pp 723–754

Sivaramakrishnan S, Seetha K, Reddy BVS (1997) Characterization of the A4 cytoplasmic male-sterile lines of sorghum using RFLP of mt DNA. Euphytica 93:301–305

Slade HB (1903) Prussic acid in sorghum. J Am Chem Soc 25:55–59

Smith CW, Frederiksen RA (2000) Sorghum: origin, history, technology and production. Wiley, New York

Snider JL, Raper RL, Schwab EB (2012) The effect of row spacing and seeding rate on biomass production and plant stand characteristics of non-irrigated photoperiod-sensitive sorghum (*Sorghum bicolor* (L.) Moench). Ind Crops Prod 37:527–535

Stephens JC, Holland RF (1954) Cytoplasmic male sterility for hybrid sorghum seed production. Agron J 46:20–23

Stewart D (2008) Lignin as a base material for materials applications: chemistry, application and economics. Ind Crops Prod 27:202–207

Sukumaran S, Xiang W, Bean SR, Pedersen JF, Kresovich S, Tuinstra MR, Tesso TT, Hamblin MT, Yu J (2012) Association mapping for grain quality in a diverse sorghum collection. Plant Genome 5:126–135. https://doi.org/10.3835/plantgenome2012.07.0016

Tang HV, Chang R, Pring DR (1998) Co-segregation of single gene associated with fertility restoration and transcript processing of sorghum mitochondrial *orf107* and *urf209*. Genetics 209:383–391

Tang CC, Sun CD, Du F, Chen F, Ameen A, Fu TC, Xie GH (2017) Effect of plant density on sweet and biomass sorghum production on semiarid marginal land. Sugar Tech 20(3):312–322

Taylor JRN, Schüssler L (1986) The protein compositions of the different anatomical parts of sorghum grain. J Cereal Sci 4:361–369

Tetreault HM, Gries T, Liu S, Toy J, Xin Z et al (2021) The sorghum (*Sorghum bicolor*) *Brown midrib30* gene encodes a chalcone isomerase required for cell wall lignification. Front Plant Sci 12(12):732307. https://doi.org/10.3389/fpls.2021.732307

Thurber CS, Ma JM, Higgins RH, Brown PJ (2013) Retrospective genomic analysis of sorghum adaptation to temperate-zone grain production. Genome Biol 14:R68

Tuinstra MR (2008) Food-grade sorghum varieties and production considerations: a review. J Plant Interact 3:69–72

Umakanth AV, Ashok Kumar A, Vermerris W, Tonapi VA (2018) Sweet sorghum for biofuel industry. In: Aruna C, Visarada KBRS, Venkatesh Bhat B, Tonapi VA (eds) Breeding sorghum for diverse end uses. Woodhead Publishing, Duxford, pp 255–270

Unkovich M, Baldock J, Forbes M (2010) Variability in harvest index of grain crops and potential significance for carbon accounting: examples from Australian agriculture. Adv Agron 105:173–219

Upadhyaya HD, Pundir RPS, Dwivedi SL, Gowda CLL, Reddy VG et al (2009) Developing a mini core collection of sorghum for diversified utilization of germplasm. Crop Sci 49:1769–1780

Vanholme R, Demedts B, Morreel K, Ralph J, Boerjan W (2010) Lignin biosynthesis and structure. Plant Physiol 153:895–905

Venkateswarlu R, Aruna C, Visarada KBRS, Venkatesh Bhat B (2018) Sorghum for animal feed. In: Aruna C, Visarada KBRS, Venkatesh Bhat B, Tonapi VA (eds) Breeding sorghum for diverse end uses. Woodhead Publishing, Duxford, pp 229–239. ISBN 9780081018804

Vermerris W (2011) Survey of genomics approaches to improve bioenergy traits in maize, sorghum and sugarcane. J Integr Plant Biol 53:105–119

Vermerris W, Saballos A (2012) Genetic enhancement of sorghum for biomass utilization. In: Paterson AH (ed) Genomics of the Saccharinae. Springer, New York, pp 391–425

Vermerris W, Saballos A, Ejeta G, Mosier NS, Ladisch MR et al (2007) Molecular breeding to enhance ethanol production from corn and sorghum stover. Crop Sci 47(S3):S142–S153. https://doi.org/10.2135/cropsci2007.04.0013IPBS

Wang D, Bean S, Mclaren J, Seib B, Madl R, Tuinstra M, Shi Y, Lenz M, Wu X, Zhao R (2008) Grain sorghum is a viable feedstock for ethanol production. J Ind Microbiol Biotechnol 35:313–320

Webster OJ, Singh SP (1964) Breeding behavior and histological structure of non-dehiscent anther character in Sorghum vulgare Pers. Crop Sci 4:656–658

Worstell JV, Kidd HJ, Schertz KF (1984) Relationship among male sterility inducing cytoplasms of sorghum. Crop Sci 24:186–189

Xin Z, Li Wang M, Barkley NA, Burow G, Franks C et al (2008) Applying genotyping (TILLING) and phenotyping analyses to elucidate gene function in a chemically induced sorghum mutant population. BMC Plant Biol 8:103. https://doi.org/10.1186/1471-2229-8-103

Xin Z, Chen J, Jiao Y, Gladman N, Hayes C, Burow G, Emendack Y, Burke JJ (2018) Registration of BTx623*ms8,* a new and easily identifiable nuclear male sterile mutant in sorghum. J Plant Registr 12:278–281. https://doi.org/10.3198/jpr2017.09.0063crgs

Yan S, Wu X, Bean SR, Pedersen JF, Tesso T, Chen YR, Wang D (2011) Evaluation of waxy grain sorghum for ethanol production. Cereal Chem 88:589–595

Yang S, Murphy RL, Morishige DT, Klein PE, Rooney WL et al (2014) Sorghum phytochrome B inhibits flowering in long days by activating expression of SbPRR37 and SbGHD7, repressors of SbEHD1, SbCN8 and SbCN12. PLoS One 9(8):e105352. https://doi.org/10.1371/journal.pone.0105352

Yu J, Zhang X, Tan T (2007) A novel immobilization method of *Saccharomyces cerevisiae* to sorghum bagasse for ethanol production. J Biotechnol 129:415–420

Zeng Y, Zhao S, Yang S, Ding S-Y (2014) Lignin plays a negative role in the biochemical process for producing lignocellulosic biofuels. Curr Opin Biotechnol 27:38–45

Zerbini E, Thomas D (2003) Opportunities for improvement of nutritive value in sorghum and pearl millet residues in South Asia through genetic enhancement. Field Crop Res 84:3–15

Photoperiod and Temperature Sensitivity in Sorghum

9

Dilara Dorter and Sevgi Marakli

Abstract

Sorghum, originating from Africa, plays a crucial role in global agriculture by providing essential nutrition for humans and animals. It is also an important plant due to its potential as a biofuel source. With a growing global population and changing climate patterns, innovative agricultural and biotechnological strategies are vital for enhancing sorghum production. The complex interplay among photoperiod sensitivity, temperature responsiveness, and rainfall adaptation in sorghum cultivars from diverse geographic regions has been commonly investigated to improve growing strategies. The International Crops Research Institute for the Semi-Arid Tropics (ICRISAT) has a rich repository of sorghum genotypes, serving as an international genebank for biotechnological research. In this chapter, growth, adaptation, and stress resistance in sorghum varieties grown under kharif and rabi seasons are discussed by using data retrieved from ICRISAT. Moreover, it is also discussed how to select the best sorghum varieties for studies related to photoperiod and temperature sensitivities.

Keywords

Abiotic stress · Germplasm · ICRISAT Genebank · Kharif season · Rabi season

D. Dorter · S. Marakli (✉)
Department of Molecular Biology and Genetics, Faculty of Arts and Sciences, Yildiz Technical University, Istanbul, Turkey
e-mail: smarakli@yildiz.edu.tr

E. Habyarimana et al. (eds.), *Omics and Biotechnological Approaches for Product Profile-Driven Sorghum Improvement*,
https://doi.org/10.1007/978-981-97-4347-6_9

9.1 Introduction

Sorghum (*Sorghum bicolor* L.) is one of the ancient crops originated from Africa. This crop is the fifth most important cereal in the world after rice, barley, wheat, and corn. Sorghum grows in regions with low rainfall and drought conditions, spanning from approximately 40°S to 45°N latitudes. Its cultivation extends from Africa to other warm-temperate regions worldwide. Africa leads in sorghum production, followed by the Americas and Asia, with key producing nations including the United States, Nigeria, Ethiopia, India, and Mexico. Sorghum thrives in soils with poor potential, shallow depths, and high clay content with a pH range between 5.5 and 8.5. Temperature preferences range from 24 °C to 27 °C after germination but frost is detrimental. Moreover, annual precipitation is between 350 and 700 mm in suitable sorghum production regions, and rainfall distribution significantly affects crop performance (Mundia et al. 2019).

Furthermore, sorghum's photoperiod sensitivity is also crucial for optimizing agricultural practices and developing more productive varieties. How cultivars grown in diverse geographic regions respond to different photoperiods is essential for adaptability and crop production. For this purpose, sorghum varieties acquired from various countries or regions possess different genotypes were evaluated in terms of morphological, physiological, and molecular manner in drought stress (Abreha et al. 2022). Other studies indicated that the majority of sorghum varieties in West Africa exhibit sensitivity to photoperiod. They initiate flowering within 20 days prior to the average time of the rainy season to end (Abdulai et al. 2012; Upadhyaya et al. 2021). Photoperiod sensitivity simultaneously optimizes the length of the vegetative phase, resulting in increased biomass production. It also serves to alleviate the effects of drought by altering late flowering, prevents attacks by birds and insects through synchronized flowering, and enhances the stability of yield across different years. Additionally, it contributes to improve grain quality by restraining mold formation through the prevention of early flowering (Abdulai et al. 2012; Guitton et al. 2018).

Sorghum plant exhibits resistance to drought and water stresses. However, different rainfall conditions together with photoperiod affect plants in various locations, and even can induce sensitivity to stress factors. Therefore, the most important criterion for success is the selection of genotypes that are appropriate for the study's objectives. As a result of the changing environment and growing population, new agricultural and biotechnology approaches have been developed for sorghum production (Xin et al. 2021; Chadalavada et al. 2021). This chapter covers detail information on sorghum production and varieties in different region in the world.

9.2 Sorghum Production

Sorghum plants are grown in low rainfall and drought regions (350–700 mm/yr) from 40°S to 45°N latitude. It is more resistant than most other cereal crops to survive high temperatures. Average temperatures between 24 °C and 27 °C are recommended for the optimum yields. On the other hand, sorghum can develop slowly in cold temperatures. The crop is mostly grown in arid environments, particularly on shallow and heavy clay soils. Production has recently shifted from the drier western production regions to the moist eastern regions. This change in the production has led to the discovery and development of cultivars that are more resistant to colder temperatures (Rakshit and Wang 2016). Changes in climate regimens affect crop variability at different sites. Therefore, it is crucial to incorporate current and accurate knowledge on climate variability and its quantitative links with agricultural processes (Chadalavada et al. 2021; Rakshit and Wang 2016).

Sorghum is extensively grown in Africa, Asia, and the southwest of the United States due to its high drought tolerance. Moreover, it is gaining popularity in other warm-temperate regions of the World (Table 9.1). The highest sorghum production was observed in Africa with 2 469,192 tons (46.8%). After Africa, the second country is America with 2 320,454 tons (34.6%) followed by Asia with 201,578 tons (15.7%) (Table 9.1).

9.3 Demands for Growth, Adaptation, and Resistance to Stress Factors

9.3.1 Soil Requirements

Poor quality and shallow soils with a high percentage of clay are typically inappropriate for growing maize but are suitable to cultivate sorghum. The optimum clay percentage is between 10 and 30% for optimum sorghum production. This plant can also grow on sandy soil unless a heavily textured subsoil is present. pH is another important criterion for plant production. Because sorghum is more tolerant of alkaline salts than other grain crops, it can easily be cultivated on soils with a pH between 5.5 and 8.5 (Chadalavada et al. 2021).

Table 9.1 Sorghum production between 2015 and 2022 all over the world (FAOSTAT 2023; Foreign Agricultural Service 2023)

Country	2015	2016	2017	2018	2019	2020	2021	2022
United States	15.16 M	12.20 M	9.19 M	9.27 M	8.67 M	9.47 M	11.37 M	4.77 M
Nigeria	7.01 M	7.56 M	6.94 M	6.80 M	6.67 M	6.36 M	6.72 M	6.74 M
Ethiopia	4.77 M	4.75 M	5.17 M	5.02 M	5.27 M	5.06 M	4.45 M	4.20 M
India	5.45 M	4.24 M	4.57 M	4.80 M	3.48 M	4.77 M	4.15 M	4.00 M
Mexico	5.20 M	5.01 M	4.85 M	4.53 M	4.35 M	4.70 M	4.84 M	4.89 M
China	2.75 M	1.95 M	2.46 M	2.91 M	3.60 M	3.55 M	3.00 M	3.00 M
Brazil	2.14 M	1.15 M	2.22 M	2.28 M	2.67 M	2.77 M	3.04 M	2.92 M
Sudan	2.74 M	6.47 M	4.16 M	5.44 M	3.71 M	2.54 M	3.53 M	5.25 M
Niger	1.92 M	1.81 M	1.95 M	2.10 M	1.90 M	2.13 M	1.20 M	1.20 M

M indicates million tons

9.3.2 Water Requirements and Drought Stress

Water stress is a condition in which a plant's water intake capacity is insufficient to encounter its water requirements. Contrarily, drought is characterized as persistent and severe water stress circumstances. Both two stresses reduce seed formation and growth rate, restrict photosynthesis activity, cause leaf drying and wilting, and thus they have a deleterious impact on plant development (Prasad et al. 2021; Bi and Wang 2022).

Sorghum is primarily cultivated in semiarid and arid regions. Approximately 60% of the sorghum is planted in some parts of Africa and up to 80% of the sorghum grown in water-scarce circumstances in America (Abreha et al. 2022). Sorghum can be cultivated in regions with limited water resources due to its ability to develop effective root systems for water absorption (Prasad et al. 2021). The water requirements of sorghum show variability depending on growth stages and climatic conditions. Particularly during the flowering and seed formation stages, the need for water increases. The development of more water-efficient and drought-resistant varieties through biotechnological research can contribute to more efficient sorghum cultivation in water-scarce regions (Abreha et al. 2022; Baloch et al. 2023; Fardin et al. 2023).

Drought decreases carbon absorption, stomatal function and cell turgidity. Therefore, this stress reduces crop growth, development, and even yield (Prasad et al. 2019, 2021). It is possible to identify the unique characteristics of sorghum underlying drought tolerance by using genetic, genomic, proteomic, and metabolomic aspects (Goche et al. 2020; Ngara et al. 2021). Gaining insight into these mechanisms linked to traits has enhanced breeding processes, providing novel drought-tolerant or heat-resistant sorghum varieties or hybrids (Getachew et al. 2023).

9.3.3 Photoperiod

Sorghum growth and development are regulated by a series of biological processes. These processes can be influenced by photoperiod. The photoperiod is a combination of the daily duration of light and dark periods, having a significant influence on the growth and development of plants. Photoperiod perception in plants is carried out by photoreceptors in leaves. A suitable photoperiod optimizes energy production through photosynthesis, thereby increasing overall productivity (Shukla et al. 2022).

Sorghum is a typical short-day plant, exhibiting different responses depending on whether the plant receives more or less light during the day. Therefore, the duration of the photoperiod is a decisive factor in determining the timing of growth and flowering. The photoperiod not only affects flowering and seed formation but also plays a role in the plant's ability to cope with stress (Dingkuhn et al. 2008). Until a sufficient night length is recognized, photoperiodism serves as a factor that inhibits genetic predisposition in flowering. Plants start the flowering process after the genetic restriction on flowering is eliminated (Chadalavada et al. 2021). In a study conducted in Mali, photoperiod modeling was performed to understand when sorghum germinates and how the light affects. For this purpose, a model was validated in terms of its capability to explain the range of genotypic responses (Dingkuhn et al. 2008). The impact of photoperiod on sorghum production plays a crucial role in flowering timing, seed formation and yield, resistance, etc. (Clerget et al. 2021; Wolabu and Tadege 2016).

Flowering Timing Sorghum is a typical short-day plant that initiates flowering after short-day lengths. Therefore, the duration of photoperiod determines the flowering timing of the plant. Precise control of photoperiod is critically important for regulating flowering timing and ensuring efficient seed formation (Guitton et al. 2018).

Seed Formation and Yield Adequate photoperiod optimizes the plant's photosynthesis activity and energy production. Consequently, sorghum with high quality and yield is produced by regulating the timing of flowering (Kamal et al. 2019).

Resistance Photoperiod can also influence the resilience of plants. Plants grown under suitable photoperiods are better adapted to stress conditions such as drought and high temperatures. This factor contributes to the overall health and productivity of the plant (Badigannavar et al. 2018; Getachew et al. 2023).

Differential Nighttime Responses Under appropriate photoperiods, plants enter a sleep-like state during the night, and efficiently use their energy. Conversely, inappropriate photoperiods may cause plants to remain awake at night, resulting in energy loss (Rai et al. 2013).

Selection of Suitable Varieties The impact of photoperiod should be taken into account when selecting suitable varieties. Sorghum varieties cultivated in different geographical regions and climatic conditions may respond differently to varying photoperiods (Shukla et al. 2022).

9.3.4 Temperature

Germination, flowering, and fruit formation phases can occur optimally within specific temperature ranges. Generally, sorghum plants develop under warm and humid conditions, but exposure to frost and extreme temperatures can negatively affect their growth (Chadalavada et al. 2021). Climate conditions affect the development and productivity of the plant in different growth stages. Countries can be categorized into three distinct groups based on climate regions. Sorghum may exhibit different responses due to its sensitivity to temperature stress in arid, temperate, subtropical, and tropical climate regions.

- Sorghum produces good quantities of products and grows well in arid environments. Plant growth can be accelerated by high temperatures.
- Sorghum may also be grown in temperate climate zones, but it may be more susceptible to temperature variations.
- Sorghum is frequently cultivated in subtropical and tropical areas. Because of the high temperature and humidity, plant diseases may be more widespread in these areas. Tillage, irrigation, and disease management are all critical parameters for efficient production.

Variability in precipitation patterns across climatic zones helps the formation of geographically distinct sorghum species. Organisms tend to evolve adaptations in response to their habitat's environmental conditions. As a result, it is suggested that sorghum plant would undergo genetic and proteomic alterations to adapt to its current environmental conditions (Shukla et al. 2022).

9.4 How Should Sorghum be Identified in Biotechnological Studies?

Sorghum is cultivated during two distinct seasons: the rainy season (kharif season, spanning from June to October) and the post-rainy season (rabi season, covering the period from October to February). High temperatures and prolonged daylight hours are characteristics of the kharif season, while lower temperatures and condensed daylight hours are characteristics of the rabi season (Upadhyaya et al. 2021). Therefore, it is crucial to develop high-yielding cultivars that are adapt to changing climatic conditions, and tolerant to biotic and abiotic threats.

9.4.1 Identify the Sorghum Genotype

ICRISAT Genebank (International Crops Research Institute for the Semi-Arid Tropics) was established by collecting sorghum samples from 94 different countries around the world in India (genebank.icrisat.org). All these samples are coded, starting from IS 1 and a total of 41,371 sorghum genotypes have been reported (ICRISAT 2023; Tonapi et al. 2021). The distribution of these collected samples in countries is indicated in Fig. 9.1.

The distribution of collected samples encompasses a wide range of highly diverse countries. Sorghum species are categorized as highly resistant, moderately resistant, susceptible, and highly susceptible based on their resistance levels in the postrainy and rainy seasons. It is observed that sorghum is quite sensitive in both seasons (Fig. 9.2).

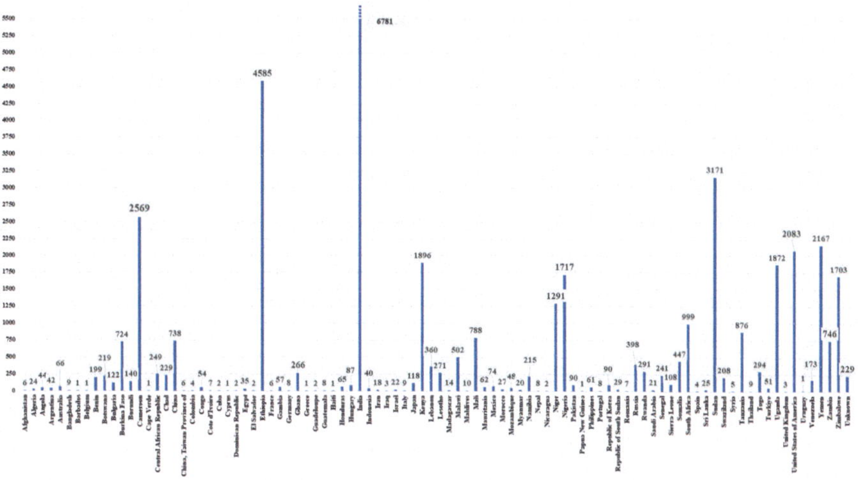

Fig. 9.1 Country-specific genotype distribution in ICRISAT Genebank

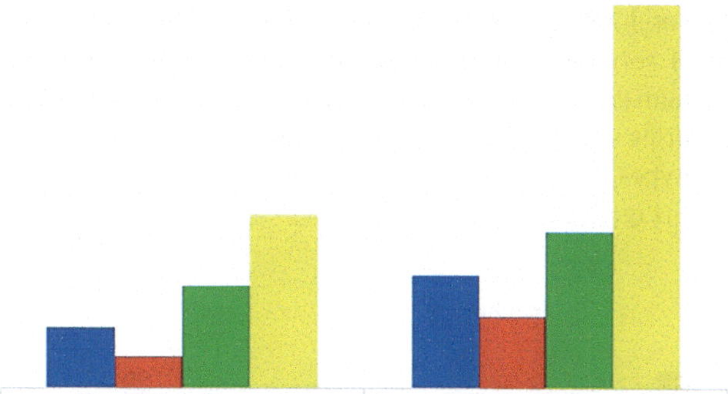

	Postrainy Season	Rainy Season
Highly resistant	1088	2047
Moderately resistant	570	1298
Suceptible	1853	2850
Highly susceptible	3150	6978

Fig. 9.2 Global distribution of sorghum's susceptibility to rainy seasons in the ICRISAT Genebank

9.4.2 Determine the Requirements Suitable for the Purpose of the Study

Sorghum genotypes with varying degrees of resistance are essential for ensuring agricultural productivity and food security in regions where sorghum is a significant crop. Here is an overview of the kharif and rabi seasons' resistance levels across different countries:

Highly Resistant Genotypes These sorghum genotypes exhibit strong resistance to the challenges posed by the rainy season, making them suitable for cultivation in regions with high rainfall.

Moderately Resistant Genotypes Genotypes in this category have a moderate level of resistance, indicating their ability to withstand some adverse weather conditions during the rainy season.

Susceptible Genotypes They are more vulnerable to the impact of heavy rainfall and may require additional management practices.

Highly Susceptible Genotypes These are the most susceptible genotypes to the rainy season's challenges and may be less suitable for cultivation in areas with intense rainfall.

Figures 9.3 and 9.4 summarize the resistance levels of various sorghum genotypes to kharif and rabi seasons, respectively, in different countries. The United States of America has a significant number of highly resistant and moderately resistant sorghum genotypes, reflecting advanced breeding programs. The highest number of sorghum genotypes is found in India. Nigeria has a significant number of sorghum genotypes with a mix of resistant and susceptible ones. Furthermore, some countries such as Algeria, Cyprus, and Mozambique have relatively few sorghum genotypes in the dataset, showing limited sorghum research or cultivation in these areas (Fig. 9.3).

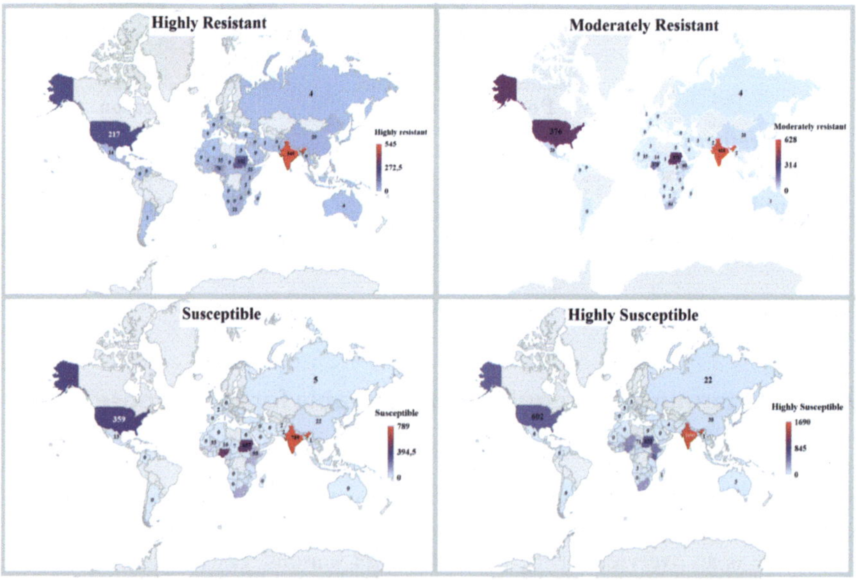

Fig. 9.3 The rain sensitivities of genotypes in the ICRISAT Genebank in the kharif season

In addition to kharif season, resistance levels of various sorghum genotypes to post-rainy/rabi season conditions in different countries are indicated in Fig. 9.4. India has a substantial number of genotypes in all categories. Some countries such as Algeria, Cyprus, and Mozambique appear to have no recorded genotypes in the dataset, possibly reflecting limited cultivation in these areas. It is essential to continue research and breeding efforts to develop more sorghum genotypes with increased resistance to post-rainy season conditions. Additionally, knowledge sharing and collaboration among countries can contribute to the development and dissemination of resilient sorghum varieties to address global food challenges.

The provided tables (Tables 9.2 and 9.3, respectively) offer valuable insights into sorghum production across four different countries during the kharif and rabi seasons. Four countries with extensive data resources were randomly selected from around the world. These tables help us understand the distribution of sorghum genotypes with varying degrees of resistance in each season. The distribution of sorghum genotypes with varying resistance levels underscores the importance of ongoing research and breeding efforts to develop resilient sorghum varieties that can thrive in diverse agricultural environments. Additionally, it emphasizes the significance of selecting appropriate genotypes based on seasonal variations and geographical factors to ensure food security and sustainable agriculture.

The United States of America exhibits a diverse range of sorghum genotypes with a significant number of moderately resistant and highly susceptible ones. India has the highest number of sorghum genotypes in this season. Moreover, Kenya, Ethiopia, and Yemen show relatively lower numbers of sorghum genotypes with varying levels of resistance.

When compared to kharif season, the number of sorghum genotypes is reduced in the United States of America. Similar to kharif, India has the highest number of

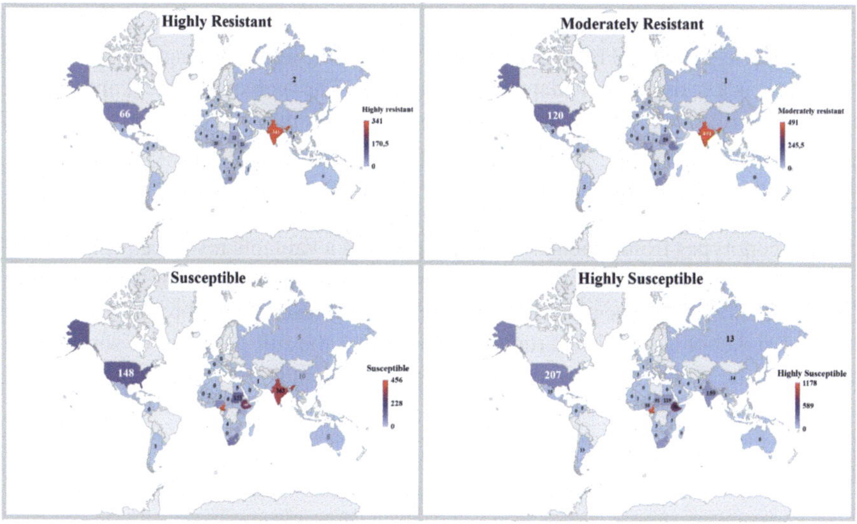

Fig. 9.4 The rain sensitivities of genotypes in the ICRISAT Genebank in the rabi season

Table 9.2 Distribution of sorghum varieties in four different countries in Kharif season (ICRISAT 2023)

Country Kharif season	Highly resistant	Moderately resistant	Susceptible	Highly susceptible
United States of America	217	376	359	602
India	545	628	789	1690
Kenya	1	17	90	604
Ethiopia	21	68	98	486
Yemen	18	12	21	44

Table 9.3 Distribution of sorghum varieties in four different countries in rabi season (ICRISAT 2023)

Country Rabi season	Highly resistant	Moderately resistant	Susceptible	Highly susceptible
United States of America	66	120	148	207
India	341	491	363	159
Kenya	5	4	19	52
Ethiopia	22	100	289	702
Yemen	1	1	11	16

sorghum genotypes in rabi season. Moreover, Kenya and Ethiopia continue to have fewer sorghum genotypes with varying levels of resistance. Yemen has also a limited number of genotypes in this season.

9.5 Conclusion

Sorghum plays a special role in the goal of sustainable agriculture and food security because it is a resistant and adaptable cereal crop. The biotechnological approaches in sorghum research include molecular markers, QTL mapping, transcriptomic and genomic studies, and genetic engineering techniques. These methods provide valuable insights into the genetic diversity and adaptive traits of sorghum (Ngara et al. 2021; Xin et al. 2021). The sorghum genebank with its extensive collection of diverse genotypes from around the world serves as a valuable resource for researchers seeking to understand the genetic underpinnings of sorghum's adaptability (Abreha et al. 2022). The continued collaboration of researchers, breeders, and agricultural stakeholders will be crucial in harnessing the full potential of sorghum to encounter the global challenges of the twenty-first century.

References

Abdulai AL, Kouressy M, Vaksmann M, Asch F, Giese M, Holger B, Abdulai CAL (2012) Latitude and date of sowing influences phenology of photoperiod-sensitive sorghums. J Agron Crop Sci 198:340. https://doi.org/10.1111/j.1439-037X.2012.00523.x

Abreha KB, Enyew M, Carlsson AS, Vetukuri RR, Feyissa T, Motlhaodi T, Nguni D, Geleta M (2022) Sorghum in dryland: morphological, physiological, and molecular responses of sorghum under drought stress. Planta 255(1):1–23. https://doi.org/10.1007/s00425-021-03799-7

Badigannavar A, Teme N, de Oliveira AC, Li G, Vaksmann M, Viana VE et al (2018) Physiological, genetic and molecular basis of drought resilience in sorghum [*Sorghum bicolor* (L.) Moench]. Indian J Plant Physiol 23:670–688

Baloch FS, Altaf MT, Liaqat W, Bedir M, Nadeem MA, Cömertpay G et al (2023) Recent advancements in the breeding of sorghum crop: current status and future strategies for marker-assisted breeding. Front Genet 14:1150616

Bi Y, Wang P (2022) Exploring drought-responsive crucial genes in Sorghum. IScience 25(11):105347. https://doi.org/10.1016/j.isci.2022.105347

Chadalavada K, Kumari BDR, Kumar TS (2021) Sorghum mitigates climate variability and change on crop yield and quality. Planta 253(5):1–19. https://doi.org/10.1007/s00425-021-03631-2

Clerget B, Sidibe M, Bueno CS, Grenier C, Kawakata T, Domingo AJ, Layaoen HL, Gutiérrez-Palacios ND, Bernal JH, Trouche G, Chantereau J (2021) Crop photoperiodism model 2.0 for the flowering time of sorghum and rice that includes daily changes in sunrise and sunset times and temperature acclimation. Ann Bot 128(1):97–113. https://doi.org/10.1093/aob/mcab048

Dingkuhn M, Kouressy M, Vaksmann M, Clerget B, Chantereau J (2008) A model of sorghum photoperiodism using the concept of threshold-lowering during prolonged appetence. Eur J Agron 28(2):74–89. https://doi.org/10.1016/j.eja.2007.05.005

FAOSTAT (2023) Food and Agriculture Organization of the United Nations (FAO). FAOSTAT Database. https://www.fao.org/faostat/en/?#data/QCL

Fardin F, Sani B, Moaveni P, Afsharmanesh G, Mozafari H (2023) Nutritional value and agronomic traits of forage sorghum under drought stress. Biocatal Agric Biotechnol 48:102624

Foreign Agricultural Service (FAS) (2023). Available from https://ipad.fas.usda.gov

Getachew F, Bayabil HK, Hoogenboom G, Kiker GA, Yu Z, Li Y (2023) Development of climate-smart sorghum ideotype for climate resilience in Ethiopia. Field Crop Res 303:109135

Goche T, Shargie NG, Cummins I, Brown AP, Chivasa S, Ngara R (2020) Comparative physiological and root proteome analyses of two sorghum varieties responding to water limitation. Sci Rep 10(1):1–18. https://doi.org/10.1038/s41598-020-68735-3

Guitton B, Théra K, Tékété ML, Pot D, Kouressy M, Témé N, Rami JF, Vaksmann M (2018) Integrating genetic analysis and crop modeling: a major QTL can finely adjust photoperiod-sensitive sorghum flowering. Field Crop Res 221(February):7–18. https://doi.org/10.1016/j.fcr.2018.02.007

ICRISAT (2023) International Crops Research Institute for the semi-arid tropics. http://genebank.icrisat.org/. Accessed 12 Jul 2023

Kamal NM, Gorafi YSA, Abdelrahman M, Abdellatef E, Tsujimoto H (2019) Stay-green trait: a prospective approach for yield potential, and drought and heat stress adaptation in globally important cereals. Int J Mol Sci 20(23):5837. https://doi.org/10.3390/ijms20235837

Mundia CW, Secchi S, Akamani K, Wang G (2019) A regional comparison of factors affecting global sorghum production: the case of North America, Asia and Africa's Sahel. Sustainability 11(7):2135

Ngara R, Goche T, Swanevelder DZH, Chivasa S (2021) Sorghum's whole-plant transcriptome and proteome responses to drought stress: a review. Life 11(7):1–25. https://doi.org/10.3390/life11070704

Prasad PVV, Djanaguiraman M, Jagadish SVK, Ciampitti IA (2019) Drought and high temperature stress and traits associated with tolerance, pp 241–265. https://doi.org/10.2134/agronmonogr58.c11

Prasad VBR, Govindaraj M, Djanaguiraman M, Djalovic I, Shailani A, Rawat N, Singla-Pareek SL, Pareek A, Vara Prasad PV (2021) Drought and high temperature stress in sorghum: physiological, genetic, and molecular insights and breeding approaches. Int J Mol Sci 22(18):9826. https://doi.org/10.3390/ijms22189826

Rai GK, Kumar RR, Bagati S (2013) Abiotic stress tolerance mechanisms in plants. In: Quinoa: botany, production and uses. CABI, Wallingford

Rakshit S, Wang Y-H (eds) (2016) The Sorghum Genome. Springer International, Cham. https://doi.org/10.1007/978-3-319-47789-3

Shukla P, Kumar A, Rakesh Kumar MKP (2022) In: Shukla P, Kumar A, Kumar R, Pandey MK (eds) Molecular response and genetic engineering for stress in plants, vol 1. IOP, Bristol. https://doi.org/10.1088/978-0-7503-4921-5

Tonapi VA, Talwar HS, Are AK, Bhat BV, Reddy CR, Dalton TJ (2021) Sorghum in the 21st century: food—fodder—feed—fuel for a rapidly changing world, pp 1–940. https://doi.org/10.1007/978-981-15-8249-3

Upadhyaya HD, Vetriventhan M, Azevedo VCR (2021) Variation for photoperiod and temperature sensitivity in the global mini core collection of sorghum. Front Plant Sci 12(June):571243. https://doi.org/10.3389/fpls.2021.571243

Wolabu TW, Tadege M (2016) Photoperiod response and floral transition in sorghum. Plant Signal Behav 11(12):e1261232. https://doi.org/10.1080/15592324.2016.1261232

Xin Z, Wang M, Cuevas HE, Chen J, Harrison M, Pugh NA, Morris G (2021) Sorghum genetic, genomic, and breeding resources. Planta 254(6):1–24. https://doi.org/10.1007/s00425-021-03742-w

Status of Sorghum Breeding in Asia

10

Niranjan Thakur, Rutwik Barmukh, Pramod Sargar, Krishna Kasanaboina, Sonal Chavan, Bruno Awio, Jieqin Li, Lihua Wang, and Ephrem Habyarimana

Abstract

This chapter offers a thorough examination of sorghum breeding in Asia, covering its history, distribution, and domestication. It addresses challenges in breeding, particularly cytoplasmic male sterility, and extensively explores the application of omics approaches such as transcriptomics, genomics, proteomics, metabolomics, and bioinformatics. Emphasis is placed on climate resilience, utilizing genomics to develop sorghum varieties adapted to biotic and abiotic stresses. We summarized biotic and abiotic stresses, along with biotechnological

N. Thakur · P. Sargar
International Crops Research Institute for the Semi-Arid Tropics, Patancheru, Hyderabad, Telangana, India

Vasantrao Naik Marathwada Krishi Vidyapeeth, Parbhani, India

R. Barmukh · E. Habyarimana (✉)
International Crops Research Institute for the Semi-Arid Tropics, Hyderabad, Telangana, India
e-mail: Ephrem.Habyarimana@icrisat.org

K. Kasanaboina · S. Chavan · B. Awio
International Crops Research Institute for the Semi-Arid Tropics, Patancheru, Hyderabad, Telangana, India

Professor Jayashankar Telangana State Agricultural University, Rajendranagar, Hyderabad, India

J. Li · L. Wang
College of Agriculture, Anhui Science and Technology University, Fengyang, China

Anhui Provincial Forage Biological Breeding International Cooperation Research Center, Fengyang, China

E. Habyarimana et al. (eds.), *Omics and Biotechnological Approaches for Product Profile-Driven Sorghum Improvement*,
https://doi.org/10.1007/978-981-97-4347-6_10

interventions to enhance important agronomic traits. This chapter also delves into modifying flowering, plant height, and the brown midrib structure's implications for animal feed. Throughout, the significance of these tools and techniques in understanding sorghum genetics and aiding breeding programs is highlighted. In conclusion, we discussed addressing current challenges in sorghum breeding in Asia and advocating ongoing research and collaboration to ensure regional food security and sustainable agriculture. Overall, it provides a comprehensive, up-to-date account of sorghum breeding, showcasing genomics and biotechnology's pivotal role in enhancing resilience and productivity.

Keywords

Biotechnology · Breeding · Genomics · Omics approaches · Sorghum

10.1 Introduction

Sorghum (*Sorghum bicolor* L. Moench) is a valuable crop with several uses that have a lasting impact on agriculture across the world. Due to its adaptability to many agroecological zones, several uses in a variety of industries—including food, livestock feed, fodder, the generation of biofuels, and its excellent nutritional value—it has been widely cultivated (Menamo et al. 2021). As a hardy and adaptable cereal crop that contributes significantly to guaranteeing food security throughout the continent, sorghum plays a significant part in Asia's agricultural tapestry (Smith and Frederiksen 2000). The main reason sorghum is grown is for its grains, which constitute a primary source of nutrition for large population worldwide, particularly in Asia and Africa (Ingle et al. 2023). These grains are an essential source of calories and nourishment, especially in areas where other cereal crops struggle owing to environmental factors. Sorghum serves as a food crop but also plays a significant part in livestock production as a necessary ingredient in animal feed and fodder (Balota et al. 2016).

The wide variety of agroecological zones found in Asia has led to the development of several sorghum ecosystems, each influenced by its unique history of domestication and adaptability (Smith and Frederiksen 2000). To fully appreciate the modern sorghum breeding in Asia, it is essential to comprehend the complex path of sorghum from its African roots to its widespread presence in Asia (Baloch et al. 2023).

In Asia, the quest for sorghum development is characterized by strong obstacles that include biotic and abiotic stresses. Abiotic challenges such as the unstable nature of drought and the rise of soil salinity collaborate with biotic stresses like the assault of diseases and the unrelenting pressure of pests (Yadav and Reddy 2016). In addition, the intentional use of cytoplasmic male sterility (CMS) in sorghum breeding provides opportunities as well as complications (Xin et al. 2016). These complex limits necessitate innovative solutions that are fully aware of local sorghum breeding specificities.

A new age of sorghum breeding began with the arrival of the genomic era, armed with powerful tools to hasten the creation of improved sorghum cultivars. A thorough investigation of the complex genetics of sorghum is made possible by transcriptomics analysis, comparative genomics, proteomics, metabolomics profiling, bioinformatics, and the use of epigenomics platforms (Varshney et al. 2018; Habyarimana et al. 2022). These methods act as vital compasses that help breeders make sense of the genetic obstacles within sorghum and decide which traits to choose and improve to get the best performance in a variety of agricultural environments. Asia's sorghum production faces a strong threat from climate change, which amplifies biotic and abiotic pressures. To increase sorghum's adaptation and resistance to the harsh effects of climate change, genomic techniques are emerging as potent tools. To ensure the sustained viability of sorghum in the face of changing environmental challenges, special attention is paid to managing biotic stresses brought on by diseases and pests (Chadalavada et al. 2021).

Agronomical trait improvement has emerged as a key goal in sorghum breeding due to the diversity of the Asian agricultural geography. With the ability to increase the yield of grains, improve grain quality, regulate flowering time, and change plant height, biotechnological interventions have become a potential method for the regulation of these characteristics (Varshney et al. 2018). Additionally, research into the brown midrib structure is an intriguing area that may have impacts on improving sorghum feed and bioenergy output (Kumar et al. 2020b).

Asia represents a dynamic synthesis of tradition and innovation, where longstanding crop domestication techniques meet state-of-the-art genomic and biotechnological methods. With its broad variety of climatic conditions and agricultural techniques, Asia is well-positioned to unleash the full genetic potential of sorghum and solidify its position as a hardy and essential crop in a constantly changing global agricultural landscape. This chapter sets out on a journey through the exciting world of sorghum breeding in Asia, exploring the multifaceted roles of sorghum as well as its cultivation, use, and potential for further improvement across these various domains. This area is characterized by a complex interaction between conventional agricultural methods, innovative biotechnological innovations, and cutting-edge genomics.

10.2 Sorghum Domestication and Distribution in Asia

10.2.1 Sorghum Domestication

Sorghum (*Sorghum bicolor*) is a grain crop that is produced around the world in tropical and subtropical climates. It is one of the most significant cereal crops in Asia as well as Africa. According to Winchell et al. (2018), sorghum was initially domesticated in northeast-central Africa between 5000 and 6000 years ago. Numerous morphological, physiological, and phenological characteristics, such as non-shattering grains, larger seeds, distinct panicles, and plant architecture, are the ways in which the domesticated phenotype may be distinguished from the wild variants. It has been determined which genes and gene families are most likely to be affected by the domestication and/or improvement processes. Notably, mutations in the shattering1 (*sh1*) gene, which presumably underwent simultaneous selection in sorghum, rice, and maize, have been linked to the loss of seed shattering (Lin et al. 2012). Other potential genes have been discovered using selection scouring and phenotype–genotype association research. These include the known domestication genes for cereals *Tb1* (*teosinte branched1*) and *ba1* (*barren stalk1*), which are involved in plant architecture (Lai et al. 2018), as well as genes linked to grain characteristics (such as *Psy1*, *phytoene synthase1*), photoperiod sensitivity (early maturity *Ma* genes), and plant height (*Dw* genes) (Mace et al. 2013). The metabolisms of starch (*ss1* and *sbe3* genes, Campbell et al. 2016), grain size and weight (Tao et al. 2017), and other major biosynthetic pathways have also been highlighted by genes.

It is unclear how sorghum domesticated and came to Asia. Sorghum, however, is thought to have initially been domesticated in Africa and subsequently brought to Asia circa 4000 BC (Fig. 10.1).

Sorghum is said to have its secondary origin in the Indian subcontinent. From the Late Harappan period, between 2000 and 1700 BC, in the Indus Valley and nearby foothills, the oldest unmistakably domesticated and grown sorghum in India is described, from where it migrated eastward. With the exception of those from Kaothe, which could be caudatums, the majority of these specimens were bicolor types (Boivin and Fuller 2009). Saraswat (1983) reported sorghum from the late Harappan Rohira around 2000–1800 BC in Sangrur, Punjab, while Kajale (1991) discovered sorghum from the earliest Savalda stages of Daimabad in Western Maharashtra dating to the early second millennium BC. Domesticated sorghum most likely traveled from India to China via maritime channels off the Asian coast as well as overland trade routes. It is possible that evolved sorghum traveled from Indo-China to China through the Mekong River or other river basins. During the Ming dynasty, travelers from China to East Africa also documented this domestication. Sorghum was brought to China by Genghis Khan on a trip to South Asia between AD 1206 and 1228, according to Hagerty (1941), and it expanded throughout the Kublai Khan era (1260–1295).

In Chinese literature from the second century BC, sorghum is described as another early crop that was grown in China. The sorghum is distinct in Indonesia and Southeast Asia as well. According to Burkill (1966), *S. propinquum* may be found from southern China via Thailand, Cambodia, Malaya, and Burma to the Philippines. They may also have a different history than the African-based varieties of sorghum, which are distinguished by their very big, loose, open panicles (Harlan and Stemler 1976).

About 10,000 years ago, in the Neolithic Age, cereals were first domesticated. Parallel selection across important crops during plant evolution resulted in various agronomic properties being altered that benefited humans; this phenomenon is known as domestication syndrome (Stetter et al. 2017). The genes that are selected in parallel always occupy a small number of crucial signaling pathway sites. Despite having various mutation types, loss or gain of function in these paired chosen genes is retained in related crop species (Lenser and Theissen 2013). Recent advances in

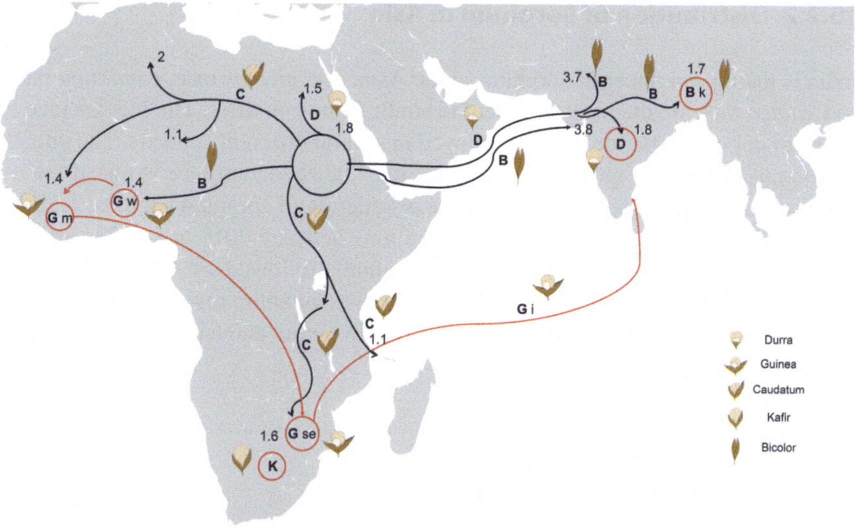

Fig. 10.1 Based on archaeobotanical and genomic data, a framework of the spread and diversification history of domesticated sorghum has been developed. Ovals with black lines represent the domestication zone for the early race bicolor and the subsequent evolution of race caudatum. Ovals with red lines show potential areas for the domestication of additional races or variations. Archaeobotanical data is used to determine trajectory; genetic evidence is used to refer to hypothetical dispersion. The numbers represent the arrival time or the evolutionary event's approximative age in years before 2000 AD. Races or varietal groupings estimated for each distribution pathway are denoted by letters on arrows as: B = bicolor race, B k = Chinese Kaoliang sorghum group within race bicolor, C = caudatum race, D = durra race, G w = race guinea, West African group, G m = race guinea, *margaritiferum* varieties (= *S. margaritiferum* Stapf.), G se = race guinea, Southeast African group, G i = race guinea, Indian group, K = race kafir (= *S. caffrorum* Snowden). (Image constructed by Niranjan Ravindra Thakur, Adapted from Fuller and Stevens 2018)

our understanding of the essential genes underpinning domestication-related features in rice, maize, and wheat (Fernie and Yan 2019) may stimulate research towards the domestication of sorghum. For instance, *SbTB1*, the sole significant gene found that controls the number of tillers in sorghum, is an ortholog of the well-known *TB1* gene in maize (Doebley et al. 1997). However, due to different selection pressure, this does not happen in certain homologs of those identified differentiated genes. Since millet *SiGC1*, the ortholog of sorghum *GC1*, already had thin glumes and grains that were simple to thresh, it did not display a parallel section in naked grains (Xie et al. 2022). It is notable that these significant domesticated genes are situated in straightforward regulatory pathways and exhibit minor pleiotropic effects, which can limit negative effects on other critical agronomic variables like the flowering time manner (Turck et al. 2008; Ge et al. 2022). Recent molecular investigations have revealed that *S. propinquum* and *S. bicolor* vary by 1.2% of nucleotides in the genome's coding regions, indicating a divergence of 1–two million years between the two sorghums (Feltus et al. 2004).

10.2.2 Distribution of Sorghum in Asia

India is the top producer of sorghum in the Asia. The *bicolor* race, *caudatum* race, and *guinea* race are the most prevalent in south Asian countries. The *bicolor* race is the most popular variation and is employed in several different contexts. The guinea race is raised for its high forage yield, whereas the caudatum race is raised for its high grain output. Table 10.1 displays the production of sorghum (in tons) and the area harvested (in hectares) in ten Asian nations between 2018 and 2022 (Statista 2023). India is Asia's top producer of sorghum, followed by China, Pakistan, Bangladesh, and Vietnam, according to the table as an entire region. Except for Sri Lanka, all nations have seen increases in sorghum output and harvested area during the past 5 years.

10.3 Constraints in Sorghum Breeding and Production in Asia

Sorghum is a hardy crop with the potential to thrive across a range of agroecological zones in Asia. However, there are several obstacles to the effective breeding and production of sorghum in Asia. According to current studies and the perspectives of industry professionals, there are a few significant obstacles that the region faces when it comes to sorghum breeding and production.

Sorghum's genetic diversity in Asia is rather low, which puts the crop's adaptation and sustainability in danger (Srinivas et al. 2019). Breeding superior cultivars with resilience to insect pests, diseases, and abiotic stresses may be limited by low genetic diversity. Furthermore, it may prevent the advancement of sorghum cultivars with desired qualities including high nutritional value and enhanced grain quality (Yadav and Reddy 2016).

10.3.1 Biotic Stress

Low yields frequently hinder the cultivation of sorghum which include factors like disease and pest. The supply of food and fodder can be impacted by the large output losses brought on by several pests. In addition, the region's sorghum production is continually hampered by the danger of new pests (Patil et al. 2017). Sorghum, however, confronts a variety of difficulties, with biotic stress factors providing a danger to its sustainability and production (Smith et al. 2015).

10.3.1.1 Insect Pests

Insect pests are one of the main causes of biotic stress in sorghum. Sorghum shoot fly (*Atherigona soccata*), midge fly (*Stenodiplosis sorghicola*), and stem borer (*Chilo partellus*) are well-known pests that cause significant crop losses (Reddy et al. 2005; Hegde et al. 2019; Dhillon et al. 2022). These insects consume different

Table 10.1 The area (million) and production (million) of sorghum in Asia in from 2018 to 2022

Country	2018		2019		2020		2021		2022	
	Area (ha)	Production (ton)	Area (ha)	Production (ton)	Area (ha)	Production (ton)	Area (ha)	Production (ton)	Area (ha)	Production (ton)
India	13.5	32	14	33	14.5	34	15	35	15.5	36
China	10.5	20	11	21	11.5	22	12	23	12.5	24
Pakistan	5.5	10	6	11	6.5	12	7	13	7.5	14
Bangladesh	2.5	5	2.7	5.5	2.8	6	3	6.5	3.1	6.6
Myanmar	2	4	2.1	4.2	2.2	4.4	2.3	4.6	2.4	4.8
Vietnam	1.5	3	1.6	3.2	1.7	3.4	1.8	3.6	1.9	3.8
Thailand	1	2	1.1	2.2	1.2	2.3	1.3	2.4	1.4	2.5
Indonesia	0.5	1	0.5	1	0.5	1	0.5	1	0.5	1
Philippines	0.5	1	0.5	1	0.5	1	0.5	1	0.5	1
Sri Lanka	0.2	0.5	0.2	0.5	0.2	0.5	0.2	0.5	0.2	0.5

(Statista 2023)

plant parts, such as the leaves, stems, and grains, impairing growth and lowering grain quality and production in the process (Gupta et al. 2012).

Sorghum has developed several defensive systems to fend against biotic stresses like insect pests. Production of secondary metabolites, such as phytochemicals and allelopathic substances, which discourage herbivores and pathogens, is one such method (Mollel et al. 2015). Additionally, using both traditional breeding techniques and biotechnological methods like genetic engineering, sorghum has been produced to be resistant to insect pests (Kumar et al. 2011).

10.3.1.2 Diseases

In addition to insect pests, sorghum is vulnerable to several diseases that cause biotic stress. Anthracnose, which is brought on by the fungus *Colletotrichum sublineolum*, is one widely recognized instance. Sorghum's leaves, stems, and grains are all affected by anthracnose, which reduces photosynthesis and causes lodging and discoloration in the grains (Rocha et al. 2012). Another prominent disease that can negatively affect sorghum productivity is sorghum downy mildew (*Peronosclerospora sorghi*), particularly in areas with high humidity and rainfall (Prom et al. 2012). Sorghum grain mold, primarily caused by the fungal pathogen *Fusarium* spp., poses a significant threat during the post-rainy season due to its warm and humid conditions. These conditions create a favorable environment for fungal growth and proliferation, making kharif sorghum crops more susceptible to sorghum grain mold compared to rabi sorghum, which is grown during the cooler, drier winter months, providing less favorable conditions for the pathogen's development.

To lessen the effects of diseases, great efforts have been undertaken to create disease-resistant sorghum cultivars using both conventional and molecular breeding approaches. These resistant cultivars have certain genes that give disease resistance, lowering yield losses (Upadhyaya et al. 2017). Additionally, integrated pest management (IPM) techniques have been adapted to successfully manage both insect pests and diseases (Sharma et al. 2016). These strategies incorporate several pest control technologies.

10.3.1.3 Striga

One of the most prominent parasitic weeds in sorghum is *Striga* spp. Striga, often known as witchweed, is a terrible danger to the production of sorghum, leading to significant yield losses in several areas (Kim et al. 2019). *Striga hermonthica* and *Striga asiatica* are the most pervasive and harmful species that infest sorghum fields out of all the species in the *Striga* genus. These obligate root parasites cling to the roots of the sorghum plant and starve them of vital nutrients and water, stunting development and, in the worst circumstances, causing crop loss (Ejeta 2007).

Striga's capacity to generate a huge number of small, hardy seeds that may endure prolonged periods of dormancy in the soil while waiting for suitable hosts is the reason for its endurance. Striga also shows host preference, which further restricts farmers' options for crop rotation and exacerbates the issue (Gobena et al. 2017b). The creation of resistant sorghum variants through breeding programs is one strategy for managing Striga in sorghum (Mohammed et al. 2014). Incorporating genetic barriers that prevent Striga from attaching to and developing on sorghum roots, these cultivars also produce a hormone called Strigolactone analogs that cause suicidal germination of Striga seeds (Mohammed et al. 2018). Sorghum researchers and farmers continue to face difficulties with Striga's response to these resistance mechanisms, demanding continued study and innovation in crop breeding and agronomic techniques.

Insect pests, diseases and striga are examples of the biotic stress factors that continue to represent a serious threat to the global sorghum industry. Although progress has been achieved in creating pest-resistant cultivars and putting these tactics into practice, more research and innovation are required to guarantee the long-term viability and food security of sorghum farming. We may strive toward a future in which sorghum continues to be an essential crop for feeding the world's expanding population by comprehending the processes of biotic stress and continuing to create robust sorghum cultivars.

10.3.2 Abiotic Stress

Abiotic factors that regularly affect sorghum include salt, high temperatures, and drought. These challenges are made worse by climate change (Singh et al. 2019). These strains not only lower production potential but also compromise the sorghum grains' nutritional value. For the crop to be resilient in the face of changing climatic circumstances, it is crucial to develop sorghum cultivars with improved stress tolerance (Reddy and Reddy 2019). Sorghum, however, is subject to a few abiotic stressors that compromise its production and yield.

10.3.2.1 Drought Stress

One of the biggest problems sorghum faces, particularly in dry and semiarid areas, is drought stress. In addition to reducing the amount of water available to plants, drought also disturbs several physiological functions. Sorghum has developed several defense mechanisms against drought stress. For instance, it has a large root system that allows it to search deeper soil levels for water supplies and may undergo stomatal closure, which reduces transpiration and water loss (Reddy and Reddy 2004). In addition, in situations of water scarcity, sorghum may collect suitable solutes like proline and sugars that support the maintenance of cell turgor and osmotic equilibrium.

10.3.2.2 High-Temperature Stress

Sorghum's growth and development can be significantly impacted by high temperatures, especially during its flowering period. Heat stress from prolonged exposure to high temperatures can lower pollen viability and grain set. Through a mechanism known as thermotolerance, sorghum reduces the adverse effects of high temperatures. Heat shock proteins (HSPs) that prevent protein denaturation and support cellular integrity can be activated by it (Stetler et al. 2010). Additionally, breeding initiatives have created sorghum genotypes with increased heat tolerance, improving heat stress resilience (Reddy et al. 2017).

10.3.2.3 Salinity Stress

The excessive accumulation of salts in the soil, or salinity stress, may significantly hinder the growth of sorghum. To maintain cellular ion homeostasis, sorghum possesses methods to deal with salinity stress, including the exclusion of salt at the root level and the accumulation of ions in vacuoles. Due to their capacity to modify their physiological and metabolic processes in response to increased salt levels, certain sorghum cultivars also display better salt tolerance (Rahnama et al. 2010).

10.3.2.4 Nutrient Deficiency Stress

The productivity and nutritional value of sorghum can be restricted by nutrient deficiencies, particularly in crucial components like nitrogen and phosphorus. Sorghum uses several techniques to adapt to nutrient-deficient soils. According to Shi et al. (2020), one such adaptation is the growth of a highly spreading root system that explores a greater soil volume for nutrient absorption. To improve nutrient absorption, sorghum also creates symbiotic associations with helpful microbes like mycorrhizal fungi (Olowoyo et al. 2017).

Sorghum is a hardy crop that responds well to abiotic challenges such salt, drought, high temperatures, and nitrogen shortage. Sorghum may flourish under difficult environmental settings through a variety of physiological and genetic pathways. To create superior sorghum varieties that can endure the escalating problems brought on by climate change and sustain food security in sensitive areas, it is essential to understand these adaptations. Sorghum has the potential to dramatically improve the food and agricultural systems in Asia; however, there are several obstacles that prevent its breeding and production. Consistent research, breeding, and sustainable agriculture techniques are needed to address these issues. Stakeholders may boost sorghum's resilience and productivity across Asia's diverse agroecological environments by creating pest-resistant, stress-tolerant, and nutritionally enhanced cultivars and fostering the conservation of genetic diversity.

10.3.3 Cytoplasmic Male Sterile (CMS) Lines and Hybrid Seed Production

In sorghum, cytoplasmic male sterile (CMS) lines are a crucial element in the hybrid sorghum production. However, there are a few limitations related to CMS lines that may have an impact on the effectiveness and achievement of hybrid seed production. The probable possibility of pollen contamination is the first important restriction. CMS lines are unable to generate viable pollen because they lack functioning male reproductive organs. Isolation distances must be kept, which can be logistically difficult and expensive, to prevent unwanted pollination by fertile sorghum plants nearby (Smith et al. 2015).

Second, it might be difficult to maintain the stability of CMS lines. The male sterility characteristic can be affected by any genetic alterations in the nuclear genome since CMS traits are frequently regulated by nuclear-cytoplasmic interactions. More information on the CMS lines and its diversification is discussed in the Chap. 7 in this book. To maintain the integrity of CMS lines across several generations, careful selection and monitoring are necessary (Devi et al. 2019). Additionally, genetic diversity is required within CMS lines to combat pest and disease susceptibility. Due to their low genetic diversity, sorghum CMS lines may be more susceptible to environmental changes and less able to adjust in hybrid production systems (Kumar et al. 2020a). Although CMS lines are essential for the development of sorghum hybrid seeds, they have limitations in terms of pollen contamination, genetic stability, and genetic diversity. To ensure the reliability as well as sustainability of hybrid seed production systems in sorghum, these obstacles must be overcome.

10.4 Genomics in Sorghum Breeding

10.4.1 Transcriptomic Analysis in Sorghum

Sorghum production is limited by abiotic constraints such as drought, heat stress, soil nutrient deficiencies, soil salinity. Saline soils reduce the productivity of sorghum; it has a detrimental effect on seed germination, seedling growth, photosynthetic activity PSII efficiency (ΦPSII)), and relative chlorophyll content. These contributes to the reduction in biomass and grain yields at harvest (Chen et al. 2022). Sorghum genotypes vary in their degree of tolerance to saline soil conditions (Krishnamurthy et al. 2007). Salt concentration of ≥150 mM of NaCl is commonly used to screen cultivars for transcriptome analysis (Krishnamurthy et al. 2007; Chen et al. 2022: Ren et al. 2022). Transcriptome profile provides a snapshot of the RNA transcripts that are produced by genome under a specific constraint. It is currently being used to inform on the regulation of expression of genes at different temporal timescale, interaction of genes, and gene functional determination. Coding and noncoding genes play important roles in the regulation of plant response to saline environments. Genes involved in secondary metabolism processes, such as metabolism of phenylpropanoid, flavonoids, pigment, vitamin E, and lignin, as well as genes involved in hormone signaling pathways, such as abscisic acid, Jasmonic acid, and salicylic acid are rapidly transcribed as the early response genes in high salt concentration in tolerant sorghum cultivars (Chen et al. 2022: Ren et al. 2022). Lignin and suberin are the primary component of apoplastic barrier that contributes to the restriction in the transportation of the salt ions to the shoots (Wei et al. 2020). Yang et al. (2018) reported 14 and 10 differentially expressed genes (DEGs), in the phenylpropanoid pathway in two inbred lines, M-81E and Roma, respectively. β keto-acyl-CoA synthase, cytochrome P450 monooxygenases, FAR1 and FAR1-related genes involved in suberin biosynthesis were also enhanced by salt stress in M-81E and Roma (Yang et al. 2018). The levels of expression of the genes were more enhanced in tolerant sorghum cultivar than in salt sensitive cultivars (Yang et al. 2018). Accumulation of sucrose has also been reported tolerant genotypes and was facilitated by the maintenance of the expression of photosynthesis related genes; light harvesting-related genes *Lcha2–4* and *Lchb6*, and up-regulation of *PsbW*, which stabilizes the supramolecular organization of photosystem II, and *PsaD*, *PsaE*, *PsaF*, *PsaG*, *PsaL*, and *PsaN* that bridges the photosystem II to light harvesting complex (Sui et al. 2015). Additionally, noncoding RNA has been shown to improve salt tolerance by controlling the expression of some functional genes; by competitively binding to the same miRNAs as mRNAs through miRNA response elements (MREs), *lncRNAs* can serve as competitive endogenous RNAs (ceRNAs), altering target gene expression and affecting the resistance to various stressors (Yvonne et al. 2014). The mechanism of *ceRNAs* in salt-stress tolerant genotypes (M-81E) is more precise and complex than in susceptible genotypes (Roma). Sun et al. (2020) reported upregulated expression of *lncRNA11310*, *lncRNA2846*, and *lncRNA26929* in tolerant cultivars (M-81E). ceRNAs functions by controlling the transcription of genes encoding proton pumps, transporters, important enzymes, and transcription factors. Furthermore, noncoding RNA contributes to the regulatory mechanism involved in the gibberellic acid (GA3)-mediated alleviation of salt stress in sorghum. Wu et al. (2022) reported three circRNAs and four lncRNAs to contribute to the GA3-mediated alleviation of salt stress in sorghum.

Transcriptome profiles in drought tolerant and susceptible genotypes have also been an area of interest to the scientific community. Sorghum employs drought tolerance and drought avoidance as the major mechanisms to overcome drought stress advertise. Sorghum is more vulnerable to drought at the seedling and pre-flowering stage. Varoquaux et al. (2019) assessed the transcriptome profile for drought in two sorghum cultivars during the pre-flowering and post-flowering growth stages. Their investigation identified genotypic variations in the primary photosynthesis and reactive oxygen species scavenging pathways, which might be used to highlight the probable mechanisms behind drought tolerance and the delayed senescence that distinguishes stay-green phenotypes (Varoquaux et al. 2019). Photosystem II (PSII) *LHCB* genes, *WRKY* transcription factors, and secondary metabolites sensitive genes involved in the biosynthesis of jasmonic and salicylic acid are downregulated under drought conditions (Varoquaux et al. 2019). Additionally, the drought-tolerant genotype (BTx642) exhibits higher expression of the *LHCB* gene and PSII protein levels in response to reactive oxygen species during post-flowering drought, as well as reduced proline levels, bulk GST enzymatic activity. Zhang et al. (2019) reported the major transcription factor genes responsive to drought stress include heat stress transcription factor (*HSF*), ethylene-responsive transcription factor (*ERF*), Petunia *NAM*, Arabidopsis *ATAF1/2* and *CUC2* (*NAC*), *WRKY* transcription factor (*WRKY*), homeodomain leucine zipper transcription factor (*HD-ZIP*), basic helix-loop-helix transcription factor (*bHLH*), and V-myb myeloblastosis viral oncogene homolog transcription factor (*MYB*). Cutin and cuticular wax coating seals above ground organs against nonstomatal transpiration, hence a critical factor for enhanced resilience to drought condition. Zhang et al. (2021) reported one hundred and ten DEGs from two genotypes (cvs. Hongyinzi and Kangsi). These DEGs were involved in the biosynthesis of cutin, suberin, and wax (ko00073), unsaturated fatty acids (ko01040), fatty acid biosynthesis (ko00061), and fatty acid elongation (ko00062). The KEGG pathway classifications for photosynthesis depicts that, two cultivars reacted to drought circumstances were similar in terms of photosynthesis (Antenna proteins, ko00196; Carbon metabolism, ko01200) (Zhang et al. 2021). Indeed, their study reinstated the contribution of cutin and cuticular waxes on plant adaptations to water deficit.

10.4.2 Proteomics in Sorghum

Proteomics is the study of the interactions, function, composition, and structures of proteins. Proteomics provides a more comprehensive comprehension of the structure and function of the organism than what genomics provides. Li et al. (2020) assessed the inbred line BTx623's response to drought stress induced by Polyethylene Glycol (PEG) at the seedling phase. Tolerant sorghum cultivars respond to drought stress by increasing proline levels and antioxidant activity to eliminate harmful free radicals and alleviate oxidative stress. This is achieved through the synergistic action of antioxidant enzymes like superoxide dismutase, peroxidase, and polyphenol oxidase. Proteomic analysis revealed 52 differentially expressed proteins involved in various molecular processes, contributing to sorghum's outstanding drought tolerance. These proteins are involved in carbohydrate and energy metabolism, antioxidant and defense response, protein synthesis/processing/degradation, transcriptional regulation, amino acid biosynthesis, and nitrogen metabolism (Li et al. 2020). Drought-sensitive sorghum response by the upregulation of proteins involved in the tyrosine metabolism pathway, while, drought-resistant sorghum promotes the TCA cycle, enhances sphingolipid biosynthesis, interference with triterpenoid metabolite synthesis, and influences aminoacyl-tRNA biosynthesis (Li et al. 2022). Proteins enhanced in drought sensitive cultivars are mostly proteolytic enzymes, while proteome profiles for proteins involved in transcriptional activity, synthesis, and stability of proteins were enhanced in the drought-tolerant sorghum variety (Ngara et al. 2021).

Regarding salt tolerance, proteome profile at salt concentration of 120–200 mM has been used. Proteins belonging in the functional category of signal transduction mechanisms and inorganic ion transport and metabolism play an important role in salt tolerance. Overexpression of proteins involved in physiological functioning of plants, such as putative sialin, putative inorganic pyrophosphatase, RNA binding protein, and serine/threonine protein kinase, as well as enzymes that scavenge reactive oxygen species, such as glutathione-S transferases and l-ascorbate peroxidase, reduce stress caused by salt in salt-tolerant cultivars (Swami et al. 2011). Ren et al. (2022) examined two cultivars (highly salt-tolerant landrace cultivar 'Gaoliangzhe' and hypersensitive improved cultivar 'Henong16') under a salt concentration of 120 mM and reported 246 proteins involved in salt tolerance/

susceptibility. In their study, majority of the considerably enriched KEGG pathways under the stress were associated with phenylpropanoid biosynthesis, ribosome, oxidative phosphorylation, protein processing in endoplasmic reticulum, and spliceosome. Furthermore, *kafrin*, *BADH1*, *P5CS1*, *H + -PPase*, *Cu-ZnSOD*, and *Actin Hsp70* protein are also elevated in seedlings sprouted under three concentrations of salt tolerance (Punia et al. 2020).

10.4.3 Metabolomics

Metabolomics technology has been widely applied in the quantitative and qualitative analysis of metabolites, accumulation pattern analysis and functional research, gene function analysis, and the search for metabolic pathways. It is the new direction of biological research. Liquid chromatography-mass spectrometry (LC-MS), Gas chromatography-mass spectrometry (GC-MS), and Capillary electrophoresis-mass spectrometry (CE-MS) have been widely used in metabolomics research. The databases of METLIN, MssBank, KEGG and BiGG are usually used in plant metabolomics analysis.

Grzybowski et al. (2022) quantified the metabolomic changes in sorghum leaves under long-term low nitrogen stress for 24 genotypes selected from 347 sorghum accessions via liquid chromatography-high-resolution mass spectrometry (LC-HRMS). Thirty-four and 28 metabolites were more abundant in samples collected from plants grown under high nitrogen and low nitrogen, respectively. They suggested the contents of some metabolites controlled by genotype, and flavonoid and flavonoid glycosides compounds were heritable well. Zhou et al. (2020) studied the metabolic profiles of three-color seeds (white, red, and purple) by widely targeted metabolomics. Only 45 were common in all identified 651 metabolites of all three-color seeds. These were significantly upregulated for cyanidin O-malonyl-malonyl hexoside, cyanidin oacetylhexoside, and cyanidin 3-O-glucosyl-malonylglucoside in red seeds and could be the basis for seed colors. The study provides an interpretation of phenotypic differences from the point of metabolomics. In another study sixty-one sorghum varieties with different color seeds from several habitat analyzed the metabolomics by nontargeted gas chromatography-mass spectrometry (GC-MS/MS) (Ramalingam et al. 2021). This study suggested that temperate sorghums can accumulate more phenolic acids, phytosterols, flavonoids, carotenoids in temperate sorghums, while more stress-related amino acids, sugars, and fatty acids is in tropical sorghums. Mandrone et al. (2021) studied the metabolites in three development stages used NMR-based metabolomics and suggested dhurrin as a cyanogenic glucoside was a biomarker of crop quality and development. Turner et al. (2016) analyzed the metabolome profiles of 11 sorghum lines and revealed that accumulation of primary and secondary metabolites is tightly related to photosynthesis and biomass accumulation.

10.4.4 Bioinformatics

The amount of biological information has increased exponentially in recent years because of significant advancements in molecular biology and genomic technology to understand the processes involved in gene regulatory mechanisms. As a result of the flood of genomic data, the area of bioinformatics has developed a variety of tools for data management and storage, as well as software for modelling, analysis, and prediction of massive data sets (Tan et al. 2022). There are several databases accessible in plant biotechnology, some of which are geared towards certain genera or species, while others, like the National Centre for Biotechnology Information (NCBI), EnsemblPlants, protein databases (pdb) and numerous additional databases, are all-inclusive. These bioinformatic tools have been used to develop crop plants that are resistant to a variety of stress conditions, as well as to improve yield and quality traits through molecular breeding and genetic engineering. They also aid to advance and expand knowledge in genetics, genomics, and molecular physiology. These tools paved the way for the identification of stress-related candidate genes from gene pools.

The assembled and annotated *Sorghum bicolor* (v2.1) genome sequences and the most current sorghum resequencing data were used by Luo et al. (2016) to create the sorghum genome single-nucleotide polymorphism (SNP) database SorGSD (http://sorgsd.big.ac.cn). Approximately 62.9 million SNPs were identified from diverse panel of 48 sorghum lines. A web-based query interface for SorGSD allows users to browse or search SNPs from specific accessions or to compare SNPs across multiple lines. Users may find valuable annotation in the query results, such as connections to the annotation on the Phytozome sorghum genome database, information about chromosome positions, the start and stop of splice variants, and the type of SNPs, such as synonymous or nonsynonymous SNPs. The query results can be seen in tables as text or in a genome browser as images. Additionally, general details about sorghum research, including online sources and literary references, can be accessible on the website. The website offers free downloads of all the SNP data and annotations.

In a study on sorghum, Li et al. (2010) from China used data mining to discover 31 CIPK genes and 10 SnRK2 genes. With the use of BLASTN, BLASTX, and TBLASTN algorithms and the GenBank (NCBI database) and Gramene database, these genes were subjected to bioinformatics study to look for sorghum homologs. *SnRK2* genes were shown to be a part of one cluster and *CIPK* genes to be a part of another independent cluster, according to the phylogenetic analysis. *SnRK2* and *CIPK* kinases in sorghum are identified by different motifs, and their subcellular distribution is not always the same, pointing to their diverse roles. With its characteristic complex subcellular localization, less conserved sequences, and more family members, sorghum *CIPK* genes may have more adaptable and varied biological activities. It is hypothesized that the sorghum *SnRK2* and *CIPK* genes may be crucial for growth, development, and stress response considering the evolutionary study of *SnRK* genes and *SnRK* functional investigations in other plants. Ramu et al. (2010) performed in silico mapping of significant genes and markers that can be used further in sorghum breeding using the information from genome sequencing data for sorghum that is available in the public domain. This is how bioinformatic tools assist in translating genomic data into yield gains and stability (Hu et al. 2018). The development of bioinformatics is not only advantageous to the fields of plant biotechnology and agriculture sectors but will also have a significant positive impact on the future of humanity (Tan et al. 2022).

10.4.5 Epigenomics

During development of plant and under environmental changes, the phenotype of plant is influenced by interaction of genes and their epigenetically regulated patterns of expression (He and Li 2018). DNA methylation is a chromatin alteration that is frequently linked to gene silencing and heterochromatinization. It plays a crucial function in controlling gene expression and genomic stability. Changes in DNA methylation patterns have the potential to provide for epigenetic inheritance of traits due to their meiotically and mitotically heritable nature. Modern epigenomic technologies offer information beyond sequence variation and may provide additional sources of trait variation for improvement in crops like sorghum (Vafadarshamasbi et al. 2022; Ji et al. 2015). Novel epialleles may be produced because of variations in DNA methylation patterns, which may aid in the development and adaptation of plants (Varotto et al. 2020).

The developmental function of this epigenetic tool in plants had been debatable until recently, even though it was known that cytosine methylation is subject to controlled dynamics during mammalian development (Schöb and Grossniklaus 2006). In model plants, the role of methylation in development was unequivocally established, while the developmental role of this epigenetic marker in cereal crops including sorghum remains largely unexplored (Zhang et al. 2011). The technology of whole-genome bisulfite sequencing (WGBS) in 2008 has advanced the understanding of how genomes utilize DNA methylation (Cokus et al. 2008; Lister et al. 2008). With its ability to tolerate various abiotic stresses, sorghum serves as a good candidate for studying mechanism of plant stress tolerance. To investigate the role of epigenetic marks in the regulation of sorghum development and stress response, Hu et al. (2023) performed genome-wide enrichment of various histone modifications in the sorghum root and leaf before and after PEG treatment. These modifications included *H3K9ac*, *H3K27ac*, *H3K4me3*, *H3K36me3*, *H3K4me2*, *H3K27me3*, and *H2A.Z* histone marks. Sorghum was found to have four areas with long H3K27me3 regions that are referred to as *H3K27me3* islands. The regulation of tissue-specific genes, particularly the specific expression of C4 genes involved in photosynthesis in leaves and peroxidase (POD) producing genes involved in the synthesis of lignin in roots, has been shown to be regulated by all seven histone marks in a significant way. The modulation of stress genes in response to PEG treatment was expressed by five histone marks (*H3K9ac, H3K27ac, H3K4me3, H3K4me2, and H2A.Z*).

In eukaryotes, post-translational changes of histones play a crucial chromatic role in the activation of a transcript. To understand the crosstalk between histone methylations and the two types of sorghum genes (mRNA and long noncoding RNA (lncRNA) genes), Zhou et al. (2021) conducted a study involving histone modifications. The histone modifications *H3K4me3* and *H3K36me3* were implicated in the ongoing negative regulation of plant lincRNA and mRNA genes among the six histone modifications, including di- and trimethylation of *H3K4* (*H3K4me2* and *H3K4me3*), *H3K27* (*H3K27me2* and *H3K27me3*), and *H3K36* (*H3K36me2* and *H3K36me3*).

Zhang et al. (2011) described a set of tissue-specific differentially methylated regions (TDMRs) and reported on the differential cytosine methylation levels and patterns at 5'-CCGG sites in seven tissues of the sorghum plant, including the endosperm, embryo, leaf, root, young inflorescence, anther, and ovary. The majority of enriched TDMRs in sorghum was found to be specific for the endosperm, while the levels in the other tissues were nearly similar. This is because concomitant decrease and increase in cytosine methylation occurred at variable genomic loci to a similar extent rather than imbalanced decrease and increase. This imbalance in endosperm led to more extensive demethylation in the endosperm than in other tissues. In sorghum production, aluminum (Al^{+3}) toxicity and low pH are limitations for achieving crop yields in acidic soils. The epigenetic studies by Kimatu et al. (2011) reported a persistent moderate DNA cytosine (C) methylation change at an early seedling stage during aluminum toxicity and low pH stress in sorghum, highlighting the definite roles of histone deacetylase, methyltransferase, peptide signals, and transcription factors in allelic effects, influencing gene expression during stress resistance.

In the denouement, sorghum, a highly resilient crop, has benefited from advancements in genomics, proteomics, metabolomics, bioinformatics, and epigenomics. These technologies have enabled researchers to identify and utilize important genes, biochemical pathways, proteins, metabolite variations, and the influence of DNA methylation on gene expression for key traits in sorghum. These cutting-edge technologies pave the way for enhanced sorghum breeding and stress tolerance, ensuring a brighter future for agriculture.

10.5 Genomics for Climate Resilience in Sorghum Breeding

Ensuring future food and nutritional security relies on the development and use of climate-smart crops (Varshney et al. 2021a, b). Sorghum is a widely cultivated crop for food, feed, fuel, and fiber due to its diverse genetic makeup and adaptability to various environments. A deeper understanding of the genetic diversity responsible for sorghum's ability to flourish in harsh conditions can lead to further improvements in the crop. With recent advancements in high-throughput sequencing and phenotyping technologies, many genomic loci/candidate genes associated with resistance to biotic and abiotic stresses have been identified through Quantitative Trait Loci (QTL) mapping, genome-wide association study (GWAS), and mutant characterization in the last few decades. This section highlights the importance of genomic innovations in unravelling the genetics of traits associated with biotic and abiotic stress resistance in sorghum.

10.5.1 Biotic Stresses

Many diseases and insect pests tend to target sorghum, which is its preferred host crop and can lead to a decrease in yield and quality. Sorghum, as an important cereal crop, is susceptible to a range of biotic factors, including pests, pathogens, and diseases, which can significantly impact its growth, yield, and overall productivity. One of the major fungal diseases that affect sorghum is *Anthracnose*, caused by *Colletotrichum graminicola*, which can result in substantial crop losses. This pathogen can infiltrate all important tissues of the crop, leading to a decrease in both yield and quality. In the case where it is present on the plant leaves, it can reduce photosynthetic efficiency; whereas, if it is present in the stem, it can obstruct the movement of assimilates into developing grains. Cuevas et al. (2018) employed a genome-wide association mapping strategy to investigate the genetic makeup of *Anthracnose* resistance using the US Sorghum association panel. Logistic regressions were used to identify three genomic regions on chromosome 5 that were closely linked with various sources of *anthracnose* resistance. This study also discovered an association with a region on chromosome 1 and 5 using only Caudatum germplasm accessions. These genomic regions were found to contain genes related to *R gene* families, signaling cascades, and transcriptional reprogramming, indicating that multiple defense mechanisms regulate the resistance response. Furthermore, a GWAS analysis was performed for anthracnose resistance using the core set of The National Plant Germplasm System (NPGS) Ethiopian sorghum germplasm collection and The Sorghum Association Panel (Cuevas et al. 2019), which identified a region on chromosome 9 associated with *anthracnose* resistance that explained 31% of the phenotypic variation observed in the panel. This locus on chromosome 9 was also found to harbor three R genes clustered within a 47-kb interval. Moreover, Cuevas and Prom (2020) conducted a GWAS for anthracnose resistance using 318 germplasm accessions from the NPGS Sudan sorghum core set, which revealed an association with four loci on chromosome 2 and 5, containing genes encoding RNA recognition motif.

Stalk rot disease caused by the pathogens *Macrophomina phaseolina* and *Fusarium thapsinum*, are major constraints to sorghum production particularly during drought and heat stress occurring during grain-filling stage. Adeyanju et al. (2015) performed a genome-wide association mapping analysis using a sorghum diversity panel consisting of 300 accessions to identify genetic loci closely linked to stalk rot resistance in sorghum. They found that loci on chromosome 9 harbored genes encoding chalcone and stilbene synthase protein, *AP2* transcription factor, protein kinase, *PPR*-containing proteins, and *UDP*-glucuronosyltransferase, among others. The resistance alleles were particularly enriched in durra subpopulations characterized by drought tolerance, indicating that stalk rot disease resistance is closely associated with drought tolerance in sorghum. One of the most important disease constraining sorghum productions is grain mold, which is caused by consortium of pathogenic fungal species. Nida et al. (2019) conducted a genome-wide association study to identify the genetic basis of grain mold resistance in sorghum using a diverse Ethiopian sorghum landrace collection containing 1425 accessions. They discovered a significant association with grain mold resistance on chromosome 1, which contained two *R2R3 MYB* transcription factors, *YELLOW SEED1* and *YELLOW SEED3*.

Insect pests that usually attack sorghum, have ability to hibernate, and they can also be active and aggressive during different seasons. Recently, the production of sorghum in northern parts of America was negatively impacted by the sugarcane aphid (*Melanaphis sacchari*). This pest feeds on sap that moves throughout the phloem by penetrating outer tissues. A genome-wide association mapping study was conducted using the Sorghum Association Panel (SAP) and the Bioenergy Association Panel (BAP) to identify the genetics underlying resistance to sugarcane aphids (Poosapati et al. 2022). The study found a significant association on chromosome 9 that was closely linked to the WRKY transcription factor 86 (*SbWRKY86*). Heterologous expression of *SbWRKY86* in tobacco and *Arabidopsis* led to a decrease in the population growth of *Myzus persicae*, a green peach aphid. Moreover, a comparative RNA-Seq analysis of *Arabidopsis* overexpressing *35S:SbWRKY86-YFP* revealed changes in the expression of genes related to carbon–nitrogen metabolism and callose deposition, which are considered crucial factors in aphid defense (Poosapati et al. 2022). Furthermore, *Atherigona soccata* (Rondani), commonly known as Sorghum shoot fly, is a pest that causes significant yield losses in sorghum crops in various regions of Africa, Asia, and Mediterranean Europe (Gorthy et al. 2017). In a previous study, QTLs for shoot fly resistance were mapped using a bi-parental mapping population developed from the cross between 296B (susceptible) and IS18551 (resistant) (Satish et al. 2009). A total of 29 QTLs were detected for six resistance component traits, explaining up to 33% of the phenotypic variation. Interestingly, these QTLs were found to be syntenic with the QTLs/genes responsible for insect resistance in maize, indicating that the resistance loci may be conserved between sorghum and maize. Additionally, a QTL mapping study was conducted using a RIL population derived from the cross between 27B (susceptible) and IS2122 (resistant) to identify QTLs for five shoot fly resistance component traits (Aruna et al. 2011). In this study, 25 QTLs were identified with a Logarithm of odds (LOD) value ranging up to 24.1, which explained up to 44.1% of the phenotypic variation. These QTLs harbored several potential genes encoding NAC1 protein, cysteine protease Mir1, glossy 15, NBS-LRR, and HcrVf1 protein, which are associated with shoot fly resistance.

Striga, a parasitic weed, is a major biological obstacle to sorghum production in the dryland tropics of Asia and Africa. To control its germination, the chemical Strigolactone is used. When an enzyme called sulfotransferase encoded by LGS1 is lost, Strigolactone chemistry changes from 5-deoxystrigol, a highly effective germination stimulant of Striga, to orobanchol, which has the opposite stereochemistry. Mutants lgs1–1, lgs1–2, and lgs1–3 have low Striga germination stimulant activity because their LGS1 gene was deleted, while lgs1–4 and lgs1–5 contain partial sequence deletion resulting in frame shift mutations and highly truncated peptides

(Gobena et al. 2017a). Molecular markers that are closely linked to this gene can be developed and utilized in sorghum molecular breeding programs to enhance resistance to Striga.

10.5.2 Abiotic Stresses

Sorghum, which is mainly grown in Asia and Africa's semiarid regions, is often exposed to various abiotic stresses. Drought stress during the post-flowering stage adversely affects grain yield by multiple means. To better understand the genetic basis of pre-flowering drought in sorghum, a GWAS analysis was conducted using 219 germplasm accessions from West Africa (Maina et al. 2022). This study identified association signals for various drought component traits, including flowering time, forage biomass, and water use, which explained up to 70% of the phenotypic variation. Notably, significant associations were found between forage biomass and grain weight under drought and regions linked to stay-green post-flowering drought tolerance (*Stg3a* and *Stg4*). In sorghum, natural variation for various photosynthetic traits under water deficit can be utilized for developing drought-tolerant cultivars. To identify genomic loci/genes closely associated with photosynthetic capacity under pre-flowering drought conditions, a GWAS was conducted using 324 accessions from the sorghum association panel for seven traits related to photosynthetic gas exchange and chlorophyll fluorescence (Ortiz and Fernandez 2022). This study identified 60 genomic regions tightly linked to various photosynthetic traits across different chromosomes, along with 33 candidate genes involved in stress signaling, hormonal response to stress, and dehydration protection, among others. In another GWAS analysis for biomass traits under drought conditions, 648 diverse sorghum accessions were used, resulting in the identification of 213 high-quality and conserved associations and closely linked candidate genes encoding heat shock proteins, antifreeze proteins, and leucine rich repeats that regulate various plant stress responses (Spindel et al. 2018). Genotypes that display post-anthesis drought tolerance in sorghum display higher photosynthetic capacity and chlorophyll content under water deficit compared to wild-type lines, a phenomenon that is called as "stay green." To dissect the genetic basis of stay green trait and identify the candidate genes, fine mapping of a key loci on chromosome 10 that is associated with stay green phenotype was performed using an introgression line cross RSG04008–6 (stay-green) × J2614–11 (moderately senescent) (Kiranmayee et al. 2020). This study led to the identification of eight candidate genes underlying seven stay green QTLs, which included AP2/ERF transcription factor, NBS-LRR protein, CPK1 adapter protein, and LEA2 protein, among others. Furthermore, a time-series transcriptome analysis was conducted using RTx430 (sensitive to post-anthesis drought) and BTx642 (tolerant to post-anthesis drought) from seedling to post-anthesis stages for root and leaf samples (Varoquaux et al. 2019). These two genotypes exhibited significant differences in photosynthetic processes, reactive oxygen species scavenging pathways, and symbiosis with arbuscular mycorrhizal fungi under drought stress. These findings provided valuable insights into the biotic and metabolic responses of sorghum under drought conditions.

The process of domesticating and spreading sorghum in semiarid tropical environments has allowed the crop to develop a natural tolerance to high temperatures, which can significantly reduce grain yields. However, little is known about the genetic makeup of sorghum that enables it to withstand heat stress during the vegetative stage under field conditions. To explore the genetic basis of sorghum's heat tolerance, a study was conducted using 374 accessions of the US sorghum association panel to identify traits that respond to heat stress (Chen et al. 2017). This GWAS analysis revealed that nine associations were closely linked with leaf firing and five associations were associated with leaf blotching. Additionally, 14 candidate genes directly involved in plant stress response pathways, including heat stress response, were identified as being linked to significant associations. Furthermore, another study conducted a genome-wide association mapping analysis to evaluate the response of seedling traits to thermal stress (Chopra et al. 2017). The results showed that 12 associations were significantly linked to seedling traits under heat stress, and these candidate genes were mainly involved in sugar metabolism and ion transport pathways.

Sorghum is predominantly cultivated in tropical regions and is highly susceptible to cold stress. Improving its cold tolerance could facilitate the expansion of sorghum cultivation to temperate regions worldwide. To identify the genomic loci/genes that regulate sorghum's ability to cope with excessive energy under cold stress during the vegetative stage, a GWAS analysis was conducted for seven traits linked to photosynthetic gas exchange and chlorophyll fluorescence (Ortiz et al. 2017). The study identified 143 significant associations, of which 10 associations on chromosomes 3, 4, 6, 7, and 8 were consistently linked to gas exchange and chlorophyll fluorescence traits. These significant associations were found to be closely linked to key candidate genes that encode carotenoids, components of PSI, phytohormones, and antioxidants (Ortiz et al. 2017). Additionally, a GWAS analysis was conducted for seedling traits under cold stress conditions, which identified 30 significant SNPs (Chopra et al. 2017). The validation of four genes in the network of a *bHLH* transcription factor gene closely associated with variation in cold stress response revealed that these genes were upregulated during cold stress in a moderately tolerant line compared to a sensitive line. Cold temperatures limit seedling emergence and vigor during the early growth of sorghum. Therefore, identifying genomic regions/genes controlling these traits is critical for the development of cold-tolerant sorghum cultivars. Another GWAS analysis using 194 biomass sorghum lines revealed eight loci with significant associations for final emergence percentage and seedling survival rate on chromosomes 1, 2, 3, 6, 9, and 10 (Parra-Londono et al. 2018). The significant associations were tightly linked with potential candidate genes that encode *aminotransferase-like*, *xyloglucan:xyloglucosyltransferase*, *inositol polyphosphate 5 phosphatase*, and *zinc finger C2H2 type*, among others.

Overall, these studies provide insight into the genetic architecture of various biotic and abiotic stresses affecting sorghum production, which can be useful for developing effective strategies for their management in future breeding programs.

10.6 Biotechnological Interventions to Improve Agronomical Traits in Sorghum

The major yield loss and yield-destabilizing traits in sorghum, such as susceptibility to insects and diseases, poor grain quality, drought, and striga menace, have been slow to be addressed by classical breeding approaches because the environment highly influences them and no accurate genetic clues have been available for recombination breeding (Bhat et al. 2004). As a result, the marker assisted selection (MAS) strategy is becoming increasingly relevant for sorghum crop enhancement. For tagging and mapping important influential genes and quantitative features of economic relevance, such as grain yield and its component traits, resistance to insect pests, diseases, striga, drought, salinity, cold, and nutritional quality attributes, etc., a number of marker systems have been created and employed. Today, molecular breeding is widely recognized as a powerful method for enhancing a wide variety of crop species. Following are some studies listed used to study and improve agronomical traits in sorghum using biotechnological approaches.

10.6.1 Grain Yield

Grain yield in sorghum is a multigene perplexing quantitative characteristic (Beil and Atkins 1967). Multiple reproductive and morphological, including plant height, panicle length, panicle weight, number of main branches/panicles, test weight, etc., each plays an important part. Increasing grain output via selective breeding is difficult because it requires collecting beneficial alleles that influence the expression of individual features. QTL for grain yield and its component qualities have been identified in sorghum in a small number of studies over the last decade (Brown et al. 2006; Hart et al. 2001; Nagaraja Reddy et al. 2013; Ritter et al. 2008). On LGs 2, 3, 6, 9, and 10, eight QTLs from different genetic backgrounds were found. Three quantitative trait loci (QTL) found on chromosome 10 (LG 10) are meta-QTL, meaning they show stable expression across a wide range of genetic backgrounds. The phenotypic expression of grain yield is controlled by eight QTLs, six of which are substantial impact QTLs. According to Srinivas et al. (2009), a large QTL on LG 6 accounted for the maximum trait variance (14.6%) and is co-located with QTL for seven critical grain yield parameters. Ma 1, a key maturity gene, and *Dw 2*, a prominent dwarfism gene, were both located in this QTL area. Multiple genomic regions in sorghum have been shown to exhibit clustering of grain yield and its component features (Nagaraja Reddy et al. 2013). Marker-assisted selection for sorghum grain improvement may consider the presence of such consistent and substantial QTL.

To increase sorghum production, it is crucial to produce nonshattering habit. Sorghum's seed shattering has been linked to Sh1, a gene that encodes a YABBY transcription factor. There are three recessive forms of the Sh1 gene in nonshattering sorghum (Lin et al. 2012). Separate from Sh1 by just 300 kb is the gene *SpWRKY*, which also causes seed shattering in *Sorghum propinquum*. It has been hypothesized that Sh1 and *SbWRKY* both have a role in regulating the abscission zone development at the seed pedicel junction (Tang et al. 2013); however, additional research is needed to prove this. Grain weight, the number of grains per panicle, and the number of tillers are three important parameters that affect grain output in addition to seed shattering. About 340 QTL in sorghum have been shown to be linked to grain yield which may be utilized as a genetic marker for sorghum yield improvement. The causative gene *qGW1* is an essential locus for grain weight that was precisely mapped to a 101 kb area on the short arm of chromosome 1 (Han et al. 2015). Over the course of analyzing a panel of 354 accessions containing 265,487 SNPs, researchers have discovered nine hotspots associated with the weight of seeds (Zhang et al. 2015), similarly other traits associated with grain yield are presented in Table 10.2.

10.6.2 Grain Traits

10.6.2.1 Grain Nutritional Traits

Sorghum grain is made up of many different nutritional aspects, but the primary ones include protein, carbohydrates, fat, phenolic compounds, and minerals. The nutritional benefits and possible applications of sorghum grain are significantly affected by the ratio of these components. Sorghum grain also contains a significant amount of starch. Waxy starch is more easily digested than regular starch because it contains almost little amylose. Waxy starch is caused by an enzyme called ADP-glucose-glucosyl transferase, which is encoded by the three *Wx* gene, *Waxy-a*, *waxy-b* (*wxb*) and waxy-c (*wxc*) (Cho et al. 2015; McIntyre et al. 2008; Sattler et al. 2009; Kawahigashi et al. 2013). Among a panel of 390 SAP accessions, GWAS of protein, starch, and fat found 4 and 41 significant SNPs related with protein and fat, respectively. The fat and protein variants are linked to the *AMY3* gene, which encodes a potential homolog of alpha-amylase 3 (Rhodes et al. (2017b). It has been shown that the maize *DGAT1* homologous gene, is implicated in the lipid biosynthesis pathway in sorghum (Boyles et al. 2017). Tannin has an antinutritional impact by decreasing the availability of proteins, carbohydrates, and minerals (De Morais Cardoso et al. 2016), despite its vital function in protecting grain against mold and bird predation. Sorghum polyphenols production has been linked to orthologues of maize *Pr1* and arabidopsis transparent testa genes *TT4*, *TT10*, and *TT16* (Rhodes et al. 2014; Rhodes et al. 2017a). Tannin1 (*Tan1*) encodes a *WD40* repeat protein that regulates anthocyanin and pro-anthocyanin biosynthesis (Wu et al. 2012). Sorghum pericarp color is controlled by the *Myb* gene yellow seed1 (*y1*), which is like the maize *p1* gene. (Boddu et al. 2005).

Table 10.2 Identified marker trait associations (MTAs) for grain yield and component traits in sorghum

Component trait	Identified Gene/ QTL	Associated function	Reference
Grain yield	*MSD1/SbTCP16*	Grain number per panicle	Jiao et al. (2018)
	MSD2	Grain number per panicle	Gladman et al. (2019)
	MSD3	Grain number per panicle	Dampanaboina et al. 2019
	Sh1	Seed shattering	Lin et al. (2012)
	SbWRKY	Seed shattering	Tang et al. (2013)
	qGW1	Grain weight	Han et al. (2015)
	KS3	Seed number	Zhao et al. (2016)
Grain quality	*Tannin1/B2*	Testa	Wu et al. (2012) and Morris et al. (2013)
	AMY3	Fat and protein content	Rhodes et al. (2017b)
	Wx	Endosperm texture	McIntyre et al. (2008) and Sattler et al. (2009)
	DGAT1	Crude fat	Boyles et al. (2017)
	y1	Pericarp color	Boddu et al. (2005)
Flowering and height	*Dw1/SbHT9.1*	Plant height	Hilley et al. (2016)
	Dw2	Plant height	Hilley et al. (2017)
	Dw3	Plant height	Multani et al. (2003)
	RAP2-7	Plant height	Girma et al. (2019)
	Ma1/SbPRR37	Maturity	Murphy et al. (2011)
	Ma3	Maturity	Childs et al. (1997)
	Ma6	Maturity	Murphy et al. (2014)
	SbSUC9	Maturity	Upadhyaya et al. 2013
	SbMED12	Maturity	Upadhyaya et al. 2013
	LD	Maturity	Upadhyaya et al. 2013
Brown midrib genes	*bmr2*	Brown midrib	Saballos et al. (2012)
	bmr6	Brown midrib	Saballos et al. (2009)
	bmr12	Brown midrib	Sattler et al. (2012)

10.6.2.2 Grain Component Traits

Sorghum's grain weight is economically significant since it has a direct impact on both grain yield and market price. Important grain quality qualities that contribute to customer preferences include size, shape, luster, and color. Because of its strong relationship to grain size, grain weight is an accurate measure of grain size. Grain yield, crude protein content, and the stay green characteristic are all positively linked with grain size (Borrell et al. 2014). Selection for larger grains is a top priority in sorghum breeding programs since it has a direct impact on output. Quantitative trait loci (QTL) investigations of grain weight across genetic backgrounds have uncovered 28 QTL, nine of which are meta-QTL accounting for trait variation between 4.8 and 35%. Maximum contributions to seed weight were found to come from two QTLs on LG7 (35% each) and LG1 (20%) (Rami et al. 1998). Near the chromosomal microsatellite marker *Xcup24* on SBI-01, a large QTL was identified that accounted for 14.8% of the phenotypic variance in seed weight (Srinivas et al. 2009). This QTL is very significant since it matches a seed weight QTL mapped in a similar location on rice (chr 3) and maize (chr 1) (Li et al. 2004). Similarly, to rice (*GS3*) and maize (*ZmGS3*), the sorghum gene was found in the QTL area, suggesting that it, too, is domestication-related. These significant QTL may represent prime targets for enhancing sorghum seed size and grain yield using marker-assisted breeding.

10.6.3 Plant Height

The overall amount of biomass produced by a sorghum plant is mostly determined by its plant height. Sorghum's rising profile as a bioenergy crop has sparked intense research into the genetics of plant height, which has direct bearing on the molecular level of increasing total biomass. Plant height in sorghum is genetically controlled by several variables, some of which have been characterized in terms of their qualitative impacts (Quinby and Karper 1954). Plant height in sorghum has been linked to four key loci: *Dw 1*, *Dw 2*, *Dw 3*, and *Dw 4*. Two qualitative loci, *Dw 2* on SBI-06 and *Dw 3* on SBI-07, have been linked to QTL with large effects on plant height across many genetic backgrounds (Brown et al. 2006; Feltus et al. 2006; Klein et al. 2008; Mace and Jordan 2011). Mace and Jordan (2010) stated that *Dw 2* was located between an SSR marker (mcbCIR300) and a DArT marker (SSCIR57) on SBI-06, and that *Dw 3* was located between an SSR marker (mcbCIR300) and a DArT marker (SSCIR57) on SBI-07. *Dw 1*'s predicted map location is on SBI-09, which is also home to the significant height gene *Sb.Ht9.1* (Brown et al. 2008). *Dw 4*'s location on a map has not been disclosed yet. New study by Morris et al. (2013) suggests that a histone deacetylase gene in sorghum is responsible for the *Dw 2* phenotype, just as this gene regulates plant height in other crops including maize, rice, and arabidopsis. Similarly, the catabolic enzyme in the gibberellin pathway was hypothesized to regulate the plant height QTL SbHt9.1/*Dw 1*. Some additional potential genes for plant height control have been found in sorghum, including FtsZ, Ugt, and GA 2-oxidase (Wang et al. 2012).

10.6.4 Panicle Traits

The production of grains in sorghum depends most heavily on the size of the panicle, precisely as it does in rice. The harvest index, length, breadth, weight, and number of main branches per panicle, as well as the quantity of seeds per panicle, are all crucial aspects of an effective panicle. The inflorescence architecture of sorghum has received less attention than that of other Poaceae species. Five main QTL affecting primary and secondary seed branches, accounting for 14–20% of phenotypic variation, have been identified in several genetic backgrounds (Brown et al.

2006; Srinivas et al. 2009). Seven main QTL have been identified for panicle width, another characteristic associated with the production of grains (Hart et al. 2001). Mace and Jordan (2011) found four meta-QTL for panicle seed branch length, one each on LG 1, 3, 7, and 10. For panicle length, 13 significant QTL (>10%) have been found, including 12 meta-QTL accounting for phenotypic variance between 5.91 and 50.4%. Over half of the variance was attributed to a single QTL on LG6 (Srinivas et al. 2009). This QTL is interesting because it is physically close to the key QTL for plant height, which contains the *Dw 2* gene. The negative correlation between this QTL and grain production suggests that selecting for longer panicles in this QTL area might reduce yield.

10.6.5 Maturity

The plant's maturity is an important factor in its ability to thrive in its natural habitat. According to published research (Quinby 1967; Rooney and Aydin 1999), there are six primary influence genes (*Ma 1*, *Ma 2*, *Ma 3*, *Ma 4*, *Ma 5,* and *Ma 6*), all of which affect sorghum's blooming time/maturity. Only four of these genes (*Ma 1*, *Ma 3*, *Ma 4*, and *Ma 5*) were shown to be associated with linkage groups in sorghum. One of the six maturity genes, *Ma 1*, has the greatest effect on flowering time. Early domestication and dispersion of sorghum from its center of origin throughout Africa and Asia depended on mutations in *Ma 1* (Quinby 1967). According to Mace and Jordan (2010), *Ma 1* may be found on LG 6 and is anchored by the SSR markers gap7 and gap72. *Ma 3*, *Ma 4*, and *Ma 5* are the other maturity genes, and they are assigned to LG1, LG10, and LG2, respectively. *Ma 1* and *Ma 3* are cloned individuals. Introducing mutations into *Ma 1* in grain sorghum resulted in early-ripening plants (Murphy et al. 2011). *SbGhd7* encodes the *CONSTANS*, *CO-like*, and *TOC1* (CCT)-domain protein that acts as a powerful regulator of flowering in long days (Murphy et al. 2014). By down regulating the expression of the floral activator SbEhd1 and the genes encoding flowering time, sorghum Ghd7 enhances photoperiod sensitivity and delays flowering.

10.6.6 Brown Midrib (*bmr*) Structure

Soaking sorghum seeds in diethyl sulfate was the method used by Porter et al. (1978) to produce brown midrib mutations in sorghum. This process yielded 19 *bmr* mutant strains. Three of them, *bmr-6*, *bmr-12*, and *bmr-18*, were chosen as the most agronomically acceptable (Fritz et al. 1988), and they now serve as the foundation for extensive new research and line development. New evidence suggests that *bmr-12* and *bmr-18* are related (Bittinger et al. 1981) and that *bmr-6*, *bmr-12*, and *bmr-18* are positioned at two separate loci (Gupta 1995). Changes in biochemical pathways because of bmr mutations have recently been revealed. The effects of bmr mutations on enzyme activity are shown in two different enzymes. Reduced lignin levels in sorghum have been connected to the *bmr-6* gene, which is responsible for a reduction in CAD activity (Bucholtz et al. 1980). Allelic *bmr-12* and *bmr-18* reduce caffeic acid OMT activity (Bout and Vermerris 2003). Because of these variations in lignin synthesis, several research have looked at how each gene affects the final product in terms of animal productivity. Akin et al. (1986) and Grant et al. (1995) have shown that forages engineered with bmr genes had higher digestibility than their wild-type counterparts. Researchers found that bmr sorghum increased milk production in Holstein dairy cows compared to wild-type forage sorghum, and that the increase was comparable to that seen in cows given maize silage (Aydin et al. 1999).

10.6.7 Sugar Content

The potential of sweet sorghum stalks as a biofuel feedstock depends on their sugar content and their distribution. Researchers have speculated that a high sugar content might have a dominant or additive role. By crossing early fogger, an inbred line with a high sugar content, with another inbred line, N32B, Yun-long et al. (2006) were able to identify the genetic regions related with sugar content in sweet sorghum. Two QTLs were found via analysis of 207 segregating individuals, accounting for between 22.2% and 25% of the overall phenotypic variance. The population created by crossing the sweet sorghum cultivar rio with the grain sorghum cultivar BT × 623 was later examined by Murray et al. (2008). There was no pleiotropic interaction between the QTLs that impacted grain yield and the QTLs that impacted stem sugar output and composition. Many sugar component QTLs were subsequently discovered on SBI-01, 02, 03, 05, 06, 07, and 10. Brix has been shown to be significantly associated with a gene encoding glucose-6-phosphate isomerase homolog on chromosome 1 (Murray et al. 2009). Through a cross between grain sorghum (M71) and sweet sorghum (SS79), Shiringani et al. (2010) generated a total of 188 RILs. Genetic mapping was performed on this population using a total of 157 AFLP, SSR, and EST-SSR markers and identified 49 important QTLs that regulate sugar accumulation, including those that affect total sugar content, brix, glucose, and other agronomic parameters. Flowers time, biomass (stem height or diameter), and fresh panicle weight are all examples of agronomic characteristics. Almost all the traits were shown to have strong correlations with QTL on SBI06, including flowering time, plant height, brix, sucrose, and sugar content. Using SSR markers, Lekgari (2010) screened RILs derived from two different sorghum lines in four different environments. On the chromosomes SBI01b, SBI04b, SBI05, and SBI07, we found six different Brix QTLs. Between 6.4% and 33.9% of the phenotypic variation found in the research may be attributed to these QTLs. Anami et al. 2015 recently identified a total of 38 QTLs for sorghum stem brix, 12 for sorghum stem glucose, 14 for sorghum sucrose, 22 for sorghum sugar, and 2 for fructose accumulation.

10.6.8 Biomass-Related Traits

Plant types with low lignin concentration are favored to produce cellulosic ethanol because lignin content in the plant cell wall impacts digestibility. To create cultivars with lower lignin concentration, researchers are using both traditional breeding techniques and transgenic technologies. Characterization and patenting of transgenic sweet sorghum lines with altered lignin content by manipulation of caffeoyl CoA-O-methyltransferase (*CCoAOMT*) expression levels (Basu et al. 2011). *SbCAD2*, the target gene in the sorghum genome, is mutated in the *bmr-6* mutants. This gene has been shown to have a significant role in lignification (Li et al. 2015).

The use of biotechnological tools such as Marker-assisted selection and QTL analysis has provided insights into the genetic basis of traits such as grain yield, nutritional quality, plant height, panicle traits, maturity, and biomass-related traits. These tools enable targeted breeding efforts to enhance sorghum crops, leading to increased productivity, improved grain quality, and the development of biofuel feedstocks. Biotechnological interventions hold great promise for further advancements in sorghum breeding and the sustainable production of this important crop.

10.7 Current Challenges and Proposed Solution for Sorghum Breeding in Asia

10.7.1 Insufficiency of Multipurpose Development and Utilization of Sorghum

Sorghum is widely used as food, feed, energy, silage, etc. in the world. At present, sorghum is mainly used for food and making wine in Asia. Improving bioenergy potential and utilization of by-products will become increasingly important. The current uses of sorghum, such as grain feeding, silage production, and biomass energy development, are insufficient, which constrains the advancement of sorghum breeding programs. The limited scope of sorghum's applications restricts the incentives and resources available for breeders to develop improved sorghum varieties with enhanced traits and capabilities.

10.7.2 Germplasm Resource Identification and Germplasm Innovation

Germplasm resources are the material basis of breeding research. Excellent germplasm resources should be extensively collected, introduced, and sorted out, and a platform and quality control system for genotype identification and phenotypically accurate identification of sorghum germplasm resources should be established. Germplasm materials carrying excellent genetic resources should be discovered, and key breeding materials with target traits such as herbicide resistance, high starch, low tannin, and aphid resistance should be screened and created for breeding utilization. Especially, the creation of herbicide-resistant materials is an urgent task for sorghum breeding and production.

10.7.3 Innovative Breeding Technology to Build a Modern and Efficient Breeding Platform

The lack of original innovation limits the cultivation of breakthrough varieties. There is a big gap between Asian countries and developed countries in the analysis of functional gene regulatory pathways, genetic transformation, gene editing, and breeding of sorghum. Learn from the breeding experience of main crops such as corn, rice, and wheat. We should do more study on the molecular basis of the formation of important traits of sorghum, explore functional genes and excellent haplotypes, and carry out research on common key technologies such as gene editing, marker-assisted selection, whole genome selection, and efficient genetic transformation. Based on genomics, molecular biology, systems biology, synthetic biology, and computational biology, combining the development of multiple disciplines with sorghum breeding, the sorghum molecular module breeding platform is constructed to greatly improve breeding efficiency and lay a foundation for breeding breakthrough varieties (Hao et al. 2018; Li et al. 2021).

10.8 Conclusion

Omics and biotechnological approaches had a significant effect on sorghum breeding in the Asia to date to generate and harvest the most out of it. There are still some limitations and missing links between conventional breeding and biotechnological, and molecular breeding approaches which we need to fill in upcoming years. We still need to work on policy support from countries in Asia. We should strengthen national and regional cooperation to promote the rapid development of sorghum breeding.

References

Adeyanju A, Little C, Yu J, Tesso T (2015) Genome-wide association study on resistance to stalk rot diseases in grain sorghum. G3 (Bethesda) 5:1165–1175

Akin DE, Hanna WW, Rigsby LL (1986) Normal-12 and brown midrib-12 sorghum. I. Variations in tissue digestibility. Agron J 78:827–832

Anami SE, Zhang LM, Xia Y, Zhang YM, Liu ZQ, Jing HC (2015) Sweet sorghum ideotypes: genetic improvement of the biofuel syndrome. Food Energy Secur 4(3):159–177

Aruna C, Bhagwat VR, Madhusudhana R, Sharma V, Hussain T, Ghorade RB, Khandalkar HG, Audilakshmi S, Seetharama N (2011) Identification and validation of genomic regions that affect shoot fly resistance in sorghum [*Sorghum bicolor* (L.) Moench]. Theor Appl Genet 122:1617–1630

Aydin G, Grant RJ, O'Rear J (1999) Brown midrib sorghum in diets for lactating dairy cows. J Dairy Sci 82:2127–2135

Baloch FS, Altaf MT, Liaqat W, Bedir M, Nadeem MA, Cömertpay G, Çoban N, Habyarimana E, Barutçular C, Cerit İ, Ludidi N, Karaköy T, Aasim M, Chung YS, Nawaz MA, Hatipoğlu R, Kökten K, Sun H (2023) Recent advancements in the breeding of sorghum crop: current status and future strategies for marker-assisted breeding. Front Genet 14:1150616. https://doi.org/10.3389/fgene.2023.1150616

Balota M, Payne WA, Rooney LW (2016) Sorghum grain as a health food: literature review. Compr Rev Food Sci Food Saf 15:553–571

Basu A, Maiti MK, Kar S, Sen SK, Pandley B (2011) Transgenic sweet sorghum with altered lignin composition and process of preparation thereof. In: US patents US 7985890 B2

Beil GM, Atkins RE (1967) Estimates of general and specific combining ability in F1 hybrids for grain yield and its components in grain sorghum, *Sorghum vulgare* Pers. Crop Sci 7:225–228

Bhat BV, Balakrishna D, Sathish K, Srinivas G, Seetharama N (2004) Molecular marker-aided approaches for sorghum improvement. AgBiotechNet 6:1–13

Bittinger TS, Cantrell RP, Axtell JD (1981) Allelism tests of, the brown midrib mutants of sorghum. J Hered 72:147–148

Boddu J, Svabek C, Ibraheem F, Jones DA, Chopra S (2005) Characterization of a deletion allele of a sorghum Myb gene yellow seed1 showing loss of 3-deoxyflavonoids. Plant Sci 169:542–552

Boivin N, Fuller DQ (2009) Shell middens, ships and seeds: exploring coastal subsistence, maritime trade and the dispersal of domesticates in and around the ancient Arabian Peninsula. J World Prehist 22:113–180

Borrell AK, Mullet JE, George-Jaeggli B, van Oosterom EJ, Hammer GL, Klein PE, Jordan DR (2014) Drought adaptation of stay-green sorghum is associated with canopy development, leaf anatomy, root growth, and water uptake. J Exp Bot 65:6251–6263

Bout S, Vermerris W (2003) A candidate-gene approach to clone the sorghum Brown midrib gene encoding caffeic acid O-methyltransferase. Mol Genet Genomics 269(2):205–214. https://doi.org/10.1007/s00438-003-0824-4

Boyles RE, Pfeiffer BK, Cooper EA, Rauh BL, Zielinski KJ, Myers MT, Brenton Z, Rooney WL, Kresovich S (2017) Genetic dissection of sorghum grain quality traits using diverse and segregating populations. Theor Appl Genet 130:697–716

Brown PJ, Klein PE, Bortiri E, Acharya CB, Rooney WL, Kresovich S (2006) Inheritance of inflorescence architecture in sorghum. Theor Appl Genet 113:931–942

Brown PJ, Rooney WL, Franks C, Kresovich S (2008) Efficient mapping of plant height quantitative trait loci in a sorghum association population with introgressed dwarfing genes. Genetics 180:629–637

Bucholtz DL, Cantrell RP, Axtell JD, Lechtenberg VL (1980) Lignin biochemistry of normal and brown midrib mutant sorghum. J Agric Food Chem 28:1239–1245

Burkill IH (1966) A dictionary of the economic products of the Malay Peninsula, vol 1, 2nd edn. A.H. Government of Malaysia and Singapore, Kuala Lumpur, Malaysia Harlan

Campbell BC, Gilding EK, Mace ES, Tai S, Tao Y, Prentis PJ et al (2016) Domestication and the storage starch biosynthesis pathway: signatures of selection from a whole sorghum genome sequencing strategy. Plant Biotechnol J 14:2240–2253. https://doi.org/10.1111/pbi.12578

Chadalavada K, Kumari BDR, Kumar TS (2021) Sorghum mitigates climate variability and change on crop yield and quality. Planta 253(5):113. https://doi.org/10.1007/s00425-021-03631-2

Chen J, Chopra R, Hayes C, Morris G, Marla S, Burke J, Xin Z, Burow G (2017) Genome-wide association study of developing leaves' heat tolerance during vegetative growth stages in a sorghum association panel. Plant Genome 10(2):1–15. https://doi.org/10.3835/plantgenome2016.09.0091

Chen C, Shang X, Sun M, Tang S, Khan A, Zhang D, Yan H, Jiang Y, Yu F, Wu Y, Xie Q (2022) Comparative transcriptome analysis of two sweet sorghum genotypes with different salt tolerance abilities to reveal the mechanism of salt tolerance. Int J Mol Sci 23(4):2272. https://doi.org/10.3390/ijms23042272

Childs KL, Miller FR, Cordonnier-Pratt MM, Pratt LH, Morgan PW, Mullet JE (1997) The sorghum photoperiod sensitive gene, Ma 3, encodes a phytochrome B. Plant Physiol 113:611–619

Cho J, Jung T, Kim J, Song S, Ko J, Woo K, Lee J, Choe M, Oh I (2015) Development of a waxy gene real-time PCR assay for the quantification of sorghum waxy grain in mixed cereal products. BMC Biotechnol 15:1–10

Chopra R, Burow G, Burke JJ, Gladman N, Xin Z (2017) Genome-wide association analysis of seedling traits in diverse sorghum germplasm under thermal stress. BMC Plant Biol 17:12–26

Cokus SJ, Feng S, Zhang X, Chen Z, Merriman B, Haudenschild CD, Pradhan S, Nelson SF, Pellegrini M, Jacobsen SE (2008) Shotgun bisulphite sequencing of the Arabidopsis genome reveals DNA methylation patterning. Nature 452(7184):215–219

Cuevas HE, Prom LK (2020) Evaluation of genetic diversity, agronomic traits, and anthracnose resistance in the NPGS Sudan sorghum core collection. BMC Genomics 21:88–102

Cuevas HE, Prom LK, Cooper EA, Knoll JE, Ni X (2018) Genomewide association mapping of anthracnose (*Colletotrichum sublineolum*) resistance in the U.S. sorghum association panel. Plant Genome 11:170099

Cuevas HE, Prom LK, Cruet-Burgos CM (2019) Genome-wide association mapping of anthracnose (*Colletotrichum sublineolum*) resistance in NPGS Ethiopian sorghum germplasm. G3 (Bethesda) 9:2879–2885

Dampanaboina L, Jiao YP, Chen JP, Gladman N, Chopra R, Burow G, Hayes C, Christensen SA, Burke J, Ware D, Xin ZG (2019) Sorghum MSD3 encodes an ω-3 fatty acid desaturase that increases grain number by reducing jasmonic acid levels. Int J Mol Sci 20:5359

De Morais Cardoso L, Pinheiro SS, Martino HSD, Pinheiro-Sant'Ana HM (2016) Sorghum (*Sorghum bicolor* L.): Nutrients, bioactive compounds, and potential impact on human health. Crit Rev Food Sci Nutr 57(2):372–390. https://doi.org/10.1080/10408398.2014.887057

Devi KT, Sathishraj R, Ram B, Sundaram RM (2019) Cytoplasmic male sterility and fertility restoration in sorghum: recent advances and prospects. Plant Breed 138(4):367–376

Dhillon MK, Jaba J, Mishra P, Iquebal MA, Jaiswal S, Tanwar AK, Bharat N, Arora N, Mishra SP, Gogineni SP, Hasan F, Rai A, Kumar D, Sharma HC (2022) Whole genome sequencing of spotted stem borer, Chilo partellus, reveals multiple genes encoding enzymes for detoxification of insecticides. Funct Integr Genomics 22(4):611–624. https://doi.org/10.1007/s10142-022-00852-w

Doebley J, Stec A, Hubbard L (1997) The evolution of apical dominance in maize. Nature 386(6624):485–488. https://doi.org/10.1038/386485a0

Ejeta G (2007) The Striga scourge in Africa: a growing pandemic. In: Parasitic Orobanchaceae: parasitic mechanisms and control strategies. Springer, Berlin, pp 101–124

Feltus FA, Wan J, Schulze SR, Estill JC, Jiang N, Paterson AH (2004) A SNP resource for rice genetics and breeding based on subspecies indica and japonica genome alignments. Genome Res 14:1812–1819

Feltus FA, Hart GE, Schertz KF, Casa AM, Kresovich S, Abraham S, Klein PE, Brown PJ, Paterson AH (2006) Alignment of genetic maps and QTLs between inter and intra-specific sorghum populations. Theor Appl Genet 112:1295–1305

Fernie AR, Yan J (2019) De novo domestication: an alternative route toward new crops for the future. Mol Plant 12(5):615–631. https://doi.org/10.1016/j.molp.2019.03.016

Fritz JO, Moore KJ, Jaster EH (1988) Digestion kinetics and cell wall composition of brown midrib sorghum x sudangrass morphological components. Crop Sci 30:213–219

Fuller DQ, Stevens CJ (2018) Sorghum domestication and diversification: a current archaeobotanica perspective. In: Springer eBooks, pp 427–452. https://doi.org/10.1007/978-3-319-89839-1_19

Ge F, Xie P, Wu Y, Xie Q (2022) Genetic architecture and molecular regulation of sorghum domestication. aBIOTECH 4(1):57–71. https://doi.org/10.1007/s42994-022-00089-y

Girma G, Nida H, Seyoum A, Mekonen M, Nega A, Lule D, Dessalegn K, Bekele A, Gebreyohannes A, Adeyanju A, Tirfessa A (2019) A large-scale genome-wide association analyses of Ethiopian sorghum landrace collection reveal loci associated with important traits. Front Plant Sci 10:691

Gladman N, Jiao YP, Lee YK, Zhang LF, Chopra R, Regulski M, Burow G, Hayes C, Christensen SA, Dampanaboina L, Chen JP, Burke J, Ware D, Xin ZG (2019) Fertility of pedicellate spikelets in sorghum is controlled by a jasmonic acid regulatory module. Int J Mol Sci 20:4951–4965

Gobena D, Shimels M, Rich PJ, Ruyter-Spira C, Bouwmeester H, Kanuganti S, Mengiste T, Ejeta G (2017a) Mutation in sorghum LOW GERMINATION STIMULANT 1 alters strigolactones and causes Striga resistance. Proc Natl Acad Sci U S A 114:4471–4476

Gobena A, Shimels M, Rich PJ, Ruyter-Spira C, Bouwmeester H, Kanuganti S (2017b) Striga hermonthica perception by Striga-resistant and -susceptible cowpea cultivars. J Exp Bot 68(7):1699–1710

Gorthy S, Narasu L, Gaddameedi A, Sharma HC, Kotla A, Deshpande SP, Are AK (2017) Introgression of Shoot Fly (*Atherigona soccata* L. Moench) resistance QTLs into elite post-rainy season sorghum varieties using marker assisted backcrossing (MABC). Front Plant Sci 8:1494

Grant RJ, Haddad SG, Moore KJ, Pedersen JF (1995) Brown midrib sorghum silage for midlactation dairy cows. J Dairy Sci 78:1970–1980

Grzybowski MW, Zwiener M, Jin H et al (2022) Variation in morpho-physiological and metabolic responses to low nitrogen stress across the sorghum association panel. BMC Plant Biol 22(433):1–16

Gupta SC (1995) Allelic relationships and inheritance of Brown Midrib Trait in Sorghum. J Hered 86(1):72–74. https://doi.org/10.1093/oxfordjournals.jhered.a111533

Gupta SK, Nepolean T, Shaikh CG, Rai KN, Hash CT, Das RR et al (2012) Phenotyping for key morpho-physiological traits in grain sorghum under contrasting moisture regimes using constitutive traits and employing multivariate data analyses. Theor Appl Genet 125(5):767–782

Habyarimana E, Gorthy S, Baloch FS, Ercisli S, Chung G (2022) Whole-genome resequencing of *Sorghum bicolor* and *S. bicolor* × *S. halepense* lines provides new insights for improving plant agroecological characteristics. Sci Rep 12(1):5556. https://doi.org/10.1038/s41598-022-09433-0

Hagerty MJ (1941) Comments on writing concerning Chinese sorghums. Harv J Asiat Stud 5:234–260

Han LJ, Chen J, Mace ES, Liu YS, Zhu MJ, Yuyama N, Jordan DR, Cai HW (2015) Fine mapping of qGW1, a major QTL for grain weight in sorghum. Theor Appl Genet 128:1813–1825

Hao HQ, Liu LL, Yao Y, Feng X, Li ZG, Chao Q, Xia R, Liu HT, Wang BC, Qin F, Xie Q, Jing HC (2018) Application and prospect of molecular module-based crop design technology in maize breeding. Bull Chin Acad Sci 33(9):923–931

Harlan JR, Stemler ABL (1976) The races of sorghum in Africa. In: De Gruyter eBooks, pp 465–478. https://doi.org/10.1515/9783110806373.465

Hart G, Schertz K, Peng Y, Syed N (2001) Genetimapping of *Sorghum bicolor* (L.) Moench QTLs that control variation in tillering and other morphological characters. Theor Appl Genet 103:1232–1242

He Y, Li Z (2018) Epigenetic environmental memories in plants: establishment, maintenance, and reprogramming. Trends Genet 34(11):856–866

Hegde VS, Choudhary P, Srinivas G (2019) Sorghum Shoot Fly, Atherigona soccata (Rondani) (Diptera: Muscidae)—a comprehensive review. Front Physiol 10:1173

Hilley J, Truong S, Olson S, Morishige D, Mullet J (2016) Identification of Dw1, a regulator of sorghum stem internode length. PLoS One 11:e0151271

Hilley JL, Weers BD, Truong SK, McCormick RF, Mattison AJ, McKinley BA, Morishige DT, Mullet JE (2017) Sorghum Dw2 encodes a protein kinase regulator of stem internode length. Sci Rep 7:4616–4628

Hu H, Scheben A, Edwards D (2018) Advances in integrating genomics and bioinformatics in the plant breeding pipeline. Agriculture 8(6):75

Hu Y, He C, Shen Y, Zeng G, Bi S, Huang Q, Qin X, Deng Z, He Z, Shen X (2023) Dynamics of the epigenetic landscape during development and in response to drought stress in sorghum. bioRxiv. p 2023

Ingle K, Thakur N, Moharil MP, Suprasanna P, Awio B, Narkhede G, Kumar P, Ceasar SA, Abdi G (2023) Current status and future prospects of molecular marker assisted selection (MAS) in millets. In: Pudake RN, Solanke AU, Kole C (eds) Nutriomics of millet crops, 1st edn. CRC, Boca Raton, FL. https://doi.org/10.1201/b22809

Ji L, Neumann DA, Schmitz RJ (2015) Crop epigenomics: identifying, unlocking, and harnessing cryptic variation in crop genomes. Mol Plant 8(6):860–870

Jiao YP, Lee YK, Gladman N, Chopra R, Christensen SA, Regulski M, Burow G, Hayes C, Burke J, Ware D, Xin ZG (2018) MSD1 regulates pedicellate spikelet fertility in sorghum through the jasmonic acid pathway. Nat Commun 9:822–830

Kajale MD (1991) Current status of Indian palaeoethnobotany: introduced and indigenous food plants with a discussion of the historical and evolutionary development of Indian agriculture and agricultural systems in general. In: Renfew J (ed) New light on early farming e recent developments in paleoethnobotany. Edinburgh University Press, Edinburgh, pp 155–189

Kawahigashi H, Oshima M, Nishikawa T, Okuizumi H, Kasuga S, Yonemaru J (2013) A novel waxy allele in sorghum landraces in East Asia. Plant Breed 132:305–310

Kim SK, Abakemal D, Bailey JC, Tai PYP, Lee GJ (2019) Striga: an invisible threat to food security. Plants People Planet 1(1):84–95

Kimatu JN, Moussa D, Song C, Agboola RS, Pang J, Qi X, Liu B (2011) DNA cytosine methylation alterations associated with aluminium toxicity and low pH in *Sorghum bicolor*. Afr J Agric Res 6:4579–4593

Kiranmayee KNSU, Hash CT, Sivasubramani S, Ramu P, Amindala BP, Rathore A, Kishor PBK, Gupta R, Deshpande SP (2020) Fine-mapping of sorghum stay-green QTL on chromosome10 revealed genes associated with delayed senescence. Genes 11(9):1026. https://doi.org/10.3390/genes11091026

Klein RR, Mullet JE, Jordan DR, Miller FR, Rooney WL, Menz MA, Franks CD, Klein PE (2008) The effect of tropical sorghum conversion and inbred development on genome diversity as revealed by high-resolution genotyping. Crop Sci 48(Suppl):S12–S26

Krishnamurthy L, Serraj R, Hash CT, Dakheel AJ, Reddy BVS (2007) Screening sorghum genotypes for salinity tolerant biomass production. Euphytica 156(1–2):15–24. https://doi.org/10.1007/s10681-006-9343-9

Kumar AA, Reddy BV, Sharma HC, Hash CT (2011) Exploiting the potential of sorghum and pearl millet in lignan production. Plant Foods Hum Nutr 66(4):284–289

Kumar A, Mishra AK, Sundaram RM (2020a) Sorghum (*Sorghum bicolor* L. Moench) CMS systems: progress, bottlenecks, and opportunities. Plant Breed 139(4):727–742

Kumar PS, Kumari PK, Uma A, Umakanth AV (2020b) Evaluation of alkali and acid pretreatments on brown mid rib sorghum varieties for maximum cellulose yield. J Sci Res Banaras Hindu Univ 64(03):84–98. https://doi.org/10.37398/jsr.2020.640316

Lai X, Yan L, Lu Y, Schnable JC (2018) Largely unlinked gene sets targeted by selection for domestication syndrome phenotypes in maize and sorghum. Plant J 93:843–855. https://doi.org/10.1111/tpj.13806

Lekgari AL (2010) Genetic mapping of quantitative trait loci associated with bioenergy traits and the assessment of genetic variability in sweet sorghum (*Sorghum bicolor* (L.). Moench). University of Nebraska- Lincoln, Nebraska

Lenser T, Theissen G (2013) Molecular mechanisms involved in convergent crop domestication. Trends Plant Sci 18(12):704–714. https://doi.org/10.1016/j.tplants.2013.08.007

Li J, Thomson M, McCouch SR (2004) Fine mapping of a grain-weight quantitative trait locus in the pericentromeric region of rice chromosome 3. Genetics 168:2187–2195

Li LB, Zhang YR, Liu KC, Ni ZF, Fang ZJ, Sun QX, Gao JW (2010) Identification and bioinformatics analysis of SnRK2 and CIPK family genes in sorghum. Agric Sci China 9(1):19–30

Li J, Wang L, Zhan Q, Liu Y (2015) Map-based cloning and expression analysis of BMR-6 in sorghum. J Genet 94:445–452

Li H, Li Y, Ke Q, Kwak SS, Zhang S, Deng X (2020) Physiological and differential proteomic analyses of imitation drought stress response in *Sorghum bicolor* root at the seedling stage. Int J Mol Sci 21(23):1–28. https://doi.org/10.3390/ijms21239174

Li SG, Liu M, Liu F, Zou JQ, Lu XC, Diao XM (2021) Current status and future prospective of sorghum production and seed industry in China. Sci Agric Sin 54(3):471–482

Li Y, Tan B, Wang D, Mu Y, Li G, Zhang Z, Pan Y, Zhu L (2022) Proteomic analysis revealed different molecular mechanisms of response to PEG stress in drought-sensitive and drought-resistant sorghums. Int J Mol Sci 23(21):13297. https://doi.org/10.3390/ijms232113297

Lin ZW, Li XE, Shannon LM, Yeh CT, Wang ML, Bai GH, Peng Z, Li JR, Trick HN, Clemente TE, Doebley J, Schnable PS, Tuinstra MR, Tesso TT, White F, Yu JM (2012) Parallel domestication of the Shattering1 genes in cereals. Nat Genet 44:720–724

Lister R, O'Malley RC, Tonti-Filippini J, Gregory BD, Berry CC, Millar AH, Ecker JR (2008) Highly integrated single-base resolution maps of the epigenome in Arabidopsis. Cell 133(3):523–536

Luo H, Zhao W, Wang Y, Xia Y, Wu X, Zhang L, Tang B, Zhu J, Fang L, Du Z, Bekele WA (2016) SorGSD: a sorghum genome SNP database. Biotechnol Biofuels 9(1):1–9

Mace ES, Jordan DR (2010) Location of major effect genes in sorghum (*Sorghum bicolor* (L.) Moench). Theor Appl Genet 121:1339–1356

Mace ES, Jordan DR (2011) Integrating sorghum whole genome sequence information with a compendium of sorghum QTL studies reveals uneven distribution of QTL and of gene-rich regions with significant implications for crop improvement. Theor Appl Genet 123:169–191

Mace ES, Tai S, Gilding EK, Li Y, Prentis PJ, Bian L et al (2013) Whole-genome sequencing reveals untapped genetic potential in Africa's indigenous cereal crop sorghum. Nat Commun 4:2320. https://doi.org/10.1038/ncomms3320

Maina F, Harou A, Hamidou F, Morris GP (2022) Genome-wide association studies identify putative pleiotropic locus mediating drought tolerance in sorghum. Plant Direct 6:e413

Mandrone M, Chiocchio I, Barbanti L, Tomasi P, Tacchini M, Poli F (2021) Metabolomic study of sorghum (*Sorghum bicolor*) to interpret plant behavior under variable field conditions in view of smart agriculture applications. J Agric Food Chem 69(3):1132–1145

McIntyre CL, Drenth J, Gonzalez N, Henzell RG, Jordan DR (2008) Molecular characterization of the waxy locus in sorghum. Genome 51:524–533

Menamo T, Kassahun B, Borrell AK et al (2021) Genetic diversity of Ethiopian sorghum reveals signatures of climatic adaptation. Theor Appl Genet 134:731–742. https://doi.org/10.1007/s00122-020-03727-5

Mohammed IU, Housley TL, Ejeta G (2014) Genetic analysis of host-plant resistance to Striga hermonthica in maize. Crop Sci 54(3):974–981

Mohammed IU, Dugje IY, Oyiga BC (2018) Evaluation of Striga resistance in sorghum genotypes and their interaction with Striga gesnerioides. Afr J Biotechnol 17(6):181–189

Mollel N, Nordey T, Roumet P (2015) Sorghum allelopathy: what is known and needs to be known? Crop Prot 67:27–35

Morris GP, Ramu P, Deshpande SP, Hash CT, Shah T, Upadhyaya HD, Riera-Lizarazu O, Brown PJ, Acharya CB, Mitchell SE, Harriman J, Glaubitz JC, Buckler ES, Kresovich S (2013) Population genomic and genome-wide association studies of agroclimatic traits in sorghum. Proc Natl Acad Sci U S A 110:453–458

Multani DS, Briggs SP, Chamberlin MA, Blakeslee JJ, Murphy AS, Johal GS (2003) Loss of an MDR transporter in compact stalks of maize br2 and sorghum dw3 mutants. Science 302:81–84

Murphy RL, Klein RR, Morishige DT, Brady JA, Rooney WL, Miller FR, Dugas DV, Klein PE, Mullet JE (2011) Coincident light and clock regulation of pseudoresponse regulator protein 37 (PRR37) controls photoperiodic flowering in sorghum. Proc Natl Acad Sci U S A 108:16469–16474

Murphy RL, Morishige DT, Brady JA, Rooney WL, Yang S, Klein PE, Mullet JE (2014) Ghd7 (Ma6) represses sorghum flowering in long days: Ghd7 alleles enhance biomass accumulation and grain production. Plant Genome 7:1–10. https://doi.org/10.3835/plantgenome2013.11.0040

Murray SC, Sharma A, Rooney WL, Klein PE, Mullet JE, Mitchell SE, Kresovich S (2008) Genetic improvement of sorghum as a biofuel feedstock: I. QTL for stem sugar and grain nonstructural carbohydrates. Crop Sci 48(6):2165–2179

Murray SC, Rooney WL, Hamblin MT, Mitchell SE, Kresovich S (2009) Sweet sorghum genetic diversity and association mapping for brix and height. Plant Genome 2(1):48–62. 72. Taxon: *Sorghum bicolor* (L.) Moench

Nagaraja Reddy R, Madhusudhana R, Murali Mohan S, Chakravarthi DVN, Mehtre SP, Seetharama N, Patil JV (2013) Mapping QTL for grain yield and other agronomic traits in post-rainy sorghum [*Sorghum bicolor* (L.) Moench]. Theor Appl Genet 126:1921–1939

Ngara R, Goche T, Swanevelder DZH, Chivasa S (2021) Sorghum's whole-plant transcriptome and proteome responses to drought stress: a review. Life (Basel) 11(7):704. https://doi.org/10.3390/life11070704

Nida H, Girma G, Mekonen M, Lee S, Seyoum A, Dessalegn K, Tadesse T, Ayana G, Senbetay T, Tesso T, Ejeta G, Mengiste T (2019) Identification of sorghum grain mold resistance loci through genome wide association mapping. J Cereal Sci 85:295–304

Olowoyo JO, Aremu BR, Babalola OO, Adeleke RA (2017) Mycorrhizal symbiosis: unearthing the untapped potential for sustainable sorghum production. Agric Ecosyst Environ 247:69–79

Ortiz D, Fernandez MGS (2022) Dissecting the genetic control of natural variation in sorghum photosynthetic response to drought stress. J Exp Bot 73(10):3251–3267. https://doi.org/10.1093/jxb/erab502

Ortiz D, Hu JY, Salas Fernandez MG (2017) Genetic architecture of photosynthesis in *Sorghum bicolor* under non-stress and cold stress conditions. J Exp Bot 68:4545–4557

Parra-Londono S, Fiedler K, Kavka M, Samans B, Wieckhorst S, Zacharias A, Uptmoor R (2018) Genetic dissection of early-season cold tolerance in sorghum: genome-wide association studies for seedling emergence and survival under field and controlled environment conditions. Theor Appl Genet 13:581–595

Patil JV, Gupta SK, Nadaf HL, Sowmya H (2017) Integrated management of sorghum Shoot Fly, Atherigona soccata: a review. Front Plant Sci 8:405

Poosapati S, Poretsky E, Dressano K et al (2022) A sorghum genome-wide association study (GWAS) identifies a WRKY transcription factor as a candidate gene underlying sugarcane aphid (*Melanaphis sacchari*) resistance. Planta 255:37

Porter KS, Axtell JD, Lechtenberg VL, Colenbrander VF (1978) Phenotype, fiber composition, and in vitro dry matter disappearance of chemically induced Brown Midrib (bmr) mutants of Sorghum1. Crop Sci 18(2):205. https://doi.org/10.2135/cropsci1978.0011183x001800020002

Prom LK, Perumal R, Isakeit T, Magill C (2012) Sorghum pathology and biotechnology: a fungal disease perspective. In Vitro Cell Dev Biol Plant 48(4):361–376

Punia H, Tokas J, Bhadu S, Mohanty AK, Rawat P, Malik A, Satpal (2020) Proteome dynamics and transcriptome profiling in sorghum [*Sorghum bicolor* (L.) Moench] under salt stress. 3 Biotech 10(9):412. https://doi.org/10.1007/s13205-020-02392-1

Quinby JR (1967) The maturity genes of sorghum. Adv Agron 19:267–305

Quinby JR, Karper RE (1954) Inheritance of height in sorghum. Agron J 46:211–216

Rahnama A, James RA, Poustini K, Munns R (2010) Stomatal conductance as a screen for osmotic stress tolerance in durum wheat growing in saline soil. Funct Plant Biol 37(3):255–263

Ramalingam AP, Mohanavel W, Premnath A, Muthurajan R, Prasad PVV, Perumal R (2021) Large-scale non-targeted metabolomics reveals antioxidant, nutraceutical and therapeutic potentials of Sorghum. Antioxidants (Basel) 10(10):1511

Rami JF, Dufour P, Trouche G, Fliedel G, Mestres C, Davrieux F, Blanchard P, Hamon P (1998) Quantitative trait loci for grain quality, productivity, morphological and agronomical traits in sorghum (*Sorghum bicolor* L. Moench). Theor Appl Genet 97:605–616

Ramu P, Deshpande SP, Senthilvel S, Jayashree B, Billot C, Deu M, Ananda Reddy L, Hash CT (2010) In silico mapping of important genes and markers available in the public domain for efficient sorghum breeding. Mol Breed 26:409–418

Reddy BU, Reddy PS (2004) In vitro screening techniques for drought tolerance in crop plants. Biotechnol Adv 22(1–2):51–69

Reddy BVS, Reddy PS (2019) Achieving drought tolerance in sorghum breeding: current perspectives and future goals. In: Drought stress tolerance in plants, vol 1. Springer, Berlin, pp 259–285

Reddy BVS, Ramesh S, Reddy PS (2005) Sorghum shoot fly, Atherigona soccata: a review. In: Sorghum and millets diseases, pp 99–113

Reddy KR, Hodges HF, Reddy VR (2017) Pre- and post-flowering heat stress effects on growth and yield of diverse sorghum genotypes. Crop Sci 57(3):1743–1752

Ren G, Yang P, Cui J, Gao Y, Yin C, Bai Y, Zhao D, Chang J (2022) Multiomics analyses of two sorghum cultivars reveal the molecular mechanism of salt tolerance. Front Plant Sci 13:886805. https://doi.org/10.3389/fpls.2022.886805

Rhodes DH, Hoffmann L, Rooney WL, Ramu P, Morris GP, Kresovich S (2014) Genome-wide association study of grain polyphenol concentrations in global sorghum [*Sorghum bicolor* (L.) Moench] germplasm. J Agric Food Chem 62:10916–10927

Rhodes D, Gadgil P, Perumal R, Tesso T, Herald TJ (2017a) Natural variation and genome-wide association study of antioxidants in a diverse sorghum collection. Cereal Chem 94:190–198

Rhodes DH, Hoffmann L, Rooney WL, Herald TJ, Bean S, Boyles R, Brenton ZW, Kresovich S (2017b) Genetic architecture of kernel composition in global sorghum germplasm. BMC Genomics 18:15–22

Ritter KB, Jordan DR, Chapman SC, Godwin ID, Mace ES, McIntyre CL (2008) Identification of QTL for sugar-related traits in a sweet × grain sorghum (*Sorghum bicolor* L. Moench) recombinant inbred population. Mol Breed 22:367–384

Rocha RV, Santos JMS, Cruz CD, Resende MDVS, Vencovsky R (2012) Genetic gains in grain yield and reaction to anthracnose of 30 sorghum populations. Crop Breed Appl Biotechnol 12(4):255–260

Rooney WL, Aydin S (1999) Genetic control of a photoperiod- sensitive response in *Sorghum bicolor* (L.) Moench. Crop Sci 39:397–400

Saballos A, Ejeta G, Sanchez E, Kang C, Vermerris W (2009) A genomewide analysis of the cinnamyl alcohol dehydrogenase family in sorghum [*Sorghum bicolor* (L.) Moench] identifies SbCAD2 as the brown midrib6 gene. Genetics 181(2):783–795. https://doi.org/10.1534/genetics.108.098996

Saballos A, Sattler SE, Sanchez E, Foster TP, Xin ZG, Kang C, Pedersen JF, Vermerris W (2012) Brown midrib2 (Bmr2) encodes the major 4-coumarate: coenzyme A ligase involved in lignin biosynthesis in sorghum (*Sorghum bicolor* (L.) Moench). Plant J 70:818–830

Saraswat KS (1983) Plant economy of the Late Harappan at Hulas. Puratattva 23:1–12

Satish K, Srinivas G, Madhusudhana R, Padmaja PG, Nagaraja Reddy R, Murali Mohan S, Seetharama N (2009) Identification of quantitative trait loci for resistance to shoot fly in sorghum [*Sorghum bicolor* (L.) Moench]. Theo Appl Genet 119:1425–1439

Sattler SE, Singh J, Haas EJ, Guo LN, Sarath G, Pedersen JF (2009) Two distinct waxy alleles impact the granule-bound starch synthase in sorghum. Mol Breed 24:349–359

Sattler SE, Palmer NA, Saballos A, Greene AM, Xin Z, Sarath G, Vermerris W, Pedersen JF (2012) Identification and characterization of four missense mutations in Brown midrib 12 (Bmr12), the Caffeic O-Methyltranferase (COMT) of Sorghum. Bioenergy Res 5(4):855–865. https://doi.org/10.1007/s12155-012-9197-z

Schöb H, Grossniklaus U (2006) The first high-resolution DNA “methylome”. Cell 126(6):1025–1028

Sharma HC, Sujàna G, Rao DM (2016) Morphological and chemical components of resistance to pod borer, Helicoverpa armigera in wild relatives of Pigeon pea. Arthropod Plant Interact 10(6):535–547

Shi Y, Gao LL, Wu Z, Zhang X, Wang MM, Zhang C et al (2020) Root anatomy and nutrient transporter expression in sorghum (*Sorghum bicolor*) lines of contrasting K+-efficiency. Plant and Soil 453(1–2):401–417

Shiringani AL, Frisch M, Friedt W (2010) Genetic mapping of QTLs for sugarrelated traits in a RIL population of *Sorghum bicolor* L. Moench. Theor Appl Genet 121(2):323–336

Singh P, Boote KJ, Kadiyala MDM, Nedumaran S, Gupta SK, Srinivas K et al (2019) An assessment of yield gains under climate change due to genetic modification of pearl millet. Sci Total Environ 664:715–726

Smith CW, Frederiksen RA (2000) Sorghum: origin, history, technology, and production. Wiley, New York, NY

Smith CW, Frederiksen RA, Onken AB (2015) Sorghum: origin, history, technology, and production. Wiley, New York, NY

Spindel J, Dahlberg J, Colgan MS, Hollingsworth J, Sievert J, Staggenborg SH, Hutmacher RB, Jansson C, Vogel JP (2018) Association mapping by aerial drone reveals 213 genetic associations for *Sorghum bicolor* biomass traits under drought. BMC Genomics 19(1):679. https://doi.org/10.1186/s12864-018-5055-5

Srinivas G, Satish K, Madhusudhana R, Reddy RN, Murali Mohan S, Seetharama N (2009) Identification of quantitative trait loci for agronomically important traits and their association with genic-microsatellite markers in sorghum. Theor Appl Genet 118:1439–1454

Srinivas G, Satish K, Madhusudhana R, Nagaraja Reddy R, Padmaja PG, Murali MS (2019) Genetic diversity in landraces and improved germplasm of sorghum. Crop Sci 59(2):607–617

Statista (2023) Sorghum production in the Asia-Pacific region in 2020, by country (in 1,000 metric tons). https://www.statista.com/statistics/681800/asia-pacific-sorghum-production-by-country/. Accessed 20 Feb 2023

Stetler RA, Gan Y, Zhang W, Liou AK, Gao Y, Cao G, Chen J (2010) Heat shock proteins: cellular and molecular mechanisms in the central nervous system. Prog Neurobiol 92(2):184–211. https://doi.org/10.1016/j.pneurobio.2010.05.002

Stetter MG, Gates DJ, Mei W, Ross-Ibarra J (2017) How to make a domesticate. Curr Biol 27(17):R896–R900. https://doi.org/10.1016/j.cub.2017.06.048

Sui N, Yang Z, Liu M, Wang B (2015) Identification and transcriptomic profiling of genes involved in increasing sugar content during salt stress in sweet sorghum leaves. BMC Genomics 16(1):534. https://doi.org/10.1186/s12864-015-1760-5

Sun X, Zheng H, Li J, Liu L, Zhang X, Sui N (2020) Comparative transcriptome analysis reveals new lncRNAs responding to salt stress in sweet sorghum. Front Bioeng Biotechnol 8:331. https://doi.org/10.3389/fbioe.2020.00331

Swami AK, Alam SI, Sengupta N, Sarin R (2011) Differential proteomic analysis of salt stress response in *Sorghum bicolor* leaves. Environ Exp Bot 71(2):321–328. https://doi.org/10.1016/j.envexpbot.2010.12.017

Tan YC, Kumar AU, Wong YP, Ling APK (2022) Bioinformatics approaches and applications in plant biotechnology. J Genet Eng Biotechnol 20(1):1–13

Tang H, Cuevas HE, Das S, Sezen UU, Zhou C, Guo H, Goff VH, Ge ZX, Clemente TE, Paterson AH (2013) Seed shattering in a wild sorghum is conferred by a locus unrelated to domestication. Proc Natl Acad Sci U S A 110:15824–15829

Tao Y, Mace ES, Tai S, Cruickshank A, Campbell BC, Zhao X et al (2017) Whole-genome analysis of candidate genes associated with seed size and weight in *Sorghum bicolor* reveals signatures of artificial selection and insights into parallel domestication in cereal crops. Front Plant Sci 8:1237. https://doi.org/10.3389/fpls.2017.01237

Turck F, Fornara F, Coupland G (2008) Regulation and identity of florigen: FLOWERING LOCUS T moves center stage. Annu Rev Plant Biol 59:573–594. https://doi.org/10.1146/annurev.arplant.59.032607.092755

Turner MF, Heuberger AL, Kirkwood JS, Collins CC, Wolfrum EJ, Broeckling CD, Prenni JE, Jahn CE (2016) Non-targeted metabolomics in diverse sorghum breeding lines indicates primary and secondary metabolite profiles are associated with plant biomass accumulation and photosynthesis. Front Plant Sci 7:953

Upadhyaya UD, Wang YH, Gowda CLL, Sharma S (2013) Association mapping of maturity and plant height using SNP markers with the sorghum mini core collection. Theor Appl Genet 126:2003–2015

Upadhyaya HD, Wang YH, Sharma R, Sharma S, Singh S, Hasenstein KH (2017) Sorghum composite collection and identification of accessions with enhanced levels of carotenoids. Euphytica 213(2):45

Vafadarshamasbi U, Mace E, Jordan D, Crisp PA (2022) Decoding the sorghum methylome: understanding epigenetic contributions to agronomic traits. Biochem Soc Trans 50(1):583–596

Varoquaux N, Cole B, Gao C, Pierroz G, Baker CR, Patel D, Madera M, Jeffers T, Hollingsworth J, Sievert J, Yoshinaga Y, Owiti JA, Singan VR, DeGraaf S, Xu L, Blow MJ, Harrison MJ, Visel A, Jansson C, Niyogi KK, Hutmacher R, Coleman-Derr D, O'Malley RC, Taylor JW, Dahlberg J, Vogel JP, Lemaux PG, Purdom E (2019) Transcriptomic analysis of field-droughted sorghum from seedling to maturity reveals biotic and metabolic responses. Proc Natl Acad Sci U S A 116:27124–27132. https://doi.org/10.1073/pnas.1907500116/-/DCSupplemental.y

Varotto S, Tani E, Abraham E, Krugman T, Kapazoglou A, Melzer R, Radanović A, Miladinović D (2020) Epigenetics: possible applications in climate-smart crop breeding. J Exp Bot 71(17):5223–5236

Varshney RK, Shi C, Thudi M, Mariac C, Wallace J, Qi P, Zhang H, Zhao Y, Wang X, Rathore A, Srivastava RK, Chitikineni A, Fan G, Bajaj P, Punnuri S, Gupta SK, Wang H, Jiang Y, Couderc M, Katta MAVSK, Paudel DR, Mungra KD, Gupta S, Kishor PBK (2018) Pearl millet genome sequence provides a resource to improve agronomic traits in arid environments. Nat Biotechnol 36:828–837

Varshney RK, Bohra A, Roorkiwal M, Barmukh R et al (2021a) Fast-forward breeding for a food-secure world. Trends Genet 37:1124–1136

Varshney RK, Bohra A, Yu J, Graner A, Zhang Q, Sorrells ME (2021b) Designing future crops: genomics-assisted breeding comes of age. Trends Plant Sci 26(6):631–649. https://doi.org/10.1016/j.tplants.2021.03.010

Wang YH, Bible P, Loganantharaj R, Upadhyaya HD (2012) Identification of SSR markers associated with height using pool-based genome-wide association mapping in sorghum. Mol Breed 30:281–292

Wei X, Yang Z, Han G, Zhao X, Yin S, Yuan F, Wang B (2020) The developmental dynamics of the sweet sorghum root transcriptome elucidate the differentiation of apoplastic barriers. Plant Signal Behav 15(3):1724465. https://doi.org/10.1080/15592324.2020.1724465

Winchell F, Brass M, Manzo A, Beldados A, Perna V, Murphy C et al (2018) On the origins and dissemination of domesticated sorghum and pearl millet across Africa and into India: a view from the Butana group of the Far Eastern Sahel. Afr Archaeol Rev 35:483–505. https://doi.org/10.1007/s10437-018-9314-2

Wu YY, Li XR, Xiang WW, Zhu CS, Lin ZW, Wu Y, Li JR, Pandravada S, Ridder DD, Bai GH, Wang ML, Trick HN, Bean SR, Tuinstra MR, Tesso TT, Yu JM (2012) Presence of tannins in sorghum grains is conditioned by different natural alleles of Tannin1. Proc Natl Acad Sci U S A 109:10281–10286

Wu Y, Liu J, Zhou G (2022) Whole-transcriptome analyses of Sorghum leaves identify key mRNAs and ncRNAs associated with GA3-mediated alleviation of salt stress. Front Plant Sci 13:1071657. https://doi.org/10.3389/fpls.2022.1071657

Xie P, Tang S, Chen C, Zhang H, Yu F, Li C, Wei H, Sui Y, Wu C, Diao X, Wu Y, Xie Q (2022) Natural variation in Glume Coverage 1 causes naked grains in sorghum. Nat Commun 13(1):1068. https://doi.org/10.1038/s41467-022-28680-3

Xin Z, Wang ML, Barkley NA, Burow G, Franks C, Pederson G, Burke J (2016) Applying genotyping (TILLING) and phenotyping analyses to elucidate gene function in a chemically induced sorghum mutant population. BMC Plant Biol 16:116

Yadav OP, Reddy BVS (2016) A comprehensive review on genetic improvement of pearl millet [*Pennisetum glaucum* (L.) R. Br.]. Crop Sci 56(1):249–267

Yang Z, Zheng H, Wei X, Song J, Wang B, Sui N (2018) Transcriptome analysis of sweet Sorghum inbred lines differing in salt tolerance provides novel insights into salt exclusion by roots. Plant and Soil 430(1–2):423–439. https://doi.org/10.1007/s11104-018-3736-0

Yun-long B, Seiji Y, Maiko I, Hong-wei C (2006) QTLs for sugar content of stalk in sweet sorghum (*Sorghum bicolor* L. Moench). Agric Sci China 5(10):736–744

Yvonne T, John R, Pier Paolo P (2014) The multilayered complexity of ceRNA crosstalk and competition. Nature 505:344–352. https://doi.org/10.1038/nature12986

Zhang M, Xu C, Von Wettstein D, Liu B (2011) Tissue-specific differences in cytosine methylation and their association with differential gene expression in sorghum. Plant Physiol 156(4):1955–1966

Zhang D, Li J, Compton RO, Robertson J, Goff VH, Epps E, Kong W, Kim C, Paterson AH (2015) Comparative genetics of seedsize traits in divergent cereal lineages represented by sorghum(Panicoidae) and rice (Oryzoidae). G3 (Bethesda) 5:1117–1128

Zhang DF, Zeng TR, Liu XY, Gao CX, Li YX, Li CH, Song YC, Shi YS, Wang TY, Li Y (2019) Transcriptomic profiling of sorghum leaves and roots responsive to drought stress at the seedling stage. J Integr Agric 18(9):1980–1995. https://doi.org/10.1016/S2095-3119(18)62119-7

Zhang X, Ni Y, Xu D, Busta L, Xiao Y, Jetter R, Guo Y (2021) Integrative analysis of the cuticular lipidome and transcriptome of *Sorghum bicolor* reveals cultivar differences in drought tolerance. Plant Physiol Biochem 163:285–295. https://doi.org/10.1016/j.plaphy.2021.04.007

Zhao J, Mantilla Perez MB, Hu J, Salas Fernandez MG (2016) Genome-wide association study for nine plant architecture traits in sorghum. Plant Genome 9:1–14

Zhou Y, Wang Z, Li Y, Li Z, Liu H, Zhou W (2020) Metabolite profiling of sorghum seeds of different colors from different sweet sorghum cultivars using a widely targeted metabolomics approach. Int J Genomics 4:6247429

Zhou C, Zhou H, Ma X, Yang H, Wang P, Wang G, Zheng L, Zhang Y, Liu X (2021) Genome-wide identification and characterization of main histone modifications in sorghum decipher regulatory mechanisms involved by mRNA and long noncoding RNA genes. J Agric Food Chem 69(7):2337–2347

Breeding Advancements of Sorghum in Europe

11

Bruno Awio, Pramod Sargar, Anisha Aghuru,
Sonal Chavan, Krishna Kasanaboina, Vinod Kumar Reddy,
Niranjan Thakur, Stanislaus Antony Ceasar,
and Ephrem Habyarimana

Abstract

The ability of sorghum to survive in harsh environments makes it an attractive alternative to maize in Europe. Sorghum is tolerant to drought and high temperatures but is susceptible to cold temperatures. The demand for this crop is expected

B. Awio · S. Chavan · K. Kasanaboina
College of Agriculture, Jayashankar Telangana State Agricultural University, Rajendranagar, Telangana, India

International Crop Research Institute for the Semi-Arid Tropics, Patancheru, Telangana, India

P. Sargar · N. Thakur
International Crop Research Institute for the Semi-Arid Tropics, Patancheru, Telangana, India

Vasantrao Naik Marathwada Agriculture University, Parbhani, Maharashtra, India

A. Aghuru
College of Agriculture, Jayashankar Telangana State Agricultural University, Rajendranagar, Telangana, India

Indian Institute of Millet Research, Hyderabad, Telangana, India

V. K. Reddy
International Crop Research Institute for the Semi-Arid Tropics, Patancheru, Telangana, India

Naini Agricultural Institute, Sam Higginbottom University of Agriculture Technology and Sciences (SHUATS), Prayagraj, Uttar Pradesh, India

S. A. Ceasar
Division of Plant Molecular Biology and Biotechnology, Department of Biosciences, Rajagiri College of Social Sciences, Cochin, Kerala, India

E. Habyarimana (✉)
International Crops Research Institute for the Semi-Arid Tropics, Hyderabad, Telangana, India
e-mail: Ephrem.Habyarimana@icrisat.org

E. Habyarimana et al. (eds.), *Omics and Biotechnological Approaches for Product Profile-Driven Sorghum Improvement*,
https://doi.org/10.1007/978-981-97-4347-6_11

to increase on the continent due to an increase in nutrition-related diseases and changing climate conditions. According to the IPCC report, the northern parts of Europe are expected to experience short-term gains from the changing climate, while the southern parts are expected to experience largely negative effects. Sorghum grows reliably well in marginalized areas. The crop also has a high nutrient density, which has various positive health effects. Broomcorn sorghum is most widely grown in Europe. The production of grain sorghum and high-biomass sorghum is also increasing. Many research projects on sorghum have been undertaken globally and in Europe to improve its adaptation to temperate environments. The collaborative sorghum conversion program that started in the 1980s between the Tropical Agriculture Research Station (TARS), USDA, ARS, S&E and Texas Agricultural Experiment Station (TAES) converted tall photoperiod-sensitive sorghum from tropical regions to dwarf photoperiod-insensitive sorghum. This allowed for desirable characteristics from tropical lines to be transferred to locally adapted sorghum varieties through conventional and modern breeding methods. Many achievements have been made in improving cold tolerance, heat tolerance, drought tolerance, perennial breeding, and nutritional enhancement. This review provides a brief history of sorghum cultivation in Europe, sorghum production and productivity status, sorghum utilization and importance, existing/existed collaborations between Europe and the rest of the continent on sorghum improvement, sorghum germplasm collection, and the extent of genetic diversity among sorghum germplasm utilized. This review provides a thesis of field and screen-house evaluations and "omic" studies conducted on four sorghum varieties (broom-corn, grain, sweet/high biomass, and forage sorghum) against some of the important temperate sorghum production constraints. The body of research included in this review serves as a baseline for further research focused on enhancing sorghum adaptability to European environments.

Keywords

QTL linkage analysis · GWAS · Genomic prediction and selection · Antioxidant levels and properties · Sorghum races · Photoperiod insensitive and sensitive sorghums · Cold tolerance · Drought tolerance · Sorghum ID

11.1 History of Sorghum Cultivation in Europe

Sorghum is classified under the genus *Sorghum*. The genus is classified into five sections: Eusorghum, Parasorghum, Chaetosorghum, Heterosorghum, and Stiposorghum (Garber 1950). The section Eusorghum, also called true sorghum, has four major species: *S. almum*, *S. halepense*, *S. propinquum*, and *Sorghum bicolor*. The latter specie consists of four subspecies: *S. bicolor* subsp. *bicolor*, *S. bicolor* subsp. *drummondii*, *S. bicolor* subsp. *verticilliflorum*, and *S. bicolor* subsp. *bicolor* (De Wet 1978). *S. bicolor* subsp. *bicolor* has the annual cultivated sorghum types and the weedy types. The subspecie bicolor was further reclassified by Harlan and Wet (1972) into five basic interfertile races (bicolor, caudatum, kafir, guinea, and dura) and ten hybrids based on panicle morphology and glume coverage on the grain. To facilitate effective sorghum improvement and utilization, the crop is classified into agronomic varieties such as grain sorghum, sweet sorghum, broomcorn sorghum, sudangrass, and sorghum-sudangrass hybrids (Berenji and Dahlberg 2004).

Sorghum originated in a semiarid tropical climate. The crop is generally sensitive to chilling temperatures. Sorghum domestication is believed to have begun in the fourth millennium BC in the Atbai region of Africa's far eastern Sahelian Belt. Sorghum impressions in its predomestication stage was un-earth in these regions in the periods of 3500–3000 BC. The crop was probably introduced to India by the second millennium BC (Boivin and Fuller 2009; Winchell et al. 2018). However, broomcorn, widely grown in Europe, was probably domesticated outside Africa. Spontaneous mutation of sorghum in different environments, coupled with repeated selection for longer panicle length, likely resulted in broomcorn-type sorghum (Doggett 1988, as cited by Berenji et al. 2011). According to the assessment of Berenji et al. (2011), the original introduction was most likely a sweet sorghum known as "sorgo," which has diffuse panicles and a height of approximately 3 m.

The first domestication of broomcorn sorghum likely occurred in China, India, or Italy. Based on carbon dating technology, broomcorn was probably domesticated in China between *ca.* 10,300 and *ca.* 8,700 calibrated years before the present (cal yr BP) (Lu et al. 2009). Other archaeological site records indicate that broomcorn was domesticated between 6100 and 5600 BC (Zhao 2011) and 5900 calibrated BC in north-eastern China (Hunt et al. 2008). From northern and eastern China, broomcorn was introduced into Central Asia in the later third millennium BC. Following its spread over the dry regions of Central Asia in the second millennium BC, the crop reached the Iranian Plateau, Southwest Asia, and then into Europe, where it became an important crop in the Mediterranean during the first century AD (Miller et al. 2016). In the Mediterranean zone, the crop is believed to have been grown more precisely in Italy. From Italy, broomcorn cultivation and corn broom manufacture spread to Spain, France, Austria, and southern Germany. At the beginning of the sixteenth century, the crop also spread into regions such as the Danube River Basin, the Austro–Hungarian Empire, and reaching the Pannonian Plain, most likely in the eighteenth century.

11.2 Past, Present, and Future of Sorghum Production and Productivity in Europe

Europe contributes ~1.4% of the total sorghum grain production in the world (FAO 2022). The production on the continent was stable from 1970 to 2008, and there was a rise in production in 2019 from 200 K tons to over 400 K tons (Fig. 11.1). Stability and the subsequent increase in production can be associated with increased productivity and area under sorghum production (Deb et al. 2004).

Greece is the leading producer of grain sorghum, followed by Hungary, Slovakia, and Austria in Europe, as per the 2020 record (FAO 2022). The least grain sorghum-producing countries are Bulgaria, Spain, Ukraine, the Russian Federation, and

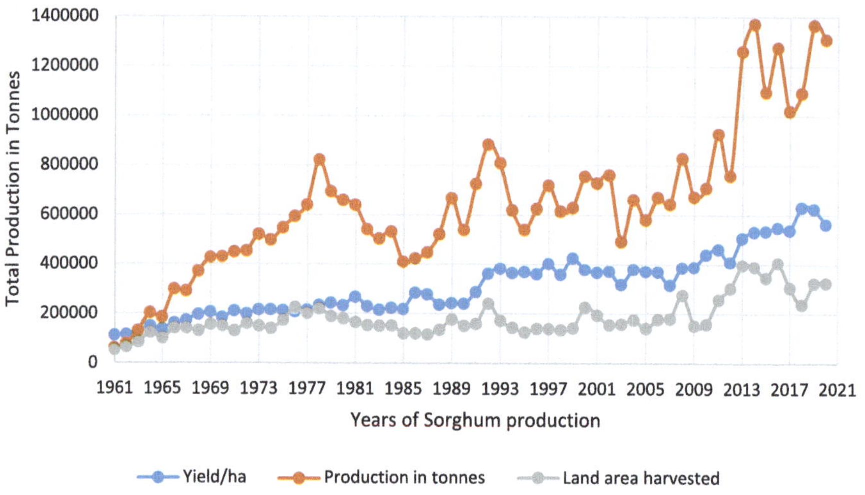

Fig. 11.1 Production/yield quantities of sorghum in Europe

Croatia (FAO 2022). Based on the 59 years of sorghum production data, the least production was recorded from 1979 to 2014, except for the production in Croatia, where a production increase was reported until 2012 (FAO 2022). The sorghum production across sorghum-producing countries from 2007 to 2020 is represented in Table 11.1. Compared to grain sorghum, broomcorn sorghum production is the highest. However, this is not always reflected in the FAO statistics. Broomcorn is the major source of brooms in Europe.

11.3 Importance of Sorghum and Objectives for Sorghum Improvement in Europe

Sorghum is an essential ingredient in animal feed and an important source of broom in Europe. It is also becoming an important source of food and bioenergy in the continent. The stems and foliage of sorghum are used to make the green chop, hay, silage, and pasture. In 2019, one farmer in south-eastern France converted 300 hectares of area under corn production for forage sorghum cultivation. Farmers in southern Hungary, part of the Carpathian Basin, and other parts of Europe with limited access to irrigation facilities would prefer sorghum production as a mitigation strategy for drought (https://phys.org/news/2019-06-sorghum-rebound-europe-climate.html). This crop is expected to be a good substitute for corn in the near future.

Grains from sorghum in Europe are ground into flour and used in the animal feed rations of monogastric animals. The nutritional composition and energy content of sorghum grain are comparable to those of maize, rice, and wheat (Jocelyne et al. 2020). In the ration formulation, sorghum can be combined with other cereals or completely substituted as an energy source. The endosperm of sorghum is 55.7–75.8% starch, and 3–3.5% nonstarchy polysaccharides such as cellulose and noncellulose. The protein content is approximately 10% (Shegro et al. 2012; Awika 2017). Due to their high tannin and phytic acid contents, red-grain sorghum is less

Table 11.1 Grain sorghum production in 1000 tons from 2009 to 2020 in sorghum-producing countries in Europe

Countries	2009	2010	2011	2012	2013	2014	2015	2016	2017	2018	2019	2020
Austria		86	81	88	110	200	222	178	217	291	315	425
Bulgaria	25	118	76	149	161	255	237	108	171	521	328	161
Croatia	1303	1155	1481	1590	1149		159	241	134			
France	203	185	181	154	179	254	178	168	207	201	269	334
Greece	25	19	6	13	10	175	290	134	138	2174	2122	2112
Hungary	89	45	194	145	169	228	182	189	289	512	1490	1284
Italy	152	169	186	100	196	226	181	193	148	187	199	230
North Macedonia	251	278	228	239	250	217	172	211	200	194	205	200
Romania	79	103	219	207	276	287	177	137	305	430	339	201
Russian Federation	11	8	50	38	143	182	160	258	85	41	81	44
Serbia	179	181	175	188	251	188	199	213	201	205	207	205
Slovakia	81	54	124	189	97	263	123	213	163	175	347	642
Spain	147	163	163	122	201	213	225	162	135	117	115	96
Ukraine	36	53	153	181	309	196	165	240	174	171	170	94
Grand total	179	151	178	192	141	28	31	35	45	312	409	407

preferred in Europe than white-grain sorghum. These antinutritional factors limit protein and carbohydrate digestibility and affect enzyme activities (Pan et al. 2019; Ali et al. 2014).

The demand for healthy, plant-based foods has led to an increase in the use of sorghum in human diets in developed countries. Sorghum has immense potential to contribute to a unique market niche among diabetic and celiac disease patients' diets. The low glycemic index of sorghum makes it ideal for diabetic patients (Gallo et al. 2021). The crop has high and diverse polyphenol contents that contribute to high antioxidant capacity and anti-inflammatory effects (Burdette et al. 2010). The gluten-free nature of sorghum, unlike wheat, makes it a desirable diet for gluten-intolerant patients (Pontieri et al. 2013). A survey of 29,212 individuals in four European countries revealed that the number of patients with celiac disease is on the rise on the continent. Gluten, a protein present in wheat, barley, and rye, has been reported to trigger an immunological response that results in the disease. A survey revealed that approximately 1% of the population has the disease, of which 2.4% of the population lives in Finland (2.0–2.8), 0.3% in Germany (0.1–0.4), and 0.7% in Italy (0.4–1.0) (Mustalahti et al. 2010).

Broomcorn comprises long, fine, elastic branches called fibers used to manufacture brooms (Berenji and Kisgeci 1996). The panicles without seeds are dried in open, sunny places (commonly in yards). When dried, panicles are grouped and bound in small sheaves with hemp cord, fiber, mulberry branches, or metal wire (Nedelcheva et al. 2007) for use as house broom, parson's broom, double broom, bettered "mihalska" broom or attic broom (Fig. 11.2) (Nedelcheva et al. 2007).

In Europe, sorghum is currently being envisioned as a source of feedstock for the renewable fuel industry. It has been estimated that approximately 121 million hectares of marginal land suitable for sweet sorghum production are on the European continent (Jiang et al. 2019). Crude oil, petroleum products, and natural gas continue to be the main sources of energy needed in Europe. However, the contribution from renewable energy sources has also continued to increase (https://ec.europa.eu/eurostat/statistics-explained/). In 2016, a body was created to promote Sorghum International Development and the National Federation for the Production of Maize and Sorghum Seeds (FNPSMS). The major objectives are to increase awareness of

Fig. 11.2 Manufactured brooms in the street market in Kyustendil (Nedelcheva et al. 2007)

sorghum's importance and bring together important players involved in the establishment of sorghum for utilization as a source of food, fodder, and energy in Europe. The group initiated a four-year promotional campaign in January 2021 entitled "Seeds for Future" (https://www.sorghum-id.com/). Organized through a grant from the European Union, the campaign targets five European countries (https://www.sorghum-id.com/). To successfully incorporate sorghum into the European farming system/environment, breeding programs should enhance the tolerance of the crop to abiotic constraints such as cold temperatures, freezing temperatures, drought, nutrient use efficiency, and others.

11.4 Collaborations on Sorghum Improvement Between Europe and Other Continents

One of the long-term collaborations between Europe and Africa was established between the French government and West African sorghum breeding programs (Ratnadass and Fleurat-Lessard 1991). The Institut de Recherches Agronomiques Tropicales et des Cultures Vivrières (IRAT) was established in 1960 to respond to Africans' demand for assistance in food crop research. This collaboration focused on improving the grain yield of tall photoperiod-sensitive sorghum genotypes locally preferred in West Africa and improving the grain yield and adaptability of grain sorghum through breeding for biotic and abiotic resistance/tolerance (Ratnadass and Fleurat-Lessard 1991). An extensive review of the nature of collaboration between France and West African countries is provided in the review document by Faye et al. (2022).

The other collaboration was the SWEETFUEL sorghum project. This was a five-year project worth £512,094,620 implemented in Europe, India, Latin America, and South Africa. The major goal was to increase sorghum-based ethanol production through crossing, advancement, and selection of new materials better suited to target regions (temperate and tropical environments). In addition to breeding for biomass yield and pertinent traits, the research also focused on agronomic methods, harvest technologies, and environmental and economic analysis, including energy balance and life cycle assessment—the European program aimed to develop high-biomass sorghum hybrids suitable for double cropping. The program revealed ten new high-biomass hybrid combinations with potential for double cropping, several inbred lines with good combining ability, and some lines with low lignin content (https://cordis.europa.eu/project/id/227422/reporting). Additionally, the initiative created new sweet sorghum hybrids with high sugar content and a productivity of up to 5–6 t ha^{-1} of soluble sugars suitable for the European environment.

The other project is being organized under the European Sorghum Satellite Research Conference "ES^{2}RC." The project's major objectives were (1) to initiate the federation of a European and West African research group (SorgNet), (2) to underline the benefits of partnerships between Europe and West Africa to address the common challenges facing these regions, and (3) to contribute to intensifying links between West African researchers and Europeans to engage the latter in the challenges that West Africa is and will face (https://www.agropolis-fondation.fr/). The meeting was held on 14 and 15 October 2021 in Montpellier, France. Two different programs, the projects "*Enhancing agricultural production and climate resilience for food security and nutrition*" and "*West Africa Breeding Networks and Extension Empowerment (ABEE)*", which are sponsored by the European Union, are being carried out at the ICRISAT Mali station and West Africa, respectively. This project aims to improve the region's food and nutrition security by promoting sorghum, millet, groundnut, and cowpea research (https://www.icrisat.org/new-eu-funded-project-to-improve-food-and-nutrition-security-in-mali-to-benefit-40000-people/).

11.5 Sorghum Germplasm Collection for Use in Breeding Programs in Europe

Globally, sorghum germplasm is conserved in about 37 gene banks. In Europe, sorghum germplasm is conserved at the NI Vavilov All-Russian Scientific Research Institute of Plant Industry (VIR) in Russia, the Laboratoire des Resources Génétiques et Amélioration des Plantes Tropicales (ORSTOM), and the Centre de Coopération Internationale en Recherche Agronomique pour le Développement (CIRAD), both of which are located in France. These gene banks have more than 3963 and 7305 sorghum accessions conserved in the respective countries (Upadhyaya et al. 2016; Venkateswaran et al. 2019). In addition, the Institute of Field and Vegetable Crops in NoviSad (Serbia) has a base collection of 365 broomcorn germplasms sourced from all over the world. Germplasm includes local landraces and breeding material (Berenji 2000, as cited by Berenji et al. 2011).

European countries also have access to sorghum accessions in other gene banks. A total of 236,617 sorghum accessions (98.3% cultivated and 1.7% constituting wild and weedy relatives) are conserved globally, and the majority of the germplasm is conserved in gene banks in Asia (39.18%), the Americas (35.72%), and Africa (16.40%) (Upadhyaya et al. 2016). The largest collection is conserved in India, followed by the United States, China, and Brazil (Fig. 11.3). The Plant Genetic Resources Conservation Unit, Southern Regional Plant Introduction Station, University of Georgia, USDA-ARS, and ICRISAT, India, conserves the largest collection (Upadhyaya et al. 2016).

11.6 Genetic Diversity of Sorghum Germplasm Used in Breeding Programs in Europe

Genetic diversity studies have been conducted to understand the level of diversity in different sorghum panels for their effective use in breeding and to comprehend crop evolution processes. A sorghum diversity panel of 338 accessions and improved breeding lines for temperate environments sourced from USDA-ARS were categorized into five geographical groups: India, the United States, South Africa, Ethiopia,

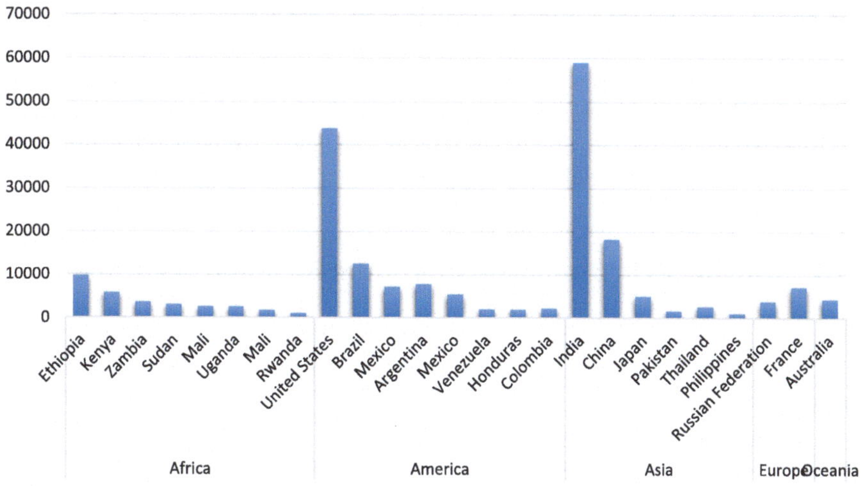

Fig. 11.3 Sorghum germplasm conservation by countries on the four continents of the world

and Nigeria, interspersed with five sorghum races (Schaffasz et al. 2019). Using the RAPD molecular marker, high genetic distance values of up to 0.8776 among accessions were reported from 46 sorghum accessions obtained from the Czech Republic plant genetic resource gene bank (Ruiz-Chután et al. 2019). The genebank conserves highly diverse germplasms with minimal inter- and intra-breeding among them. The results of the study is in line with the mandate of gene banks, which are designed to store and conserve germplasm for future use. Here, the genetic resources were grouped into five categories: cluster I included materials from Russia, Germany, Egypt, Turkey, and China; cluster II included materials from Russia; cluster III included materials from Russia, Argentina, Turkey, Yemen, the United States, Nigeria, South Africa, Nigeria, and Moldova; cluster IV included materials from Russia, Ethiopia and Sudan; and cluster V included materials from Russia (Ruiz-Chután et al. 2019). The results of these studies demonstrated the impact of spatial adaptation on the genetics of sorghum. Here, geographic barriers and distance prevent gene flow between populations, leading to the local accumulation of genetic differences among populations. In a more recent study, a whole-genome SNP sequence of 3333 diverse Ugandan *S. bicolor* germplasm accessions was used. These genetic materials were collected by Plant Genetic Resources Centre at the Uganda National Genebank. Chakrabarty et al. (2022) reported that germplasms from lowlands (central, eastern, northern, and north-western regions) were genetically distinct from those from highlands (south-western region). However, more outcrossing occurred in the materials from the former than in those from the latter. The Ugandan germplasms were found to adequately represent the diversity in the global collection, indicating the potential for the collection to be used to address global breeding constraints.

In broomcorn, genetic variability within and between clusters is comparatively high (Hu et al. 2009; Zhu et al. 2020). Nevertheless, its diversity is less than that of non-broomcorn sorghum (Hu et al. 2009). However, the narrow genetic diversity of the germplasm collection is conserved at the Institute of Field and Vegetable Crops in NoviSad (Serbia). This is because, in the process of collection, the broomcorn panicle type genotypes were selected (Berenji et al. 2011). The collection, according to Berenji et al. (2011), represents a distinct and unique broomcorn germplasm that is not found anywhere else in the globe.

11.7 Advances in Broomcorn Sorghum Breeding in Europe

Broomcorn differs from non-broomcorn sorghum in terms of the average panicle length, top primary branch length, node number of the panicle rachis, and ratio of the top primary branch length to the panicle length. The average length in broomcorn for these parameters ranges from 24.48 to 52.51 cm, 15.54 to 42.90 cm, 2.6 to 3.8, and 0.64 to 0.80, respectively, while non-broomcorn, panicle length were 36.49 cm, 10.12 cm, 7.5 cm, and 0.28 cm, respectively (Zhu et al. 2020). Genetic improvement in broomcorn has focused on improving plant characteristics, panicle characteristics, seed characteristics, hybrid cultivar breeding, and disease resistance (Berenj and Sikora).

Broomcorn in Europe is categorized into three major groups: European dwarf type, Western dwarf type, and standard brown corn. Dwarfism occurred through spontaneous mutations and selection within the tall (standard) types of broomcorn (Berenj and Sikora 2002). Dwarf broomcorn cultivars are preferred for broom manufacturing in Europe. It is more tolerant to dry weather conditions and produces finer fiber for broom making than standard broomcorn cultivars. Their harvest index is most desirable for broom making (Berenj and Sikora 2002). The Institute of Field and Vegetable Crops in Novi Sad (Yugoslavia) released three dwarf varieties of broomcorn, 'Backibiser', 'Neoplanta', and 'Panonija'. However, these varieties have negative panicle exertion, which hinders the application of machine harvesting in this crop. The positive relationship between plant height and peduncle length has led to the crop being primarily manually harvested.

Through breeding, more than 11 varieties adapted to a wide range of climate conditions have been developed (Sikora and Berenji 2009). The institute released four improved varieties, Sava, Tisa, Jumak, and Jantarith, with higher yields and better-quality broomcorn brushes (Sikora and Berenji 2009). The panicle yields of broomcorn are influenced by precipitation and the maximum air temperature during the vegetative stage. The differences between the maximum and minimum temperatures, growing degree days, and the sum of precipitation are also important parameters for increasing panicle yield (Sikora et al. 2018). Sikora et al. (2018) categorized these genotypes based on their environmental suitability. Cultivars 'Sava', 'Jumak', and 'Deer 418' are adapted to hot and dry environments (i.e., environments with high precipitation and the sum of precipitation at the vegetative growth stage), while cultivars 'Prima', 'Tan Sava', and 'Sz. Szlovak' are suitable for environments of cold and wet environments. The cultivar 'Sz.1023' is adapted to environments with a minimum average temperature at the vegetative stage and average relative humidity. The 'Reform' and 'Neoplanta Plus' cultivars were adapted to intermediate conditions. They strive and are stable in performance under wet, hot, and dry environments.

The cultivars 'Reform' and 'Neoplanta plus' were selected for early maturity and fiber fineness (Table 11.2). These cultivars preserved the good agrotechnological traits of former cultivars (Sikora and Berenji 2009). The 'Sz.1023', 'Prima', and 'Jumak' varieties combine the length of the panicle, length of the fiber, coefficient of fiber fineness, and weight of the threshed panicle desired by the industry (Table 11.2) (Sikora et al. 2018). In Serbia, only 'Reform' and 'Prima' are grown on a large scale among the released varieties (Berenji et al. 2011).

Furthermore, resistance to biotic constraints is also important in broomcorn breeding. Diseases of major importance are anthracnose, maize dwarf mosaic virus (MDMV), *Fusarium* spp., anthracnose stalk rot (*Colletotrichum graminicola* [Ces] G.W. Wils.), and *Hadrotrichum sorghi* (Berenj and Sikora 2002). The breeding program in Serbia has screened broomcorn sorghum available in its collection for disease resistance. Some effective sources of resistance to anthracnose and partial resistance to Fusarium were identified (Berenji et al. 2011). However, no significant level of MDMV resistance has been identified in broomcorn genotypes. Breeding efforts using grain sorghum lines (Krish) source of resistance to MDMV to improve resistance in broomcorn sorghum are underway.

Other areas for further investment included broomcorn hybrid development, seedless hybrid broomcorn development, concurrent improvement of broomcorn for panicle, seed, and stem characteristics for use as brooms and bioenergy

Table 11.2 Agrotechnological traits of broomcorn cultivars grown in Europe

	Traits											
Cultivars	PHG (cm)	SHG (cm)	PLG (cm)	PDL (cm)	FLG (cm)	LSL (cm)	UTP (g)	TRP (g)	GWG (g)	TPR (g)	FNO	FFI (g/m)
Jumak	158	65	93	38	65	41	66	26	41	37	56	508
Sava	175	78	97	44	65	42	84	31	54	35	67	518
Neoplanta+	195	100	95	52	61	43	73	24	50	31	64	446
Reform	173	84	88	25	64	40	62	22	40	35	55	496
Prima	153	60	98	51	73	40	65	26	39	39	56	520
Tan sava	189	90	100	28	71	42	74	26	48	35	60	488
Sz. Szlovak	150	52	98	40	69	39	56	23	33	40	51	497
Sz. 1023	153	57	96	16	80	39	68	30	39	42	54	553
Dia	150	52	98	40	69	40	62	23	39	37	53	480
Szilard	143	57	89	34	65	40	62	21	41	33	52	437
Deer 418	219	148	71	24	48	42	59	15	44	24	52	472
Average	169	76	93	36	66	41	66	24	43	35	56	492

PHG is plant height, SHG is stalk height, PLG is the length of the panicle, PDL is the length of the peduncle, FLG is the length of the fiber, LSL is the length of the top leaf sheath, UTP is the weight of unthreshed panicle, TRP is the weight of threshed panicle, GWG is grain weight per panicle, TPR is threshed panicle ratio, FNO is the ratio of fiber to panicle length, and FFI is coefficient of fiber fineness

production, elimination of red coloration of panicles, and increasing genetic diversity. The diversity may be increased through modern breeding approaches such as TILLING and genome editing.

11.8 Advances in Grain Sorghum Breeding in Europe

Some studies on the improvement of sorghum cold tolerance, drought tolerance, sorghum perenniality, and antioxidant levels have been conducted in Europe. Breeding for early season and terminal cold stress has been emphasized in temperate environments. The effects of temperature on germination and seedling emergence in temperate environments have been estimated using base temperature and the thermal time requirements of cultivars. Cultivars with low base temperatures and long thermal times are desirable for early sowing in temperate environments. Early emergence combined with high vigor under suboptimal temperature facilitates early sowings and is the major determinant of potential population establishment. Researchers have conducted germplasm evaluation for cold tolerance in natural and controlled temperate environments. Their studies revealed weak correlations for cold tolerance parameters in germplasms screened under controlled and field conditions (Parra-Londono et al. 2018; Schaffasz et al. 2019). This difference has been associated with variations in soil temperature, among other factors, under field conditions (Parra-Londono et al. 2018). Relatively large genetic variability in cold tolerance parameters (leaf greenness, root dry weight, and shoot dry weight) has been reported in Europe (Patanè et al. 2006; Parra-Londono et al. 2018; Schaffasz et al. 2019), indicating the potential for improving cold tolerance through breeding. Genotypes possessing these traits respond to variance in temperate environmental conditions, but the variance is not large enough to mask the genotypic expression for cold tolerance. Schaffasz et al. (2019) reported a reduction in shoot dry weight under severe cold conditions in different environments. According to their evaluation, P1602736, SC614, and SC1201 were relatively adaptable to cold temperate environments. Parra-Londono et al. (2018) reported that more than 50% of 194 genotypes exhibit stable cold tolerance characteristics in the field, but these genotypes experienced greater genotype-by-environment interactions under lower temperatures in a controlled chamber. Sorghum germplasm sourced from tropical environments, particularly in isolated highlands, has been reported to represent a good source of diverse cold-tolerant genotypes for temperate environments (Chakrabarty et al. 2022).

There is, however, very scanty information on sorghum terminal drought tolerance in Europe. Terminal cold-tolerant genotypes are grouped based on their reproductive parameters. Chakrabarty et al. (2021) identified ten genotypes with better tolerance to cold stress based on their study conducted in Germany, Mexico, and America. The genotypes had greater seed numbers, seed weights, thousand kernel weights, and panicle harvest index. The genotypes identified for juvenile cold stress in the above studies are good sources of tolerance for improving the productivity and adaptation of sorghum in temperate environments.

Flowering and maturity of sorghum in temperate environments are substantially delayed in cooler compared with those in warmer environments. Schaffasz et al. (2019) evaluated a set of 338 diverse sorghum germplasms. They reported that 11% of the materials at Rauischholzhausen and 6% at Poel vs. 2% at Gross-Gerau were photoperiod sensitive. The genotypes also showed reduced dry matter content. Genotypes with a delayed flowering period extending into extremely cold conditions are often male sterile and have poor seed set (Schaffasz et al. 2019). Cold stress also results in lower thousand kernel weight, causing lower grain yield (Chakrabarty et al. 2021). However, early-maturing genotypes can escape terminal drought stress conditions. Therefore, most tropical sorghum genotypes that are photoperiod sensitive and are cultivated in temperate environments have genes associated with photoperiod insensitivity and early maturity introgressed into their genetic background (Quiles-Belén and Sotomayor-Ríos 1983). The derived lines are used in temperate environments to improve locally adapted cultivars.

Grain sorghum has a very narrow genetic base for overwintering. The ability of sorghum to overwinter can minimize weed growth, prevent soil erosion, protect biodiversity and ecosystem services, and reduce seed costs and energy expenditures from tillage (Piper and Kulakow 1994). This endeavor has been made possible by efforts to take advantage of the perennial nature of *S. halepense*. The rhizomes of the specie extend to a depth of 40–50 cm below ground level, a depth adequate for the rhizomes to survive freezing temperatures (Warwick et al. 1986). Piper and Kulakow (1994) evaluated tetraploid *S. bicolor*, F_1, BC_1, and BC_2 of crosses between tetraploid *S. bicolor* and *S. halepense* and revealed no compromise in the allocation of assimilates for rhizome and seed production within any hybrid combination. Ninety-seven derived lines from these populations were evaluated in Italy by Habyarimana and published in (Habyarimana et al. 2018). The authors identified high biomass and grain yielding dual-purpose sorghum genotypes with high winter survival rates and fiber mass fractions. The top ten and two percent selections from the whole SB x SH population outperformed both the best check hybrid and the commercial forage hybrid in terms of biomass production. Nonetheless, a number of studies are necessary, including the need to address the practical objective of combining the perennial growth habit and seed yield, as well as the question of how much rhizome formation will permit overwintering and how much seed yield is sufficient (Piper and Kulakow 1994).

Breeding for improving phytochemical levels has also become important in developed countries. High genetic variability for these secondary metabolites has been reported in sorghum (Dykes and Rooney 2007; Alfieri et al. 2019). This trait is highly heritable and mostly controlled by a few genes (Wu et al. 2012; Habyarimana and Lopez-Cruz 2019). The amounts of these secondary metabolites, phenols (FEN), tannins (TAN), and flavonoids (FLA), are relatively greater in wild sorghum species (Wu et al. 2012) and their associated hybrids than in cultivated sorghum (Habyarimana and Lopez-Cruz 2019). The cross of *S. bicolor* x *S. halepense* has been evaluated in Europe and is considered a good source of antioxidant variability for breeding. Incorporation of these materials in breeding will require different breeding strategies. For instance, prior to breeding for antioxidant production, undesirable levels can be culled using multivariate techniques or an appropriate breeding scheme (such as tandem selection) (Habyarimana and Lopez-Cruz 2019).

11.9 Advances in High Biomass, Fiber, and Sweet Sorghum Breeding in Europe

Braconnier et al. (2014) proposed an ideotype suited for biofuel-type sorghum cultivars for the European environment. These genotype(s) exhibit high biomass (height >3 m), low-temperature tolerance, water deficit tolerance, adaptation to late flowering, photosensitivity, quality biomass, lodging tolerance, and high soluble sugar content in stalks (Braconnier et al. 2014). Biofuel sorghum in Europe is grown in the summer season (April/May sowing). However, for double cropping, breeding efforts target sowing in September/October, just after winter (Braconnier et al. 2014). Early-sowing sweet sorghum types have been found to have enhanced sugar accumulation and competition for carbohydrates for grain filling and sugar accumulation in stems in photosensitive sweet sorghum genotypes (Braconnier et al. 2014). However, early-season sowing of sorghum is hindered by the suboptimal temperatures experienced in Europe. Sweet sorghum, in particular, is relatively more sensitive to cold stress than high biomass or fiber sorghum (Saita et al. 2011). Through research, genotypes with low base temperatures and short thermal times, such as 'PR811F' and 'PR849', have been identified (Saita et al. 2011). Patanè et al. (2021) reported the seed germination rates of the cultivars 'Padana 4', 'PR811F', 'PSE24213', 'PR849', and 'Zerberus' were greater than 80% at 8 °C. Investigations revealed that photochemical quenching (Havaux 1989), energy for light dissipation (φDI0, DIo/CS), and the polyphasic transient phases of Chl-a fluorescence (Zegada-Lizarazu et al. 2016) are strong indicators of cold tolerance in sweet sorghum. The cold tolerance of 'Bulldozer', a cultivar grown in Europe with high germination, seed emergence, stem elongation, and high biomass under suboptimal temperatures, has been partially attributed to its efficient photosynthetic system. In this genotype, photosystem II (PSII) is more resistant to photoinactivation and has a more effective photochemical regulatory mechanism for controlling incoming light energy. The genotype effectively dissipates surplus energy and divert received energy into other energy sinks (Zegada-Lizarazu et al. 2016). These genotypes can be used in breeding programs to improve cold tolerance in sweet and high-biomass sorghum genotypes.

Plant height, stem thickness, brix content, and biomass quality are important bioenergy traits for sorghum breeding. Schaffasz et al. (2019) reported wide genotypic variability for bioenergy-associated traits in a 338-diversity sorghum panel across both warmer and colder European environments. Generally, landraces, inbred, and hybrids are important genetic sources for biofuel production. Hybrids followed by landraces were found superior to inbred lines for aboveground biomass production, plant height, lignin content, and cellulose content in Italy (Habyarimana et al. 2016). Landraces, inbred, and hybrids have similar content for nonstructural carbohydrates and plant maturity (Habyarimana et al. 2016). Through the

SWEETFEUL project, large amounts of breeding materials were developed for European environments. In the project, more than ten male high-biomass sorghum parents with dry biomass yields ranging from 18 to 26 t ha^{-1} and a short cycle time (less than 110 days from sowing to harvest) appropriate for double cropping were reported (Braconnier et al. 2014). The program also developed several inbred lines with good combining ability and identified some lines with low lignin content. The initiative also created several new sweet sorghum hybrids with high sugar content and a productivity of up to 5–6 t ha^{-1} of soluble sugars suitable for the European environment. Furthermore, through the SWEETFUEL project, many genetic materials from Asian, American, and African breeding programs were developed that are freely accessible and represent a great genetic resource for biofuel improvement in Europe (Fig. 11.4).

Sorghum grown in summer in the Mediterranean region are exposed to drought constraints. The seedling and post-flowering stages are the most sensitive to soil moisture deficits (Prasad et al. 2019; Bibi et al. 2012). Sorghum produces lower biomass under water stress conditions (Habyarimana et al. 2002). Indirect selection for high biomass under moisture stress has been possible by selecting taller stature, stay-green trait, and selection of medium to long growth cycle genotypes (Habyarimana et al. 2004), high root and shoot biomass (Habyarimana et al. 2002). The selection of high aboveground biomass production under drought conditions was found transferable under irrigated management (Habyarimana et al. 2002).

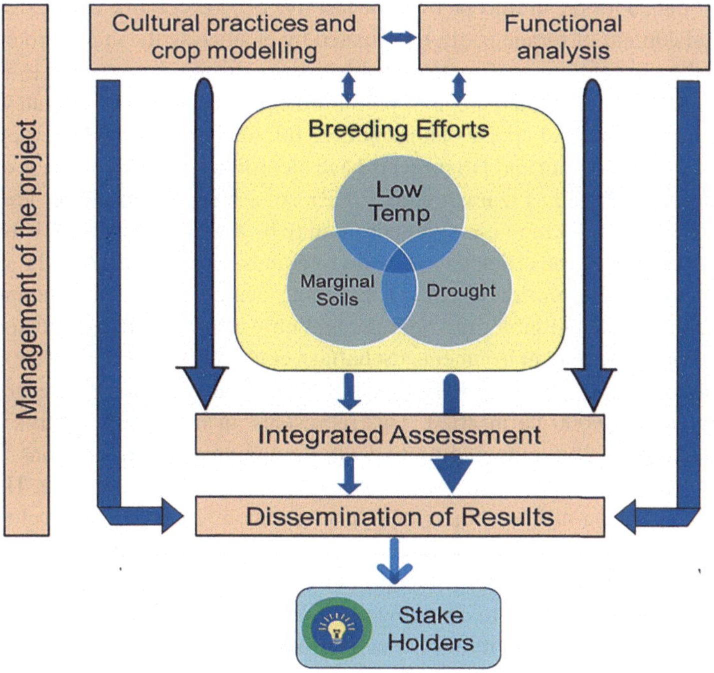

Fig. 11.4 Organization of the works of the SWEETFEUL project

Several drought-tolerant sorghum biofuel-type genotypes have been identified (Habyarimana et al. 2002, 2004). These genetic resources are recommended for production under moisture-stress conditions and as parental materials to improve locally adapted cultivars.

11.10 Advances in Forage Sorghum Improvement

Europe imports a significant amount of animal feed from the Americas (San Martin et al. 2021; Díaz de Otálora et al. 2022). Locally grown fodder crops, such as maize, grass, and other roughage crops, contribute about 50% of the dry matter supplied to cattle (Karlsson et al. 2021). Green forage crops, such as lucerne and other legumes, and annual and perennial grasses are also grown (Ergon et al. 2018). In waterlogged, acidic, and cold soils as well as in situations of water and wind erosion, fodder grasses serve as the foundation for fodder phytocenoses (Kosolapov et al. 2021). However, animal husbandry cannot advance without a well-managed supply of fodder (Kosolapov et al. 2021).

Climate change could pose serious obstacles to the production of grassland-based fodder throughout Europe. Maize is commonly used as silage in Germany and Northwest Europe, and the area under maize cultivation is rapidly expanding toward the north due to subsidies promoting biogas production. However, there has been no significant change in the digestibility, crude protein, or metabolizable energy of maize forage in Europe over the last five decades (Taube et al. 2020). The nutritional content of forage is often evaluated by estimating the crude protein content and the concentration of metabolizable energy. It may be possible to address these obstacles and take advantage of possibilities in the context of climate change by selecting the optimal species and mixtures for sowing grasslands (Ergon et al. 2018). Various European and US projects have identified sorghum forage as a promising feedstock (Zegada-Lizarazu et al. 2013). As a biofuel and fodder crop, sorghum, a versatile and climate-resilient crop, may be a good substitute for maize. It has superior nutrient use efficiency, drought tolerance, and resistance to the western corn root worm (*Diabrotica virgifera*). Breeding evidence for cold tolerance and forage quality was negligible for sorghum, and only a few reports on this topic are available for European environments (Schaffasz et al. 2019; Niu et al. 2022).

The area of forage sorghum in Europe has been steadily growing in recent years and surpassed 200,000 ha in 2020. This represents an increase of approximately 16–20% in the EU28 (compared to 2019) in the major producing nations: Russia (almost 40% of soil, +6% compared to 2019), France (+25%), Ukraine, Hungary (+85%), and Italy (+10%) (https://european-seed.com/2020/07/the-development-of-sorghum-forage-in-europe-a-new-perspective-for-farmers/). This increasing area substantiates the need for future selection of germplasms with tolerance to biotic and abiotic stresses to adapt to changing environmental conditions, with an emphasis on improving forage quality for the economic viability of sorghum growing and expansion in Europe.

11.11 Sorghum Breeding Advancements in Europe Using Genomic and Other "Omics" Technologies

11.11.1 Genome Wide Association Studies (GWAS) and Quantitative Trait Loci (QTL) Analysis for Gene Discovery

Genomic regions for resistance and tolerance to abiotic and biotic stresses were reported in many sorghum breeding studies. Programs have been evaluating several germplasms and populations to dissect the genetic mechanisms controlling the stresses. Such germplasms that have been evaluated, included the 194 biomass lines from KWS company (Einbeck, Germany) (Fiedler et al. 2016; Parra-Londono et al. 2018), 36 sorghum accessions from temperate environments (Maulana et al. 2017), a broad diversity panel of 338 genotypes representing photoperiod-sensitive temperate conversion sorghum lines and tropical conditions (Schaffasz et al. 2019), and 3333 Ugandan germplasm materials (Chakrabarty et al. 2021). The detection of QTL with moderate to large effect sizes can be accomplished using QTL linkage analysis (Singh and Singh 2015). Several populations, including F_2, $F_{2:3}$, near isogenic lines (NILs), and others, can be used, although recombinant inbred lines (RILs) are the most useful and informative for QTL detection (Singh and Singh 2015). Fiedler et al. (2016) used two separate $F_{2:3}$ populations that were segregated for cold tolerance and had relatively dense linkage maps (255 markers and 544 markers) in the populations. Their study revealed, on average, 34 significant QTL associated with cold tolerance traits (dry matter growth rate, leaf appearance rate, chlorophyll content (SPAD), and fluorescence (Fv0/Fm0 and ΦPSII)). The SBI-01 and SBI-02 chromosomes included more QTL (hotspots) than did the other eight sorghum chromosomes. Promising QTL were found in genomic regions with candidate genes implicated in the C-repeat binding pathway or encoding cold-shock proteins (on chromosomes SBI-09, SBI-06, SBI-04, SBI-03, SBI-02, and SBI-01).

Association mapping panels used in GWAS offer the opportunity to map QTL with a much smaller effect size that contributes to a quantitative trait (Singh and Singh 2015). Natural populations, breeding lines, and lines produced from multiple crosses are used for QTL detection (Singh and Singh 2015). This method harnesses the power of historical and recent recombination to locate QTL associated with traits present in a population. Due to the smaller recombination blocks, the markers identified with significant associations with the trait in a population are expected to be much closer to the gene(s) of interest than are those identified via conventional QTL linkage mapping (Singh and Singh 2015). Schaffasz et al. (2019) used 1507 SNPs from 338 germplasm collections. They reported three significant cold tolerance-associated SNPs in the field environment and 12 SNPs in a controlled environment. These SNPs were associated with the emergence percentage and shoot dry weight, root dry weight, emergence index, and leaf greenness, respectively. Parra-Londono et al. (2018) identified 21 SNPs significantly associated with the final emergence percentage under late-sown fields and cold environments (survival rates at 10.7 °C and 12.3 °C). According to their study, the three SNPs associated with the final emergence percentage and survival rate were most closely related to genes associated with *serine/threonine kinase (Sb02g033290)*, zinc finger domain with *ubiquitin transferase* activity (*Sb02g033390*), and two-pore calcium channel protein 1 (*Sb03g031110*). In the most recent study, Chakrabarty et al. (2022), using a set of 3333 Ugandan diverse sorghum germplasms and a total of 34,469 SNP markers, identified three SNPs on chromosomes Sb02, Sb06, and Sb09, and these SNPs were consistently associated with cold tolerance traits at the juvenile growth stage in three environments (two fields and a screen-house environment). The SNPs were in the regions of nine previously identified cold tolerance-associated QTL. The QTL was associated with six genes involved in various forms of cold acclimatization and tolerance identified in sorghum, maize, and rice. Chakrabarty et al. (2021) reported 51 SNP markers associated with reproductive cold tolerance parameters. The traits and associated SNPs included the panicle harvest index (5 associations), seed number (21), seed yield (9), thousand kernel weight (9), date to flower (1), and plant height (10). Among these associated SNPs, SNP marker Sb10-5394955 on chromosome Sb10 showed multiple cold-related traits; associations to panicle harvest index, seed yield, and thousand kernel weight. The SNP showed relatively weak LD with adjacent SNPs on either side ($R^2 < 0.8$). The SNP is close to genes *Sobic.010G067300*, and *leucine-rich repeat receptor-like protein kinase*. Furthermore, Chakrabarty et al. (2021) found no relationship between SNPs associated with plant height and flowering date, indicating that cold tolerance at the reproductive stage can easily be bred in both plant statures. Marker trait associations of agronomic and bioenergy-related traits were also assessed using the same germplasm collection evaluated by Schaffasz et al. (2019). Forty-six significant SNPs were found to be associated with agronomic parameters. In Europe's warmer regions, four SNPs were linked to the number of days to flowering. In contrast, more SNPs were linked to cold tolerance at the reproductive stage in cooler temperate zones. While three SNPs were linked to stem diameter, more SNP-trait associations were observed for dry matter content in cooler environments than in warmer ones. The percent brix content was linked to three major SNPs. However, most of these marker-trajectory correlations in sorghum and other crops are in limited use in downstream marker-assisted breeding due to, among other factors, the difficulty of reliably detecting the same QTL across various genetic backgrounds and environments. Indeed, QTL linkage mapping and association genetics have been ineffective in mapping rare variants associated with economically important traits. Since quantitative traits are usually affected by many genes with small effect sizes, the benefits of using marker-assisted selection can be limited by the proportion of genetic variance explained by QTL.

11.11.2 Genomic Selection

Meissen et al. (2001) introduced genomic selection as a unique breeding approach to overcome the drawbacks of phenotypic selection and restrictions imposed by QTL linkage mapping and association genetics. Here, a large set of molecular markers, i.e., more than 40,000 polymorphic SNP markers distributed all over the crop chromosomes, are used in establishing a genomic prediction model to predict the candidates' selection performance. Genomic selection combines molecular and phenotypic data in a training population to obtain the genomic estimated breeding values (GEBVs) of individuals in a testing population that has been genotyped but not phenotyped. Several statistical genetic models for genomic prediction have been developed (Merrick and Carter 2021). These genomic prediction models and selection methods are under investigation in many crop species to improve selection for complex traits and oligogenic traits. The effectiveness of a prediction model is determined by the amount of phenotypic variance that can be explained by the model and root mean square error, among other factors. In the comparison of G-BLUP, which incorporates only the genomic matrix, fitting a combined pedigree-genomic matrix with the ideal weight was observed to consistently yield the greatest gains in predictive ability and resulted in the greatest decreases in prediction bias in sorghum (Velazco et al. 2019b). The incorporation of plant height as a covariant in the prediction model improved the accuracy of the selection of candidates that produce high grain yield in sorghum. The study by Velazco et al. (2019a) suggested that predicting low heritable traits can be improved when an indirect, easily measurable trait is included in selecting a specific low heritable trait. Plant height of sorghum is a proxy trait for grain yield improvement (George-Jaeggli 2009). As in other crops, sorghum individuals' genomic prediction accuracy decreases as the genomic relationship between the training population and the predicted individuals increases (Hunt et al. 2018). Therefore, successful prediction and selection will be achieved when genotype training, validation, and test sets exhibit some relationships. Many genomic selection and prediction successes have been attained from populations with some genomic relationships, such as inbred selection from within families (Crossa et al. 2017). Further efforts are being made to account for nonadditive genomic effects and the influence of genotype × environment interactions (Burgueño et al. 2012; Lopez-Cruz et al. 2015; Cuevas et al. 2017). Depending on the crop, a genomic prediction accuracy of 0.2–0.5 is considered satisfactory for selection when compared to that of a marker-assisted selection breeding program (Heffner et al. 2010, 2011). The application of genomic selection for sorghum improvement is limited in Europe. In a study by Habyarimana and Lopez-Cruz (2019), genomic selection models were applied (GBLUP, BRR, Bayesian LASSO, and Bayes) in a diverse panel of *S. bicolor* lines, landraces, and RILs (*S. bicolor* × *S. halepense*) to improve antioxidant-associated traits (phenols, flavonoids, total antioxidant capacity, and condensed tannins). The accuracies of the four genomic selection models were comparable across antioxidant-associated traits, with the accuracy assessed using a validation set ranging from 0.49 to 0.58. These accuracy levels are deemed sufficient to support breeding for improved antioxidant levels while enabling significant genetic benefits per unit of time and expense.

11.12 Conclusion and Recommendation

The value of sorghum in Europe is increasing due to its benefits. As a crop that can withstand heat and drought, its significance is growing in light of impending climate change. The northern parts of Europe are expected to experience short-term gains (such as higher crop yields and the expansion of forests) from the changing climate. In contrast, the southern parts are expected to experience largely negative effects (such as declines in agricultural productivity and water scarcity) (Bednar-Friedl et al. 2022). It is anticipated that maize yields will decline by up to 50% when a 3 °C increase in temperature occurs, especially in southern Europe (Bednar-Friedl et al. 2022). Wheat yields in northern Europe may increase if the temperature rises below 2 °C (Bednar-Friedl et al. 2022). Compared to maize, sorghum and pearl millet are considered less drought sensitive. To increase their ability to adapt to drought and heat, sorghum and pearl millet limit the rate of leaf area expansion and transpiration (Choudhary et al. 2020).

In addition, sorghum has a variety of positive health effects. High antioxidant levels improve cholesterol profiles, and these compounds are safe food sources for individuals with celiac disease. Compared to cereals such as rice and wheat, sorghum grains have a rich mineral content, high fiber content, and moderate digestibility. Thus, sorghum meals are advised for those with diabetes, jaundice, and obesity. Compared to other cereals, sorghums with high tannin levels lower the incidence of several cancers. Policosanol-containing sterols found in waxy sorghum control the production of endogenous cholesterol and cholesterol absorption. These health benefits associated with sorghum have increased demand and consumption in developed countries.

Furthermore, sorghum can contribute to providing solutions to the increasing demand for energy in Europe. Among the crops for producing ethanol, sweet sorghum has the highest energy yield (Bassam 2010). The sugars from sorghum biomass are used in making bioethanol. Moreover, lingo-cellulosic materials are convertible into other energy products, such as bio-crude oil, gasoline, electricity, pulp for paper, compost, and charcoal pellets (Bassam 2010). Sorghum will soon contribute more to Europe's food, feed, and energy sectors. To achieve these goals, Sorghum—International Development (Sorghum ID) was created to promote the interest of sorghum and its hybrid genetics for several uses. To encourage European sorghum production (grain, silage, and energy) and develop hybrid genetics, the associations bring together various organizations and businesses from many European countries (the European Union and Eastern Europe) along the sorghum value chain. The action plan is based on (1) representing and promoting sorghum interests vis-à-vis national and European institutions, consistent with the other major crops; (2) putting into practice promotion strategies that highlight the sorghum crop's production-related economic and technical advantages; (3) promoting sorghum's potential applications and markets for human food, animal feed, and non-food products; and (4) assembling and mobilizing possible parties to create synergies and gain momentum over time.

References

Ali MA, Ali A, Ahmed MI, Hassan SW, Khan SR, Abid AA (2014) Phosphorus effect on growth and yield parameters on mung bean. Sci Int 26(4):1821–1824

Alfieri M, Cabassi G, Habyarimana E, Quaranta F, Balconi C, Redaelli R (2019) Discrimination and prediction of polyphenolic compounds and total antioxidant capacity in sorghum grains. J Near Infrared Spectrosc 27(1):46–53. https://doi.org/10.1177/0967033518825351

Awika JM (2017) Sorghum: its unique nutritional and health-promoting attributes. In: Gluten-free ancient grains: cereals, pseudocereals, and legumes: sustainable, nutritious, and health-promoting foods for the 21st century. Elsevier, Amsterdam, pp 21–54. https://doi.org/10.1016/B978-0-08-100866-9.00003-0

Bassam EN (2010) Handbook of bioenergy crops; a complete reference to species, development and applications. Earthscan, London. www.earthscan.co.uk

Bednar-Friedl B, Biesbroek R, Schmidt DN, Alexander P, Børsheim KY, Carnicer J, Georgopoulou E, Haasnoot M, Le Cozannet G, Lionello P, Lipka O, Möllmann C, Muccione V, Mustonen T, Piepenburg D, Whitmarsh L (2022) Europe. In: Pörtner H-O, Roberts DC, Tignor M, Poloczanska ES, Mintenbeck K, Alegría A, Craig M, Langsdorf S, Löschke S, Möller V, Okem A, Rama B (eds) Climate change 2022: impacts, adaptation and vulnerability. Cambridge University Press, Cambridge, pp 1817–1927. https://doi.org/10.1017/9781009325844.015

Berenji J, Kisgeci J (1996) Broomcorn-classical example of industrial use of sorghum. In: 1st European seminar on sorghum for energy and industry, pp 43–48

Berenji JB (2000) Genetic resources in broomcorn. u: Abstracts of the Maize and Sorghum EUCARPIA XVIIth International Conference on Maize and Sorghum Genetics and Breeding at the end of the 20th Century. Belgrade, str, 10

Berenj J, Sikora V (2002) Trends and achievements in broomcorn breeding. Breed Prod 30:81–88

Berenji J, Dahlberg J (2004) Perspectives of sorghum in Europe. J Agron Crop Sci 190:332–338. www.blackwell-synergy.com

Berenji J, Dahlberg J, Sikora V, Latković D (2011) Origin, history, morphology, production, improvement, and utilization of broomcorn [Sorghum bicolor (L.) Moench] in Serbia 1. Econ Bot 65(2):190–208

Bibi A, Sadaqat HA, Tahir MHN, Akram HM (2012) Screening of sorghum (sorghum bicolor Var Moench) for drought tolerance at seedling stage in polyethylene glycol. Int J Anim Plant Sci 22(3):671–678

Boivin N, Fuller DQ (2009) Shell middens, ships and seeds: exploring coastal subsistence, maritime trade and the dispersal of domesticates in and around the ancient Arabian Peninsula. J World Prehist 22(2):113–180. https://doi.org/10.1007/s10963-009-9018-2

Braconnier S, Schaffert R, Parella R, Borges-Damascenao C, Srinivasa Rao P, Rao P, Basavaraj G, Reddy B, Zacharias A, Reinhardt G, Rettenmaier N, Detzel A, Keller H, Köppen S, Cornelius C, Monti A, Zegada Lizarazu W, Amaducci S, Fracasso A, Bursztyn M (2014) Summary report of the SWEETFUEL project. www.sweetfuel-project.eu

Burdette A, Garner PL, Mayer EP, Hargrove JL, Hartle DK, Greenspan P (2010) Anti-inflammatory activity of select sorghum (Sorghum bicolor) Brans. J Med Food 13(4):879–887

Burgueño J, de los Campos G, Weigel K, Crossa J (2012) Genomic prediction of breeding values when modeling genotype × environment interaction using pedigree and dense molecular markers. Crop Sci 52(2):707–719. https://doi.org/10.2135/cropsci2011.06.0299

Chakrabarty S, Kravcov N, Schaffasz A, Snowdon RJ, Wittkop B, Windpassinger S (2021) Genetic architecture of novel sources for reproductive cold tolerance in sorghum. Front Plant Sci 12:772177. https://doi.org/10.3389/fpls.2021.772177

Chakrabarty S, Mufumbo R, Windpassinger S, Jordan D, Mace E, Snowdon RJ, Hathorn A (2022) Genetic and genomic diversity in the sorghum gene bank collection of Uganda. BMC Plant Biol 22(1):378. https://doi.org/10.1186/s12870-022-03770-y

Choudhary S, Guha A, Kholova J, Pandravada A, Messina CD, Cooper M, Vadez V (2020) Maize, sorghum, and pearl millet have highly contrasting species strategies to adapt to water stress and climate change-like conditions. Plant Sci 295:110297. https://doi.org/10.1016/j.plantsci.2019.110297

Crossa J, Pérez-Rodríguez P, Cuevas J, Montesinos-López O, Jarquín D, de los Campos G, Burgueño J, González-Camacho JM, Pérez-Elizalde S, Beyene Y, Dreisigacker S, Singh R, Zhang X, Gowda M, Roorkiwal M, Rutkoski J, Varshney RK (2017) Genomic selection in plant breeding: methods, models, and perspectives. In: Trends in plant science, vol 22. Elsevier, Amsterdam, pp 961–975. https://doi.org/10.1016/j.tplants.2017.08.011

Cuevas J, Crossa J, Montesinos-López OA, Burgueño J, Pérez-Rodríguez P, de los Campos G (2017) Bayesian genomic prediction with genotype × environment interaction kernel models. G3 Genes Genomes Genet 7(1):41–53

De Wet JMJ (1978) Special paper: systematics and evolution of sorghum sect. Sorghum (gramineae). Am J Bot 65(4):477–484. https://doi.org/10.1002/j.1537-2197.1978.tb06096.x

Deb UK, Bantilan MC, Roy AD, Rao PP (2004) Global sorghum production scenario. In: Sorghum genetic enhancement: research process, dissemination and impacts. Patancheru, International Crops Research Institute for the Semi-Arid Tropics, pp 21–38

Díaz de Otálora X, Dragoni F, Del Prado A, Estellés F, Wilfart A, Krol D, Balaine L, Anestis V, Amon B (2022) Identification of representative dairy cattle and fodder crop production typologies at regional scale in Europe. Agron Sustain Dev 42(5):94

Doggett H (1988) Sorghum, 2nd Edition. Longman Scientific and Technical, Essex, UK. SÁVOLY, Ferenc: Az Alföld fásításától és öntözésétől a mezőgazdaság terén várható bioklimatikus értéknövelésről. Erdészeti Lapok 59, 1920, no. 17–18, pp. 387–408

Dykes L, Rooney LW (2007) Phenolic compounds in cereal grains and their health benefits. Cereal Foods World 52(3):105–111. https://doi.org/10.1094/CFW-52-3-0105

Ergon Å, Seddaiu G, Korhonen P, Virkajärvi P, Bellocchi G, Jørgensen M, Østrem L, Reheul D, Volaire F (2018) How can forage production in Nordic and Mediterranean Europe adapt to the challenges and opportunities arising from climate change? Eur J Agron 92:97–106

Faye JM, Diatta C, Maina F, Morris GP (2022) Past, present, and future of West African sorghum improvement: building a roadmap for climate-adaptive, farmer-adopted varieties. In: Kane NA, Foncéka D, Dalton TJ (eds) Crop adaptation and improvement for drought-prone environments. https://kstatelibraries.pressbooks.pub/feed-the-future-innovation. https://creativecommons.org/licenses/by-sa/4.0/. Creative Commons Attribution-ShareAlike 4.0 International License

Food and Agriculture Organization (FAO) (2022) The State of Food and Agriculture 2022, Rome

Fiedler K, Bekele WA, Matschegewski C, Snowdon R, Wieckhorst S, Zacharias A, Uptmoor R (2016) Cold tolerance during juvenile development in sorghum: a comparative analysis by genomewide association and linkage mapping. Plant Breed 135(5):598–606. https://doi.org/10.1111/pbr.12394

Gallo LRDR, Reis CEG, Mendonça MA, Da Silva VSN, Pacheco MTB, Botelho RBA (2021) Impact of Gluten-Free sorghum bread genotypes on glycemic and antioxidant responses in healthy adults. Foods 10(10):2256. https://doi.org/10.3390/foods10102256

Garber ED (1950) Cytotaxonomic studies in the genus Sorghum. Univ Calif Publ Bot 23:283–361

George-Jaeggli B (2009) Physiology and genetics of height-yield associations in sorghum. https://www.researchgate.net/publication/43510354

Habyarimana E, Lopez-Cruz M (2019) Genomic selection for antioxidant production in a panel of sorghum bicolor and s. bicolor × s. halepense lines. Gene 10(11):841. https://doi.org/10.3390/genes10110841

Habyarimana E, Laureti D, di Fonzo N, Lorenzoni C (2002) Biomass production and drought resistance at the seedling stage and in field conditions in sorghum. Maydica 47:303–309. https://www.researchgate.net/publication/261357448

Habyarimana E, Laureti D, de Ninno M, Lorenzoni C (2004) Performances of biomass sorghum [Sorghum bicolor (L.) Moench] under different water regimes in Mediterranean region. Ind Crop Prod 20(1):23–28. https://doi.org/10.1016/j.indcrop.2003.12.019

Habyarimana E, Lorenzoni C, Marudelli M, Redaelli R, Amaducci S (2016) A meta-analysis of bioenergy conversion relevant traits in sorghum landraces, lines and hybrids in the Mediterranean region. Ind Crop Prod 81:100–109. https://doi.org/10.1016/j.indcrop.2015.11.051

Habyarimana E, Lorenzoni C, Redaelli R, Alfieri M, Amaducci S, Cox S (2018) Towards a perennial biomass sorghum crop: A comparative investigation of biomass yields and overwintering of Sorghum bicolor x S. halepense lines relative to long term S. bicolor trials in northern Italy. Biomass Bioenergy 111:187–195. https://doi.org/10.1016/j.biombioe.2017.03.004

Harlan JR, Wet JMJ (1972) A simplified classification of cultivated sorghum 1. Crop Sci 12(2):172–176

Havaux M (1989) Fluorimetric determination of the genetic variability existing for chilling tolerance in sweet sorghum and Sudan grass. Plant Breed 102:327–332

Heffner EL, Lorenz AJ, Jannink J-L, Sorrells ME (2010) Plant breeding with genomic selection: gain per unit time and cost. Crop Sci 50:1681–1690

Heffner EL, Jannink J-L, Sorrells ME (2011) Genomic selection accuracy using multifamily prediction models in a wheat breeding program. Plant Genome 4:65–75

Hu X, Wang J, Lu P, Zhang H (2009) Assessment of genetic diversity in broomcorn millet (Panicum miliaceum L.) using SSR markers. J Genet Genomics 36(8):491–500. https://doi.org/10.1016/S1673-8527(08)60139-3

Hunt HV, van der Linden M, Liu X, Motuzaite-Matuzeviciute G, Colledge S, Jones MK (2008) Millets across Eurasia: chronology and context of early records of the genera Panicum and Setaria from archaeological sites in the old world. Veg Hist Archaeobotany 17(suppl. 1):5–18. https://doi.org/10.1007/s00334-008-0187-1

Hunt CH, van Eeuwijk FA, Mace ES, Hayes BJ, Jordan DR (2018) Development of genomic prediction in sorghum. Crop Sci 58(2):690–700. https://doi.org/10.2135/cropsci2017.08.0469

Jiang D, Ma T, Ding F, Fu J, Hao M, Wang Q, Chen S (2019) Mapping global environmental suitability for sorghum bicolor (L.) Moench. Energies 12(10):1928. https://doi.org/10.3390/en12101928

Jocelyne RE, Béhiblo K, Ernest AK (2020) Comparative study of nutritional value of wheat, maize, sorghum, millet, and fonio: some cereals commonly consumed in Côte d'Ivoire. Eur Sci J 16(21):118

Karlsson JO, Parodi A, Van Zanten HH, Hansson PA, Röös E (2021) Halting European Union soybean feed imports favours ruminants over pigs and poultry. Nat Food 2(1):38–46

Kosolapov VM, Cherniavskih VI, Kostenko SI (2021) Fundamentals for forage crop breeding and seed production in Russia. Vavilovskii Zhurnal Genet Selektsii 25(4):401–407. https://doi.org/10.18699/VJ21.044

Lopez-Cruz M, Crossa J, Bonnett D, Dreisigacker S, Poland J, Jannink JL, Singh RP, Autrique E, de los Campos, G. (2015) Increased prediction accuracy in wheat breeding trials using a marker × environment interaction genomic selection model. G3 Genes Genomes Genet 5(4):569–582. https://doi.org/10.1534/g3.114.016097

Lu H, Zhang J, Liu K-B, Wu N, Li Y, Zhou K, Ye M, Zhang T, Zhang H, Yang X, Shen L, Xu D, Li Q (2009) Earliest domestication of common millet (Panicum miliaceum) in East Asia extended to 10,000 years ago. PNAS 106(8):7367–7372. https://doi.org/10.1073/pnas.0900158106

Maulana F, Weerasooriya D, Tesso T (2017) Sorghum landrace collections from cooler regions of the world exhibit magnificent genetic differentiation and early season cold tolerance. Front Plant Sci 8:756. https://doi.org/10.3389/fpls.2017.00756

Meissen THE, Hayes BJ, Goddard ME (2001) Prediction of total genetic value using genome-wide dense marker maps. Genetics 157:1819–1829

Merrick LF, Carter AH (2021) Comparison of genomic selection models for exploring predictiveability of complex traits in breeding programs. Plant Genome 14:e20158. https://doi.org/10.1002/tpg2.20158

Miller NF, Spengler RN, Frachetti M (2016) Millet cultivation across Eurasia: origins, spread, and the influence of seasonal climate. The Holocene 26(10):1566–1575. https://doi.org/10.1177/0959683616641742

Mustalahti K, Catassi C, Reunanen A, Fabiani E, Heier M, McMillan S, Murray L, Metzger MH, Gasparin M, Bravi E, Mki M (2010) The prevalence of celiac disease in Europe: results of a centralized, international mass screening project. Ann Med 42(8):587–595. https://doi.org/10.3109/07853890.2010.505931

Nedelcheva AM, Dogan Y, Guarrera PM (2007) Plants traditionally used to make brooms in several European countries. J Ethnobiol Ethnomed 3:20. https://doi.org/10.1186/1746-4269-3-20

Niu H, Han Y, Ping J, Wang Y, Lv X, Chu J (2022) Genome wide association analysis of acid detergent fiber content of 206 forage sorghum (Sorghum bicolor (L.) Moench) accessions. Genet Resour Crop Evol 69(5):1941–1951

Pan Y, Cheng J, Sun D (2019) Cold plasma-mediated treatments for shelf life extension of fresh produce: a review of recent research developments. Compr Rev Food Sci Food Saf 18(5):1312–1326. https://doi.org/10.1111/1541-4337.12474

Parra-Londono S, Fiedler K, Kavka M, Samans B, Wieckhorst S, Zacharias A, Uptmoor R (2018) Genetic dissection of early-season cold tolerance in sorghum: genome-wide association studies for seedling emergence and survival under field and controlled environment conditions. Theor Appl Genet 131(3):581–595. https://doi.org/10.1007/s00122-017-3021-2

Patanè C, Cavallaro V, Avola G, D'Agosta G (2006) Seed respiration of sorghum [Sorghum bicolor (L.) Moench] during germination as affected by temperature and osmoconditioning. Seed Sci Res 16(4):251–260. https://doi.org/10.1017/SSR2006259

Patanè C, Cosentino SL, Cavallaro V, Saita A (2021) Screening for cold tolerance during germination within sweet and fiber sorghums [sorghum bicolor (L.) moench] for energy biomass. Agronomy 11(4):620. https://doi.org/10.3390/agronomy11040620

Piper JK, Kulakow PA (1994) Seed yield and biomass allocation in Sorghum bicolor and F1 and backcross generations of S. bicolor x S. halepense hybrids. Can J Bot 72:468–474

Pontieri P, Mamone G, de Caro S, Tuinstra MR, Roemer E, Okot J, de Vita P, Ficco DBM, Alifano P, Pignone D, Massardo DR, del Giudice L (2013) Sorghum, a healthy and gluten-free food for celiac patients as demonstrated by genome, biochemical, and immunochemical analyses. J Agric Food Chem 61(10):2565–2571. https://doi.org/10.1021/jf304882k

Prasad PVV, Djanaguiraman M, Jagadish SVK, Ciampitti IA (2019) Drought and high temperature stress and traits associated with tolerance. In: Ciampitti IA, Prasad PVV (eds) Sorghum: a state of the art and future perspectives, vol 58. ASA, CSSA, SSSA, Madison, pp 245–265

Quiles-Belén A, Sotomayor-Ríos A (1983) The sorghum conversion program (No. 1976-2017-3580)

Ratnadass A, Fleurat-Lessard F (1991) Improvements in the practical methods of assessment of losses caused by insects in grain stored at the village level in Tropical Africa. In: Fleurat-Lessard F, Ducom P (eds) 5th International working conference on stored-product protection, Bordeaux, France, pp 1681–1691

Ruiz-Chután JA, Salava J, Janovská D, Žiarovská J, Kalousová M, Fernández E (2019) Assessment of genetic diversity in sorghum bicolor using RAPD markers. Genetika 51(3):789–803. https://doi.org/10.2298/GENSR1903789R

Saita A, Patane C, Guarnaccia P (2011) Genotypes screening for cold tolerance during germination in sorghum for energy biomass. European Biomass Conference and Exhibition, pp 550–555

San Martin D, Orive M, Iñarra B, García A, Goiri I, Atxaerandio R, Urkiza J, Zufia J (2021) Spent coffee ground as second-generation feedstuff for dairy cattle. Biomass Convers Biorefinery 11:589–599

Schaffasz A, Windpassinger S, Friedt W, Snowdon R, Wittkop B (2019) Sorghum as a novel crop for Central Europe: using a broad diversity set to dissect temperate-adaptation. Agronomy 9(9):535

Shegro A, Shargie NG, van Biljon A, Labuschagne MT (2012) Diversity in starch, protein and mineral composition of sorghum landrace accessions from Ethiopia. J Crop Sci Biotechnol 15(4):275–280. https://doi.org/10.1007/s12892-012-0008-z

Sikora V, Berenji J (2009) Razvoj sortimenta sirka metla a u Institutu za ratarstvo i povrtarstvo Novi Sad. Strucni Rad Expert Note, pp 363–368

Sikora V, Stojanović A, Brdar-Jokanović M, Kiprovski B, Mutavdžić B, Ugrenović V, Tóth Š (2018) Broomcorn [Sorghum bicolor (L.) MOCNCH] panicle yield as affected by environmental variables and agro-technological traits. In Pak J Bot 50(2):545–552

Singh D, Singh K (2015) Marker-assisted plant breeding: principles and practices. Springer, Cham

Taube F, Vogeler I, Kluß C, Herrmann A, Hasler M, Rath J, Loges R, Malisch CS (2020) Yield progress in forage maize in NW Europe—breeding progress or climate change effects? Front Plant Sci 11:1214

Upadhyaya HD, Vetriventhan M, Deshpande S (2016) Sorghum germplasm resources characterization and trait mapping. Springer, Cham, pp 77–94. https://doi.org/10.1007/978-3-319-47789-3_4

Velazco JG, Jordan DR, Mace ES, Hunt CH, Malosetti M, van Eeuwijk FA (2019a) Genomic prediction of grain yield and drought-adaptation capacity in sorghum is enhanced by multi-trait analysis. Front Plant Sci 10:997. https://doi.org/10.3389/fpls.2019.00997

Velazco JG, Malosetti M, Hunt CH, Mace ES, Jordan DR, van Eeuwijk FA (2019b) Combining pedigree and genomic information to improve prediction quality: an example in sorghum. Theor Appl Genet 132(7):2055–2067. https://doi.org/10.1007/s00122-019-03337-w

Venkateswaran K, Elangovan M, Sivaraj N (2019) Origin, domestication and diffusion of sorghum bicolor. In: Breeding sorghum for diverse end uses. Elsevier, Amsterdam, pp 15–31. https://doi.org/10.1016/B978-0-08-101879-8.00002-4

Warwick SI, Phillips D, Anderew C (1986) Rhizome depth: the critical factor in winter survival of Sorghum halepense (L.) Pers. (Johnson grass). Weed Res 26:381–387

Winchell F, Brass M, Manzo A, Beldados A, Perna V, Murphy C, Stevens C, Fuller DQ (2018) On the origins and dissemination of domesticated sorghum and pearl millet across Africa and into India: a view from the Butana Group of the Far Eastern Sahel. Afr Archaeol Rev 35(4):483–505. https://doi.org/10.1007/s10437-018-9314-2

Wu Y, Li X, Xiang W, Zhu C, Lin Z, Wu Y, Li J, Pandravada S, Ridder DD, Bai G, Wang ML, Trick HN, Beane SR, Tuinstra MR, Tesso TT, Yu J (2012) Presence of tannins in sorghum grains is conditioned by different natural alleles of Tannin1. Proc Natl Acad Sci USA 109(26):10281–10286. https://doi.org/10.1073/pnas.1201700109

Zegada-Lizarazu W, Parrish D, Berti M, Monti A (2013) Dedicated crops for advanced biofuels: consistent and diverging agronomic points of view between the USA and the EU-27. Biofuels Bioprod Biorefin 7(6):715–731

Zegada-Lizarazu W, Fernando Luna D, Monti A (2016) Differential characteristics of photochemical acclimation to cold in two contrasting sweet sorghum hybrids. Physiol Plant 157(4):479–489. https://doi.org/10.1111/ppl.12430

Zhao Z (2011) New archaeobotanic data for the study of the origins of agriculture in China. Curr Anthropol 52(4):659308

Zhu M, Chen J, Yuyama N, Luo L, Xiao X, Lv Y, Liu Y, Cai H (2020) Genetic diversity and population structure of broomcorn sorghum investigated with simple sequence repeat markers. Trop Plant Biol 13(1):62–72. https://doi.org/10.1007/s12042-019-09251-1

Marker-Assisted Breeding in Major Insect Pest of Sorghum Crop

12

Dhanyakumar Onkarappa, Arunsaikumar Karrem, Murugan Marimuthu, and Jagdish Jaba

Abstract

The fundamental concept behind marker-assisted breeding revolves around the identification and utilization of genetic markers linked to biotic stress resistance traits. These markers act as guideposts within the plant's DNA, allowing breeders to precisely identify the presence of desirable traits without the need for extensive field testing. While the use of resistant germplasms in breeding programs targeting improved sorghum resistance to various biotic stresses has been relatively limited, it presents a significant opportunity for expansion and enhancement in sorghum pest management strategies. To seize this opportunity, continued research endeavors are imperative. By delving deeper into the identification

D. Onkarappa
International Crops Research Institute for the Semi-Arid Tropics, Hyderabad, Telangana, India

Department of Entomology, Tamil Nadu Agricultural University, Coimbatore, Tamil Nadu, India

A. Karrem
International Crops Research Institute for the Semi-Arid Tropics, Hyderabad, Telangana, India

Department of Agricultural Entomology, University of Agricultural Sciences, Raichur, Karnataka, India

M. Marimuthu
Department of Entomology, Tamil Nadu Agricultural University, Coimbatore, Tamil Nadu, India

J. Jaba (✉)
International Crops Research Institute for the Semi-Arid Tropics, Hyderabad, Telangana, India
e-mail: Jagdish.Jaba@icrisat.org

E. Habyarimana et al. (eds.), *Omics and Biotechnological Approaches for Product Profile-Driven Sorghum Improvement*,
https://doi.org/10.1007/978-981-97-4347-6_12

and integration of resistant traits, there exists the potential for substantial progress in sorghum breeding programs. The incorporation of marker-assisted breeding techniques into these programs aims to elevate crop productivity and ensure food security while simultaneously reducing the dependence on chemical pesticides. This approach offers a more sustainable and environmentally friendly method for managing pests in sorghum cultivation.

Keywords

Biotic stress · Molecular breeding · Molecular markers · Sorghum

12.1 Introduction

Sorghum (*Sorghum bicolor*) is an often cross-pollinated diploid crop species ($2n = 20$), belonging to the family Poaceae and subfamily, Panicoideae. Sorghum is a C_4 metabolism plant that has fifth most important cereal next to wheat, rice, maize, and barley (Gibson 2009). It is the staple food for over 750 million people in the world. It is widely grown for food, fodder, forage, and fuel in the semiarid tropics, especially in Asia, Africa, the United States of America, and Australia. In Asia, the production of sorghum amounted to approximately 9,201,578 tons in 2020, cultivated across an area of 6,945,276 hectares. India, being the third-largest sorghum producer globally, cultivated sorghum on a vast expanse of 4.7 million hectares, resulting in the production of 5.5 million metric tons of grain in the year 2020 (http://data.un.org/Data.aspx?d=FAO&f=itemCode%3A83).

In mixed crop-livestock systems, sorghum serves as a valuable source of stover for dairy and draught animals. Since sorghum has the same energy content as maize and other cereals, it can replace maize for making better silage. Being coarse and fast-growing in the warm season, sorghum provides livestock with feed in midsummer when no other source is available (Juerg et al. 2009). Recently, sorghum has gained greater significance due to the increasing demand for its use as poultry feed and for bioethanol production from sweet sorghum stalks (Rao et al. 2009). However, there has been an apparent increase in usage and production of sorghum. Despite its adaptability and utility, sorghum faces significant challenges due to various biotic stresses, including insect pests, nematodes, and weeds. These biotic stresses can lead to substantial yield losses and impact the overall productivity of sorghum cultivation.

In each year, the biotic stress causes either individually or in combination with other factors have shown result in substantial economic yield losses in sorghum production (Mengistu et al. 2019). Currently, more than 150 insect species have been recorded in sorghum crops with some of the most destructive being are Sorghum shoot fly *Atherigona soccata*, aphid species such as *Rhopalosiphum maidis* and *Melanaphis sacchari*, and Sorghum midge, *Stenodiplosis sorghicola*, stem borer, *Chilo partellus* which causes the reduction in production(Kumari et al. 2021; Vashisth et al. 2022). Weed infestations in sorghum fields have a significant detrimental effect on yield. Among the common weeds, wild Striga stands out as a major hindrance to sorghum productivity, particularly in Africa. In severely infested soils, Striga spp. can lead to crop yield reductions of up to 100% (Kavuluko et al. 2021). The utilization of modern genetically modified approaches to identify and transfer candidate genes for enhanced productivity and resistance to biotic stress holds the potential to significantly improve overall sorghum performance (Hariprasanna and Rakshit 2016).

Utilizing chemical methods to control infestations is neither cost-effective for small-scale farmers nor environmentally safe. In contrast, host plant resistance (HPR) offers a more dependable option for smallholder farmers who cannot afford expensive chemicals. HPR represents a complex trait resulting from interactions among various component traits, both morphological and biochemical, which confer resistance to biotic stress (Sharma et al. 2003; Dhillon and Sharma 2004; Riyazaddin et al. 2015; Arora et al. 2021). Numerous genotypes exhibiting resistance to biotic stresses have been identified, and some of them are actively promoted for inclusion in resistance breeding programs (Baloch et al. 2023).

The advancement of breeding-based technologies aimed at addressing biotic stress has encountered challenges due to the intricate nature of resistance traits and the difficulty in transferring resistance to offspring through conventional breeding methods. Furthermore, as a quantitatively inherited trait heavily reliant on environmental factors, conventional breeding has not provided a significant success in managing biotic stress (Aruna et al. 2011). Another complicating factor arises from difficulties in crossing wild genotypes with domesticated resistant varieties, as this often results in undesirable agronomic traits and complicates the process of generating improved resistant progeny (Satish et al. 2012).

Marker-assisted selection (MAS) serves as a platform for the genetic manipulation of identified quantitative trait loci (QTL). The molecular dissection of these QTLs aids in the development of markers that are specific to particular traits. However, it is worth noting that the markers identified within the QTLs are susceptible to recombination, which can be a drawback. This potential recombination may result in inconsistencies between genotypic and phenotypic measurements. Genic markers mitigate the risk of genotypic and phenotypic variation and contribute to the fine mapping of candidate genes. This, in turn, offers insights into the molecular mechanisms associated with trait resistance. Furthermore, the identification of specific genes within QTLs, accomplished through expression and transformation studies, provides a robust pathway toward trait resistance (Baloch et al. 2023).

The availability of the sorghum genome greatly facilitates the identification of genes responsible for conferring resistance to biotic stress, especially when compared to other similar crops (Satish et al. 2012). Given the economic impact of biotic stress, it becomes essential to comprehend the genetic factors that underlie the inheritance of resistance. Such understanding provides valuable insights for molecular biologists to develop advanced technologies for managing biotic stress. Therefore, the discovery and utilization of novel genes responsible for host plant resistance are of paramount importance in the quest to develop sorghum cultivars that are resistant to biotic stress in particular to insects, plant parasitic nematodes (PPN) and weed.

12.2 Sorghum Germplasm and Its Collection

The cultivated sorghum, scientifically known as *S. bicolor* subsp. *bicolor*, is traditionally categorized into five major races, which are *bicolor*, *guinea*, *caudatum*, *kafir*, and *durra*, along with ten possible hybrid groups resulting from interactions among these races (Harlan and De Wet 1972; Harlan 2011). Within the broader *S. bicolor* species, there exist numerous "wild" subspecies, each with its own distinct races, many of which carry potentially valuable traits. However, the classification of sorghum into various races is primarily based on a relatively small number of genes governing morphological characteristics. In contrast, traits crucial for adapting to specific habitats, which exhibit extensive variability across sorghum germplasm and are quantitatively inherited. Additionally, closely related accessions may exhibit significant field-level variations in appearance, while some seemingly similar accessions may actually originate from vastly different genetic backgrounds. Therefore, relying solely on a limited set of discrete morphological characteristics for classifying germplasm accessions may not offer an accurate representation of the genetic diversity present within germplasm of sorghum (Varshney and Tuberosa 2007).

In breeding programs different germplasm collections has expand significantly, often becoming too extensive for a single breeder's practical assessment. To address this issue, a systematic approach has introduced the concept called a "core collection." The core collection's primary aim is to encapsulate the genetic diversity within the larger germplasm collection while reducing the number of individual accessions. This streamlined core collection serves as a practical resource for breeders searching for new traits. It is easily screened for desirable characteristics, with the potential for integration into breeding programs. Additionally, related accessions from the larger collection can be examined. Developing an informative core collection comes with its own challenges. Core collections must strike a balance, being compact enough for practical utility while retaining a substantial portion of the genetic diversity in the species represented in the comprehensive collection.

12.3 Biotic Stress in Sorghum

Biotic stress in sorghum crops refers to the adverse effects caused by various living organisms, such as insects, nematodes, and weeds. These biotic stresses alone or in combination cause significant impact on sorghum crop production and lead to economic losses to the small land holding farmers (Mengistu et al. 2019).

A broad spectrum of insect pests inflicts considerable harm to sorghum crops, beginning with seedling pests, root borers, stem borers, and leaf feeders, as well as sucking pests and ear head pests (Table 12.1) (Kumari et al. 2021). *Striga* spp., a parasitic weed, is a widespread issue affecting sorghum yields across all continents. It can cause substantial reductions in crop yields, with the potential to lead to a 100% decrease in production (Kavuluko et al. 2021).

Table 12.1 Major biotic stress causing economic yield loss in sorghum

Sr. no	Common name	Scientific name	Order	Family
Insects				
1	Shoot fly	*Atherigona soccata* Rondani	Diptera	Muscidae
2	Stem borer	*Chilo partellus* Swinhoe	Lepidoptera	Pyralidae
3	Midge	*Stenodiplosis sorghicola* Coquillett	Diptera	Cecidomyiidae
4	White grub	*Holotrichia consanguinea* Hope	Coleoptera	Scarabaeidae
5	Armyworm	*Mythimna separata* Walker	Lepidoptera	Noctuidae
6	Cutworm	*Agrotis ipsilon* Ochsenheimer	Lepidoptera	Noctuidae
7	Grasshopper	*Hieroglyphus* sp	Orthoptera	Acrididae
8	Pyrilla	*Pyrilla perpusilla* Linnaeus	Hemiptera	Lophopidae
9	Shoot bug	*Peregrinus maidis* Ashmead	Hemiptera	Delphacidae
10	Earhead caterpillars	*Helicoverpa armigera* Hubner *Eublemma* spp.	Lepidoptera	Noctuidae
11	Fall Army Worm	*Spodoptera frugiperda*	Lepidoptera	Noctuidae
Weeds				
1	Witchweed	*Striga* Spp.	Lamiales	Orobanchaceae
Nematode				
1	Dagger	*Xiphinema americanum*	Dorylaimida	Longidoridae
2	Root knot	*Meloidogyne* spp.	Tylenchida	Heteroderidae
3	Stubby root	*Paratrichodorus minor*	Triplonchida	Trichodoridae
4	Stunt	*Tylenchorhynchus* spp	Tylenchida	Belonolaimidae
5	Needle	*Longidorus africanus*	Dorylaimida	Longidoridae
6	Sting	*Belonolaimus longicaudatus*	Tylenchida	Belonolaimidae

(Source: Claflin (1983); Sharma and Nwanze (1997))

Plant parasitic nematodes are posing a detrimental impact on sorghum crop production (Traore et al. 2012). In the early stages of root infection, the affected sorghum plants exhibit stunted growth and delayed flowering, ultimately resulting in reduced overall harvest yields (Little et al. 2018). Sorghum crops are susceptible to a wide array of plant parasitic nematodes, including the Dagger nematode, Root knot nematode, Stubby root nematode, Stunt nematode, Needle nematode, and Sting nematode (Table 12.1).

12.4 Molecular Markers in Sorghum

Molecular markers play a crucial role in identifying and tracking genetic factors associated with biotic stress resistance in crops like sorghum. These markers allow breeders to select and develop stress-resistant varieties more efficiently. These molecular markers are valuable tools for sorghum breeders and geneticists working to enhance resistance to biotic stresses. They are employed to identify specific DNA sequences that are associated with particular traits or genes of interest. They enable the precise selection of desirable traits and the development of biotic stress-resistant sorghum varieties, ultimately contributing to improved crop productivity and sustainability.

The restriction fragment length polymorphism (RFLP) is the earliest types of molecular markers system that has been initially widely used in sorghum. It identifies differences in DNA sequences that result in variation in the lengths of DNA fragments produced by restriction enzymes and it has been widely used in genetic mapping and phylogenetic studies (Tao et al. 1998). Thereafter, more marker systems emerged for sorghum, including amplified fragment length polymorphism (AFLP), have been used to generate some selective amplification of a subset of genomic DNA fragments and they are often used for fingerprinting and genetic diversity assessment (Wu et al. 2006). In later, randomly amplified polymorphic DNA (RAPD) markers are generated by random amplification of genomic DNA segments using short, arbitrary primers, but they are simple to use and lack reproducibility as compared to other markers.

In recent days, single sequence repeats (SSRs) are widely used because they have short repetitive DNA sequences, such as dinucleotide or trinucleotide repeats and they are highly polymorphic and commonly used for genetic diversity analysis, mapping, and marker-assisted breeding. Similarly, single-nucleotide polymorphisms (SNP) are also widely used in sorghum crop and it is the most abundant type of molecular marker, representing single base-pair variations in the DNA sequence. It is valuable for high-throughput genotyping, association studies, and genomic analysis (Haussmann et al. 2002; Bowers et al. 2003). Commonly used marker systems, such as SSR and SNP, heavily rely on sequence information, resulting in a significant cost per data point. These limitations pose a substantial challenge to breeding programs, making it difficult to attain a favorable return on investment that would support the routine use of marker-assisted breeding, particularly for quantitative traits. Diversity Arrays Technology (DArT) offers a cost-effective marker technology that functions based on hybridization principles, providing a high level of multiplexing without the need for sequence data dependency (Mace et al. 2008).

In sorghum, each of these marker systems has found application for a range of purposes, encompassing QTL mapping, genetic diversity assessment, and fingerprinting. The availability of the sorghum genome sequence opens up an opportunity to expand the repertoire of SSR molecular markers. The genome sequence itself is estimated to harbor approximately 71,000 SSRs (Baloch et al. 2023). Identifying and harnessing novel genes associated with host plant resistance are very importance in the pursuit of developing sorghum cultivars that exhibit robust resilience to biotic stresses. By delving into the exploration of new genetic factors, we can pave the way for the creation of sorghum varieties that not only withstand biotic stressors but also thrive in challenging agricultural conditions, ultimately increase its crop yield and ensuring food security in sorghum growing regions. The discovery and incorporation of these genes hold the key to enhancing sorghum's natural defense mechanisms, making it a more reliable and sustainable crop option for small-scale farming communities.

12.5 Marker Assisted and Its Utility in Plant Breeding Against Biotic Stress

Numerous marker systems have been developed and utilized to pinpoint and map genes and quantitative characteristics (QTL) that hold considerable economic significance. These encompass traits like resistance to insect pests and nematodes, tolerance to Striga. Marker-assisted selection (MAS) plays a pivotal role in breeding programs aiming to enhance these traits. This process involves several crucial steps, including the identification of desirable genetic markers associated with target traits, genotyping of plant materials to detect these markers, and the subsequent selection of individuals with the desired genetic profile. The utilization of MAS not only expedites the breeding process but also enhances its precision by allowing breeders to make informed decisions based on the genetic makeup of the plants, ultimately leading to the development of improved sorghum cultivars that better meet the diverse needs of farmers and consumers.

12.5.1 Marker-Assisted Breeding Against Insects

Marker-assisted breeding against insect pests is an evolving field of agricultural research and breeding strategies aimed at mitigating the detrimental impact of insect infestations on crop production. Sorghum, a staple crop in many regions, is particularly susceptible to insect attacks, with a range of species posing significant threats. Among these pests, major lepidopterous insects like the stem borer (*Chilo partellus*) and the fall armyworm (*Spodoptera frugiperda*) cause substantial damage. Additionally, dipterans such as gall midges (*Stenodiplosis sorghicola*), shoot flies (*Atherigona soccata*), and hemipteran insects like the green aphid (*Schizaphis graminum*) are known as some of the most destructive pests in sorghum cultivation. The development of marker-assisted breeding techniques in sorghum aims to combat these challenges by identifying and utilizing genetic markers associated with insect resistance traits, ultimately leading to the creation of insect-resistant (Okosun et al. 2021).

The presence of two distinct genetic mechanisms, namely antixenosis and antibiosis, that contribute to the resistance observed in the sorghum line ICSV745 × 90,562 against the midge. Antixenosis involves the sorghum plant's ability to deter or repel the midge from settling on or infesting the crop, thus reducing its susceptibility. On the other hand, antibiosis signifies the sorghum plant's inherent ability to negatively affect the midge's biology or survival upon infestation, further enhancing the resistance level (Tao et al. 2003).

Marker-assisted breeding techniques harness the synergy of molecular genetics and traditional plant breeding. The incorporation of marker-assisted breeding and map-based cloning approaches has propelled in the quest to identify and characterize genetic loci associated with resistance to green aphid (*S. graminum*). Various studies have collectively reported the registration of 36 quantitative trait loci (QTLs) linked to resistance against *S. graminum*. These QTLs denote critical genomic regions to identify genetic markers that significantly contribute to develop new cultigens the sorghum plant against *S. graminum* infestations (Wu and Huang 2008; Punnuri et al. 2013).

Among the dipterous insect, sorghum shoot fly (*Atherigona soccata*) is recognized as a significant and damaging pest of sorghum. Its larvae have a penchant for severing the new shoot tips, resulting in symptoms akin to a "dead heart" in the sorghum plant. Many attempts have been made and identified polymorphisms within sorghum that confer resistance to shoot flies. These genetic variations have been instrumental in pinpointing specific genetic loci responsible for enhancing sorghum's resistance. Employing a range of molecular techniques, including SSRs, SNPs, and GWAS approaches, numerous molecular traits, and resistance loci has been uncovered in different sorghum genotypes and cultivars. Presently, some notable resistant germplasms have been utilized among these are SFCR 151, IS 1057, PS 30715–1, IS 1071, IS 1082, IS 4663, IS 1096, IS 2312, IS 2146, IS 18369, IS 2205, IS 18551, and IS 2394, all of which have shown promise in resistance against *A. soccata* (Chamarthi et al. 2011; Kumari et al. 2021) (Table 12.2).

The marker-assisted breeding is to identify and utilize genetic markers associated with insect resistance traits. These markers serve as signposts in the plant's DNA, enabling breeders to pinpoint the presence of desirable traits without the need for extensive field testing. The utilization of resistant germplasms in breeding programs aimed at enhancing sorghum's resistance to several insect-pests has been somewhat limited. This is an opportunity for substantial expansion and improvement in sorghum pest management strategies. The identification and incorporation of resistant traits are potential for significant advancements in sorghum breeding programs, making it a more sustainable and environmentally friendly approach to insect management in sorghum cultivation.

Some of the plant parasitic nematodes (PPN) play a significant role in causing damage to plant roots, leading to adverse effects on crop yields. Of these nematodes, the Southern root-knot nematode stands out for its notorious potential to inflict harm upon plant roots. Recent endeavors in this field have taken a noteworthy turn and actively pursued the identification of Single Nucleotide Polymorphism (SNP) markers. This marks a fresh and innovative approach aimed at comprehensively addressing and understanding the complexities surrounding the impact of PPN on plant roots and overall crop productivity (Little et al. 2018).

12.6 Conclusion

In the sorghum crop, developing resistance in host plants against biotic stresses presents a competitive challenge. This challenge arises from the intricacies associated with variations in pest genotypes, the complex genetic control involved, and the difficulties in effective phenotyping for these stresses. Despite these challenges, significant progress has been made in sorghum crop improvement, transitioning from conventional methods to utilizing advanced molecular tools to enhance resistance to biotic stress. The integration of genetic maps, as exemplified by certain

Table 12.2 Insect pest resistance loci in sorghum (Source: Baloch et al. 2023)

Insect	Biotype	Population	Linked Marker	QTL gene	Chromosome	Reference
Green Aphid	I	F2, 371	Starssbnm93-Starssbem286	Qstsgrsbi09ii	SBI-09	Punnuri et al. (2013)
			Starssbem296-Starssbnm93	Qstsgrsbi09i	SBI-09	
			Starssbnm102-Starssbem298	Qstsgrsbi09iii	SBI-09	
	I, K	RIL	Xtxp12	VIS- GBK1	SBI-04	Nagaraj et al. (2005)
			Xtxp335	IS-GBK5	SBI-05	
			Xtxp43	SPA-I2	SBI-05	
			Xtxp335	SPA-K2	SBI-05	
			Xcup20	VIS-GBK2	SBI-04	
			Sb1–10	VIS-GBI8	SBI-04	
	I	F2	Xtxp230-Xtxp67	QSsgr-09–02	SBI-09	Wu and Huang (2008)
			Xtxp358-Xtxp289	QSsgr-09–01	SBI-09	
	E	F2:3	Xtxp289-Xtxp358	QSsgr-09–01	SBI-09	Wu et al. (2007)
		F3	pSB347	Ssg1	SBI-09	Katsar et al. (2002)
	C, E	F3	pSB262-pSB089	Ssg3	SBI-05	
	C	F3	CSU030	Ssg2	SBI-08	
	C,I,E, K	F3	pSB443	Ssg4	SBI-03	
	K, I	RIL	B18–885	–	SBI-01	Agrama et al. (2002)
			OPB12–795	–	SBI-05	

(continued)

Table 12.2 (continued)

Insect	Biotype	Population	Linked Marker	QTL gene	Chromosome	Reference
Shoot-fly		RIL	Xtxp99-Xcup57	QTdu.dsr-7 QTdl. dsr-10.1 QTdu. dsr-10 QTdl. dsr-10.2 4.2	SBI-07 SBI-10 SBI-10 SBI-10	Aruna et al. (2011)
			Xtxp320-Xgap1			
			Xcup16-Xtxp320			
			Trit-Xtxp320			
			Xtxp248–Xtxp316		SBI-01	Apotikar et al. (2011)
			Xnhsbm1013-Xnhsbm1044	QTdu.dsr-10.1 QTdl.dsr-1.2 QTdl. dsr-6 QGs.dsr-5	SBI-10 SBI-01 SBI-06 SBI-05	Satish et al. (2009)
			Xtxp302-Dsenhsbm80			
			Xtxp274-Xtxp317			
			Xtxp30-Xtxp6			
		RIL	XnhsbmSFC34	Cysteine protease Mir1	SBI-10	Satish et al. (2012)
	F1 and F2		Ch8_20519857-Ch8_20217087	qRE8.1	SBI-02 SBI-06	Pandian et al. (2020)
			Ch4_48400958-Ch4_46023941	qVI4.0		

studies, holds great promise for developing more efficient breeding systems in sorghum. This is crucial for optimizing marker-assisted selection (MAS) breeding strategies aimed at enhancing host plant resistance to diverse biotic stresses. As genotyping tools become increasingly available and affordable, prioritizing trait phenol-typing becomes imperative. The future success in developing MAS for host plant resistance to biotic stresses relies on the quality of data generation and the judicious use of genomic tools to establish trait-marker associations, ensuring their effective deployment in breeding programs.

References

Agrama H, Widle G, Reese J et al (2002) Genetic mapping of QTLs associated with greenbug resistance and tolerance in *Sorghum bicolor*. Theor Appl Genet 104:1373–1378

Apotikar DB, Venkateswarlu D, Ghorade RB et al (2011) Mapping of shoot fly tolerance loci in sorghum using SSR markers. J Genet 90:59–66. https://doi.org/10.1007/s12041-011-0046-1

Arora N, Mishra SP, Nitnavare RB, Jaba J, Kumar AA, Bhattacharya J, Sohu RS, Sharma HC (2021) Morpho-physiological traits and leaf surface chemicals as markers conferring resistance to sorghum shoot fly (Atherigona soccata Rondani). Field Crop Res 261:1–12. ISSN 0378-4290

Aruna C, Bhagwat VR, Madhusudhana R et al (2011) Identification and validation of genomic regions that affect shoot fly resistance in sorghum [*Sorghum bicolor* (L.) Moench]. Theor Appl Genet 122:1617–1630

Baloch FS, Altaf MT, Liaqat W et al (2023) Recent advancements in the breeding of sorghum crop: current status and future strategies for marker-assisted breeding. Front Genet 14:1150616

Bowers JE, Abbey C, Anderson S et al (2003) A high-density genetic recombination map of sequence-tagged sites for sorghum, as a framework for comparative structural and evolutionary genomics of tropical grains and grasses. Genetics 165:367–386

Chamarthi SK, Sharma HC, Sahrawat KL et al (2011) Physico-chemical mechanisms of resistance to shoot fly, Atherigona soccata in sorghum, *Sorghum bicolor*: physico-chemical mechanisms of resistance in sorghum to shoot fly. J Appl Entomol 135:446–455. https://doi.org/10.1111/j.1439-0418.2010.01564.x

Claflin LE (1983) Plant-parasitic nematodes affecting sorghum. In: Sorghum root and stalk rots, p 53

Dhillon MK, Sharma PD (2004) Studies on biology and behavior of Earias vittella (Lepidoptera: Noctuidae) for mechanisms of resistance in different cotton genotypes. Crop Prot 23:235–241

Gibson DJ (2009) Grasses and grassland ecology. Oxford University Press, New York

Hariprasanna K, Rakshit S (2016) Economic importance of sorghum. In: Rakshit S, Wang Y-H (eds) The sorghum genome. Springer International, Cham, pp 1–25

Harlan JR (2011) Origins of African plant domestication. Walter de Gruyter, Berlin

Harlan JR, De Wet JMJ (1972) A simplified classification of cultivated sorghum 1. Crop Sci 12:172–176. https://doi.org/10.2135/cropsci1972.0011183X001200020005x

Haussmann B, Hess D, Seetharama N et al (2002) Construction of a combined sorghum linkage map from two recombinant inbred populations using AFLP, SSR, RFLP, and RAPD markers, and comparison with other sorghum maps. Theor Appl Genet 105:629–637

Juerg B, Thompson W, Rooney W, Bean B (2009) Management of biomass and sweet sorghum in the southwest US. Abstract. In: International annual meetings. ASA-CSSA-SSSA. 1–5 Nov, Pittsburgh, PA

Katsar CS, Paterson AH, Teetes GL, Peterson GC (2002) Molecular analysis of sorghum resistance to the greenbug (Homoptera: Aphididae). J Econ Entomol 95:448–457

Kavuluko J, Kibe M, Sugut I et al (2021) GWAS provides biological insights into mechanisms of the parasitic plant (Striga) resistance in sorghum. BMC Plant Biol 21:392. https://doi.org/10.1186/s12870-021-03155-7

Kumari A, Goyal M, Kumar R, Sohu RS (2021) Morphophysiological and biochemical attributes influence intra-genotypic preference of shoot fly [Atherigona soccata (Rondani)] among sorghum genotypes. Protoplasma 258:87–102

Little CR, Bandara AY, Todd TC, Perumal R (2018) Sorghum diseases and their management in cultivation: stalk, root and other diseases. In: Achieving sustainable cultivation of sorghum, vol 1. Burleigh Dodds Science Publishing, Cambridge, pp 447–484

Mace ES, Xia L, Jordan DR et al (2008) DArT markers: diversity analyses and mapping in *Sorghum bicolor*. BMC Genomics 9:26. https://doi.org/10.1186/1471-2164-9-26

Mengistu G, Shimelis H, Laing M, Lule D (2019) Assessment of farmers' perceptions of production constraints, and their trait preferences of sorghum in Western Ethiopia: implications for anthracnose resistance breeding. Acta Agric Scand B Soil Plant Sci 69(3):241–249. https://doi.org/10.1080/09064710.2018.1541190

Nagaraj N, Reese JC, Tuinstra MR et al (2005) Molecular mapping of sorghum genes expressing tolerance to damage by greenbug (Homoptera: Aphididae). J Econ Entomol 98:595–602

Okosun OO, Allen KC, Glover JP, Reddy GV (2021) Biology, ecology, and management of key sorghum insect pests. J Integr Pest Manag 12:4

Pandian BA, Varanasi A, Vennapusa AR et al (2020) Characterization, genetic analyses, and identification of QTLs conferring metabolic resistance to a 4-hydroxyphenylpyruvate dioxygenase inhibitor in sorghum (*Sorghum bicolor*). Front Plant Sci 11:596581

Punnuri S, Huang Y, Steets J, Wu Y (2013) Developing new markers and QTL mapping for greenbug resistance in sorghum [*Sorghum bicolor* (L.) Moench]. Euphytica 191:191–203

Rao SP, Rao SS, Seetharama N et al (2009) Sweet sorghum for biofuel and strategies for its improvement. International Crops Research Institute for the Semi-Arid Tropics, Hyderabad

Riyazaddin M, Kavi Kishor PB, Ashok Kumar A et al (2015) Mechanisms and diversity of resistance to sorghum shoot fly, a therigona soccata. Plant Breed 134:423–436. https://doi.org/10.1111/pbr.12276

Satish K, Srinivas G, Madhusudhana R et al (2009) Identification of quantitative trait loci for resistance to shoot fly in sorghum [*Sorghum bicolor* (L.) Moench]. Theor Appl Genet 119:1425–1439

Satish K, Madhusudhana R, Padmaja PG et al (2012) Development, genetic mapping of candidate gene-based markers and their significant association with the shoot fly resistance quantitative trait loci in sorghum [*Sorghum bicolor* (L.) Moench]. Mol Breed 30:1573–1591

Sharma HC, Nwanze KF (1997) Insect pests of sorghum: biology, extent of losses, and economic thresholds. In: Plant resistance to insects in sorghum, pp 9–23

Sharma HC, Taneja SL, Rao NK, Rao KP (2003) Evaluation of sorghum germplasm for resistance to insect pests. International Crops Research Institute for the Semi-Arid Tropics, Hyderabad

Tao YZ, Jordan DR, Henzell RG, McIntyre CL (1998) Construction of a genetic map in a sorghum recombinant inbred line using probes from different sources and its comparison with other sorghum maps. Aust J Agr Res 49:729–736

Tao YZ, Hardy A, Drenth J et al (2003) Identifications of two different mechanisms for sorghum midge resistance through QTL mapping. Theor Appl Genet 107:116–122

Traore M, Lompo F, Thio B et al (2012) Influence of crop rotation, fertilization and tillage on populations of plant parasitic nematodes of sorghum (*Sorghum bicolor* (L.) Moench). Biotechnol Agron Soc Environ 16:59–66

Varshney RK, Tuberosa R (eds) (2007) Genomics-assisted crop improvement. Springer, Dordrecht

Vashisth S, Jagdish J, Sharma SP, Sharma HC (2022) Biochemical mechanisms of induced resistance to Chilo partellus in sorghum. Int J Pest Manag. https://doi.org/10.1080/09670874.2022.203686

Wu Y, Huang Y (2008) Molecular mapping of QTLs for resistance to the greenbug *Schizaphis graminum* (Rondani) in *Sorghum bicolor* (Moench). Theor Appl Genet 117:117–124

Wu YQ, Huang Y, Tauer CG, Porter DR (2006) Genetic diversity of sorghum accessions resistant to greenbugs as assessed with AFLP markers. Genome 49:143–149. https://doi.org/10.1139/g05-095

Wu X, Zhao R, Bean SR et al (2007) Factors impacting ethanol production from grain sorghum in the dry-grind process. Cereal Chem 84:130–136. https://doi.org/10.1094/CCHEM-84-2-0130

Mitigating Abiotic Stress Through the Application of Genomic and Breeding Strategies in Sorghum

13

Lorenzo Stagnati, Alessandra Lanubile, Matteo Busconi, and Adriano Marocco

Abstract

Resistance to environmental stress, as well as enhanced yield performance and stability, and grain and stem qualities, has been improved in sorghum to be used as food, feed, and biomass for energy production. Transcriptome research on sorghum response to drought stress, temperature extremes, salinity, and heavy metals revealed the enrichment of core genes involved in basic and critical functions (such as reproductive development, leaf and seed development, cell differentiation, and chloroplast organization), as well as in dispensable genes involved in adaptive biological processes (such as secondary metabolic processes, RNA processing, and amino acid transport). The powerful approaches to uncover the complex traits of abiotic stress responses using genome-wide single nucleotide polymorphism (SNP) markers are genome-wide association study (GWAS), quantitative trait loci (QTLs) mapping, and genomic selection (GS). In addition, the comparative genomics analysis is adopted to explore the genomic variation between sweet and grain sorghums. The genome-editing technologies and the mutant libraries provide efficient resources to identify mutations allowing the rapid introduction into elite germplasm. The development of next generation and third-generation sequencing technologies as well as high-throughput phenomics will allow more effective exploitation of large-scale breeding populations.

L. Stagnati · A. Lanubile · M. Busconi · A. Marocco (✉)
Department of Sustainable Crop Production, Università Cattolica del Sacro Cuore, Piacenza, Italy
e-mail: adriano.marocco@unicatt.it

E. Habyarimana et al. (eds.), *Omics and Biotechnological Approaches for Product Profile-Driven Sorghum Improvement*,
https://doi.org/10.1007/978-981-97-4347-6_13

Keywords

Drought stress · High temperature stress · Cold stress · Salinity stress · Sorghum tolerance · Genomic technologies · Breeding approaches

13.1 Introduction

The species *Sorghum bicolor* (L.) Moench is classified in the family *Poaceae*, subfamily *Panicoideae*, tribe *Andropogoneae*, subtribe *Sorghinae.* The species are annually cultivated, wild, and weedy sorghums in addition to two rhizomatous taxa: *S. halepense* (L.) Pers. and *S. propinquum* (Kunth) Hitch. *S. bicolor* is subsequently divided in three subspecies: *S. bicolor* subsp. *bicolor*, *S. bicolor* subsp. *drummondii*, and *S. bicolor* subsp. *verticilliflorum* (Wiersema and Dahlberg 2007). Cultivated sorghums belongs to *S. bicolor* subsp. *bicolor* (Harlan and de Wet 1972), while House (1985) further identified five different basic races: bicolor, caudatum, durra, guinea, and kafir.

All sorghums (*Sorghum bicolor* subsp. *bicolor*) are diploids and have $2n = 20$ chromosomes. The genus *Sorghum* comprises a high genetic diversity as seen for variation of color and shape (Uptmoor et al. 2003; Assar et al. 2005), therefore offering the potential for crop improvement and increased productivity. Commercial cultivars of *S. bicolor* (L.) Moench are classified into the following agronomic variants: (a) grain sorghum cultivars with high grain yield cultivated as food and feed crop; (b) fiber sorghum, cultivars with a high content of fiber and potentially used as a fiber or energy crop; (c) broom sorghum, cultivars characterized by an inflorescence with long, elastic branches and used to produce brooms; (d) sweet sorghum, cultivars potentially used as energy and/or food crop because of sugary and juicy stalk; and (e) biomass sorghum, cultivars with high ligno-cellulosic biomass yield, generally used as energy crop.

Sorghum is a C4 plant with many appealing features: (a) it adapts well to different environmental conditions; (b) it has a short life cycle that can be completed in few months; (c) it has a high efficiency in the use of water and nutrients; (d) the cultivation and harvest can be highly mechanized; and (e) sweet sorghum cultivars can produce and accumulate high levels of sugar in the stalk and starch in seed's endosperm, while the bagasse obtained from sorghum has an interesting biological value when used as forage as well as good combustion when used for cogeneration.

Breeding recently aims at improving sorghum to serve as food, feed, or fuel (Vermerris et al. 2007), according to the local necessities of the countries. To this aim, characters as yield performance and stability, resistance to pests and pathogens, and grain and stem qualities, among others must be improved.

Sorghum survives with less than 300 mm of water supply during a cropping season of 100 days, but it responds favorably to irrigation or additional rainfall. To achieve good yields of 50–100 t/ha of total above ground biomass (fresh weight) sorghum, "the camel" among crops, requires between 550 and 800 mm of water. Even though sorghum is a dry-land crop, an early-season drought can block growth before panicle initiation while the plant remains vegetative, rebooting leaf production and flowering when more favorable conditions occur again, while a mid-season drought stops leaf development. In drought conditions, soluble sugars are responsible of maintaining a reasonable turgor level; it is reported that the level of reducing sugars and proteins can indicate drought tolerance capacity in different sorghum genotypes (Erdei et al. 2009).

Sun radiation and water are efficiently transformed by sorghum (radiation use efficiency, RUE = 3.7 g/m^2 of dry matter per MJ/m^2 of absorbed photosynthetically active radiation) into dry matter (water use efficiency, WUE = 193 mm of water/kg of dry matter produced or 5.2 g of dry matter/kg of water consumed) (Khawaja 2014). When water is scarce, RUE may be significantly lower. However, RUE and water consumption correlate and, interestingly, stressed plants use available water more efficiently than unstressed plants (Dercas and Liakatas 2007). Consequently, sorghum represents an alternative crop system under reduced irrigation conditions (Vasilakoglou et al. 2011). Because of its tropical origin, sorghum is sensitive to cold and breeding for chilling tolerance would help expand sorghum cultivation into colder regions.

Even though sorghum is a fundamental crop in many areas of the world, breeding and genomic studies have lagged behind other cereal crops. Scientific research, boosted by the recent interest for the crop, has advanced and resulted in the sorghum genome sequencing more than 10 years ago (Paterson et al. 2009), the development of several Sorghum Association Panels (SAPs) (Casa et al. 2008; Upadhyaya et al. 2009; Morris et al. 2013), the establishment of mutant libraries, the completion of sorghum pangenome, the availability of a gene expression atlas, and the development of an online database (Xin et al. 2008; Shakoor et al. 2014; Makita et al. 2015; Jiao et al. 2016; Addo-Quaye et al. 2018; Tao et al. 2021), thus introducing sorghum to a new era. The vast knowledge accumulated during the years can make sorghum a critical crop for addressing global food and energy security in a changing global climate.

13.2 Genetic Basis of Traits for Adaptation to Climate Change and Abiotic Stresses

During cultivation, sorghum can be affected by different abiotic stresses as temperature extremes, drought, salinity, and presence of heavy metals in the soil. With the aim of understanding the development of sorghum and its response to the environment, transcriptomic researches have made available spatial-temporal datasets under different stresses and many gene expression profiles have been published (Chopra et al. 2015; Gelli et al. 2017; Marla et al. 2017). In the case of drought tolerance, dhurrin, heat shock proteins (HSPs), and antifreeze proteins have been found as relevant (Hayes et al. 2015; Spindel et al. 2018), while carotenoids, phytohormones, thioredoxin, components of photosystem I, and antioxidants play important roles in chilling tolerance (Ortiz et al. 2017). The colocalization of chilling tolerance loci with grain tannin content and dwarfing genes suggest that focusing on non-tannin and dwarfing alleles in breeding plans for short-cycle grain sorghum resulted in unintentional selection for chilling sensitivity (Marla et al. 2019). Sorghum growth and photosynthesis are impacted by heat stress where transport proteins, transcription factors (TFs), and HSPs are key players of sorghum response (Chen et al. 2017; Chopra et al. 2017).

13.2.1 Genes Involved in Drought Stress Tolerance

Drought stress can cause severe yield losses in sorghum and mechanisms of plant response have been investigated for both pre-flowering (before anthesis) and the post-flowering (during grain filling). Pre-flowering drought stress have an impact on leaf growth causing different symptoms (leaf rolling, leaf erectness, reduced height) and on flower development resulting in delayed flowering, panicle blasting, saddle effect as well as a reduction of panicle size (Tuinstra et al. 1996); while post-flowering drought stress directly impact on grain yield (Rakshit et al. 2016).

The stay-green trait is associated to post-flowering drought tolerance because stay-green genotypes are able to retain chlorophyll thus being able to continue the photosynthetic process for longer time compared to "senescent" genotypes (Thomas and Howarth 2000). In response to drought, a high number of differentially expressed genes have been highlighted by comparative transcriptome studies carried out with sorghum genotypes with contrasting WUE. Transcriptomic analysis of sorghum seedlings subjected to drought stress identified the early and late responsive genes for drought tolerance, as well as genes expressed only in drought-tolerant genotypes (Pasini et al. 2014; Fracasso et al. 2017; Abdel-Ghany et al. 2020). Mild and severe drought stress resulted in the upregulation of two ethylene-responsive transcription factor (ERFs) while re-watering resulted in the down-regulation of ERF (Zhang et al. 2019). In other studies, mild drought resulted in strong upregulation of genes encoding for the mitochondrial transcription termination factor (mTERF) family, anion-transporting ATPase family proteins, and LEA (late embryogenesis abundant) proteins, hydroxyproline-rich glycoprotein family proteins; while under severe drought, proteins with putative homology to abscisic acid-insensitive 2 (ABI2) and mannosyl-transferase were among the most abundant (Devnarain et al. 2019). Moreover, many sets of differentially expressed gene products have been found as enhanced by drought stress, such as LEA proteins, delta 1-pyrroline-5-carboxylate synthase (P5CS2), high-affinity K^+ transporter 1 (HKT1), proteins associated with response to abscisic acid (ABA), dehydration-responsive element-binding transcription factor (DREB1A), RING finger 1 (SDIR1), CBL-interacting serine/threonine-protein kinase 1 (CIPK1), and trehalose-6-phosphate synthase (TPS) (Johnson et al. 2014, 2015). In a stay-green line, *P5CS2* showed an increased expression that was associated with higher levels of proline. This gene has been localized in the *Stg1* QTL region (Johnson et al. 2015), which was previously identified as stay-green QTL (Subudhi et al. 2000; Xu et al. 2000). Additionally, drought stress resulted in the upregulation of important TFs families including heat stress transcription factor (HSF), ERFs, Petunia NAM, Arabidopsis ATAF1/2 and CUC2 (NAC), WRKY transcription factor (WRKY), and homeodomain leucine zipper transcription factor (HD-ZIP) (Varoquaux et al. 2019). Genes encoding for HSPs, LEAs, chaperones, aquaporins, and expansins might play important roles in sorghum drought tolerance (Varoquaux et al. 2019). WRKY transcription factor gene was highly over-expressed in sorghum under severe stress (Devnarain et al. 2016); in roots, WRKYs were found to be largely inhibited under pre-flowering stress but not under post-flowering drought stress (Varoquaux et al. 2019). Genes encoding DREBs, AP2/ERF (APETALA2-Ethylene Responsive Factors), and MYB TFs are among the early response genes in sorghum when plants are subjected to PEG stress (Sharoni et al. 2011; Xu et al. 2011; Abdel-Ghany et al. 2020; Mathur et al. 2020). In a stay-green sorghum line investigated by Johnson et al. (2015), the DREB1A transcription factor gene was expressed at higher levels; it is interesting to note that the expression of sorghum DREB2 in rice improved both tolerance and yield under drought stress (Bihani et al. 2011).

The above-mentioned genes can represent relevant targets in breeding plans aimed at improving drought tolerance in sorghum and other cereals because of their general upregulation during drought stress. Further integrations with the third-generation and single-cell RNA sequencing will significantly enhance the accuracy and specificity of transcriptome sequencing.

13.2.2 Genes Associated with High Temperature Stress Tolerance

Global warming and extreme weather conditions are challenging worldwide agriculture and there is an urgent necessity to increase high temperature (HT) tolerance in crops due to the rising temperature (Kotak et al. 2007; Djanaguiraman et al. 2014). Similarly to other cereals, the reproductive stage is more sensitive to HT stress than the vegetative period, resulting in severe yield losses. In sorghum, the periods between 10 days before anthesis and 5 days after anthesis were most sensitive to HT which causes the maximum decreases in floret fertility and, as consequence, a reduction in seed set (Prasad et al. 2021). In sorghum, HT tolerance includes the accumulation of HSPs (Gurley 2000), biological membrane stability under HT through adjustments of membrane lipid composition, and changes in fatty acid unsaturation levels (Sung et al. 2003), and production of antioxidants (Prasad et al. 2021; Xin et al. 2021). The roles under HT stress of a basic helix-loop-helix (bHLH) transcription factor and phytochrome interacting factor 4 (PIF4) were found in sorghum (Proveniers and van Zanten 2013). The *ARF* (auxin response factor) genes were both constitutively upregulated after HT stress and induced during HT stress (Chen et al. 2019a).

13.2.3 Genes Associated with Both Drought and High Temperature Stress Tolerance

In real-field conditions, crops are often subjected to a combination of HT and drought stresses. Under such stress condition, sorghum, which is among the most

HT tolerant crops, is able to manifest little damages (Chen et al. 2017). For this reason sorghum is a powerful source to seek HT tolerant genes that can be isolated and subsequently used to increase HT tolerance in crops that can be affected by HT injury. The screening of a sorghum mutant library to describe HT injury has described several phenotypes including plant death, leaf rolling, leaf bleach/blotch, leaf firing, panicle blasting, reduction in pollen production as well as a reduction in seed set and size (Xin et al. 2021).

Plants that experience drought and HT conditions try to survive activating various signaling pathways that induce the expression of several trans-acting factors which trigger the expression of drought and HT-responsive genes. Mitogen-activated protein kinase (MAPK/MPKs), Ca-dependent protein kinases (CDPKs), sugars (as signaling molecules), nitric oxide (NO), and phytohormone are the main signaling molecules that modulate the expression of many genes implicated in stress-response mechanism (Ahmad et al. 2012). The gene drought-hypersensitive mutant1 (DSM1) is induced by drought to confer drought tolerance through the MAPK pathway. Activated stress-responsive genes detoxify the reactive oxygen species (ROS), reactivate the structure of proteins and the activity of enzymes, and maintain the cellular homeostasis, which ultimately leads to tolerance against abiotic stresses (Ciarmiello et al. 2011). HT and drought stress are able to induce HSFs that have a fundamental role in HT stress response regulating the expression of HSPs. For example, high-level expression of HSF1 was observed during HT stress, while HSF5, HSF6, HSF10, HSF13, HSF19, HSF23, and HSF25 were upregulated during drought (Nagaraju et al. 2015; Sudhakar et al. 2016).

Additionally, the heterotrimeric G-protein complexes of sorghum, such as the Rho-type GTPase-activating protein 1 (RGA1) includes cis-regulatory elements such as ABA, ARE, GT-1 boxes, LTR, and MeJAE, which are thought to have a role in abiotic stress signaling. Moreover, it is reported that drought situations result in the upregulation for RGA1 transcription, while the same is downregulated under HT stress. These observations remark the crucial role of G-protein complexes in drought and HT stress tolerance in sorghum (Prasad et al. 2021).

13.2.4 Identification of Genes for Improving Cold Stress Tolerance

Cold temperature, which generally occurs at the beginning of the growing season, can severely affect sorghum during early stages of plant development with negative consequences on emergence, plantlets vigor, and general metabolism (Natoli et al. 2002).

Tissue-specific expression studies of sodium proton antiporter genes (*SbNHX* and *SbNHE*) showed that they were highly expressed under cold stress in roots, embryos, stems, and leaves (Hima Kumari et al. 2018). TFs belonging to the nuclear factor Y (NFY) family are implicated in several developmental processes including abiotic stress tolerance (Siriwardana et al. 2016). In the stem, genes encoding for *SbNFY-A*, *SbNFY-B*, and *SbNFY-C* subunits were expressed under cold stress

(Maheshwari et al. 2019), thus helping the elucidation of the evolution of diverse NFY members and their roles during cold stress responses (Maharajan et al. 2021).

13.2.5 Identification of Genes for Salinity, Heavy Metal, and Alkali Stress Tolerance

Among the 1125 million hectares of worldwide agricultural land affected by salinity, 76 million hectares are a consequence of human-induced salinization and sodification (Hossain 2019): In such unhappy global scenario it mandatory to breed salt-tolerant varieties in sorghum as in other crops.

The Nigerian landrace Mota Maradi is tolerant to salinity, while the reference sorghum line BTx623 is susceptible. By a comparative analysis of the 2 strains grown in soil treated with 100 mM NaCl at 10 days after germination, significant differences on physiological performance, particularly on growth and biomass gain, photosynthetic rate, and the accumulation of Na^+, K^+, proline, and sucrose were shown. Differentially expressed genes highlighted by transcriptome profiling of the leaves, leaf sheaths, stems, and roots included ion transporters such as HKT, NHX, AKT (Arabidopsis K^+ Transporter), HAK5 (High Affinity K^+ Uptake Transporter 5), and KUP3 (K^+ Uptake Transporter 3), and were likely responsible for the differences in Na^+ and K^+ accumulation. Meanwhile, differentially expressed genes involved in photosynthesis, cellular growth, signaling, and ROS scavenging were also highlighted between Mota Maradi and BTx623 (Ukwatta et al. 2021). The HKT family gene of sorghum (*SbHKT1;4*) was highly induced in both shoot and root tissues of a salt-tolerant genotype under high salt stress (200 mM NaCl). Functional characterization of SbHKT1;4 in mutants of *Saccharomyces cerevisiae* proved that SbHKT1;4 mediates Na^+ and K^+ uptake under high salt stress by maintaining optimal Na^+ and K^+ balance and enhancing plant growth in the salt-tolerant genotype under salt stress (Wang et al. 2014a). Nutrient transporters are relevant in abiotic stress tolerance in sorghum; among the sulfate transporter (*SULTR*) family genes, *SbSULTR1;2*, *SbSULTR1;3*, *SbSULTR3;3*, and *3;5* were highly induced in the root of sorghum after 24 h under 200 μM Cd stress (Akbudak et al. 2018a). *SbSULTR2;2*, *SbSULTR3;2*, *SbSULTR3;3*, and *SbSULTR3;5* were highly expressed in the leaf of sorghum after 6 h under 200 μM Cd.

When plants experience salt stress or other abiotic stresses, there is the accumulation of the amino acid proline which functions as an osmolyte. One of the key enzymes in proline biosynthesis is P5CS. Su et al. (2011) reported two P5CS genes (*SbP5CS1* and *SbP5CS2*) on chromosomes 3 and 9, respectively; these genes were found to be expressed in sorghum root and leaf tissues at 4 and 8 h after high salt treatment (250 mM NaCl); breeders can focus on these 2 genes in order to develop salt stress tolerant varieties. In addition, in the sorghum genome, four LEA3 genes (*SbLEA3;1*, *SbLEA3;2*, *SbLEA3;3*, and *SbLEA3;4*) were identified. The expression pattern analysis of LEA3 genes in seeds (at early and maturity stage) and leaves under high salt stress (150 mM NaCl) (Dalal et al. 2013) revealed that *SbLEA3;1* and *SbLEA3;3* were highly expressed in the leaf at the early stage of seed development and elevated significantly at maturity when plants experience salt stress. The stress-associated protein (SAP) family gene (*SbSAP14*) has been found to be highly expressed in the root of sorghum at 2 and 4 h under high salt stress (250 mM) (Wang et al. 2013): when the gene is overexpressed in transgenic rice, it enhanced seed germination and seedling survival rate under high salt stress through the induction of antioxidant genes (*ascorbate peroxidase 2* (*APX2*), *catalase B* and *C* (*CatB* and *CatC*), and *superoxide dismutase* (*SodA1*)) (Wang et al. 2013).

Abiotic and biotic stresses are able to induce TFs like the DREB proteins. Under Cd stress it was found that *SbDREB2B* was highly induced in root while *SbDREB2D* was highly expressed in both leaf and root tissues (Akbudak et al. 2018b), suggesting the relevance of this gene in response to heavy metals even though further functional studies are needed to better clarify and understand the role of *DREB2* genes. Critical components in various stress signal transduction pathways are the CIPKs. Thirty-two CIPK (*SbCIPK1* to *SbCIPK32*) genes have been discovered in sorghum genome and CIPK expression patterns have been investigated in leaves of plants subjected to alkali stress finding that six *SbCIPK*s were highly expressed (Guo et al. 2015).

13.3 Genomic and Genetic Resources

13.3.1 Sorghum Genomes and Pangenome

Fifteen years ago, the first sorghum genome sequence was concluded in an elite grain sorghum inbred line, BTx623 (Paterson et al. 2009). It has been employed as the reference for genome-wide analyses to detect genetic variations from other sorghum accessions. It includes 58.8% retrotransposons, 8.7% DNA transposons, and about 34,129 annotated genes. For comparative genomics, the sequencing by the whole-genome shotgun strategy of two sweet sorghum lines (Keller and E-tian) and one grain sorghum line (Ji2731) was carried out and compared with BTx623 (Zheng et al. 2011). These four genomes presented a large number of genetic variations, including 1,057,018 SNPs, 99,948 insertion/deletion mutations (Indels), 16,487 present/absent variations (PAVs), and 17,111 copy number variations (CNVs) (Zheng et al. 2011; Mace et al. 2013). The sweet and grain sorghums showed variations in 1442 genes. Consequently, 44 sorghum accessions, including a wild relative (*S. propinquum*) were sequenced (Mace et al. 2013) pinpointing new genetic potential among the sorghum accessions native from the African continent.

Combining methods of deep sequencing, genetic linkage analysis, and transcriptome data, a high-quality reference genome sequence (version 3) was constructed (McCormick et al. 2018). By means of new technologies, a chromosome-level de novo assembly was produced from the sorghum inbred line Tx430 (Deschamps et al. 2018) and in the sweet sorghum Rio (Cooper et al. 2019). Despite the differences in agronomic characteristics, sweet and grain sorghum genomes are highly similar in structure (Cooper et al. 2019). Sweet sorghums are enhanced with genes involved in starch and sucrose metabolism along with lignin and coumarin biosynthesis pathways.

The pangenome results from the de novo assembly of genome sequences belonging to many genetic diverse lines (Wang et al. 2021a). The sorghum pangenome derives from de novo assembly of 16 cultivated sorghum genomes and its wild relative. This pangenome sequence has 955 Mb and includes 44,079 gene families, 0.3–0.5 million insertions/deletions, and 15,293,465 SNPs have been found (Ruperao et al. 2021; Tao et al. 2021). Within this sorghum pangenome sequence, the genes are distinguished as either core genes observed in at least 15 genomes, shell genes described in 2–14 genomes, or cloud genes present in only one genome. Core genes constitute only 36% (15,867) of the total gene families, followed by shell genes (63.6%, 28,026), and cloud genes (0.4%, 186). The sorghum pangenome shows a percentage of dispensable genes (shell plus cloud) more elevated to than reported in *Oryza sativa* (54%), *Glycine max* (49%), and *Brachypodium distachyon* (45%), pointing out that this species may have a greater genetic variation (Xin et al. 2021).

RNA-seq data and functional enrichment analyses reveal that core genes are enriched with basic and critical functions (such as the developing process, RNA processing, reproductive system development, leaf and seed development, cell differentiation, and chloroplast organization), whereas dispensable genes are endorsed with adaptive biological processes (such as secondary metabolic processes, cellular metabolic processes, and amino acid transport). The pangenome sequence supplies the sorghum community with a solid genetic foundation for gaining trait biology knowledge, representing a platform for exploiting genetic diversity aimed to future sorghum advance and gene discovery. Recently, an online resource for the research community has been released reporting various genetic information and community advancement (https://www.sorghumbase.org; Gladman et al. 2022).

13.3.2 Sorghum Germplasm and Genetic Diversity

With the intensifying of international cooperation and germplasm trade, in association with the regional genomic resource employment, a higher number of accessions are being exploited in sorghum breeding and genetic research. Based on the shape of spikelets/panicles, the sorghum accessions can be classified into five basic races and ten intermediate types (Dillon et al. 2007): *Bicolor* is the most primitive of the five races, including primitive forage sorghums with sweet stems; *Caudatum* is of most recent origin and used for beer brewery; *Kafir* is insensitive to photoperiod and most male-sterile lines were originated from it; *Guinea* is the oldest group and mostly photoperiod sensitive; and *Durra* is the most relevant sorghum type in Ethiopia, also widely used in breeding programs in Asia (Cuevas et al. 2017).

Northeast Africa, a special area extending from Ethiopia to Sudan, is the region of domestication and diversification of sorghum (Wendorf et al. 1992). Hence, Ethiopia and Sudan hold the oldest wild types and richest diversities.

The International Crops Research Institute for the Semi-Arid Tropics (ICRISAT, http://www.icrisat.org) and the United States Department of Agriculture's National Plant Germplasm System (USDA-NPGS, http://www.ars-grin.gov) are the 2 most considerable sorghum germplasm banks storing more than 41,000 accessions, including historic accessions, wild relatives, landraces, and improved breeding lines. In addition, there are about 30 other institutions, such as the Australian Tropical Crops and Forages Genetic Resources Centre, which contains the largest collection of Australian wild sorghums, the National Bureau of Plant Genetic Resources in India with about 20,000 collections (www.nbpgr.ernet.in), and the Institute of Crop Germplasm Resources in China, which houses about 16,874 collections (https://www.cgris.net/cgris_english.htm).

Among the >41,000 accessions stored in USDA-NPGS, 7217 accessions originated from Ethiopia as the largest group, followed by Sudan (2552) (Dahlberg et al. 2004). Germplasm resources from these two regions are significantly different. Five ancestral populations are included in a representative Sudan core collection, whereas the Ethiopian core set includes 11 ancestral populations, and the pairwise genetic distance among the Sudan core set accessions is bigger than that in Ethiopia, suggesting that the Sudan collection shows genetically higher diversity (Cuevas et al. 2017; Cuevas and Prom 2020).

A high level of genetic variation has also been reported in local germplasm from Ethiopia (Cuevas et al. 2017; Girma et al. 2019), Sudan (Cuevas and Prom 2020), and Nigeria (Olatoye et al. 2018).

Sweet sorghum is characterized by rich stem juice and sugar and has polyphyletic origin, indeed is not phylogenetically separated from the grain sorghum (Murray et al. 2009; Wang et al. 2009; Burks et al. 2015). By a population genomic work, it was highlighted that stem juiciness trait has multiple origins and mutations in the *Dry* gene, which controls the pithy/juicy stem trait and codes for a NAC transcription factor, and represented important milestones in the origin of sweet sorghum (Zhang et al. 2018).

The major problems in germplasm collections are the lack of phenotypic and genotypic data, the accessions redundancy, and the underutilization of wild sorghum. The first problem can be solved by decreasing sequencing costs and rapid development of phenomics. The accession redundancy requires maintenance of genotypes that do not contribute to panel diversity. Instead, wild sorghum has great potential in future sorghum breeding (Galaihalage et al. 2020).

13.3.3 Diversity Panels

The World Core Collections (WCC) was assembled at ICRISAT and consists of 2247 landraces clustered based on plant height and flowering time (Grenier et al. 2001a, b). From WCC a mini-core collection of 242 accessions was gained, based on phenotypic distance, for easier use in breeding (Upadhyaya et al. 2009, 2021). The Sorghum Conversion Program (SCP) was created by the USDA to insert new genetic diversity into modern cultivars (Hao et al. 2021). BTx406, which possesses recessive alleles for photoperiod insensitivity and dwarfism at the *Ma1* and *Dw1-Dw3* genes, was employed to change tropical late-flowering sorghums to photoperiod insensitive, early maturing, and short stature phenotypes. Both new grain sorghum varieties and germplasm for genetic research were developed by SCP (Hao et al. 2021). Hundreds of accessions have been exploited for association mapping in sorghum. In this regard, SAP represents the first panel assembled, deriving from the combination of 228 SCP tropical lines and 149 breeding lines (Casa et al. 2008). Combining 3 different panels, SAP, mini-core collection, and sorghum reference set, Morris et al. (2013) constructed a panel of 971 accessions to explore sorghum diversity. A further SCP panel of 700 lines was established to investigate variation in dhurrin synthesis and catabolism genes (Hayes et al. 2015). To improve sweet sorghum breeding program, several panels consisting only on sweet or biomass accessions have been constructed, such as the sweet sorghum panel and the bioenergy association panel (Murray et al. 2009; Wang et al. 2009; Brenton et al. 2016). Many regional genomic resources have also been used, like the Nigerian Diversity Population consisting of 516 accessions (Maina et al. 2018), a panel of 421 Senegalese sorghum landraces (Faye et al. 2019), a complex set of 1425 Ethiopian landrace accessions (Girma et al. 2019; Nida et al. 2019), and the Sudan core collection of 318 accessions (Cuevas and Prom 2020).

13.3.4 NAM, MAGIC, and TILLING Populations

The nested association mapping (NAM) population was first settled in maize and achieved rapid popularity due to its incredible power for scrutinizing the genomic architecture of phenotypic variations and identifying small effect loci (Hao et al. 2021). To date, only one true NAM population has been described in sorghum (Bouchet et al. 2017). An elite line, RTx430, which has been extensively exploited for breeding programs, was used as a common parent. As alternate parents, 10 different genotypes representing global sorghum diversity were also chosen, and 2214 lines from 10 recombinant inbred line (RIL) families formed the final NAM population.

For dissecting the genomic structure and improving breeding populations, multiparent advanced generation intercross (MAGIC) populations also represent a valid option. In sorghum, using 19 diverse founder lines intercrossed randomly through a genetic male sterility system, the first MAGIC population was obtained, arising 1000 inbred lines utilized for gene mapping and breeding (Ongom and Ejeta 2018).

Targeting Induced Local Lesions In Genomes (TILLING) has been successfully applied to cereal crops (Xin et al. 2008). By mutagenizing the sorghum reference BTx623, 2 mutant populations were created with ethyl methane sulfonate (EMS) (Xin et al. 2008; Addo-Quaye et al. 2018), carrying 1600 and 4800 mutant individuals, respectively. These mutant lines represent useful tools for forward and reverse genetic research.

13.4 Strategies of Molecular Marker Assisted Selection

13.4.1 Genome-Wide Association Study (GWAS) in Sorghum

GWAS is a powerful approach using genome-wide SNP markers for the dissection of the genetic basis of agronomical traits by employing the various association mapping panels mentioned above. Morris et al. (2013) constructed a widely used panel genotyped using genotyping-by-sequencing (GBS) and resulting in 265,487 SNPs. The genetic basis of several plant traits, grain yield components, and cold and heat stress tolerance traits were dissected (Chen et al. 2017; Chopra et al. 2017; Ortiz et al. 2017).

Spindel et al. (2018) performed a high-density phenomics GWAS to uncover the genetic basis of pre-flowering and post-flowering drought tolerance in sorghum. The experiment was conducted over a growing season in California using a collection of 648 diverse sorghum lines; phenotypes were measured once per week by using drones. The advantage of aerial drones is that they can cover large areas in a reduced amount of time, thus allowing all genotypes under examination to be measured simultaneously without being impeded by the development of the plants as in the case of classical manual phenotyping. From seedling emergence until 50% average flowering, irrigations were avoided in order to impose a pre-flowering water stress. In the post-flowering drought treatments, irrigations were terminated until 50% flowering. Drone phenotyping allowed the collection of biomass, height, and leaf area data for individual field plots. Phenotyping produced around 171,000 datapoints that, coupled with 183,989 GBS markers, revealed 213 high-quality, replicated, and conserved GWAS associations. About 48 were considered exceptionally strong candidate genes which include HSPs, cytochromes, antifreeze proteins, and proteins characterized by the presence of repeated domains as leucine rich repeats, armadillo repeats, ankyrin repeats, and WD40 which are considered relevant in plant response to stresses. Among the best candidate genes highlighted by the study, many encode proteins with DNA binding domains suggesting potential transcriptional regulators or genes encoding for kinases and phosphatases with potential role in gene regulation.

In addition, genotype-by-environment (G × E) interaction in sorghum have also been studied. The knowledge at genomic level of traits for adaptation to environment would help predict the performance of sorghum in different environments. In fact, the data obtained demonstrate that most of the known genetic loci could only explain a small part of phenotypic variation. Environmental variables for drought and Al toxicity explained a substantial portion of SNP variation in a panel of 1943 georeferenced sorghum landraces (Lasky et al. 2015). A significant proportion of nucleotide variation reflecting adaptation along the precipitation gradient was found in Nigerian sorghum landraces: genes controlling variation in morphology and flowering time play essential roles in drought adaptation (Olatoye et al. 2018).

Faye et al. (2019) characterized the genomic regions shaped by climate adaption using 213,916 SNPs in 421 Senegalese sorghum landraces. Both floral regulatory pathways and stay-green contributed to climate adaption in Sahel and Soudan zones.

Recently, the GWAS for seed mass variation in a set of 1901 georeferenced sorghum landraces showed that this trait was adaptive to precipitation gradients (Wang et al. 2020a). The results suggest that knowledge at SNP-level on adaptation could be used to predict adaption traits under a specific environment.

13.4.2 Quantitative Trait Loci (QTL) Mapping

Many traits of agronomic relevance, including yield, biotic, and abiotic stress responses, have a polygenic nature and are controlled by multiple QTL. Genetic linkage maps have been constructed using amplified fragment length polymorphism (AFLPs), restriction fragment length polymorphism (RFLPs), diversity arrays technology (DArTs), randomly amplified polymorphism DNA (RAPDs), simple sequence repeats (SSRs), and SNPs (Prasad et al. 2021). To date, several QTL have been identified in sorghum that are involved in drought, HT, cold, and salinity tolerance (Table 13.1).

13.4.2.1 Identification of QTL for Drought and High Temperature Tolerance

Many important plant and root traits for drought tolerance are controlled by multiple genes located within QTL. Overall, 12 traits (WUE, transpiration ratio, transpiration, total number of green leaves, stomatal density, stay-green, rate of leaf senescence, leaf drying score, green leaf area, CO_2 assimilation ratio, chlorophyll fluorescence, and chlorophyll content) were associated with drought tolerance within 245 QTL (Prasad et al. 2021). The identification of the QTL associated with these traits is highly important for the development of drought-tolerant sorghum cultivars.

Drought-tolerant sorghum genotypes have been identified, which include 00MN7645 (Sukumaran et al. 2016), QL41 (Tao et al. 2000), B35 (Crasta et al. 1999; Subudhi et al. 2000; Xu et al. 2000), BTx642 (Harris et al. 2007), SC-56 (Kebede et al. 2001), and E-36-1 (Haussmann et al. 2002). These genotypes have been used as sources of drought tolerance genes and to develop recombinant inbred line (RIL), near-isogenic line (NIL), and mapping populations for the identification of QTL.

A small number of QTL have been identified under pre-flowering drought stress in sorghum (Prasad et al. 2021). Initial research used RILs obtained by crossing two contrasting genotypes, Tx7078 (pre-flowering tolerant, post-flowering susceptible) and B35 (pre-flowering susceptible, post-flowering tolerant). The RILs were genotyped with RFLP and RAPD markers, and six QTL specific for pre-flowering drought tolerance were identified (Tuinstra et al. 1996). The expressions of these QTL were correlated with the severity of the drought stress levels. Kebede et al. (2001) identified four QTL for pre-flowering drought tolerance with RILs obtained from the cross SC56 × Tx7000: it was possible to find that tolerant parent TX7000 contributed an allele linked to pre-flowering drought tolerance to the PfrG QTL. Such QTL contributed 15–37.7% of the phenotypic variations in 2 different

Table 13.1 Quantitative trait loci (QTLs) mapped for drought, cold, and salinity tolerance in sorghum

Type of stress	Type of population	Cross	Number of QTLs	Chromosome/ linkage group	References
Drought pre-flowering	98 RILs	Tx7078 × B35	6	D, F, M	Tuinstra et al. (1996)
	125 RILs	SC56 × Tx7000	4	A, G, J	Kebede et al. (2001)
	70 RILs	Tx7078 × B35	3	9, 10	Kapanigowda et al. (2014)
	96 landraces		3	3, 8, 9	Abou-Elwafa and Shehzad (2018)
Drought post-flowering	98 RILs	Tx7078 × B35	16	B, C, E, F, I	Tuinstra et al. (1997)
	96 RILs	Tx430 × B35	9	A, B, D, G, I, J	Crasta et al. (1999)
	98 RILs	Tx7000 × B35	4	A, D, J	Subudhi et al. (2000)
	160 RILs	QL39 × QL41	10	A, B, C, G, I	Tao et al. (2000)
	98 RILs	Tx7000 × B35	4	A, D, J	Xu et al. (2000)
	125 RILs	Tx7000 × SC56	12	A, B, C, D, E, F, G, J	Kebede et al. (2001)
	226 RILs	IS9830 × E36-1 N13 × E36-1	3	A, E, G	Haussmann et al. (2002)
	NILs	RTx7000 × BTx642	4	3, 2, 5	Harris et al. (2007), Borrell et al. (2014a, b)
	245 RILs	M35-1 × B35	7	9, 10	Reddy et al. (2014)
	436 RILs	Tx436 × 00MN7645	4	2, 3, 4, 6	Sukumaran et al. (2016)
	208 RILs	CK60 × San Chi San	9	1, 6, 7, 8, 9	Gelli et al. (2017)
	152 $F_{2:4}$ recombinants	RSG04008-6 × J2614-11	7	10L	Kiranmayee et al. (2020)
Cold	153 RILs	Shan Qui Red × SRN39	5	1, 2, 3, 4, 7	Knoll et al. (2008)
	171 RILs	RTx430 × PI610727	14	1, 2, 4, 7, 9	Burow et al. (2011)
Salinity	181 RILs	Shihong137 × L-Tian	29	1, 2, 3, 4, 6, 7, 8, 9, 10	Wang et al. (2014b)

environments, suggesting strong genotype × environment interaction at this locus. Additionally, the QTL co-localized with other quantitative loci associated to agronomic traits such as flowering time, stay-green, plant height, and lodging resistance (Sanchez et al. 2002). A different study, conducted on 70 RILs, was able to highlight 3 QTL in chromosomes 9 and 10, which account for 17–21% of variance for the transpiration ratio associated with pre-flowering drought tolerance (Kapanigowda et al. 2014). The QTL influencing the transpiration ratio co-segregated with agricultural traits, which indicates that the introgression of these QTL through marker-assisted selection (MAS) could confer tolerance to pre-flowering drought.

Abou-Elwafa and Shehzad (2018) evaluated 96 sorghum accessions under optimum and drought stressed conditions. The drought stress began 7th week after sowing and was maintained until the 16th week. They identified 3 drought-responsive QTL (Xtxp69 on chromosome 3, SbAGA01 on chromosome 8, and SbAGB03 on chromosome 9), which are associated with the drought-resistance linked phenotypic traits and 15 drought-responsive genes.

The post-flowering drought and HT response in sorghum is linked to the stay-green trait which positively affects yield improvement. The main studies on QTL mapping used the tolerant genotype B35 and the results obtained are reported below:

(a) A set of 98 RILs was obtained from crossing TX7078 × B35, genotyped using 170 RAPD and RFLP markers to identify 3 QTL for seed weight stability, 5 for yield stability, 6 for stay-green, and 2 for grain yield under post-flowering stress (Tuinstra et al. 1997).
(b) Seven QTL for stay-green and 2 for seed maturity were identified in 96 RILs obtained from the cross B35 × Tx430 using 142 RFLP markers (Crasta et al. 1999). The contribution to phenotypic variation of 7 stay-green QTL ranged from 11.6% to 28.6%, while the seed maturity QTL (TXS1299) contributed to 47.5%, and QTL (UMC27) contributed to 20.3% of the phenotypic variability (Crasta et al. 1999).
(c) Four and 3 QTL for stay-green and chlorophyll content, respectively, were identified in 98 RILs of sorghum derived from B35 × Tx7000, using 91 RFLP, SSR, and RAPD markers (Subudhi et al. 2000). Among 4 stay-green QTL, 1 QTL (Stg2) has been identified as a key contributor explaining the maximum amount of phenotypic variation (49.8%).
(d) Xu et al. (2000) reported a multiple environment testing under drought stress for identifying QTL associated with stay-green: 15 and 9 QTL were identified for stay-green and chlorophyll content, respectively, in 98 RILs of sorghum derived from B35 × Tx7000, using 162 RFLP markers.

Additional QTL for stay-green have been identified in other sorghum genotypes:

(a) The 125 RILs derived from SC56 × Tx7000 and genotyped with 170 RFLP allowed the identification of QTL controlling stay-green (14 QTL), pre-flowering drought tolerance (7 QTL), flowering time (5 QTL), plant height (6 QTL), and lodging tolerance (3 QTL) (Kebede et al. 2001). The phenotypic variance explained by individual QTL ranged from 9.9% to 35.8%.
(b) Tao et al. (2000) identified 10 QTL for stay-green in 160 RILs from the cross QL39 × QL41, genotyped with 281 RFLP and 25 SSR markers at various sites with different climatic conditions.

(c) Haussmann et al. (2002) used 2 sets of RILs, developed from the crosses IS9830 × E36-1 and N13 × E36-1, to detect 10 QTL for stay-green under drought stress.
(d) Borrell et al. (2014a) reported the enhanced effect of stay-green QTL on grain yield in 160 RILs from the cross BQL39 × BQL41 under post-anthesis drought stress. Mapping of the stay-green QTL on chromosome 10 from an introgression line cross, RSG04008-6 (stay-green) × J2614-11 (moderately senescent), revealed the presence of several genes associated with delayed leaf senescence (Kiranmayee et al. 2020).

Several QTL linked to chlorophyll SPAD were identified including SPAD readings at booting (SPADB) and maturity (SPADM), and they were associated with stay-green ratings (Xu et al. 2000). For instance, Reddy et al. (2014) identified 7 QTL in 245 RILs: major QTL on linkage group 9 and 10 accounted for 15% of the SPADB variation and 14.1% of the SPADM variation, respectively. QTL for chlorophyll content measured at different stages of plant growth were also found (Sukumaran et al. 2016; Gelli et al. 2017). In another study, a QTL for chlorophyll concentration in the flag leaf, which explained 13% of phenotypic variance, was detected on chromosome 4 (Kapanigowda et al. 2014). QTL for chlorophyll content SPAD readings overlapped with QTL for stay-green under drought stress conditions suggesting the possibility of developing drought-tolerant sorghum cultivars combining these important traits (Xu et al. 2000; Harris et al. 2007; Borrell et al. 2014b).

Few sorghum lines (RTx430, BTx3197, RTx7000, and B35) have been studied for HT stress. The cultivars possessing stay-green traits appeared to be HT tolerant, suggesting a possible relationship between drought and HT responses (Prasad et al. 2021). Twenty-six QTL have been found to be associated with five heat tolerance traits (shoot length, shoot fresh weight, leaf firing, leaf blotching, and chlorophyll content). The main QTL identified for leaf firing needs to be validated in different genetic backgrounds to assess the co-segregation pattern.

13.4.2.2 Identification of QTL for Improving Cold Stress Tolerance

Emergence, seedling vigor, and general metabolism are affected by cold stress. Knoll et al. (2008) reported 2 QTL for seed germination in 153 RILs derived from a cross between the cold tolerant Shan Qui Red and the cold sensitive SRN39 genotypes, using 56 SSR, 67 RAPD, and 17 RFLP markers. Moreover, a total of 14 QTL distributed over 5 chromosomes (1, 2, 4, 7, and 9) for percentage of seed germination (5 QTL), field emergence (8 QTL), and seedling vigor (1 QTL) were identified in 171 RILs derived from cold sensitive RTx430 and cold tolerant PI610727 genotypes, using 141 SSR markers (Burow et al. 2011). The QTL Germ 12–2.1 for seed germination showed the highest LOD score, and it is the most promising QTL for improving seed germination and field emergence under cold stress.

13.4.2.3 Identification of QTL for Salinity Stress Tolerance

Few studies focused on QTL mapping of traits for salinity tolerance. Wang et al. (2014b) discovered 5 QTL at germination stage (3 for germination percentage and

2 for germination vigor) and 19 QTL at seedling stage (4 for shoot height, 3 each for root length and shoot dry weight, 2 each for shoot fresh weight, root fresh weight, total fresh weight and total dry weight, and 1 for root dry weight) in 181 RILs of sorghum derived from Shihong137 × L-Tian, using 616 SSR markers. The LOD score and phenotypic variation explained for each QTL ranged from 2.5% to 5.86% and 5.3% to 21.9%, respectively. Recently, the same authors found 53 QTL for plant height, stem diameter, total biomass, stem fresh weight, juice weight, and Brix under salt stress using the same RILs of sorghum and SSR markers (Wang et al. 2020b). Phenotypic variation explained for each QTL ranged from 4% to 20%. They reported that six QTL (qTB6, qSFW9, qJW9, qBrix2, qBrix10, and qSTI-Brix9) might be the main effect QTL for salt stress tolerance of sorghum (Wang et al. 2020b). These 6 QTL mapped in a cluster on chromosome 7 and they could be further investigated both for improving salt stress tolerance by MAS, and for isolation of genes for salt tolerance at seedling stage of sorghum.

As salinity tolerance is closely related to production, it is important to map QTL in all growth stages (from seedling to yield) (Huang 2018). Although such QTL analysis has not yet been reported in sorghum, some studies have been conducted in barley (Sbei et al. 2014), rice (Puram et al. 2018), and maize (Luo et al. 2017), and they provide methods for evaluating salt tolerance at vegetative and reproductive stages of the sorghum crop.

13.4.3 Genomic Selection (GS)

The newly developed GS technology utilizes information from whole genome-wide marker data regardless of whether they are associated with target traits or not. Thus, for GS there is no need to find out the QTL related to target traits and to collect phenotypic data, thereby reducing costs and time (Xu et al. 2020). In addition, by exploiting the genotype-to-phenotype relationship at the whole-genome level, major and minor gene effects can be captured (Guo et al. 2019)

In sorghum, GS has mainly focused on a variety of training populations, such as natural population, testcross hybrids, and mixed populations. Yu et al. (2016) chose a set of 962 biomass sorghum accessions that was characterized with 340,496 GBS SNPs. The overall diversity was represented by a core set of 299 accessions subsequently phenotyped and characterized for biomass yield and other related traits. The data originated by the 299 sorghums of the training set were used for cross-validation with multiple analytical methods indicating high prediction accuracy for biomass yield. The previous findings were subsequently confirmed with a 200-accession validation set chosen from the reference set. Moreover, an independent population was used to investigate the potential to apply the prediction model to broader genetic contexts. By this project, it is demonstrated that it is possible to develop a cost-effective strategy to assess the valuable germplasm maintained in gene banks.

Dos Santos et al. (2020) evaluated five genomic prediction models: Bayesian network (BN), Pleiotropic Bayesian network (PBN), Dynamic Bayesian network (DBN), multi-trait GBLUP (MTr-GBLUP), and multi-time GBLUP (MTi-GBLUP)

Models, in a diversity panel constituted by 869 biomass sorghum lines. The panel was genotyped with 100,435 SNP markers, and phenotyped for plant height (PH) with 2 measurements per week from 30 to 120 days after planting (DAP) and for end-of-season dry biomass yield (DBY). The results suggest that 45 DAP is an optimal developmental stage to set a two-level indirect selection framework for biomass sorghum allowing the indirect selection for PH and DBY at the end of the season on the basis of lines ranking at 45 DAP.

A panel of 380 biomass sorghum genotypes composed of *S. bicolor* landraces and lines and *S. bicolor* × *S. halepense* advanced inbred lines was used to demonstrate the feasibility of genomic selection. The selection index, including plant height, aboveground dry biomass yield, and dry mass fraction of fresh mass material, was the most promising. Moreover, the work analyzed the possibility to use the information from the sorghum crops grown from seed to predict the performance of the populations regrown from the belowground rhizomes (Habyarimana et al. 2020).

Over the past two decades, several different statistical models and machine learning methods have been proposed for GS, and RR-BLUP (Ride-Regression Best Linear Unbiased Prediction) models seem to be the more popular. Simulated traits from diversity panels in maize and sorghum were evaluated using RR-BLUP models that include fixed-effect covariates tagging peak GWAS signals (Rice and Lipka 2019). It was possible to quantify the ability of such covariates to increase GS prediction accuracy in the RR-BLUP model under a wide variety of genetic architectures and genomic backgrounds. The authors recommend that before the implementation of the genome selection model within a sorghum breeding program it should be better to explore the performance of the model on a trait-by-trait basis.

Genomic prediction accuracy is affected by marker density: depending on the trait, the increase in accuracy reaches a plateau as the population size and marker density increase (Hao et al. 2021).

Genomic selection has been considered as a promising tool to address many issues through its adoption in sorghum breeding. Fernandes et al. (2018) compared genomic selection methods that used correlated traits to help predict biomass yield in sorghum and founding trait-assisted genomic selection as the best approach. Correlated traits included moisture, plant height measured at monthly intervals between planting and harvest, and the area under the growth progress curve. A new strategy was proposed called trait-assisted GS, in which correlated traits are used along with marker data in the validation population to predict a focal trait. Trait-assisted GS increased prediction accuracy of yields prediction in the validation population by up to 50% because of the use of plant height in both the training and validation populations. Therefore, trait-assisted GS is an ideal approach when the focal trait is correlated with other traits that can be measured earlier or in a cost-saving manner.

From the analysis of sorghum reference genomes and a collection of 354 diverse sorghum accessions it was possible to identify more than 2 million SNPs, and a subsequent association analysis for relevant agronomic traits was able to identify 398 associated SNPs (Ruperao et al. 2021). Under drought conditions, a genome expression analysis was conducted resulting in the identification of 1788 genes that were functionally linked to the cell membrane, catalytic activity, stimulus response, various metabolic processes, and cellular and biological regulation. Among the genes identified from the drought-condition expression analysis, it was possible to observe that 79 were absent from the reference genome assembly. More such research analyses are required to strengthen the sorghum pangenome assembly for increased trait association and its use in breeding program.

13.4.4 Comparative Genomics in Sorghum

Genomic variation between sweet and grain sorghums has been analyzed by applying comparative genomics. The genomes of two sweet sorghums, Keller and E-Tian and of a grain sorghum, Ji2731, were recently resequenced identifying a huge amount of SNPs, INDELS, PAVs, and CNVs. Furthermore, nearly 1500 genes differentiating sweet and grain sorghums were identified, some of which involved in lignin and coumarin synthesis and starch and sucrose metabolism as outlined in Sect. 13.3.1 (Zheng et al. 2011).

By comparing two different genomes, Rio (sweet sorghum) and BTx623 (grain sorghum), Cooper et al. (2019) identified a high presence of loss of function and nonsynonymous mutations in genes involved in sugar metabolism. These results represent an important starting point to understand the molecular basis of sweet sorghum-associated characters. No comparative genomics studies on forage and broom sorghum types have been reported up to now. The understanding of the genetics underlying phenotypic divergence among sorghum types is expected to provide useful information for sorghum breeding for different end uses.

Genetic gains in this crop can benefit from wild relatives such as *S. halepense*. Habyarimana et al. (2022) reported that a panel of 172 populations of *S. bicolor* and *S. bicolor* × *S. halepense* (SbxSh) advanced lines generated a total of 567,046,841 SNPs; 91,825,474 INDELS; 1,532,171 SVs; and 4,973,961 CNVs. The SbxSh materials accumulated more variants and mutations with powerful effects on genetic differentiation. Out of a total of 5548 genes specific to SbxSh and mapped to the biological process, 34 genes were root-specific while 2 genes, *Root Primordium Defective* 1 (*RPD*1) and *Retarded Root Growth* (*RRG*), were found to have a key role in the growth and development of the root.

In a framework of climate change resilience, agroecological practices, and a more general sustainable agriculture, the concept of developing perennial crops is of particular relevance. In order to dissect the architecture of perennialism and other sustainability related traits the use of genomic information from wild relatives of cultivated species can boost the study of many relevant adaptation aspects.

13.4.5 Genome Editing and Mutagenesis

In the past 10 years, genome-editing technology, based on Clustered Regularly Interspaced Short Palindromic Repeats (CRISPR)/CRISPR-associated (Cas) system, has advanced rapidly (Jinek et al. 2012). Genome editing allows the simultaneous introduction of mutations of any type into genomes at pre-determined positions promising the transformation of breeding in agriculture (Chen et al. 2019b; Maharajan et al. 2022). Several studies of genome editing have been reported on sorghum. The activity of Cas9/single guide gRNA in sorghum has been demonstrated in callus obtained from immature embryos using as reporter the gene encoding for red fluorescence protein (Jiang et al. 2013). One study has reported the application of CRISPR/Cas9 that targeted the centromere-specific histone 3 (*SbCENH3*) gene, while the inheritance of the CRISPR-induced mutations has yet to be demonstrated (Che et al. 2018). The *k1C* gene family encodes for α-kafirins and a targeted mutagenesis with a single guide RNA to modify the conserved region of *k1C* genes resulted in the development of sorghums with a higher increased protein digestibility and lysine content in grain (Li et al. 2018). Char et al. (2020) developed a CRISPR/Cas9 system customized for *Agrobacterium*-mediated transfer. The efficiency of the system was proved with the targeted mutagenesis of *SbFT* and *SbGA2ox5 genes which, respectively,* underly QTLs for flowering time and plant height and since the Cas9/sgRNA was still present, additional mutations can arise in subsequent generations. The *SbFT* mutant plants exhibited significant difference in flowering time.

Although different transformation techniques have been tested to apply genome editing in sorghum, only genotypes RTx430 and P898012 have reasonable transformation efficiency (Gurel et al. 2012; Liu and Godwin 2012; Liu et al. 2019). Therefore, the development of transformation methods, genotype independent, is still the main obstacle to applying genome-editing technologies in sorghum. Although genome editing can introduce nearly any type of mutations in a precise location of a genome, it is still critical the knowledge of the target gene to be edited and, ideally, of the desired mutations in the gene in order to achieve the expected outcome quickly. Despite the recent development of improved editing technologies, the editing of multiple genes with multiple alterations in a genome in order to determine the impact of these alterations on plant health, productivity, and quality is still challenging.

On the other hand, reverse and forward genetics can be used to identify causal mutations using mutant libraries derived from induced mutagenesis. Given that each induced mutant line carries thousands of mutations (Jiao et al. 2016), the desired traits can be identified by screening a limited number of lines and in a timely effective frame (Jiao et al. 2016; Addo-Quaye et al. 2018) using BSAseq (Bulked Segregant Analysis) (Wang et al. 2021b). Even though BSAseq workflow in sorghum was originally developed in bulked F2 populations, the availability of next-generation sequencing technologies has made this approach more affordable for identifying causative mutations that can represent promising targets for genome editing. However, induced mutagenesis also introduces a high amount of undesired background mutations (Addo-Quaye et al. 2018; Jiao et al. 2016). These mutations need to be removed before the introduction of such mutants in breeding projects. Even though recurrent backcrosses can be used to eliminate background mutations, several generations of backcross are necessary to remove approximately 90% of the unlinked background mutations. Unfortunately, there are no effective genetic method currently available to remove the unwanted linked mutations. The advantage of genome editing over classical and untargeted mutagenesis is the possibility to introduce precise modifications with few or no offsite mutations, thus allowing a rapid introduction of superior traits into elite germplasm for breeding.

13.5 Breeding for Drought and High Temperature Stress Tolerance

Breeding and improvement of crops gained a great advantage by adopting and integrating the technologies of DNA markers, MAS, and genetic engineering, which provide more detailed foreground and background selection approaches, together with the discovery and use of more different genetic resources. The development of more advanced next-generation and third-generation sequencing approaches, GWAS, GS, and genome editing, as well as noninvasive high-throughput phenomics, has completely reshaped the toolkits available to breeders for crop improvement. It is now possible for a more effective and precise exploitation of the natural or artificially generated genetic variability, the phenotyping, and selection of broad breeding populations in a new and more efficient manner (Maharajan et al. 2021; Yang et al. 2021). Concerning drought, breeding for tolerance to water scarcity followed two fundamental approaches: (1) a direct selection setting studies where the stressing conditions are uniform and predictable; (2) an indirect selection based on the selection of the best genotypes in controlled stress environments that can be close or distant from target locations. However, in consequence of the high genotype-by-environment interaction associated to many quantitative physiological traits as stress tolerance, indirect selection methods are the preferred selection approaches for evaluating plant water status and function within single environments (Singh et al. 2020). Breeding programs for selecting drought tolerance in sorghum aim at the development and identification of new cultivated varieties characterized by a high WUE able to maximize grain production from a given quantity of water (Khawaja 2014; Belete 2018). A high WUE in sorghum is related to different developmental traits and, concerning this aspect, significant genetic variations are present among different genotypes: (1) the root apparatus is a fundamental organ for developing drought tolerant varieties that should be included in any breeding program aimed at this. In this perspective, the knowledge of root variations among sorghum genotypes should be deeply studied and understood; (2) the sorghum evapotranspiration under standard and, in particular, stressed field conditions can be reduced by the presence of epicuticle waxes on leaves and stalks; (3) the stay-green phenotype, the capacity of the plant to maintain green leaves and photosynthetic capacity especially under stressing conditions, represents the most critical drought tolerance trait in sorghum (Jordan et al. 2012; Borrell et al. 2000, 2014a). In recent years, sorghum hybrids, presenting the stay-green trait, have been developed and successfully cultivated.

Improving for drought tolerance in future can take advantage of sorghum genetic variability across the globe and of studies that identified and reported more than 240 QTLs for water scarcity tolerance. Of these, the biggest number is related to physiological and growth factor traits, including the stay-green phenotype. The identification of a QTL is the preliminary step for the identification of DNA marker–trait associations and the further development of marker-assisted/marker-based approaches for the selection of segregating generations derived from breeding programs for drought-based traits. A further boosting of breeding for drought tolerance can be derived by exploiting the validated QTL for drought and adopting high-throughput phenomics platforms.

Breeding programs will be boosted by the development of reverse genetics tools. Li et al. (2019) overexpressed in maize two sorghum LRR-receptor-like kinase genes (*SbER*1-1 and *SbER*2-1) under a ubiquitin promoter of maize, in order to improve maize drought tolerance. The authors, increasing drought stress severity, observed a gradual increase in the expression of *SbER*1-1 and *SbER*2-1 in leaves and stems, while not in roots. The two transgenic lines showed delayed leaf senescence, an increased photosynthetic rate, and more lignin content compared to wild type under water stress. In *Arabidopsis* and maize, the expression of *SbER2-1* conferred increased drought tolerance, particularly regarding WUE. The expression of *mannitol-1-phosphate dehydrogenase* (*mtlD*) gene in transgenic sorghum lines exhibits a high root growth, an increment in soluble sugar content, and a reduction of malondialdehyde content under water scarcity conditions, showing increased tolerance to drought (Maheswari et al. 2017).

Increasing HT tolerance in sorghum is a complex process because it is influenced by both the environment and the interaction of different component traits and by the confusing effect of drought and HT. Despite the two traits being independent, their separation is challenging resulting that, traditionally, selection for drought tolerance was accompanied by the co-selection for HT tolerance. Variability in seedlings HT tolerance was studied in sorghum (Prasad et al. 2021). So far, a positive correlation between grain yield and HT has been considered as an initial selection standard for HT tolerance. New possibilities for capturing 3D images and IR images on temperature of canopy and tolerance traits (shoot length, shoot fresh weight, chlorophyll content, leaf firing, leaf blotching) may be provided by the availability of high-throughput and precision phenomics platforms. The analysis of these traits can increase the understanding of the genetic base of HT tolerance in sorghum in order to plan more appropriate breeding programs. Only few sorghum lines (RTx430, BTx3197, RTx7000, and B35) have been considered for HT stress. Lines, B35 and BTx3197, were used as heat tolerance sources in sorghum breeding (Ashok Kumar et al. 2013; Prasad et al. 2021).

The integration of phenomics, genomics, and models of phenology incorporating drought and HT tolerance in breeding plans represent a new opportunity for the introgression and validation of the QTL for the traits related to stress tolerance (Singh et al. 2014; Prasad et al. 2021).

Acknowledgments The study is part of the NODES project which has received funding from the MUR–M4C2 1.5 of PNRR with grant agreement no. ECS00000036.

References

Abdel-Ghany SE, Ullah F, Ben-Hur A, Reddy AS (2020) Transcriptome analysis of drought-resistant and drought-sensitive sorghum (*Sorghum bicolor*) genotypes in response to peg-induced drought stress. Int J Mol Sci 21(3):772. https://doi.org/10.3390/ijms21030772

Abou-Elwafa SF, Shehzad T (2018) Genetic identification and expression profiling of drought responsive genes in sorghum. Environ Exp Bot 155:12–20

Addo-Quaye C, Tuinstra M, Carraro N, Weil C, Dilkes BP (2018) Whole-genome sequence accuracy is improved by replication in a population of mutagenized sorghum. G3 (Bethesda) 8(3):1079–1094. https://doi.org/10.1534/g3.117.300301

Ahmad P, Bhardwaj R, Tuteja N (2012) Plant signalling under abiotic stress environment. In: Ahmad P, Prasad MNV (eds) Environmental adaptations and stress tolerance of plants in the era of climate change. Springer, New York, pp 297–323

Akbudak MA, Filiz E, Kontbay K (2018a) DREB2 (dehydration-responsive element-binding protein 2) type transcription factor in sorghum (*Sorghum bicolor*): genome-wide identification, characterization and expression profiles under cadmium and salt stresses. 3 Biotech 8:1–6. https://doi.org/10.1007/s13205-018-1454-1

Akbudak MA, Filiz E, Kontbay K (2018b) Genome-wide identification and cadmium induced expression profiling of sulfate transporter (*SULTR*) genes in sorghum (*Sorghum bicolor* L.). Biometals 31:91–105. https://doi.org/10.1007/s10534-017-0071-5

Ashok Kumar A, Sharma HC, Sharma R, Blummel M, Sanjana Reddy P, Reddy BVS (2013) Phenotyping in sorghum [*Sorghum bicolor* (L.) Moench]. In: Phenotyping for plant breeding. Springer, New York, pp 73–109

Assar AHA, Uptmoor R, Abdelmula AA, Salih M, Ordon F, Friedt W (2005) Genetic variation in sorghum germplasm from Sudan, ICRISAT and USA assessed by simple sequence repeats (SSRs). Crop Sci 45(4):1636. https://doi.org/10.2135/cropsci2003.0383

Belete T (2018) Breeding for resistance to drought: a case in sorghum (*Sorghum bicolor* (L.) Moench). J Agric For Meteorol Res 1:1–10

Bihani P, Char B, Bhargava S (2011) Transgenic expression of sorghum *DREB2* in rice improves tolerance and yield under water limitation. J Agric Sci 149(1):95. https://doi.org/10.1017/S0021859610000742

Borrell AK, Hammar GL, Henzell RG (2000) Does maintaining green leaf area in sorghum improve yield under drought II. Dry matter production and yield. Crop Sci 40:1037–1048. https://doi.org/10.2135/cropsci2000.4041037x

Borrell AK, Mullet JE, George-Jaeggli B, van Oosterom EJ, Hammer GL, Klein PE, Jordan DR (2014a) Drought adaptation of stay-green sorghum is associated with canopy development, leaf anatomy, root growth, and water uptake. J Exp Bot 65(21):6251–6263. https://doi.org/10.1093/jxb/eru232

Borrell AK, van Oosterom EJ, Mullet JE, George-Jaeggli B, Jordan DR, Klein PE, Hammer GL (2014b) Stay-green alleles individually enhance grain yield in sorghum under drought by modifying canopy development and water uptake patterns. New Phytol 203(3):817–830. https://doi.org/10.1111/nph.12869

Bouchet S, Olatoye MO, Marla SR, Perumal R, Tesso T, Yu J, Tuinstra M, Morris GP (2017) Increased power to dissect adaptive traits in global sorghum diversity using a nested association mapping population. Genetics 206:573–585

Brenton ZW, Cooper EA, Myers MT, Boyles RE, Shakoor N, Zielinski KJ, Rauh BL, Bridges WC, Morris GP, Kresovich S (2016) A genomic resource for the development, improvement, and exploitation of sorghum for bioenergy. Genetics 204:21–33

Burks PS, Kaiser CM, Hawkins EM, Brown PJ (2015) Genome wide association for sugar-yield in sweet sorghum. Crop Sci 55:2138–2148

Burow G, Burke JJ, Xin Z, Franks CD (2011) Genetic dissection of early-season cold tolerance in sorghum (*Sorghum bicolor* (L.) Moench). Mol Breed 28:391–402. https://doi.org/10.1007/s11032-010-9491-4

Casa AM, Pressoir G, Brown PJ, Mitchell SE, Rooney WL, Tuinstra MR, Franks CD, Kresovich S (2008) Community resources and strategies for association mapping in sorghum. Crop Sci 48(1):30–40. https://doi.org/10.2135/cropsci2007.02.0080

Char SN, Wei J, Mu Q, Li X, Zhang ZJ, Yu J, Yang B (2020) An Agrobacterium-delivered CRISPR/Cas9 system for targeted mutagenesis in sorghum. Plant Biotechnol J 18(2):319–321. https://doi.org/10.1111/pbi.13229

Che P, Anand A, Wu E, Sander JD, Simon MK, Zhu W, Sigmund AL, Zastrow-Hayes G, Miller M, Liu D, Lawit SJ, Zhao ZY, Albertsen MC, Jones TJ (2018) Developing a flexible, high-efficiency Agrobacterium-mediated sorghum transformation system with broad application. Plant Biotechnol J 16(7):1388–1395. https://doi.org/10.1111/pbi.12879

Chen J, Chopra R, Hayes C, Morris G, Marla S, Burke J, Xin Z, Burow G (2017) Genome-wide association study of developing leaves' heat tolerance during vegetative growth stages in a sorghum association panel. Plant Genome 10(2). https://doi.org/10.3835/plantgenome2016.09.0091

Chen D, Wang W, Wu Y, Xie H, Zhao L, Zeng Q, Zhan Y (2019a) Expression and distribution of the auxin response factors in *Sorghum bicolor* during development and temperature stress. Int J Mol Sci 20:4816

Chen KL, Wang YP, Zhang R, Zhang H, Gao C (2019b) CRISPR/Cas genome editing and precision plant breeding in agriculture. Annu Rev Plant Biol 70:667–697. https://doi.org/10.1146/annurev-arplant-050718-100049

Chopra R, Burow G, Hayes C, Emendack Y, Xin Z, Burke J (2015) Transcriptome profiling and validation of gene based single nucleotide polymorphisms (SNPs) in sorghum genotypes with contrasting responses to cold stress. BMC Genomics 16:1040–1050

Chopra R, Burow G, Burke JJ, Gladman N, Xin Z (2017) Genome-wide association analysis of seedling traits in diverse sorghum germplasm under thermal stress. BMC Plant Biol 17:12–26

Ciarmiello LF, Woodrow P, Fuggi A, Pontecorvo G, Carillo P (2011) Plant genes for abiotic stress. In: Shanker A, Venkateswarlu B (eds) Abiotic stress plants mechanisms and adaptation. IntechOpen, London, pp 283–308. https://www.intechopen.com/chapters/18407

Cooper EA, Brenton ZW, Flinn BS, Jenkins J, Shu SQ, Flowers D, Luo F, Wang YS, Xia P, Barry K, Daum C, Lipzen A, Yoshinaga Y, Schmutz J, Saski C, Vermerris W, Kresovich S (2019) A new reference genome for *Sorghum bicolor* reveals high levels of sequence similarity between sweet and grain genotypes: implications for the genetics of sugar metabolism. BMC Genomics 20:420–432

Crasta O, Xu W, Rosenow D, Mullet J, Nguyen H (1999) Mapping of post-flowering drought resistance traits in grain sorghum: association between QTLs influencing premature senescence and maturity. Mol Gen Genet 262(3):579–588. https://doi.org/10.1007/s004380051120

Cuevas HE, Prom LK (2020) Evaluation of genetic diversity, agronomic traits, and anthracnose resistance in the NPGS Sudan sorghum core collection. BMC Genomics 21:88–102

Cuevas HE, Rosa-Valentin G, Hayes CM, Rooney WL, Hoffmann L (2017) Genomic characterization of a core set of the USDANPGS Ethiopian sorghum germplasm collection: implications for germplasm conservation, evaluation, and utilization in crop improvement. BMC Genomics 18:108–124

Dahlberg JA, Burke JJ, Rosenow DT (2004) Development of a sorghum core collection: refinement and evaluation of a subset from Sudan. Econ Bot 58:556–567

Dalal M, Kumar GS, Mayandi K (2013) Identification and expression analysis of group 3 LEA family genes in sorghum [*Sorghum bicolor* (L.) Moench]. Acta Physiol Plant 35:979–984. https://doi.org/10.1007/s11738-012-1123-7

Dercas N, Liakatas A (2007) Water and radiation effect on sweet sorghum productivity. Water Resour Manag 21(9):1585–1600. https://doi.org/10.1007/s11269-006-9115-2

Deschamps S, Zhang Y, Llaca V et al (2018) A chromosome-scale assembly of the sorghum genome using nanopore sequencing and optical mapping. Nat Commun 9:4844

Devnarain N, Crampton BG, Chikwamba R, Becker JV, O'Kennedy MM (2016) Physiological responses of selected African sorghum landraces to progressive water stress and re-watering. S Afr J Bot 103:61–69. https://doi.org/10.1016/j.sajb.2015.09.008

Devnarain N, Crampton BG, Olivier N, Van der Westhuyzen C, Becker JV, O'Kennedy MM (2019) Transcriptomic analysis of a *Sorghum bicolor* landrace identifies a role for beta-alanine betaine biosynthesis in drought tolerance. S Afr J Bot 127:244–255. https://doi.org/10.1016/j.sajb.2019.08.049

Dillon SL, Shapter FM, Robert HJ, Cordeiro G, Izquierdo L, Lee SL (2007) Domestication to crop improvement: genetic resources for *Sorghum* and *Saccharum* (*Andropogoneae*). Ann Bot 100:975–989

Djanaguiraman M, Prasad PVV, Murugan M, Perumal R, Reddy UK (2014) Physiological differences among sorghum (*Sorghum bicolor* L. Moench) genotypes under high temperature stress. Environ Exp Bot 100:43–54. https://doi.org/10.1016/j.envexpbot.2013.11.013

Dos Santos JPR, Fernandes SB, McCoy S, Lozano R, Brown PJ, Leakey ADB, Buckler ES, Garcia AAF, Gore MA (2020) Novel Bayesian networks for genomic prediction of developmental traits in biomass sorghum. G3 (Bethesda) 10:769–781

Erdei E, Pepo P, Boros N, Toth S, Szabo B (2009) Morphological and biochemical indicators of drought tolerance in sweet sorghum (*Sorghum dochna* L.). Cereal Res Commun 37:157–160

Faye JM, Maina F, Hu ZB, Fonceka D, Cisse N, Morris GP (2019) Genomic signatures of adaptation to Sahelian and Soudanian climates in sorghum landraces of Senegal. Ecol Evol 9:6038–6051

Fernandes SB, Dias KOG, Ferreira DF, Brown PJ (2018) Efficiency of multi-trait, indirect, and trait-assisted genomic selection for improvement of biomass sorghum. Theor Appl Genet 131:747–755

Fracasso A, Magnanini E, Marocco A, Amaducci S (2017) Real-time determination of photosynthesis, transpiration, water-use efficiency and gene expression of two *Sorghum bicolor* (Moench) genotypes subjected to dry-down. Front Plant Sci 8:1–12

Galaihalage KSA, Myrans H, Norton SL, Gleadow R, Furtado A, Henry RJ (2020) Wild sorghum as a promising resource for crop improvement. Front Plant Sci 11:1108. https://doi.org/10.3389/fpls.2020.01108

Gelli M, Konda AR, Liu K, Zhang C, Clemente TE, Holding DR, Dweikat IM (2017) Validation of QTL mapping and transcriptome profiling for identification of candidate genes associated with nitrogen stress tolerance in sorghum. BMC Plant Biol 17:123–140

Girma G, Nida H, Seyoum A, Mekonen M, Nega A, Lule D, Dessalegn K, Bekele A, Gebreyohannes A, Adeyanju A, Tirfessa A, Ayana G, Taddese T, Mekbib F, Belete K, Tesso T, Ejeta G, Mengiste T (2019) A large-scale genome-wide association analyses of Ethiopian sorghum landrace collection reveal loci associated with important traits. Front Plant Sci 10:691–705

Gladman N, Olson A, Wei S, Chougule C, Lu Z, Tello-Ruiz M, Meijs I, Van Buren P, Jiao Y, Wang B, Kumar V, Kumari S, Zhang L, Burke J, Chen J, Burow G, Hayes C, Emendack Y, Xin Z, Ware D (2022) SorghumBase: a web-based portal for sorghum genetic information and community advancement. Planta 255:35. https://doi.org/10.1007/s00425-022-03821-6

Grenier C, Bramel-Cox PJ, Hamon P (2001a) Core collection of sorghum: I. Stratification based on eco-geographical data. Crop Sci 41:234–240

Grenier C, Hamon P, Bramel-Cox PJ (2001b) Core collection of sorghum: II. Comparison of three random sampling strategies. Crop Sci 41:241–246

Guo M, Liu Q, Yu H, Zhou T, Zou J, Zhang H, Bian M, Liu X (2015) Characterization of alkali stress-responsive genes of the CIPK Family in sweet Sorghum [*Sorghum bicolor* (L.) Moench]. Crop Sci 55:1254–1263. https://doi.org/10.2135/cropsci2013.08.0520

Guo TT, Yu XQ, Li XR, Zhang HZ, Zhu CS, Flint-Garcia S, McMullen MD, Holland JB, Szalma SJ, Wisser RJ, Yu JM (2019) Optimal designs for genomic selection in hybrid crops. Mol Plant 12:390–401

Gurel S, Gurel E, Miller TI, Lemaux PG (2012) Agrobacterium-mediated transformation of *Sorghum bicolor* using immature embryos. Methods Mol Biol 847:109–122. https://doi.org/10.1007/978-1-61779-558-9_10

Gurley WB (2000) HSP101: a key component for the acquisition of thermotolerance in plants. Plant Cell 12(4):457–460. https://doi.org/10.1105/tpc.12.4.457

Habyarimana E, Lopez-Cruz M, Baloch FS (2020) Genomic selection for optimum index with dry biomass yield, dry mass fraction of fresh material, and plant height in biomass sorghum. Genes 11:61–76

Habyarimana E, Gorthy S, Baloch FS et al (2022) Whole-genome resequencing of *Sorghum bicolor* and *S. bicolor* × *S. halepense* lines provides new insights for improving plant agroecological characteristics. Sci Rep 12:5556. https://doi.org/10.1038/s41598-022-09433-0

Hao H, Li Z, Leng C, Cheng LC, Hong Luo H, Liu Y, Wu X, Liu Z, Shang L, Jing HC (2021) Sorghum breeding in the genomic era: opportunities and challenges. Theor Appl Genet 134:1899–1924. https://doi.org/10.1007/s00122-021-03789-z

Harlan JR, de Wet JMJ (1972) A simplified classification of cultivated sorghum. Crop Sci 12:172–176

Harris K, Subudhi P, Borrell A, Jordan D, Rosenow D, Nguyen H, Klein P, Klein R, Mullet J (2007) Sorghum stay-green QTL individually reduce post-flowering drought-induced leaf senescence. J Exp Bot 58(2):327–338. https://doi.org/10.1093/jxb/erl225

Haussmann B, Mahalakshmi V, Reddy B, Seetharama N, Hash C, Geiger H (2002) QTL mapping of stay-green in two sorghum recombinant inbred populations. Theor Appl Genet 106(1):133–142. https://doi.org/10.1007/s00122-002-1012-3

Hayes CM, Burow GB, Brown PJ, Thurber C, Xin ZG, Burke JJ (2015) Natural variation in synthesis and catabolism genes influences dhurrin content in sorghum. Plant Genome 8:1–9

Hima Kumari P, Anil Kumar S, Ramesh K, Sudhakar Reddy P, Nagaraju M, Bhanu Prakash A, Shah T, Henderson A, Srivastava RK, Rajasheker G, Chitikineni A (2018) Genome-wide identification and analysis of *Arabidopsis* sodium proton antiporter (NHX) and human sodium proton exchanger (NHE) homologs in *Sorghum bicolor*. Genes 9:236. https://doi.org/10.3390/genes9050236

Hossain MS (2019) Present scenario of global salt affected soils, its management and importance of salinity research. Int Res J Biol Sci 1:1–3. https://scirange.com/abstract/irjbs.2019.1.3

House LR (1985) A guide to sorghum breeding, 2nd edn. International Crops Research Institute for the Semi-Arid Tropics, Andhra Pradesh

Huang RD (2018) Research progress on plant tolerance to soil salinity and alkalinity in sorghum. J Integr Agric 17:739–746. https://doi.org/10.1016/S2095-3119(17)61728-3

Jiang W, Zhou H, Bi H, Fromm M, Yang B, Weeks DP (2013) Demonstration of CRISPR/Cas9/sgRNA-mediated targeted gene modification in Arabidopsis, tobacco, sorghum and rice. Nucleic Acids Res 41:e188

Jiao Y, Burke J, Chopra R, Burow G, Chen J, Wang B, Hayes C, Emendack Y, Ware D, Xin Z (2016) A sorghum mutant resource as an efficient platform for gene discovery in grasses. Plant Cell 28(7):1551–1562. https://doi.org/10.1105/tpc.16.00373

Jinek M, Chylinski K, Fonfara I, Hauer M, Doudna JA, Charpentier E (2012) A programmable dual-RNA-guided DNA endonuclease in adaptive bacterial immunity. Science 337(6096):816–821. https://doi.org/10.1126/science.1225829

Johnson SM, Lim FL, Finkler A, Fromm H, Slabas AR, Knight MR (2014) Transcriptomic analysis of *Sorghum bicolor* responding to combined heat and drought stress. BMC Genomics 15(1):456

Johnson SM, Cummins I, Lim FL, Slabas AR, Knight MR (2015) Transcriptomic analysis comparing stay-green and senescent *Sorghum bicolor* lines identifies a role for proline biosynthesis in the stay-green trait. J Exp Bot 66(22):7061–7073. https://doi.org/10.1093/jxb/erv405

Jordan DR, Hunt CH, Cruickshank AW, Borrell AK, Henzell RG (2012) The relationship between the stay-green trait and grain yield in elite sorghum hybrids grown in a range of environments. Crop Sci 52:1153–1161. https://doi.org/10.2135/cropsci2011.06.0326

Kapanigowda MH, Payne WA, Rooney WL, Mullet JE, Balota M (2014) Quantitative trait locus mapping of the transpiration ratio related to preflowering drought tolerance in sorghum (*Sorghum bicolor*). Funct Plant Biol 41(10–11):1049–1065. https://doi.org/10.1071/Fp13363

Kebede H, Subudhi PK, Rosenow DT, Nguyen HT (2001) Quantitative trait loci influencing drought tolerance in grain sorghum (*Sorghum bicolor* L. Moench). Theor Appl Genet 103(2–3):266–276. https://doi.org/10.1007/s001220100541

Khawaja C (2014) Energy sorghum—an alternative energy crop. A handbook, WIP—renewable energies, Munich, Germany. ISBN: 978-3-936338-31-7

Kiranmayee KNSU, Hash CT, Sivasubramani S, Ramu P, Amindala BP, Rathore A, Kishor PBK, Gupta R, Deshpande SP (2020) Fine-mapping of sorghum stay-green QTL on chromosome10 revealed genes associated with delayed senescence. Genes 11:1026

Knoll J, Gunaratna N, Ejeta G (2008) QTL analysis of early-season cold tolerance in sorghum. Theor Appl Genet 116:577–587. https://doi.org/10.1007/s00122-007-0692-0

Kotak S, Larkindale J, Lee U, von Koskull-Doring P, Vierling E, Scharf KD (2007) Complexity of the heat stress response in plants. Curr Opin Plant Biol 10(3):310–316. https://doi.org/10.1016/j.pbi.2007.04.011

Lasky JR, Upadhyaya HD, Ramu P, Deshpande S, Hash CT, Bonnette J, Juenger TE, Hyma K, Acharya C, Mitchell SE, Buckler ES, Brenton Z, Kresovich S, Morris GP (2015) Genome-environment associations in sorghum landraces predict adaptive traits. Sci Adv 1:e1400218

Li A, Jia S, Yobi A, Ge Z, Sato SJ, Zhang C, Angelovici R, Clemente TE, Holding DR (2018) Editing of an alpha-kafirin gene family increases, digestibility and protein quality in sorghum. Plant Physiol 177(4):1425–1438. https://doi.org/10.1104/pp.18.00200

Li H, Han X, Liu X, Zhou M, Ren W, Zhao B, Zhao J (2019) A leucine-rich repeat-receptor-like kinase gene *SbER2-1* from sorghum (*Sorghum bicolor* L.) confers drought tolerance in maize. BMC Genomics 20:737. https://doi.org/10.1186/s12864-019-6143-x

Liu G, Godwin ID (2012) Highly efficient sorghum transformation. Plant Cell Rep 31(6):999–1007. https://doi.org/10.1007/s00299-011-1218-4

Liu G, Li J, Godwin ID (2019) Genome editing by CRISPR/Cas9 in sorghum through biolistic bombardment. Methods Mol Biol 1931:169–183. https://doi.org/10.1007/978-1-4939-9039-9_12

Luo M, Zhao Y, Zhang R, Xing J, Duan M, Li J, Wang N, Wang W, Zhang S, Chen Z, Zhang H (2017) Mapping of a major QTL for salt tolerance of mature field-grown maize plants based on SNP markers. BMC Plant Biol 17:1. https://doi.org/10.1186/s12870-017-1090-7

Mace ES, Tai SS, Gilding EK, Li YH, Prentis PJ, Bian LL, Campbell BC, Hu WS, Innes DJ, Han XL, Cruickshank A, Dai CM, Frère C, Zhang HK, Hunt CH, Wang XY, Shatte T, Wang MM, Su Z, Li J, Lin XZ, Godwin ID, Jordan DR, Wang J (2013) Whole genome sequencing reveals untapped genetic potential in Africa's indigenous cereal crop sorghum. Nat Commun 4:2320–2328

Maharajan T, Ajeesh Krishna TP, Kiriyanthan RM, Ignacimuthu S, Ceasar SA (2021) Improving abiotic stress tolerance in sorghum: focus on the nutrient transporters and marker-assisted breeding. Planta 254:90. https://doi.org/10.1007/s00425-021-03739-5

Maharajan T, Krishna TPA, Rakkammal K, Ceasar SA, Ramesh M (2022) Application of CRISPR/Cas system in cereal improvement for biotic and abiotic stress tolerance. Planta 256:106. https://doi.org/10.1007/s00425-022-04023-w

Maheshwari P, Kummari D, Palakolanu SR, Nagasai Tejaswi U, Nagaraju M, Rajasheker G, Jawahar G, Jalaja N, Rathnagiri P, Kavi Kishor PB (2019) Genome-wide identification and expression profile analysis of nuclear factor Y family genes in *Sorghum bicolor* L.(Moench). PLoS One 14:e0222203. https://doi.org/10.1371/journal.pone.0222203

Maheswari M, Varalaxmi Y, Yadav SK, Jyothilakshmi N, Vanaja M, Venkateswarlu B (2017) Enhanced tolerance of transgenic sorghum expressing *mtlD* gene to water-deficit stress. Indian J Biotechnol 16:63–67

Maina F, Bouchet S, Marla SR, Hu ZB, Wang JN, Mamadou A, Abdou M, Saidou AA, Morris GP (2018) Population genomics of sorghum (*Sorghum bicolor*) across diverse agroclimatic zones of Niger. Genome 61:223–232

Makita Y, Shimada S, Kawashima M, Kondou-Kuriyama T, Toyoda T, Matsui M (2015) MOROKOSHI: transcriptome database in *Sorghum bicolor*. Plant Cell Physiol 56(1):e6. https://doi.org/10.1093/pcp/pcu187

Marla SR, Shiva S, Welti R, Liu SZ, Burke JJ, Morris GP (2017) Comparative transcriptome and lipidome analyses reveal molecular chilling responses in chilling-tolerant sorghums. Plant Genome 10:1–16

Marla SR, Burow G, Chopra R, Hayes C, Olatoye MO, Felderhoff T, Hu Z, Raymundo R, Perumal R, Morris GP (2019) Genetic architecture of chilling tolerance in sorghum dissected with a nested association mapping population. G3 Bethesda 9(12):4045–4057. https://doi.org/10.1534/g3.119.400353

Mathur S, Priyadarshini SS, Singh V, Vashisht I, Jung K-H, Sharma R, Sharma MK (2020) Comprehensive phylogenomic analysis of ERF genes in sorghum provides clues to the evolution of gene functions and redundancy among gene family members. 3 Biotech 10(3):1–16. https://doi.org/10.1007/s13205-020-2120-y

McCormick RF, Truong SK, Sreedasyam A, Jenkins J, Shu SQ, Sims D, Kennedy M, Amirebrahimi M, Weers BD, McKinley B, Mattison A, Morishige DT, Grimwood J, Schmutz J, Mullet JE (2018) The *Sorghum bicolor* reference genome: improved assembly, gene annotations, a transcriptome atlas, and signatures of genome organization. Plant J 93:338–354

Morris GP, Ramu P, Deshpande SP, Hash CT, Shah T, Upadhyaya HD, Riera-Lizarazu O, Brown PJ, Acharya CB, Mitchell SE, Harriman J, Glaubitz JC, Buckler ES, Kresovich S (2013) Population genomic and genome-wide association studies of agroclimatic traits in sorghum. Proc Natl Acad Sci U S A 110(2):453–458

Murray SC, Rooney WL, Hamblin MT, Mitchell SE, Kresovich S (2009) Sweet sorghum genetic diversity and association mapping for brix and height. Plant Genome 2:48–62

Nagaraju M, Reddy SP, Kumar AS, Srivastava KR, Kavi KPB, Rao DM (2015) Genome-wide scanning and characterization of *Sorghum bicolor* L. heat shock transcription factors. Curr Genomics 16:279–291

Natoli A, Gorni C, Chegdani F, Ajmone Marsan P, Colombi C, Lorenzoni C, Marocco A (2002) Identification of QTLs associated with sweet sorghum quality. Maydica 47:311–322

Nida H, Girma G, Mekonen M, Lee S, Seyoum A, Dessalegn K, Tadesse T, Ayana G, Senbetay T, Tesso T, Ejeta G, Mengiste T (2019) Identification of sorghum grain mold resistance loci through genome wide association mapping. J Cereal Sci 85:295–304

Olatoye MO, Hu ZB, Maina F, Morris GP (2018) Genomic signatures of adaptation to a precipitation gradient in Nigerian sorghum. G3 (Bethesda) 8:3269–3281

Ongom PO, Ejeta G (2018) Mating design and genetic structure of a multi-parent advanced generation intercross (MAGIC) population of sorghum (*Sorghum bicolor* (L.) Moench). G3 (Bethesda) 8:331–341

Ortiz D, Hu JY, Salas Fernandez MG (2017) Genetic architecture of photosynthesis in *Sorghum bicolor* under non-stress and cold stress conditions. J Exp Bot 68:4545–4557

Pasini L, Bergonti M, Fracasso A, Marocco A, Amaducci S (2014) Microarray analysis of differentially expressed mRNAs and miRNAs in young leaves of sorghum under dry-down conditions. J Plant Physiol 171:537–548

Paterson AH, Bowers JE, Bruggmann R, Dubchak I, Grimwood J, Gundlach H, Haberer G, Hellsten U, Mitros T, Poliakov A, Schmutz J, Spannagl M, Tang H, Wang X, Wicker T, Bharti AK, Chapman J, Feltus FA, Gowik U, Grigoriev IV, Lyons E, Maher CA, Martis M, Narechania A, Otillar RP, Penning BW, Salamov AA, Wang Y, Zhang L, Carpita NC, Freeling M, Gingle AR, Hash CT, Keller B, Klein P, Kresovich S, McCann MC, Ming R, Peterson DG, Mehboobur R, Ware D, Westhoff P, Mayer KF, Messing J, Rokhsar DS (2009) The *Sorghum bicolor* genome and the diversification of grasses. Nature 457(7229):551–556. https://doi.org/10.1038/nature07723

Prasad VBR, Govindaraj M, Djanaguiraman M, Djalovic I, Shailani A, Rawat N, Singla-Pareek SL, Pareek A, Vara Prasad PV (2021) Drought and high temperature stress in sorghum: physiological, genetic, and molecular insights and breeding approaches. Int J Mol Sci 22:9826. https://doi.org/10.3390/ijms22189826

Proveniers MCG, van Zanten M (2013) High temperature acclimation through PIF4 signaling. Trends Plant Sci 18:59–64

Puram VR, Ontoy J, Subudhi PK (2018) Identification of QTLs for salt tolerance traits and pre-breeding lines with enhanced salt tolerance in an introgression line population of rice. Plant Mol Biol Rep 36:695–709. https://doi.org/10.1007/s11105-018-1110-2

Rakshit S, Swapna M, Dalal M, Sushma G, Ganapathy KN, Dhandapani A, Karthikeyan M, Talwar HS (2016) Post-flowering drought stress response of post-rainy sorghum genotypes. Indian J Plant Physiol 21:8–14. https://doi.org/10.1007/s40502-015-0187-y

Reddy NRR, Ragimasalawada M, Sabbavarapu MM, Nadoor S, Patil JV (2014) Detection and validation of stay-green QTL in postrainy sorghum involving widely adapted cultivar, M35-1 and a popular stay-green genotype B35. BMC Genomics 15(1):1–16. https://doi.org/10.1186/1471-2164-15-909

Rice B, Lipka AE (2019) Evaluation of RR-BLUP genomic selection models that incorporate peak genome-wide association study signals in maize and sorghum. Plant Genome 12(1):180052

Ruperao P, Thirunavukkarasu N, Gandham P, Selvanayagam S, Govindaraj M, Nebie B, Manyasa E, Gupta R, Das RR, Odeny DA, Gandhi H, Edwards D, Deshpande SP, Rathore A (2021) Sorghum pan-genome explores the functional utility for genomic-assisted breeding to accelerate the genetic gain. Front Plant Sci 12:666342. https://doi.org/10.3389/fpls.2021.666342

Sanchez A, Subudhi P, Rosenow D, Nguyen H (2002) Mapping QTLs associated with drought resistance in sorghum (*Sorghum bicolor* L. Moench). Plant Mol Biol 48(5):713–726. https://doi.org/10.1023/A1014894130270

Sbei H, Sato K, Shehzad T, Harrabi M, Okuno K (2014) Detection of QTLs for salt tolerance in Asian barley (*Hordeum vulgare* L.) by association analysis with SNP markers. Breed Sci 64:378–388. https://doi.org/10.1270/jsbbs.64.378

Shakoor N, Nair R, Crasta O, Morris G, Feltus A, Kresovich S (2014) A Sorghum bicolor expression atlas reveals dynamic genotype specific expression profiles for vegetative tissues of grain, sweet and bioenergy sorghums. BMC Plant Biol 14(1):35. https://doi.org/10.1186/1471-2229-14-35

Sharoni AM, Nuruzzaman M, Satoh K, Shimizu T, Kondoh H, Sasaya T, Choi I-R, Omura T, Kikuchi S (2011) Gene structures, classification and expression models of the AP2/EREBP transcription factor family in rice. Plant Cell Physiol 52(2):344–360. https://doi.org/10.1093/pcp/pcq196

Singh P, Nedumaran S, Traore PCS, Boote KJ, Rattunde HFW, Prasad PVV, Singh NP, Srinivas K, Bantilan MCS (2014) Quantifying potential benefits of drought and heat tolerance in rainy season sorghum for adapting to climate change. Agric For Meteorol 185:37–48. https://doi.org/10.1016/j.agrformet.2013.10.012

Singh RK, Ashish Prasad A, Muthamilarasan M, Parida SK, Prasad M (2020) Breeding and biotechnological interventions for trait improvement: status and prospects. Planta 252:54. https://doi.org/10.1007/s00425-020-03465-4

Siriwardana CL, Gnesutta N, Kumimoto RW, Jones DS, Myers ZA, Mantovani R, Holt BF III (2016) Nuclear factor Y, subunit A (NF-YA) proteins positively regulate flowering and act through flowering locus T. PLoS Genet 12:e1006496. https://doi.org/10.1371/journal.pgen.1006496

Spindel JE, Dahlberg J, Colgan M, Hollingsworth J, Sievert J, Staggenborg SH, Hutmacher R, Jansson C, Vogel JP (2018) Association mapping by aerial drone reveals 213 genetic associations for *Sorghum bicolor* biomass traits under drought. BMC Genomics 19:679–696

Su M, Li XF, Ma XY, Peng XJ, Zhao AG, Cheng LQ, Chen SY, Liu GS (2011) Cloning two P5CS genes from bioenergy sorghum and their expression profiles under abiotic stresses and MeJA treatment. Plant Sci 181:652–659. https://doi.org/10.1016/j.plantsci.2011.03.002

Subudhi PK, Rosenow DT, Nguyen HT (2000) Quantitative trait loci for the stay green trait in sorghum (*Sorghum bicolor* L. Moench): consistency across genetic backgrounds and environments. Theor Appl Genet 101:733–741. https://doi.org/10.1007/s001220051538

Sudhakar RP, Srinivas RD, Sivasakthi K, Bhatnagar MP, Vadez V, Sharma KK (2016) Evaluation of sorghum [*Sorghum bicolor* (L.)] reference genes in various tissues and under abiotic stress conditions for quantitative real-time PCR data normalization. Front Plant Sci 7:529

Sukumaran S, Li X, Li X, Zhu C, Bai G, Perumal R, Tuinstra MR, Prasad PV, Mitchell SE, Tesso TT (2016) QTL mapping for grain yield, flowering time, and stay-green traits in sorghum with genotyping-by-sequencing markers. Crop Sci 56(4):1429–1442. https://doi.org/10.2135/cropsci2015.02.0097

Sung D-Y, Kaplan F, Lee K-J, Guy CL (2003) Acquired tolerance to temperature extremes. Trends Plant Sci 8(4):179–187. https://doi.org/10.1016/S1360-1385(03)00047-5

Tao Y, Henzell R, Jordan D, Butler D, Kelly A, McIntyre C (2000) Identification of genomic regions associated with stay green in sorghum by testing RILs in multiple environments. Theor Appl Genet 100(8):1225–1232. https://doi.org/10.1007/s001220051428

Tao Y, Luo H, Xu J, Cruickshank A, Zhao X, Teng F, Hathorn A, Wu X, Liu Y, Shatte T, Jordan D, Jing H, Mace E (2021) Extensive variation within the pan-genome of cultivated and wild sorghum. Nat Plants 7(6):766–773. https://doi.org/10.1038/s41477-021-00925-x

Thomas H, Howarth CJ (2000) Five ways to stay green. J Exp Bot 51:329–337. https://doi.org/10.1093/jexbot/51.suppl_1.329

Tuinstra MR, Grote EM, Goldsbrough PB, Ejeta G (1996) Identification of quantitative trait loci associated with pre-flowering drought tolerance in sorghum. Crop Sci 36:1337–1344. https://doi.org/10.2135/cropsci1996.0011183X003600050043x

Tuinstra MR, Grote EM, Goldsbrough PB, Ejeta G (1997) Genetic analysis of post-flowering drought tolerance and components of grain development in *Sorghum bicolor* (L.) Moench. Mol Breed 3:439–448. https://doi.org/10.1023/A:1009673126345

Ukwatta J, Pabuayon ICM, Park J, Chen J, Chai X, Zhang H, Zhu JK, Xin Z, Shi H (2021) Comparative physiological and transcriptomic analysis reveals salinity tolerance mechanisms in *Sorghum bicolor* (L.) Moench. Planta 254:98. https://doi.org/10.1007/s00425-021-03750-w

Upadhyaya HD, Pundir RPS, Dwivedi SL, Gowda CLL, Reddy VG, Singh S (2009) Developing a mini core collection of sorghum for diversified utilization of germplasm. Crop Sci 49(5):1769–1780. https://doi.org/10.2135/cropsci2009.01.0014

Upadhyaya HD, Vetriventhan M, Azevedo VCR (2021) Variation for photoperiod and temperature sensitivity in the global mini core collection of Sorghum. Front Plant Sci 12:571243

Uptmoor R, Wenzel W, Friedt W, Donaldson G, Ayisi K, Ordon F (2003) Comparative analysis on the genetic relatedness of *Sorghum bicolor* accessions from Southern Africa by RAPDs, AFLPs and SSRs. Theor Appl Genet 106(7):1316–1325. https://doi.org/10.1007/s00122-003-1202-7

Varoquaux N, Cole B, Gao C, Pierroz G, Baker CR, Patel D, Madera M, Jeffers T, Hollingsworth J, Sievert J (2019) Transcriptomic analysis of field-droughted sorghum from seedling to maturity reveals biotic and metabolic responses. Proc Natl Acad Sci U S A 116(52):27124–27132. https://doi.org/10.1073/pnas.1907500116

Vasilakoglou I, Dhima K, Karagiannidis N, Gatsis T (2011) Sweet sorghum productivity for biofuels under increased soil salinity and reduced irrigation. Field Crop Res 120(1):38–46. https://doi.org/10.1016/j.fcr.2010.08.011

Vermerris W, Saballos A, Ejeta G, Mosier NS, Ladisch MR, Carpita NC (2007) Molecular breeding to enhance ethanol production from corn and sorghum stover. Crop Sci 47(3):143–153. https://doi.org/10.2135/cropsci2007.04.0013IPBS

Wang ML, Zhu CS, Barkley NA, Chen ZB, Erpelding JE, Murray SC, Tuinstra MR, Tesso T, Pederson GA, Yu JM (2009) Genetic diversity and population structure analysis of accessions in the US historic sweet sorghum collection. Theor Appl Genet 120:13–23

Wang Y, Zhang L, Zhang L, Xing T, Peng J, Sun S, Chen G, Wang X (2013) A novel stress-associated protein SbSAP14 from *Sorghum bicolor* confers tolerance to salt stress in transgenic rice. Mol Breed 32:437–449. https://doi.org/10.1007/s11032-013-9882-4

Wang TT, Ren ZJ, Liu ZQ, Feng X, Guo RQ, Li BG, Li LG, Jing HC (2014a) *SbHKT1; 4*, a member of the high-affinity potassium transporter gene family from *Sorghum bicolor*, functions to maintain optimal Na+/K+ balance under Na+ stress. J Integr Plant Biol 56:315–332. https://doi.org/10.1111/jipb.12144

Wang H, Chen G, Zhang H, Liu B, Yang Y, Qin L, Chen E, Guan Y (2014b) Identification of QTLs for salt tolerance at germination and seedling stage of *Sorghum bicolor* L. Moench. Euphytica 196(1):117–127. https://doi.org/10.1007/s10681-013-1019-7

Wang JN, Hu ZB, Upadhyaya HD, Morris GP (2020a) Genomic signatures of seed mass adaptation to global precipitation gradients in sorghum. Heredity 124:108–121

Wang H, Wang R, Liu B, Yang Y, Qin L, Chen E, Zhang H, Guan Y (2020b) QTL analysis of salt tolerance in *Sorghum bicolor* during whole-plant growth stages. Plant Breed 139:455–465. https://doi.org/10.1111/pbr.12805

Wang B, Jiao Y, Chougule K et al (2021a) Pan-genome analysis in sorghum highlights the extent of genomic variation and sugarcane aphid resistance genes. bioRxiv. 2021.01.03.424980

Wang L, Lu Z, Regulski M, Jiao Y, Chen J, Ware D, Xin Z (2021b) BSAseq: an interactive and integrated web-based workflow for identification of causal mutations in bulked F2 populations. Bioinformatics 37(3):382–387. https://doi.org/10.1093/bioinformatics/btaa709

Wendorf F, Close AE, Schild R, Wasylikowa K, Housley RA, Harlan JR, Królik H (1992) Saharan exploitation of plants 8,000 years BP. Nature 359:721–724

Wiersema JH, Dahlberg J (2007) The nomenclature of *Sorghum bicolor* (L.) Moench (*Gramineae*). Taxon 56:944

Xin Z, Wang ML, Barkley NA, Burow G, Franks C, Pederson G, Burke J (2008) Applying genotyping (TILLING) and phenotyping analyses to elucidate gene function in a chemically induced sorghum mutant population. BMC Plant Biol 8:103. https://doi.org/10.1186/1471-2229-8-103

Xin Z, Wang M, Cuevas HE, Chen J, Harrison M, Pugh NA, Morris G (2021) Sorghum genetic, genomic, and breeding resources. Planta 254:114. https://doi.org/10.1007/s00425-021-03742-w

Xu W, Subudhi PK, Crasta OR, Rosenow DT, Mullet JE, Nguyen HT (2000) Molecular mapping of QTLs conferring stay-green in grain sorghum (*Sorghum bicolor* L. Moench). Genome 43:461–469. https://doi.org/10.1139/g00-003

Xu ZS, Chen M, Li LC, Ma YZ (2011) Functions and application of the AP2/ERF transcription factor family in crop improvement. J Integr Plant Biol 53(7):570–585. https://doi.org/10.1111/j.1744-7909.2011.01062.x

Xu YB, Liu XG, Fu JJ, Wang HW, Wang JK, Huang CL, Prasanna BM, Olsen MS, Wang GY, Zhang AM (2020) Enhancing genetic gain through genomic selection: from livestock to plants. Plant Commun 1:100005

Yang Y, Mumtaz Ali Saand MA, Huang L, Abdelaal WB (2021) Applications of multi-omics technologies for crop improvement. Front Plant Sci 12:563953

Yu XQ, Li XR, Guo TT, Zhu CS, Wu YY, Mitchell SE, Roozeboom KL, Wang DH, Wang ML, Pederson GA, Tesso TT, Schnable PS, Bernardo R, Yu JM (2016) Genomic prediction contributing to a promising global strategy to turbocharge gene banks. Nat Plants 2:16150

Zhang LM, Leng CY, Luo H, Wu XY, Liu ZQ, Zhang YM, Zhang H, Xia Y, Shang L, Liu CM, Hao DY, Zhou YH, Chu CC, Cai HW, Jing HC (2018) Sweet sorghum originated through selection of *Dry*, a plant-specific NAC transcription factor gene. Plant Cell 30:2286–2307

Zhang R, Zhou Y, Yue Z, Chen X, Cao X, Ai X, Jiang B, Xing Y (2019) The leaf-air temperature difference reflects the variation in water status and photosynthesis of sorghum under waterlogged conditions. PLoS One 14(7):e0219209

Zheng LY, Guo XS, He B, Sun LJ, Peng Y, Dong SS, Liu TF, Jiang S, Ramachandran S, Liu CM, Jing HC (2011) Genome-wide patterns of genetic variation in sweet and grain sorghum (*Sorghum bicolor*). Genome Biol 12:R114

Omics Approaches for Improving Sorghum Value Chains

14

Pradeep Ruperao and Srikanth Bollam

Abstract

Sorghum (*Sorghum bicolor*) holds global significance as a versatile crop, serving diverse purposes like food, animal feed, and more. It is the fifth most crucial grain crop, originating from Africa, and yields around 60 million tons annually. With advancements in sequencing and omics tools, we have seen improved crop enhancement and food security. The progress in high-tech experimental and data processing tools has greatly advanced genomic research in sorghum. By combining artificial intelligence and machine learning with omics methods, we can speed up genetic advancements in crops. This involves gaining profound insights in to the genetic and molecular underpinnings of key traits, allowing for targeted improvements. In this chapter, we delve into the different omics approaches used to enhance sorghum genetics.

Keywords

Sorghum omics · Genomics · Proteomics · Metabolomics · Transcriptomics · Interactomics · Epigenomics · Hormonomics · QTL-omics · Bioinformatics · Artificial intelligence · Machine learning · Deep learning

P. Ruperao (✉) · S. Bollam (✉)
International Crops Research Institute for the Semi-Arid Tropics, Patancheru, India
e-mail: Pradeep.ruperao@icrisat.org; Srikanth.bollam@icrisat.org

E. Habyarimana et al. (eds.), *Omics and Biotechnological Approaches for Product Profile-Driven Sorghum Improvement*,
https://doi.org/10.1007/978-981-97-4347-6_14

14.1 Introduction

Sorghum is a small-grained cereal that many people rely on as food in Africa and Asia. There are different kinds of sorghum, and they all come from three subspecies: *arundinaceum*, *bicolor*, and *drummondii*. All the cultivated sorghum variations were partitioned into five groups/races such as, *bicolor*, *durra*, *caudatum*, *guinea*, and *kafir* (Harlan and Wet 1972). Sorghum is versatile crop and is grown for different end uses worldwide, including grain sorghum, sweet sorghum, forage sorghum, and broom sorghum (Batey 2017). Sweet sorghum, for instance, is a natural type that is different from regular sorghum. Such major phenotypic variations are due to the variations (including SNPs and indels) at the genome level in the sorghum varieties. The genetic structural alterations include the presence-absence variations (PAVs) and copy number variations (CNVs).

Recent advances in sequencing technologies have boosted research efforts on crop improvement programs and advancements in science of food (Varshney et al. 2009). This opens new possibilities for using these technologies to make crops better. There is a term called "Foodomics" that uses different omics technologies to improve food and nutrition (Capozzi and Bordoni 2013). Combining these omics technologies helps make crops more nutritious and deal with issues of food security (Tian et al. 2016).

"Omics" covers a bunch of different areas like genomics, transcriptomics, metabolomics, proteomics, and bioinformatics (Nayak et al. 2021) (Fig. 14.1). Bringing all these omics technologies together helps us understand how biological systems work in a broader sense (Pazhamala et al. 2021). This chapter talks about these different omics technologies and how they can be used to make crops better, with a focus on sorghum.

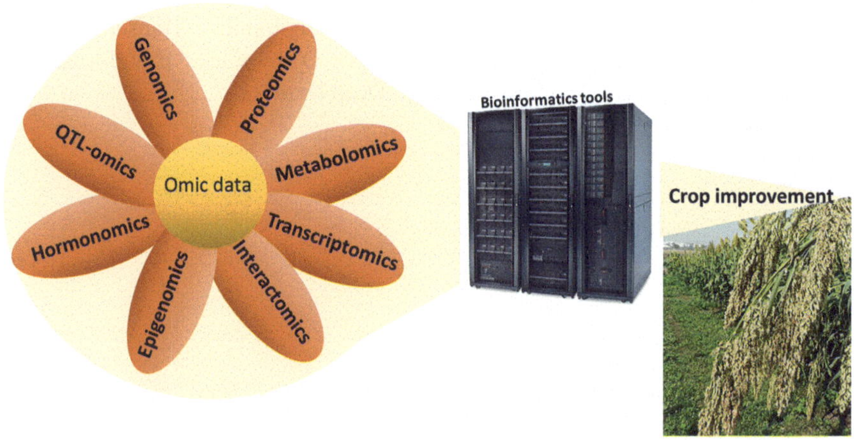

Fig. 14.1 Integrated omics approaches for functional foods for crop improvement through bioinformatics tools

14.2 Genomics

Major germplasm repositories for sorghum are housed in two primary gene banks: the International Crops Research Institute for the Semi-Arid Tropics (ICRISAT) and the United States Department of Agriculture-National Plant Germplasm System (USDA-NPGS), which collectively harbors over 41,000 diverse sorghum accessions. Additional collections exist in Australia, India, and China. Based on morphological and physiological traits, these lines were categorized into five basic and ten intermediate races (Dillon et al. 2007). ICRISAT researchers have defined a World Core Collection (WCC) panel and a smaller Mini-Core Collection (MCC) panel, the latter comprising 242 accessions (Upadhyaya et al. 2009). Combining these with the Sorghum Association Panel (SAP) and a reference set, a diverse panel was constituted with 971 sorghum genotypes (Morris et al. 2013).

Understanding the complete set of genes/genome in sorghum is pivotal for identifying key genes responsible for agronomic traits. The availability of genome assemblies for three sorghum lines (Deschamps et al. 2018; Cooper et al. 2019; Chen et al. 2022) and pangenomes of diverse sorghum panels (Jensen et al. 2020; Ruperao et al. 2021) contributes significantly to sorghum genomic resources. Thousands of sorghum genetic stocks with whole genome sequences serve as core genomic resources, aiding in the development of molecular markers and identification of quantitative trait loci (QTLs) for target traits, facilitating genomics-assisted breeding and genomic selection (Varshney et al. 2005).

Advancements in DNA technologies have transformed plant breeding from conventional to molecular methods (Moose and Mumm 2008), incorporating DNA markers, marker-assisted selections (MAS), gene editing, and genetic engineering (Bruce 2012). Second- and third-generation sequencing technologies have significantly bolstered genomic tools like genome-wide association studies (GWAS), genomic selection (GS), genome editing (GE), and molecular modules (MMs), alongside high-throughput phenomics. Unlike MAS, GS utilizes genome-wide marker information, eliminating the need for associated phenotypes and target trait QTLs in the breeding population (Meuwissen et al. 2001). In sorghum, GS leverages both training and test populations to predict genomic estimated breeding values (GEBVs) for the identification of the selection population, focusing on training models with natural populations, mixed populations, and testcross hybrids (Yu et al. 2016; Habyarimana et al. 2020; Dos Santos et al. 2020).

Genomics offers valuable insights into the genetic and genomic mechanisms of sorghum germplasm, aiding in the in-depth understanding of domestication and diversification processes. This knowledge facilitates the incorporation of genetic diversity from wild progenitors into improved varieties through breeding selection. Diversity analysis in a wide range of sorghum accessions helps identify candidate genomic regions, and the compilation of sorghum populations for functional genomics, diverse germplasm panels, and mapping populations allows for exploration and editing of genes controlling specific traits.

14.3 Transcriptomics

Transcriptomes serve as a dynamic measure of differentially expressed genes in a particular tissue at different stages, with quantification achieved through molecular tools. The comprehensive analysis of expressed RNAs, particularly mRNAs translated into proteins, plays a crucial role in understanding the regulatory functions of noncoding sequences during normal plant growth (Szymański et al. 2003; Guleria et al. 2011). Various techniques are employed for transcriptomic studies, including Northern blotting, hybridization methods, expressed sequence tags (ESTs), reverse transcriptase quantitative PCR (RT-qPCR), serial analysis of gene expression (SAGE), RNA microarray, and RNA sequencing (Lowe et al. 2017). Among these, microarrays and RNA-seq are commonly employed in plant transcriptome studies, with a specific focus on sorghum, where investigations often focus on drought stress, RNA profiling, plant tissues, and water stress treatments (Cheng et al. 1993; Wood and Goldsbrough 1997; Hamza et al. 2016).

Numerous transcriptomic studies in sorghum have delved into growth and development (Kebrom and Mullet 2016; McCormick et al. 2018), revealing differentially expressed genes associated with specific traits. For instance, the transcriptome analysis of Della stem samples, collected from floral initiation stage to 30 days post-grain maturity, identified 200 differentially expressed genes crucial for stem composition and biomass accumulation (McKinley et al. 2016). To comprehensively understand the entire expressed gene landscape, a broader analysis involving 47 samples from roots, leaves, stems, seeds, and panicles was conducted, contributing to the annotation of the reference genome (McCormick et al. 2018). Transcriptomic investigations have also explored gene expression profiles under abiotic and biotic stresses, highlighting the distinct differences between samples (Gelli et al. 2017; Marla et al. 2017). Notably, bulk sample transcriptomes from 400 leaf and root samples revealed 3977 genes related with drought stress, exhibited variations in photosynthesis, pathways associated with reactive oxygen species (ROS) scavenging, and their symbiotic relationships with arbuscular mycorrhizal (AM) fungi under drought stress, these differences suggest a potential area to focus for strengthening the sorghum breeding efforts (Varoquaux et al. 2019).

14.4 Proteomics

Proteomics, focusing on protein structure, function, and interactions, offers a superior platform for large-scale analysis of expressed proteins in cells responding to environmental stimuli compared to genomics (Blackstock and Weir 1999). Categorized into gel-based and nongel-based techniques, proteomics employs mass spectrometry and bioinformatics analysis for protein structure and functional characterization (Monteoliva and Albar 2004). Despite earlier advancements in plant proteomics, primarily focusing on abiotic stress responses (Kosová et al. 2011; Barkla et al. 2013; Ahmad et al. 2016), proteomic studies in sorghum remain limited, highlighting knowledge gaps in sorghum proteomes in response to various stresses, including water deprivation (Jedmowski et al. 2014; Goche et al. 2020; Li et al. 2020).

To address this gap, advanced techniques such as liquid chromatography coupled with mass spectrometry (LC-MS) and crystallography are utilized for detecting protein expression and structure. A proteomic study on sorghum leaves under drought stress, employing two-dimensional difference gel electrophoresis (2D DIGE) analysis, revealed higher levels of proteolytic enzymes in the drought-sensitive variety compared to the stabilized protein levels in the drought-tolerant sorghum variety (Jedmowski et al. 2014). Another study, analyzing the root proteome of two sorghum varieties in response to drought stress using gel-free isobaric tags for relative and absolute quantitation (iTRAQ) technology, identified 237 and 187 drought-sensitive and tolerant-responsive proteins, with 51 proteins common to both varieties (Goche et al. 2020). The unique proteins displayed distinct expression dynamics, contributing to the differential performance of the two varieties under stress conditions. The study highlighted that the drought-tolerant sorghum variety (SA1441) responded to water deficits by increasing protein levels related to transcription, protein synthesis, protease inhibition, and signaling, whereas the drought-susceptible accession (ICSB338) increased the accumulation of proteolytic enzymes while downregulating metabolism and protein synthesis (Goche et al. 2020). Further functional studies are essential to unravel the significance of these proteins in stress response.

14.5 Metabolomics

Metabolomics, the study of small molecules or metabolites in organisms, involves the detection and quantification of specific metabolites within a sample. Utilizing tools such as gas chromatography-mass spectrometry (GC-MS), liquid chromatography-mass spectrometry (LC-MS), nuclear magnetic resonance (NMR), and near-infrared spectrometry (NIR), metabolomics provides a comprehensive approach for quantifying biologically active compounds, fingerprinting, and profiling.

This emerging technology plays a crucial role in understanding how genetic variability influences phenotypic variations in crop plants, offering a mechanistic interpretation of plant responses at the molecular level (Feussner and Polle 2015). Metabolites, being regulated by gene expression and formation of transcripts, provide a functional approach complementary to genomic and transcriptomic strategies, with implications for identifying targets in plant breeding (Bino et al. 2004; Fernie and Schauer 2009). Applications of metabolomics include characterizing biochemical variation within species, assessing plant responses to environmental conditions, and understanding responses to both biotic and abiotic stresses (Kusano et al. 2015; Heuberger et al. 2014; Scandiani et al. 2015; Ganie et al. 2015).

In sorghum, leaf metabolites exhibit significant variation among grain and biomass forms, impacting morphological, physiological, and cell-wall structural traits (Turner et al. 2016). A study by Manuele et al. in the Emilia-Romagna region, Italy, utilized NMR-based metabolomics to assess 12 sorghum crop fields at seedling, advanced vegetative, and ripening stages, correlating agronomic assessment and metabolomic data to grain yield (Mandrone et al. 2021). Another investigation highlighted the metabolic responses to low nitrogen, revealing altered metabolic profiles and reduced defense responses, indicating potential interactions among biotic and abiotic stresses affecting the plant's metabolic capacity for defense (Sheflin et al. 2019a).

14.6 Interactomics

Interactomics, a field focused on studying a comprehensive array of molecular interactions within biological samples, encompasses protein-protein interactions (PPIs) and interactions between small molecules and proteins. Such interactome studies are crucial for understanding the intricate organization of cellular interactions (Koh et al. 2012). While various methods exist for identifying PPIs, yeast-2-hybrid, and mass spectrometry (MS) have significantly contributed to these investigations (Uetz et al. 2000; Koh et al. 2012). PPIs, forming protein interaction networks (PIN), play essential roles in diverse cellular processes, facilitating the transfer of information and the transport of molecules in living cells (Ashburner et al. 2000).

Interactome datasets, generated through experimental datasets, analysis, and visualization using systems biology tools, contribute to online data resources (Salwinski et al. 2004; Butland et al. 2005; Yilmaz et al. 2022). Integrating biological systems, including interactomes and protein-protein interaction networks, provides critical insights into the regulation of plant morphological, physiological, developmental, stress resistance, and pathological mechanisms (Zhang et al. 2018). Another widely studied interactome is the DNA-protein interactome, encompassing gene-regulatory networks, where regulatory proteins, target genes, and transcription factors form intricate networks, defining both direct and indirect genetic interactions. Leveraging interactome datasets effectively uncovers gene networks associated with specific traits, enabling accelerated breeding technologies.

In contrast to current marker-assisted breeding approaches, which focus on improving individual genes, interactome-based breeding has the potential for broader impact by modifying entire gene networks. Such studies allow the dissection of pathways related to domestication, facilitating the modification of favorable alleles associated with them for faster breeding advancements (Fernie and Yan 2019). Notable examples of pathway-driven improvements include 'Golden rice', where the biosynthetic pathway for β-carotene was modified to enhance vitamin A content in rice (Paine et al. 2005), and 'C4 Rice', an initiative transferring C4 genes and regulatory pathways from maize and sorghum to rice for improved crop yield (Wang et al. 2017). Identifying various networks regulating agronomic traits is crucial for future interactome-based breeding programs, promising rapid improvements in productivity and resilience in food crops.

14.7 Epigenomics

Epigenomics, the comprehensive study of biochemical variations in histone proteins, nuclear DNA, and noncoding RNAs, plays a pivotal role in understanding cellular processes such as chromosome inactivation, cell differentiation, development, and senescence (Kumar 2018). Key epigenetic modifications, including histone modification and DNA methylation, intricately govern gene expression, chromatin structure, and genome stability, particularly influencing gene silencing and heterochromatin. Detection methods for DNA methylation involve bisulfite conversion, restriction enzymes with methylation-sensitive and insensitive properties, and affinity-based immunoprecipitation (Bock et al. 2010; Morrison et al. 2021). Whole-genome bisulfite DNA sequencing (WGBS) technology stands out as a gold-standard technique, generating comprehensive DNA methylation maps (Cokus et al. 2008).

In recent times, epigenomics has gained substantial attention for unraveling gene function regulation in various species. High-throughput genomic assays have allowed reading and writing epigenomes with heightened resolution, facilitating the characterization and functional interrogation of plant epigenome variations. These tools reveal the consequences of environmental cues on gene expression regulation, potentially triggering phenotypic alterations. The reversibility of epigenetic changes makes manipulation a promising strategy for breeding and therapeutic applications (Shin et al. 2022). Owing to its heritability, modifications in DNA methylation patterns offer epigenetic inheritance in various traits, advancing genome annotations and helping discover regulatory components underlying key traits. Epigenomic resources, extending beyond transcript variations, serve as prospective sources of variation for improving specific traits in different crop species.

Although sorghum epigenetic research lags behind other crops like rice and maize, comparative analyses have identified homologous genes associated with methylation regulatory enzymes in sorghum (Vafadarshamasbi et al. 2022). Under limited phosphorus availability conditions, sorghum exhibited significant modifications in DNA methylation profiles, uncovering unique shifts related to candidate genes and regulatory motifs influencing root structure architecture. The sorghum methylome, once better understood, holds the potential to enhance our knowledge of the molecular basis of traits and improve sorghum performance through epigenome-wide prediction and association studies.

14.8 Hormonomics

Hormonomics represents an approach that concurrently monitors multiple regulated plant hormones, providing dynamic and critical data on the physiological and developmental environments of plant cells at the organ level. Plant hormones, essential components in metabolic and signaling networks, encompass well-known entities such as auxins, cytokinins, gibberellins, salicylic acid, abscisic acid, jasmonates, brassinosteroids, strigolactones, and ethylene (Waters et al. 2017). Genomic analyses have revealed that allelic variants in genes regulate a myriad of biological phenomena, encompassing growth, development, production, and responses to various stresses (Gilding et al. 2013; Kawa et al. 2021).

In sorghum, hormones serve as stress relievers, with cytokinin contributing to stress alleviation through increased biosynthesis (Markovich et al. 2017). Advanced methods, including artificial intelligence (AI), high-throughput analysis, and gene editing, are making significant contributions to hormonomics. AI-based artificial neural networks and decision tree methods have been developed to facilitate hormone activities, requiring high-throughput data for model training (Aasim et al. 2023). The generation and analysis of high-throughput quantitative data for measuring phytohormones in sorghum are actively practiced to explore chemical diversity and hormone sensitivity (Sheflin et al. 2019b). CRISPR/Cas technology is also being leveraged not just for delivering gene-modified components into the plant cells, but also for regenerating lost functional activity from the genetically engineered cells through plant tissue culture technics (Liu et al. 2019; Che et al. 2022; Elkonin et al. 2023).

Plant hormones can exist in inactive forms, conjugated with several metabolites like sugars and amino acids, and can detach to become activated. Generating and interpreting data for such dynamic stages aids in identifying valuable biomarkers that monitor the developmental and physiological phases of plants under unstable conditions (Hirayama and Mochida 2023).

Hormonomics studies contribute to the concurrent analysis of several plant hormones, exploring plant-microbe interactions, investigating hormonomic methodologies, implementing AI-based real-time plant hormone monitoring, and facilitating gene discovery and allele mining.

14.9 QTL-omics

The comprehensive analysis of structural and functional gene characterizations underlying quantitative traits fall under the purview of QTL-omics. This expansive dataset, facilitated by QTLomics, has been instrumental in mapping and characterizing QTLs in sorghum (El Mannai et al. 2012; Pandian et al. 2020; Kumar et al. 2023). The availability of a high-quality genome assembly, enriched with well-annotated genome sequences and resequencing data, serves as a valuable resource for the development of numerous genetic markers, subsequently utilized in QTL mapping and molecular breeding programs. Various marker assays, such as SSR (Baggett et al. 2021) and SNP markers (Burow et al. 2019; Senthilvel et al. 2019; Enyew et al. 2022), including the KASP assay (Burow et al. 2019), Infinium BeadChip (Senthilvel et al. 2019; Prince et al. 2020), and Axiom (O'Connell et al. 2022), have been developed in sorghum. With the advent of emerging sequencing technologies, cost-effective and efficient sequencing-based genotyping approaches, such as RADseq (restriction site associated DNA sequencing), GBS (genotyping by sequencing), and WGRS (whole-genome resequencing), have also been developed (Christensen et al. 2015).

QTL mapping using biparental mapping populations, with low-density markers, tends to yield a lower resolution of QTLs compared to GWAS, which leverage high-density markers for enhanced mapping resolution (Tao et al. 2020). Innovative approaches, such as combining GWAS with GBS, multiparent advanced generation intercross (MAGIC) populations, and trait mapping by pooled sequencing techniques like BSR-Seq, Indel-Seq, Seq-BSA, and QTL-Seq, are being employed (Pandey et al. 2016). Systematic integration of various omics resources developed in sorghum holds the potential to expedite crop improvement.

There are several factors influence QTLomics, including genetic recombination, linked QTLs, QTL allele effects, phenotypic precision, genotyping inaccuracies, and environmental cues (Francia et al. 2005; Collard et al. 2005). Furthermore, populations selected from diverse gene pools may exhibit varying expression levels in different environments, and recombination between QTL-associated genes may modify the introgressed segments. The primary challenge for QTLomics lies in integrating data from whole-genome sequencing, RNA sequencing, epigenomics, and phenotypic evaluation to achieve more accurate outcomes while addressing genotyping errors and preventing the overestimation of QTL effects (Kumawat et al. 2016).

14.10 Bioinformatics

The objective of plant "omics" is to unravel biological processes in plants at the genetic and molecular levels, facilitating the efficient utilization of plants as biological resources for crop improvement. A pivotal tool in this pursuit is genomics, which plays a crucial role in identifying genes, understanding gene functions, and deciphering their responses to environmental stimuli. Gene expression data, coupled with informatics tools, can provide insights into plant health and trait value, aiding in decision-making of breeding management systems. The burgeoning "omic" programs spawn extensive datasets necessitating storage, processing, and dissemination within the research community. This data encompasses sequence information, markers, maps, mutations, structures, functions, pathways, and more. Plant bioinformatics emerges as a vital discipline, managing, curating, processing, mining, and presenting key information meaningfully to the scientific community.

Genomic information has myriad applications, prompting the integration of "omics" data into large-scale functional genomics projects. A comprehensive understanding of systems biology facilitates modelling changes in gene expression, networks, protein interactions, and exploration of metabolite pools within a given sample. Bioinformatics, as an interdisciplinary field merging biotechnology, computer science, and information technology, has rapidly progressed in managing genomic, transcriptomic, and proteomic datasets. During the genomic revolution's early stages, a primary challenge for bioinformatics was the storage, management, and retrieval of massive genomic sequence data. Over time, computational "omic" resources and large databases were developed to meet this challenge.

Bioinformatics actively engages in the analysis and interpretation of various biological data types, contributing to tool development, algorithm creation, and statistical modules for handling large datasets. In the context of sorghum, bioinformatics efforts have been pivotal in genome and pangenome assembly, providing additional genetic information. Databases like Phytozome (Goodstein et al. 2012) and the National Center for Biotechnology Information (NCBI) (Sayers et al. 2022) have played a crucial role in making sorghum genomic resources accessible. These resources have facilitated the identification of genes associated with different traits in sorghum. Table 14.1 (adapted from Ngara et al. 2021) presents a comprehensive list of tools available for various "omics" areas applicable in crop breeding programs.

14.11 Artificial Intelligence and Machine Learning

The conventional breeding processes, traditionally time-consuming and labor-intensive, are now undergoing a revolutionary transformation through the integration of AI to expedite plant breeding cycles. This acceleration, known as speed breeding, involves optimizing crop growth in controlled climate conditions. The synergy of AI tools with marker-assisted selection and gene editing facilitates the selection and manipulation of crops with desirable traits, thereby enhancing crop improvement (Rai 2022). AI tools, when integrated with complex omics data, have the potential to simulate biological systems. This enables the prediction of plant responses to varying climate conditions, thereby informing future breeding cycles. Speed breeding techniques have already been successfully employed in various crops, including *Glycine max* L (Pierre Roumet 1997), sorghum (Forster 2014), rice (Collard et al. 2017), *Pisum sativum* (Mobini and Warkentin 2016), and more (Rai 2022).

Ongoing developments aim to further incorporate AI into crop improvement. Advanced computational and bioinformatics tools, coupled with applications of AI and machine learning, facilitate deep learning of biological data for a comprehensive understanding. This includes crop prediction modeling for genomic selection (Fig. 14.2). The inherently nonlinear and nondeterministic nature of genotype, environment, and interaction data makes classical statistical methods less efficient. Nonlinear machine learning techniques, such as neural networks, partial least square regression, and support vector machines, prove more adept at handling large and complex datasets in plant breeding. Machine learning applications span various areas, from classifying plant genotypes with molecular markers to modeling and predicting quantitative plant characteristics and interpreting complex relationships among plant traits.

Machine learning algorithms are instrumental in efficiently processing high-throughput phenotyping data, enabling accurate research outcomes. Numerous statistical and machine learning models, including rrBLUP, BayesA/B, RKHS, and random forest, have been developed for genomic selection (Heslot et al. 2012; Yu et al. 2016; Dos Santos et al. 2020). The integration of AI components with crop improvement extends to machine learning, nanotechnology, biosensors like lidar, and statistical data analysis packages, enabling precise trait assessments for crop improvement programs.

Table 14.1 Databases and bioinformatics tools supporting the sorghum "omics" research (adapted from (Ngara et al. 2021)

Database	Website	Features
Genomic resources/tools		
NCBI	https://ncbi.nlm.nih.gov	Plant genomics resource
Phytozome	https://phytozome.jgi.doe.gov	Plant genomics resource
Gramene	https://www.gramene.org	Grass genomics resource
GreenPhylDB	http://www.greenphyl.org	Comparative and functional genomics in plants
PGSB PlantsDB	http://pgsb.helmholtz-muenchen.de/plant/plantsdb.jsp	Comparative genomics in plants
PlantGDB	http://www.plantgdb.org	Tools and resources for plant genomics
SorghumFDB	http://structuralbiology.cau.edu.cn/sorghum/index.html	Sorghum functional genomics with network analysis
SorGSD	http://sorgsd.big.ac.cn	Sorghum SNP data
DNApod	http://tga.nig.ac.jp/dnapod	Genome-wide DNA polymorphism datasets of plants
Transcriptomic resources		
Grassius	https://grassius.org/links.php	Grass transcription factors and gene promoters
GreenCircRNA	http://greencirc.cn	Plant circular RNAs (circRNAs)
KAAS	www.genome.jp/tools/kaas/	Pathway enrichment analyses of transcripts to classify spatial and temporal pathways
miRbase	http://www.mirbase.org	Plant microRNA (miRNA) data
Morokoshi transcriptome database	http://matsui-lab.riken.jp/morokoshi/Home.html	Sorghum transcriptome data
PlantTFDB	http://planttfdb.gao-lab.org/index.php	Plant transcription factors
PSRN	http://syslab5.nchu.edu.tw	Plant stress-specific transcriptome data
psRNATarget	http://plantgrn.noble.org/psRNATarget/	Potential miRNA sorghum target predictions
PtRFdb	http://www.nipgr.ac.in	Plant transfer RNA-derived fragments (tRFs) data
psRobot; Plant small RNA analysis toolbox	http://omicslab.genetics.ac.cn/psRobot/	Potential miRNA sorghum target predictions
UEA sRNA workbench	http://srna-workbench.cmp.uea.ac.uk	Various smallRNA (sRNA) tools
Proteomic resources		
CropPal	https://crop-pal.org	Protein subcellular location
ProtAnnDB	http://www.polebio.lrsv.ups-tlse.fr/ProtAnnDB/	Protein annotation
ExPASy	https://www.expasy.org/	Bioinformatics resources for proteomics

(continued)

Table 14.1 (continued)

Database	Website	Features
Uniprot	https://www.uniprot.org	Protein sequence and functional information
Gene ontology/metabolic pathways		
AgriGo	http://bioinfo.cau.edu.cn/agriGO/	Gene ontology of plant and agricultural species
KEGG	https://www.genome.jp/kegg/	Gene functional information
SorghumCyc	http://pathway.gramene.org/gramene/sorghumcyc.shtml	Metabolic pathways in sorghum

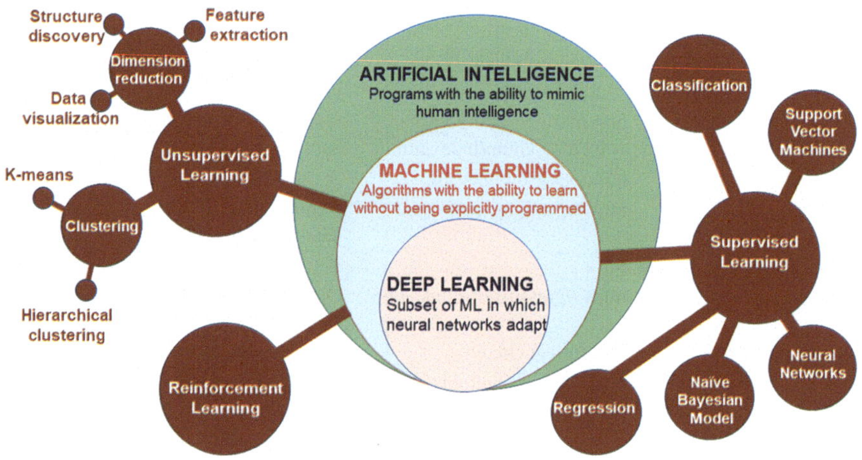

Fig. 14.2 Artificial intelligence methods and submethodologies

In addition to omics approaches, genome editing tools such as RNAi, CRISPR/Cas9, and TALENs are revolutionizing crop improvement. In sorghum plant breeding, advancements include the development of a sorghum panicle detection and counting pipeline was designed by using unmanned aerial system pictures, combining image segmentation and processing through convolutional neural networks (Lin and Guo 2020). The US Department of Agriculture and the Edward Buckler team are utilizing machine learning to detect climate adaptations of 1000 species, including maize, sorghum, and sugar cane (Beans 2020).

14.12 Conclusion

In the face of an impending global population surge and shifting climatic conditions, sorghum emerges as a pivotal crop for future agricultural sustainability. Recent advancements in the genetic and genomic stocks of sorghum have positioned it alongside major cereals like wheat, rice, and maize. However, the imperative to enhance sorghum characteristics, encompassing yield (both grain and fodder), quality (lysin, amylopectin, iron, zinc, and sugar content), and traits demanding substantial inputs (such as resistance to biotic and abiotic stresses, nutrient utilization, and water use efficiency), is critical for ensuring sustainable sorghum cultivation. To propel sorghum breeding programs toward developing cultivars that are both nutritionally enriched and climate-resilient, a holistic approach is essential. Integrating the identification of genes and genomic regions associated with pivotal agronomic, adaptive, and quality traits with high-throughput omics technologies, advanced phenotyping tools, and genomics-assisted breeding strategies is imperative. This integrated approach is crucial for discerning genetic variability, superior alleles, and their regulatory networks. By harmonizing these diverse elements, sorghum breeding programs can be accelerated, fostering the rapid development of cultivars that not only meet nutritional demands but also exhibit resilience to the challenges posed by a changing climate. This concerted effort is indispensable for securing the future sustainability of sorghum cultivation.

References

Aasim M, Ali SA, Altaf MT, Ali A, Nadeem MA, Baloch FS (2023) Artificial neural network and decision tree facilitated prediction and validation of cytokinin-auxin induced in vitro organogenesis of sorghum (Sorghum bicolor L.). Plant Cell Tissue Organ Cult 153:611–624

Ahmad P, Abdel Latef AAH, Rasool S, Akram NA, Ashraf M, Gucel S (2016) Role of proteomics in crop stress tolerance. Front Plant Sci 7:1336. https://doi.org/10.3389/fpls.2016.01336

Ashburner M, Ball CA, Blake JA, Botstein D, Butler H, Cherry JM et al (2000) Gene ontology: tool for the unification of biology. Nat Genet 25:75556. https://doi.org/10.1038/75556

Baggett JP, Tillett RL, Cooper EA, Yerka MK (2021) De novo identification and targeted sequencing of SSRs efficiently fingerprints Sorghum bicolor sub-population identity. PLoS One 16:248213. https://doi.org/10.1371/journal.pone.0248213

Barkla BJ, Vera-Estrella R, Pantoja O (2013) Progress and challenges for abiotic stress proteomics of crop plants. Proteomics 13:401. https://doi.org/10.1002/pmic.201200401

Batey I (2017) The diversity of uses for cereal grains. In: Cereal grains: assessing and managing quality, 2nd edn. Elsevier, Amsterdam. https://doi.org/10.1016/B978-0-08-100719-8.00003-6

Beans C (2020) Crop researchers harness artificial intelligence to breed crops for the changing climate. Proc Natl Acad Sci USA 117:27066–27069. https://doi.org/10.1073/pnas.2018732117

Bino RJ, Hall RD, Fiehn O, Kopka J, Saito K, Draper J et al (2004) Potential of metabolomics as a functional genomics tool. Trends Plant Sci 9:418–425. https://doi.org/10.1016/j.tplants.2004.07.004

Blackstock WP, Weir MP (1999) Proteomics: quantitative and physical mapping of cellular proteins. Trends Biotechnol 17:121–127. https://doi.org/10.1016/S0167-7799(98)01245-1

Bock C, Tomazou EM, Brinkman A, Müller F (2010) Genome-wide mapping of DNA methylation: a quantitative technology comparison. Nat Biotechnol 28:1106–1114

Bruce TJA (2012) GM as a route for delivery of sustainable crop protection. J Exp Bot 63:537–541. https://doi.org/10.1093/jxb/err281

Burow G, Chopra R, Hughes H, Xin Z, Burke J (2019) Marker assisted selection in sorghum using kasp assay for the detection of single nucleotide polymorphism/insertion deletion. In: Methods in molecular biology. Springer, Cham. https://doi.org/10.1007/978-1-4939-9039-9_6

Butland G, Peregrin-Alvarez JM, Li J, Yang W, Yang X, Canadien V et al (2005) Interaction network containing conserved and essential protein complexes in Escherichia coli. Nature 433:531–537. https://doi.org/10.1038/nature03239

Capozzi F, Bordoni A (2013) Foodomics: a new comprehensive approach to food and nutrition. Genes Nutr 8:1–4. https://doi.org/10.1007/s12263-012-0310-x

Che P, Wu E, Simon MK, Anand A, Lowe K, Gao H et al (2022) Wuschel2 enables highly efficient CRISPR/Cas-targeted genome editing during rapid de novo shoot regeneration in sorghum. Commun Biol 5:344. https://doi.org/10.1038/s42003-022-03308-w

Chen Y, Zhang Y, Wang H, Sun J, Ma L, Miao F et al (2022) A high-quality genome assembly of Sorghum dochna. Front Genet 13:844385. https://doi.org/10.3389/fgene.2022.844385

Cheng Y, Weng J, Joshi CP, Nguyen HT (1993) Dehydration stress-induced changes in translatable RNAS in sorghum. Crop Sci 33:1397–1400. https://doi.org/10.2135/cropsci1993.0011183X003300060053x

Christensen KD, Dukhovny D, Siebert U, Green RC (2015) Assessing the costs and cost-effectiveness of genomic sequencing. J Pers Med 5. https://doi.org/10.3390/jpm5040470

Cokus SJ, Feng S, Zhang X, Chen Z, Merriman B, Haudenschild CD et al (2008) Shotgun bisulphite sequencing of the Arabidopsis genome reveals DNA methylation patterning. Nature 452. https://doi.org/10.1038/nature06745

Collard BCY, Jahufer MZZ, Brouwer JB, Pang ECK (2005) An introduction to markers, quantitative trait loci (QTL) mapping and marker-assisted selection for crop improvement: the basic concepts. Euphytica 142. https://doi.org/10.1007/s10681-005-1681-5

Collard BCY, Beredo JC, Lenaerts B, Mendoza R, Santelices R, Lopena V et al (2017) Revisiting rice breeding methods – evaluating the use of rapid generation advance (RGA) for routine rice breeding. Plant Prod Sci 20:337–352. https://doi.org/10.1080/1343943X.2017.1391705

Cooper EA, Brenton ZW, Flinn BS, Jenkins J, Shu S, Flowers D et al (2019) A new reference genome for Sorghum bicolor reveals high levels of sequence similarity between sweet and grain genotypes: implications for the genetics of sugar metabolism. BMC Genomics 20:420. https://doi.org/10.1186/s12864-019-5734-x

Deschamps S, Zhang Y, Llaca V, Ye L, Sanyal A, King M et al (2018) A chromosome-scale assembly of the sorghum genome using nanopore sequencing and optical mapping. Nat Commun 9. https://doi.org/10.1038/s41467-018-07271-1

Dillon SL, Shapter FM, Henry RJ, Cordeiro G, Izquierdo L, Lee LS (2007) Domestication to crop improvement: genetic resources for sorghum and saccharum (andropogoneae). Ann Bot 100. https://doi.org/10.1093/aob/mcm192

Dos Santos JPR, Fernandes SB, McCoy S, Lozano R, Brown PJ, Leakey ADB et al (2020) Novel Bayesian networks for genomic prediction of developmental traits in biomass sorghum. G3 Genes Genomes Genet 10. https://doi.org/10.1534/g3.119.400759

El Mannai Y, Shehzad T, Okuno K (2012) Mapping of QTLs underlying flowering time in sorghum [sorghum bicolor (L.) Moench]. Breed Sci 62. https://doi.org/10.1270/jsbbs.62.151

Elkonin LA, Gerashchenkov GA, Borisenko NV, Kenzhegulov OA, Sarsenova K, Rozhnova NA et al (2023) Development of sorghum mutants with improved in vitro protein digestibility by CRISPR/Cas9 editing of kafirin genes. Crop J. https://doi.org/10.1016/j.cj.2023.02.005

Enyew M, Feyissa T, Carlsson AS, Tesfaye K, Hammenhag C, Seyoum A et al (2022) Genome-wide analyses using multi-locus models revealed marker-trait associations for major agronomic traits in Sorghum bicolor. Front Plant Sci 13. https://doi.org/10.3389/fpls.2022.999692

Fernie AR, Schauer N (2009) Metabolomics-assisted breeding: a viable option for crop improvement? Trends Genet 25. https://doi.org/10.1016/j.tig.2008.10.010

Fernie AR, Yan J (2019) De novo domestication: an alternative route toward new crops for the future. Mol Plant 12. https://doi.org/10.1016/j.molp.2019.03.016

Feussner I, Polle A (2015) What the transcriptome does not tell - proteomics and metabolomics are closer to the plants' patho-phenotype. Curr Opin Plant Biol 26. https://doi.org/10.1016/j.pbi.2015.05.023

Forster BP (2014) Accelerated plant breeding. CABI Rev. https://doi.org/10.1079/PAVSNNR20149043

Francia E, Tacconi G, Crosatti C, Barabaschi D, Bulgarelli D, Dallaglio E et al (2005) Marker assisted selection in crop plants. Plant Cell Tissue Organ Cult 82. https://doi.org/10.1007/s11240-005-2387-z

Ganie AH, Ahmad A, Pandey R, Aref IM, Yousuf PY, Ahmad S et al (2015) Metabolite profiling of low-P tolerant and low-P sensitive maize genotypes under phosphorus starvation and restoration conditions. PLoS One 10. https://doi.org/10.1371/journal.pone.0129520
Gelli M, Konda AR, Liu K, Zhang C, Clemente TE, Holding DR et al (2017) Validation of QTL mapping and transcriptome profiling for identification of candidate genes associated with nitrogen stress tolerance in sorghum. BMC Plant Biol 17. https://doi.org/10.1186/s12870-017-1064-9
Gilding EK, Frère CH, Cruickshank A, Rada AK, Prentis PJ, Mudge AM et al (2013) Allelic variation at a single gene increases food value in a drought-tolerant staple cereal. Nat Commun 4. https://doi.org/10.1038/ncomms2450
Goche T, Shargie NG, Cummins I, Brown AP, Chivasa S, Ngara R (2020) Comparative physiological and root proteome analyses of two sorghum varieties responding to water limitation. Sci Rep 10. https://doi.org/10.1038/s41598-020-68735-3
Goodstein DM, Shu S, Howson R, Neupane R, Hayes RD, Fazo J et al (2012) Phytozome: a comparative platform for green plant genomics. Nucleic Acids Res 40. https://doi.org/10.1093/nar/gkr944
Guleria P, Mahajan M, Bhardwaj J, Yadav SK (2011) Plant small RNAs: biogenesis, mode of action and their roles in abiotic stresses. Genomics Proteomics Bioinformatics 9. https://doi.org/10.1016/S1672-0229(11)60022-3
Habyarimana E, Lopez-Cruz M, Baloch FS (2020) Genomic selection for optimum index with dry biomass yield, dry mass fraction of fresh material, and plant height in biomass sorghum. Gene 11. https://doi.org/10.3390/genes11010061
Hamza NB, Sharma N, Tripathi A, Sanan-Mishra N (2016) MicroRNA expression profiles in response to drought stress in Sorghum bicolor. Gene Expr Patterns 20. https://doi.org/10.1016/j.gep.2016.01.001
Harlan JR, Wet JMJ (1972) A simplified classification of cultivated sorghum1. Crop Sci 12:172–176. https://doi.org/10.2135/CROPSCI1972.0011183X001200020005X
Heslot N, Yang HP, Sorrells ME, Jannink JL (2012) Genomic selection in plant breeding: a comparison of models. Crop Sci 52. https://doi.org/10.2135/cropsci2011.06.0297
Heuberger AL, Broeckling CD, Kirkpatrick KR, Prenni JE (2014) Application of nontargeted metabolite profiling to discover novel markers of quality traits in an advanced population of malting barley. Plant Biotechnol J 12. https://doi.org/10.1111/pbi.12122
Hirayama T, Mochida K (2023) Plant hormonomics: a key tool for deep physiological phenotyping to improve crop productivity. Plant Cell Physiol 63. https://doi.org/10.1093/pcp/pcac067
Jedmowski C, Ashoub A, Beckhaus T, Berberich T, Karas M, Brüggemann W (2014) Comparative analysis of sorghum bicolor proteome in response to drought stress and following recovery. Int J Proteom 2014. https://doi.org/10.1155/2014/395905
Jensen SE, Charles JR, Muleta K, Bradbury PJ, Casstevens T, Deshpande SP et al (2020) A sorghum practical haplotype graph facilitates genome-wide imputation and cost-effective genomic prediction. Plant Genome 13:e20009. https://doi.org/10.1002/tpg2.20009
Kawa D, Taylor T, Thiombiano B, Musa Z, Vahldick HE, Walmsley A et al (2021) Characterization of growth and development of sorghum genotypes with differential susceptibility to Striga hermonthica. J Exp Bot 72. https://doi.org/10.1093/jxb/erab380
Kebrom TH, Mullet JE (2016) Transcriptome profiling of tiller buds provides new insights into phyb regulation of tillering and indeterminate growth in Sorghum. Plant Physiol 170. https://doi.org/10.1104/pp.16.00014
Koh GCKW, Porras P, Aranda B, Hermjakob H, Orchard SE (2012) Analyzing protein-protein interaction networks. J Proteome Res 11. https://doi.org/10.1021/pr201211w
Kosová K, Vítámvás P, Prášil IT, Renaut J (2011) Plant proteome changes under abiotic stress - contribution of proteomics studies to understanding plant stress response. J Proteome 74. https://doi.org/10.1016/j.jprot.2011.02.006
Kumar S (2018) Epigenomics of plant responses to environmental stress. Epigenomes 2. https://doi.org/10.3390/epigenomes2010006
Kumar N, Boatwright JL, Brenton ZW, Sapkota S, Ballén-Taborda C, Myers MT et al (2023) Development and characterization of a sorghum multi-parent advanced generation intercross

(MAGIC) population for capturing diversity among seed parent gene pool. G3 Genes Genomes Genet. https://doi.org/10.1093/g3journal/jkad037
Kumawat G, Gupta S, Ratnaparkhe MB, Maranna S, Satpute GK (2016) QTLomics in soybean: a way forward for translational genomics and breeding. Front Plant Sci 7. https://doi.org/10.3389/fpls.2016.01852
Kusano M, Yang Z, Okazaki Y, Nakabayashi R, Fukushima A, Saito K (2015) Using metabolomic approaches to explore chemical diversity in rice. Mol Plant 8. https://doi.org/10.1016/j.molp.2014.11.010
Li H, Li Y, Ke Q, Kwak SS, Zhang S, Deng X (2020) Physiological and differential proteomic analyses of imitation drought stress response in sorghum bicolor root at the seedling stage. Int J Mol Sci 21. https://doi.org/10.3390/ijms21239174
Lin Z, Guo W (2020) Sorghum panicle detection and counting using unmanned aerial system images and deep learning. Front Plant Sci 11. https://doi.org/10.3389/fpls.2020.534853
Liu G, Li J, Godwin ID (2019) Genome editing by CRISPR/Cas9 in sorghum through biolistic bombardment. In: Methods in molecular biology. Springer, Cham. https://doi.org/10.1007/978-1-4939-9039-9_12
Lowe R, Shirley N, Bleackley M, Dolan S, Shafee T (2017) Transcriptomics technologies. PLoS Comput Biol 13. https://doi.org/10.1371/journal.pcbi.1005457
Mandrone M, Chiocchio I, Barbanti L, Tomasi P, Tacchini M, Poli F (2021) Metabolomic study of sorghum (Sorghum bicolor) to interpret plant behavior under variable field conditions in view of smart agriculture applications. J Agric Food Chem 69. https://doi.org/10.1021/acs.jafc.0c06533
Markovich O, Steiner E, Kouřil Š, Tarkowski P, Aharoni A, Elbaum R (2017) Silicon promotes cytokinin biosynthesis and delays senescence in arabidopsis and sorghum. Plant Cell Environ 40. https://doi.org/10.1111/pce.12913
Marla SR, Shiva S, Welti R, Liu S, Burke JJ, Morris GP (2017) Comparative transcriptome and lipidome analyses reveal molecular chilling responses in chilling-tolerant sorghums. Plant Genome 10. https://doi.org/10.3835/plantgenome2017.03.0025
McCormick RF, Truong SK, Sreedasyam A, Jenkins J, Shu S, Sims D et al (2018) The sorghum bicolor reference genome: improved assembly, gene annotations, a transcriptome atlas, and signatures of genome organization. Plant J. https://doi.org/10.1111/tpj.13781
McKinley B, Rooney W, Wilkerson C, Mullet J (2016) Dynamics of biomass partitioning, stem gene expression, cell wall biosynthesis, and sucrose accumulation during development of sorghum bicolor. Plant J 88. https://doi.org/10.1111/tpj.13269
Meuwissen THE, Hayes BJ, Goddard ME (2001) Prediction of total genetic value using genome-wide dense marker maps. Genetics 157. https://doi.org/10.1093/genetics/157.4.1819
Mobini SH, Warkentin TD (2016) A simple and efficient method of in vivo rapid generation technology in pea (Pisum sativum L.). In Vitro Cell Dev Biol Plant 52:530–536. https://doi.org/10.1007/s11627-016-9772-7
Monteoliva L, Albar JP (2004) Differential proteomics: an overview of gel and non-gel based approaches. Brief Funct Genomic Proteomic 3. https://doi.org/10.1093/bfgp/3.3.220
Moose SP, Mumm RH (2008) Molecular plant breeding as the foundation for 21st century crop improvement. Plant Physiol 147. https://doi.org/10.1104/pp.108.118232
Morris GP, Ramu P, Deshpande SP, Hash CT, Shah T, Upadhyaya HD et al (2013) Population genomic and genome-wide association studies of agroclimatic traits in sorghum. Proc Natl Acad Sci USA. https://doi.org/10.1073/pnas.1215985110
Morrison J, Koeman JM, Johnson BK, Foy KK, Beddows I, Zhou W et al (2021) Evaluation of whole-genome DNA methylation sequencing library preparation protocols. Epigenetics Chromatin 14. https://doi.org/10.1186/s13072-021-00401-y
Nayak SN, Aravind B, Malavalli SS, Sukanth BS, Poornima R, Bharati P et al (2021) Omics technologies to enhance plant based functional foods: an overview. Front Genet 12. https://doi.org/10.3389/fgene.2021.742095
Ngara R, Goche T, Swanevelder DZH, Chivasa S (2021) Sorghum's whole-plant transcriptome and proteome responses to drought stress: a review. Life 11. https://doi.org/10.3390/life11070704

O'Connell A, Deo J, Deomano E, Wei X, Jackson P, Aitken KS et al (2022) Combining genomic selection with genome-wide association analysis identified a large-effect QTL and improved selection for red rot resistance in sugarcane. Front Plant Sci 13. https://doi.org/10.3389/fpls.2022.1021182

Paine JA, Shipton CA, Chaggar S, Howells RM, Kennedy MJ, Vernon G et al (2005) Improving the nutritional value of golden rice through increased pro-vitamin A content. Nat Biotechnol 23. https://doi.org/10.1038/nbt1082

Pandey MK, Roorkiwal M, Singh VK, Ramalingam A, Kudapa H, Thudi M et al (2016) Emerging genomic tools for legume breeding: current status and future prospects. Front Plant Sci 7. https://doi.org/10.3389/fpls.2016.00455

Pandian BA, Varanasi A, Vennapusa AR, Sathishraj R, Lin G, Zhao M et al (2020) Characterization, genetic analyses, and identification of QTLs conferring metabolic resistance to a 4-hydroxyphenylpyruvate dioxygenase inhibitor in sorghum (sorghum bicolor). Front Plant Sci 11. https://doi.org/10.3389/fpls.2020.596581

Pazhamala LT, Kudapa H, Weckwerth W, Millar AH, Varshney RK (2021) Systems biology for crop improvement. Plant Genome 14. https://doi.org/10.1002/tpg2.20098

Pierre Roumet FM (1997) Germination of immature soybean seeds to shorten reproductive cycle duration. Crop Sci 37:521–525. https://doi.org/10.2135/cropsci1997.0011183X003700020035x

Prince SJ, Vuong TD, Wu X, Bai Y, Lu F, Kumpatla SP et al (2020) Mapping quantitative trait loci for soybean seedling shoot and root architecture traits in an inter-specific genetic population. Front Plant Sci 11. https://doi.org/10.3389/fpls.2020.01284

Rai KK (2022) Integrating speed breeding with artificial intelligence for developing climate-smart crops. Mol Biol Rep 49:11385–11402. https://doi.org/10.1007/s11033-022-07769-4

Ruperao P, Thirunavukkarasu N, Gandham P, Selvanayagam S, Govindaraj M, Nebie B et al (2021) Sorghum pan-genome explores the functional utility for genomic-assisted breeding to accelerate the genetic gain. Front Plant Sci 12. https://doi.org/10.3389/FPLS.2021.666342

Salwinski L, Miller CS, Smith AJ, Pettit FK, Bowie JU, Eisenberg D (2004) The database of interacting proteins: 2004 update. Nucleic Acids Res 32. https://doi.org/10.1093/nar/gkh086

Sayers EW, Bolton EE, Brister JR, Canese K, Chan J, Comeau DC et al (2022) Database resources of the national center for biotechnology information. Nucleic Acids Res 50. https://doi.org/10.1093/nar/gkab1112

Scandiani MM, Luque AG, Razori MV, Casalini LC, Aoki T, O'Donnell K et al (2015) Metabolic profiles of soybean roots during early stages of Fusarium tucumaniae infection. J Exp Bot 66. https://doi.org/10.1093/jxb/eru432

Senthilvel S, Ghosh A, Shaik M, Shaw RK, Bagali PG (2019) Development and validation of an SNP genotyping array and construction of a high-density linkage map in castor. Sci Rep 9. https://doi.org/10.1038/s41598-019-39967-9

Sheflin AM, Chiniquy D, Yuan C, Goren E, Kumar I, Braud M et al (2019a) Metabolomics of sorghum roots during nitrogen stress reveals compromised metabolic capacity for salicylic acid biosynthesis. Plant Direct 3. https://doi.org/10.1002/pld3.122

Sheflin AM, Kirkwood JS, Wolfe LM, Jahn CE, Broeckling CD, Schachtman DP et al (2019b) High-throughput quantitative analysis of phytohormones in sorghum leaf and root tissue by ultra-performance liquid chromatography-mass spectrometry. Anal Bioanal Chem 411. https://doi.org/10.1007/s00216-019-01658-9

Shin H, Choi WL, Lim JY, Huh JH (2022) Epigenome editing: targeted manipulation of epigenetic modifications in plants. Genes Genom 44:307–315. https://doi.org/10.1007/s13258-021-01199-5

Szymański M, Barciszewska MZ, Zywicki M, Barciszewski J (2003) Noncoding RNA transcripts. J Appl Genet 44:1–19

Tao Y, Zhao X, Wang X, Hathorn A, Hunt C, Cruickshank AW et al (2020) Large-scale GWAS in sorghum reveals common genetic control of grain size among cereals. Plant Biotechnol J 18. https://doi.org/10.1111/pbi.13284

Tian J, Bryksa BC, Yada RY (2016) Feeding the world into the future – food and nutrition security: the role of food science and technology. Front Life Sci 9. https://doi.org/10.1080/21553769.2016.1174958

Turner MF, Heuberger AL, Kirkwood JS, Collins CC, Wolfrum EJ, Broeckling CD et al (2016) Non-targeted metabolomics in diverse sorghum breeding lines indicates primary and secondary metabolite profiles are associated with plant biomass accumulation and photosynthesis. Front Plant Sci 7. https://doi.org/10.3389/fpls.2016.00953

Uetz P, Glot L, Cagney G, Mansfield TA, Judson RS, Knight JR et al (2000) A comprehensive analysis of protein-protein interactions in Saccharomyces cerevisiae. Nature 403. https://doi.org/10.1038/35001009

Upadhyaya HD, Pundir RPS, Dwivedi SL, Gowda CLL, Reddy VG, Singh S (2009) Developing a mini core collection of sorghum for diversified utilization of germplasm. Crop Sci 49. https://doi.org/10.2135/cropsci2009.01.0014

Vafadarshamasbi U, Mace E, Jordan D, Crisp PA (2022) Decoding the sorghum methylome: understanding epigenetic contributions to agronomic traits. Biochem Soc Trans 50. https://doi.org/10.1042/BST20210908

Varoquaux N, Cole B, Gao C, Pierroz G, Baker CR, Patel D et al (2019) Transcriptomic analysis of field-droughted sorghum from seedling to maturity reveals biotic and metabolic responses. Proc Natl Acad Sci USA 116:27124–27132. https://doi.org/10.1073/pnas.1907500116

Varshney RK, Graner A, Sorrells ME (2005) Genomics-assisted breeding for crop improvement. Trends Plant Sci 10. https://doi.org/10.1016/j.tplants.2005.10.004

Varshney RK, Nayak SN, May GD, Jackson SA (2009) Next-generation sequencing technologies and their implications for crop genetics and breeding. Trends Biotechnol 27. https://doi.org/10.1016/j.tibtech.2009.05.006

Wang P, Khoshravesh R, Karki S, Tapia R, Balahadia CP, Bandyopadhyay A et al (2017) Re-creation of a key step in the evolutionary switch from C3 to C4 leaf anatomy. Curr Biol 27. https://doi.org/10.1016/j.cub.2017.09.040

Waters MT, Gutjahr C, Bennett T, Nelson DC (2017) Strigolactone signaling and evolution. Annu Rev Plant Biol 68. https://doi.org/10.1146/annurev-arplant-042916-040925

Wood AJ, Goldsbrough PB (1997) Characterization and expression of dehydrins in water-stressed sorghum bicolor. Physiol Plant 99. https://doi.org/10.1034/j.1399-3054.1997.990120.x

Yilmaz M, Paulic M, Seidel T (2022) Interactome of arabidopsis thaliana. Plants 11. https://doi.org/10.3390/plants11030350

Yu X, Li X, Guo T, Zhu C, Wu Y, Mitchell SE et al (2016) Genomic prediction contributing to a promising global strategy to turbocharge gene banks. Nat Plants 2. https://doi.org/10.1038/nplants.2016.150

Zhang Y, Swart C, Alseekh S, Scossa F, Jiang L, Obata T et al (2018) The extra-pathway interactome of the TCA cycle: expected and unexpected metabolic interactions. Plant Physiol 177. https://doi.org/10.1104/pp.17.01687

New Breeding Trends in Sorghum

15

Dinakaran Elango, Wanyan Wang, Neethu Francis, Debamalya Chatterjee, Ann Murithi, Visalakshi Chandra, Theivasigamani Parthasarathi, Einstein Mariya David, Vanitha Jayaraman, Kamaleeswari Govindarajan, Bonti Gogoi, Somashekhar Punnuri, Mahendar Thudi, Mahalingam Govindaraj, Ashok Kumar Are, Yinping Jiao, and Surinder Chopra

D. Elango (✉) · A. Murithi
Department of Agronomy, Iowa State University, Ames, IA, USA
e-mail: delango@iastate.edu

W. Wang
Ecosystem Science and Management, Penn State University, University Park, PA, USA

N. Francis
Department of Millets, Tamil Nadu Agricultural University, Coimbatore, India

D. Chatterjee · S. Chopra
Department of Plant Science, Penn State University, University Park, PA, USA

V. Chandra
ICAR-Central Tuber Crops Research Institute, Trivandrum, Kerala, India

T. Parthasarathi · E. M. David
VIT School of Agricultural Innovations and Advanced Learning (VAIAL), Vellore Institute of Technology, Vellore, India

V. Jayaraman · K. Govindarajan
Faculty of Agriculture, SRM Institute of Science and Technology, Chennai, India

B. Gogoi
Assam Agricultural University, Jorhat, Assam, India

S. Punnuri · M. Thudi
Agriculture Research Station, Fort Valley State University, Fort Valley, GA, USA

M. Govindaraj
Harvest Plus, Alliance CIAT-Bioversity International, Hyderabad, India

A. K. Are
International Crops Research Institute for the Semi-Arid Tropics, Hyderabad, India

Y. Jiao
Department of Plant and Soil Science, Institute of Genomics for Crop Abiotic Stress Tolerance (IGCAST), Texas Tech University, Lubbock, TX, USA

E. Habyarimana et al. (eds.), *Omics and Biotechnological Approaches for Product Profile-Driven Sorghum Improvement*,
https://doi.org/10.1007/978-981-97-4347-6_15

Abstract

Sorghum, a versatile cereal crop, serves as a staple food for millions and is increasingly recognized for its resilience to climate adversities. This chapter provides an encompassing overview of the current status of sorghum breeding, discussing the achievements, challenges, and future prospects. Historically, breeding efforts have primarily targeted yield improvement, biotic and abiotic stress resistance, and quality enhancement. With the advent of molecular breeding tools, there has been a paradigm shift toward genomics-assisted breeding, which holds promise in accelerating the development of superior sorghum varieties. We also shed light on the integration of high-throughput phenotyping and advanced genomics, which are playing pivotal roles in understanding the genetic architecture of complex traits. The future of sorghum breeding is anticipated to be dominated by precision breeding, where the amalgamation of CRISPR/Cas9 genome editing, artificial intelligence, and digital agriculture will redefine the conventional breeding pipelines. Challenges persist, particularly in bridging the gap between research advancements and their translation to smallholder farmers' fields. It is imperative for the global community to foster collaborations, ensuring the benefits of advanced breeding technologies reach all, sustaining food security and agricultural sustainability in the face of a changing climate.

Keywords

Agricultural sustainability · AI · Digital agriculture · Food security · Genetic architecture · Genomics-assisted breeding · High-throughput phenotyping · Smallholder farmers

15.1 Introduction

Sorghum (*Sorghum bicolor* L. Moench) is a versatile crop ranked fifth in importance among the cereals, pivotal to the livelihoods of millions across the globe (Reddy et al. 2012a, b). With its origins rooted in the African continent (Harlan and de Wet 1971), it has since spread across varied geographies, thanks to its adaptability to diverse climatic conditions (de Wet 1978). Sorghum being the staple food for more than 500 million people, particularly in the arid and semi-arid tropics of Africa and Asia, remains an indispensable component in ensuring food security and nutrition (Rooney 1996; Elango et al. 2020b). Historically, sorghum has been cultivated for multiple purposes, ranging from food, feed, and fodder to alcoholic beverages (House 1985). Its resilience, attributed to the C_4 photosynthetic pathway and other plant characteristics including transpiration efficiency and advantageous root architecture (Djanaguiraman et al. 2023), allows it to efficiently use water and withstand high temperatures, making it an apt candidate for regions with challenging environmental conditions (Krishnamurthy et al. 2014; Elango 2020; Elango et al. 2020b). However, sorghums are very sensitive to low temperature stresses (Singh 1985; Krishnamurthy et al. 2014; Cloutier et al. 2021; Elango et al. 2022).

Breeding efforts in sorghum were primarily focused on increasing grain quality, yield, and resilience to biotic and abiotic stresses (House 1985; Reddy et al. 2012a). Conventional breeding techniques, which relied heavily on hybridization followed by phenotypic selection, laid the foundation for the development of several elite sorghum varieties. However, with the burgeoning global population, set to breach the 9 billion mark by 2050, there is an escalating demand to expedite breeding processes and enhance genetic gains. Emerging technological innovations over the past two decades have revolutionized the sphere of sorghum breeding. The complete sequencing of the sorghum genome in 2009 was a landmark achievement, unveiling a wealth of genetic information (Paterson et al. 2009). The advanced breeding line BTx623 released by Texas A&M University, extensively used in sorghum breeding and genomics research, was used for sequencing. Because of the compact genome (730 Mb), sorghum is a desirable model crop for functional genomic studies of grasses. Consequently, genomics-assisted breeding (GAB), which amalgamates molecular markers with traditional breeding, has been instrumental in accelerating the breeding cycles and enhancing the precision of selection.

Genome-wide association studies (GWAS) have been used extensively for the past several years to investigate the genetic architecture of complex human diseases. In recent times, rapid advancements in sequencing technologies leading to the widespread availability of large-scale SNP markers have made GWAS a more popular choice in crop improvement. Genomic regions behind traits associated with abiotic and biotic stresses like tolerance to aluminium stress (Caniato et al. 2011; Hufnagel et al. 2018; Too et al. 2018), stalk rot resistance (Adeyanju et al. 2015), epicuticular wax (EW) formation (Elango et al. 2020b), environmental adaptative traits (Lasky et al. 2015), and agroclimatic adaptative traits (Morris et al. 2013) were successfully delineated in sorghum through GWAS. Apart from these, genomic regions for grain quality (Sukumaran et al. 2012), seed size (Zhang et al. 2015), and grain polyphenol concentrations (Rhodes et al. 2014) were also identified. Over the years, sorghum research community around the globe has developed various mapping panels; notable panels are the Sorghum Association Panel (SAP) (Casa et al. 2008), the minicore panel from International Crops Research Institute for the Semi-Arid Tropics (ICRISAT) (Upadhyaya et al. 2009), Bio-energy Association Panel (BAP) (Brenton et al. 2016), Queensland diversity panel (Mace et al. 2019), Nigerian diversity population (Maina et al. 2018), world core collection (Grenier et al. 2001), nested association mapping (NAM) panel (Bouchet et al. 2017), and CIRAD (the French agricultural research and international cooperation organization) landraces (Deu et al. 2006), of which most of the panels were sequenced and the high-density sequence reads are available for public use (Boyles et al. 2019). Further, Jiao et al. (2016) developed the EMS (ethyl methanesulfonate) based mutant library in sorghum for characterizing mutations for various traits of interest. These panels and sequencing libraries were invaluable resources for improving sorghums for years to come.

Furthermore, the integration of high-throughput phenotyping platforms and advanced computational tools has empowered breeders to dissect the genetic architecture of complex traits more comprehensively (Watanabe et al. 2017; Ortiz et al.

2018; Wang et al. 2018; Zhou et al. 2019; Bao et al. 2019). While advancements in genomics and biotechnology offer a promising trajectory, the translation of these innovations from research labs to farmers' fields remains a significant bottleneck (Elango and Kawarazuka 2019). Additionally, socioeconomic constraints, policy limitations, and the digital divide further exacerbate the challenges in harnessing the full potential of modern breeding tools. The main aim of this chapter is to discuss about the recent developments in sorghum breeding and outline the challenges ahead in the face of climate change to deliver improved cultivars for wider adaptation.

15.2 CRISPR in Sorghum Crop Improvement

CRISPR has emerged as a powerful tool for genome editing with immense potential for crop improvement (Table 15.1). However, limited progress has been made in the utilization of CRISPR techniques to improve agronomic traits, yield, and quality of sorghum. The major limitations in gene editing of sorghum are: (1) delivery of the gene construct into the plant cell, and (2) regenerating a fertile plant from the edited cell through tissue culture. Early evidence of the stable transformation of sorghum

Table 15.1 List of successful sorghum gene-editing experiments using CRISPR techniques

Gene name	Purpose	Delivery method	Performed by
Wuschel2 (Wus2)	Direct somatic embryogenesis and regeneration to reduce the duration of the tissue culture cycle	*Agrobacterium* transformation	Che et al. (2022)
SbBADH2	Increased aroma of seeds and leaves	*Agrobacterium* transformation	Zhang et al. (2022)
LIGULELESS-1 (LG1)	Develop an easily scorable phenotype that can used for genome editing protocol improvements in sorghum	Biolistic bombardment	Brant et al. (2021)
Cinnamyl alcohol dehydrogenase (CAD)	To develop a biolistic bombardment methodology to produce stable edited plants of the sorghum inbred line, Tx430	Biolistic bombardment	Liu et al. (2019)
Phytoene desaturase (PDS)	To develop a biolistic bombardment methodology to produce stable edited	Biolistic bombardment	Liu et al. (2019)
Sorghum centromere-specific histone H3 (Sb-CENH3)	To develop a high efficiency *Agrobacterium* transformation system for sorghum. *Sb-CENH3* knockouts can be used to developed haploid inducers in the crop	*Agrobacterium* transformation	Che et al. (2018)
An out-of-frame red fluorescence protein gene (DsRED2)	To demonstrate Cas9/sgRNA activity in sorghum immature embryos	*Agrobacterium* transformation	Jiang et al. (2013)

proliferating embryo was demonstrated using *Agrobacterium tumefaciens* binary vector, *Y158*-mediated transformation (Jiang et al. 2013). Later, a ternary vector or cohabitating vector system was used, and regeneration frequencies ranging from 6% to 29% were achieved in the sorghum Tx430 line (Che et al. 2022). In recent years, biolistic bombardment showed success in the stable transformation (Liu et al. 2019; Brant et al. 2021). The recent report of *Wuschel2* (*Wus2*)-enabled improvement of transformation efficiency, as well as CRISPR/Cas-targeted genome editing frequency, opened a new avenue to improve homology-directed repair efficiency to bypass the less reliable canonical nonhomologous end joining (NHEJ) chromosomal repair-mediated editing in sorghum (Zhang et al. 2022).

15.3 Genomic Selection

The need for rapid improvement has never been more critical, given the challenges posed by climate change. Genomic selection (GS), with its ability to predict the performance of untested genotypes using dense molecular markers, offers promise in this context (Crossa et al. 2017). GS has now provided a better alternative of a genome-based approach for combining desirable traits with fewer field trials (Bernardo and Yu 2007; Abberton et al. 2016; Bernardo 2016; Hickey et al. 2017). GS differs from marker-assisted selection (MAS), wherein only a few markers associated with target traits are utilized. Instead, GS uses all available marker data to predict breeding values, allowing breeders to make selections without phenotypic data of the test cross, thereby shortening the breeding cycle (Meuwissen et al. 2001).

Modern genotyping platforms have made it feasible to genotype thousands of sorghum accessions rapidly. Such platforms are instrumental in GS, providing the dense molecular data required for accurate predictions (Mace and Jordan 2011). Various models, such as the ridge-regression best linear unbiased predictor (RR-BLUP) and Bayesian approaches, have been tested for sorghum. These models help in estimating marker effects and predicting the genetic value of individuals (Cuevas et al. 2016). GS has shown promise for complex traits in sorghum, such as drought tolerance and grain yield. Early studies indicate that the prediction accuracy for these traits is competitive, making GS a viable strategy for their improvement. By eliminating the need for phenotyping in certain breeding stages, GS can significantly reduce the duration of the breeding cycle (Heffner et al. 2010). GS allows breeders to make simultaneous selections for multiple traits, leading to improved genetic gains per unit of time (Technow et al. 2012). Though GS requires initial investments in genotyping, it can lead to cost savings in the long run by reducing the need for extensive field trials (Poland and Rutkoski 2016). GS can be seamlessly integrated with other modern breeding technologies like high-throughput phenotyping and CRISPR-based gene editing to further enhance the breeding efficiency (Xu et al. 2017). While challenges remain in the adoption and optimization of GS for sorghum, the potential benefits are significant. With continued research and collaboration, GS stands to revolutionize sorghum crop improvement, paving the way for more resilient and high-yielding varieties.

15.4 Microbiome-Associated Breeding

Plants are associated with vast microbes, either at the surfaces of plant tissues like root (rhizosphere), leaf (phyllosphere) or pollen and nectar (anthosphere), or within plant organs (endosphere), which together consist of plant microbiome (Hardoim et al. 2008; Vorholt 2012; Lundberg et al. 2013; Remus-Emsermann and Schlechter 2018). Plant microbiome has been known as the second genome of plants, and plays an important role in plant growth, health, and defense. There is evidence that microorganisms help plants by increasing nutrient availability and producing phytohormones to promote plant growth (Stringlis et al. 2018a, b, 2019). Nitrogen fixation is a well-studied example of how various microbes such as *Frankia*, *Rhizobium*, *Pseudomonas*, and *Azotobacter* fix nitrogen by converting nitrogen to nitrate thus promoting plant growth (Mahmud et al. 2020). For assimilation of other nutrients, for instance, phosphate, *Trichoderma* fungi were shown to facilitate phosphate acquisition by secreting chelating metabolites that solubilize phosphate which can be absorbed directly by plants (Altomare et al. 1999). In Arabidopsis, endophytic symbiont *Colletotrichum tofieldiae* was found to facilitate phosphate translocation in phosphate deficient conditions (Hiruma et al. 2016). Additionally, a wide array of microbes, including both bacteria and fungi, can produce indole acetic acid (IAA) or auxin-like compounds which act as phytohormones to promote plant growth (Duca et al. 2014; Garnica-Vergara et al. 2016).

In agriculture, biotic stresses are the main reason for crop yield loss, increasing the cost of management and control. There is evidence demonstrating that microbes could have multiple functions to buffer stresses imposed on the plants. Many studies showed that they can produce *1-aminocyclopropane-1-carboxylic acid (ACC) deaminase* to reduce plant endogenous ethylene, which is highly accumulated in plants under stress conditions and inhibit plant growth (Glick et al. 2007). In maize, the root system produces benzoxazinoids, attracting the rhizobacterium *Pseudomonas putida* which primes the plants with jasmonic acid (JA)-dependent inducible systemic resistance (ISR) against pathogen *Colletotrichum graminicola* (Neal and Ton 2013). Similarly, when leguminous plants are under nitrogen deficient condition, flavonoids are largely produced to attract microbes with nitrogen fixing ability (Hassan and Mathesius 2012). Kwak et al. (2018) compared the rhizosphere microbiomes of a resistant and a susceptible tomato variety. They found that a *flavobacterial* genome was enriched in resistant variety microbiota and the transplantation of rhizosphere microbiota from resistant plants confers the susceptible plants with the resistance to pathogen *Ralstonia solanacearum*.

Research on association of microbes with sorghum has revealed that microbes associated with sorghum have plant growth-promoting (PGP) functions. For example, *Azospirillum brasilense* harbors multiple PGP effects including phytohormone activity, nitrogen fixation, and proliferation of the root system (Steenhoudt and Vanderleyden 2000; Lucy et al. 2004; Bashan and de-Bashan 2010). Genera *Pantoea* and *Rhizobia* possessing several PGP features including IAA production were isolated from sweet sorghum root microbiome (Mareque et al. 2015). Genera *Enterobacter* and *Klebsiella* exhibited high nitrogenase activity in sweet sorghum plants (Pedersen et al. 1978; Mareque et al. 2015). Genera *Massilia* and *Bacillus* are featured endophytes within plant tissues, characterized with effects on host protection and growth promotion, both of which were found enriched in sorghum root microbiome (Santoyo et al. 2016; Mareque et al. 2018). In Fitzpatrick's study on root microbiome across 30 angiosperm plant species, evidence showed that there is a positive correlation between relative abundance of *Streptomyces* and increased drought tolerance of the host plants (Fitzpatrick et al. 2018). Later a sorghum root microbiome study identified an increased *Streptomyces* colonization in root and rhizosphere under drought stress and the colonization is associated with promoted root development (Naylor and Coleman-Derr 2018). Moreover, a study from Cloutier et al. (2020), identified certain bacterial compositions were highly abundant in sorghum rhizosphere under frost stress. The newly emerging technologies like metagenomics and proteomics will make it possible to screen for more potential microbes with PGP function on sorghum, to help in the pursuit of more sustainable agroecosystems.

15.5 Metabolite-Assisted Breeding: Sorghum Phytoalexins

Metabolomics is an emerging approach used for crop improvement. In sorghum, it has been utilized predominantly to enhance abiotic and biotic stress tolerance responses and nutritional profiling. Phytoalexins are stress-induced secondary metabolites that play a crucial role in regulating plants' responses, often enhancing their ability to cope with biotic and abiotic stresses. The term "phytoalexin" was initially coined by Müller and Börger in 1940 following their observation of this phenomenon in late blight disease affected potato tubers. Among *poaceae* family, sorghum produces novel phytoalexins (Ibraheem et al. 2010). Sorghum synthesizes the flavonoid class antifungal phytolaexins in reaction to the infection of *Colletotrichum sublineola* (Nicholson et al. 1987), known as 3-deoxyanthocyanidins (3-DAs); these compounds arise from a colorless precursor flavan-4-ol. Apigeninidin and luteolinidin are the main 3-DA family compounds identified in sorghum. The luteolinidin, apigeninidin, and their methyl and acyl derivatives, such as luteolinidin-5-glucoside, 7-*O*-methyl-apigeninidin, luteolinidin-5-glucoside, 5-methoxyluteolinidin, and a recently discovered novel pyrano-apigeninidin-4-vinyl phenol, are among the seven antifungal 3-DA compounds in sorghum (Awika and Rooney 2004; Bai et al. 2014). The synthesis of 3-DAs occurs at the site of pathogen entry (Snyder and Nicholson 1990). Initially colorless, the antifungal compound's color intensifies 30 h post infection as it disperses into the fungus's cytoplasm, triggering the surrounding cells to produce more red-colored antifungal compounds, ultimately killing the pathogen. These 3-DA family compounds are crucial in arresting the growth of fungal hyphae, thereby conferring resistance to anthracnose disease (*Colletotrichum sublineola*) in sorghum (Ibraheem et al. 2010). Compared to susceptible genotypes, resistant genotypes produce 3-DAs more quickly and at a higher rate (Basavaraju et al. 2009; Ibraheem et al. 2010). Functional copies of the *yellow seed1* (*y1*) gene are responsible for the production of 3-DAs in sorghum (Boddu et al. 2005; Ibraheem et al. 2010, 2015). Recently, 3-DAs were shown to have toxic effects on aphids (Kariyat et al. 2019) and fall armyworm (Chatterjee et al. 2023) growth and mortality. Hence, 3-DAs could be exploited as a potential bio-fungicide or bio-pesticide to control the various pests and pathogens in sorghum and other crops.

Apart from their antifungal properties, 3-DAs have many positive effects on human health. Due to their strong antioxidant and phase II enzyme activity, they can restrict the proliferation of cancerous cells in human (Yang et al. 2009; Zhu et al. 2017; Xiong et al. 2019). Due to their high stability under elevated temperatures and acidic conditions, they are also suitable as natural food colorants (Awika and Rooney 2004; Awika 2008, 2011; Yang et al. 2014). In Africa, red pigment compounds are extracted from sorghum sheaths to be utilized as colorant in cheese. Recent studies have shown that these sheath extracts are rich in 3-DAs and exhibit significant antioxidant activities (Kayodé et al. 2011, 2012). With the growing consumer preference for natural and organic food colorants, the characterization and quantification of 3-DA variations in genetic resources of sorghum could facilitate the identification of 3-DA rich lines, which can be further utilized to enhance both plant resilience and human health.

15.6 Breeding for Climate Adaptive Traits: Epicuticular Wax

Crops exhibit distinctive morpho-adaptive traits that confer resilience to climatic stresses. In breeding for climate resilience, traits like deep root system, stomatal density, leaf orientation, size, and thickness are frequently explored. Moreover, EW, which forms the outermost barrier between plants and their environment, is a novel trait that can contribute significantly to plants' adaption to stresses. EW has been demonstrated to confer tolerance against a variety of abiotic stresses like cold, physical leaf injury, air pollution, strong UV rays, heat, and drought (Thomas and Barber 1974; Eglinton and Hamilton 1967; Percy and Baker 1990; Reicosky and Hanover 1978; Jefferson et al. 1989; Blum 1975; Bengston et al. 1978). It can also provide protection from insect and pathogen damage (Edwards 1982; El-Otmani et al. 1989).

Sorghum variants for EW production were first reported in 1941 as bloomless types (no visible EW), bloom types (visible EW), and sparse bloom types (reduced visible EW) (Ayyangar and Ponnaiya 1941). Allelism tests suggested that homozygous recessive genes, designated as *bmbm* for bloomless and *hh* for sparse bloom, determined these phenotypes. Further allelism analysis among the sorghum mutants revealed bi-allelic (*bm* and *bm2*) and tri-allelic (*h*, *h2* and *h3*) regulation of bloomless and sparse bloom phenotypic expression, respectively (Peterson et al.1982). Madhusudhana and Patil (2013) identified the significant QTL *Dw4*, affecting bloom production in sorghum, during a plant height study involving a cross between 296B and IS 18551. Burow et al. (2009) discovered the *BLMC* locus, crucial for high wax production, in a mutant population from BTx 623 and KFS 2021. Mizuno et al. (2013) located a genomic region linked to bloom inhibition, with the *Sb06g023280* gene's inversion identified as the cause, a gene belonging to the *ABC transporter family*. Recent GWAS identified genetic associations of EW trait in sorghum (Elango et al. 2020b). EW is another potential trait that could be exploited in breeding drought and heat tolerant cultivars.

15.7 High-Throughput Phenotyping Platforms

With the evolution of precision agriculture and the advent of modern breeding technologies, high-throughput phenotyping (HTP) and unmanned aerial systems (UAS) have emerged as transformative tools. In sorghum breeding, these technologies are unlocking new potentials by offering rapid, scalable, and nondestructive assessment of diverse phenotypic traits, thereby expediting the breeding cycle. HTP and UAS tools offer innovative solutions to bridge the genotype-phenotype gap, essential for effective breeding interventions. The precise data from these technologies ensures the selection of elite genotypes with higher accuracy, leading to improved genetic gains. Although initial investments are substantial, the long-term benefits of reduced field trials and rapid selections make HTP and UAS cost-effective tools.

HTP platforms, equipped with advanced sensors and imaging technologies, facilitate the rapid collection of phenotypic data from large sorghum populations. Some of the widely used HTP platforms employed in sorghum research are ground-based phenotyping platforms (phenomobile, phenospex, and automated greenhouses and growth chambers), UAS or drones, remote sensing platforms, and integrated phenotyping platforms. Phenomobile is a mobile platform equipped with a range of sensors, including RGB cameras, hyperspectral cameras, and LIDAR. It is used to capture plant height, biomass, and canopy spectral reflectance among other traits (White et al. 2012). Phenospex system offers a range of sensors for plant scanning, including 3D laser scanning and multispectral imaging, helping in traits related to plant architecture and color (Fiorani and Schurr 2013). Automated greenhouses and growth chambers are combined with conveyor systems and imaging stations, and allow for detailed phenotyping of early-stage sorghum plants (Araus and Cairns 2014). Satellites and manned aircrafts equipped with advanced sensors can cover vast areas, providing data on larger field trials or even capturing phenotypic variation at the landscape level. Software platforms like *LemnaGrid*, *PlantCV*, and *Ilastik* aid in processing the vast amount of image data generated by HTP platforms. These tools extract meaningful traits from raw images, turning them into actionable phenotypic data (Fahlgren et al. 2015). Some systems integrate sensor data from multiple sources (e.g., ground-based sensors, drones, and satellite imagery) to provide a holistic view of plant performance in various environments (Andrade-Sanchez et al. 2014; Barker et al. 2016).

Ground-based phenotyping systems have been employed in sorghum research to assess traits such as plant height, biomass, and leaf architecture (Salas Fernandez et al. 2017; Xiang et al. 2019; Young et al. 2019). These platforms, by providing real-time data, enable breeders to make timely decisions, especially in early breeding cycles. Drones, equipped with multispectral and hyperspectral cameras, offer a bird's-eye view of sorghum fields. They capture data on canopy temperature, chlorophyll content, and overall plant vigor, crucial for assessing drought tolerance and health. UAS allows breeders to evaluate large field trials, ensuring the collection of consistent data across vast spatial scales (Shakoor et al. 2016). The integration of UAS with machine learning and data analytics tools provides nuanced insights, aiding in the prediction of grain yield, disease resistance, and other essential traits. The integration of HTP and UAS in sorghum breeding represents a significant leap toward precision breeding. As these technologies continue to evolve, they promise to revolutionize sorghum breeding, ensuring the development of resilient, high-yielding varieties optimized for diverse agroecological challenges.

15.8 Breeding for Biotic and Abiotic Stress Tolerance

Drought tolerance is a consequence of heritable morphological and anatomical traits such as deeper root system, leaf EW, physiological responses (osmotic adjustment), and stay-green traits as well as adaptive mechanisms (Tari et al. 2013). Stay green is a post-flowering drought adaptation trait expressed by delayed leaf senescence due to improved water balance in the plant (Borrell et al. 2014) and functional stay green trait is considered important for improving sorghum's drought tolerance (Christopher et al. 2016). Sorghum is moderately tolerant to salinity, and sorghum cultivars show differential salt tolerance at different growth stages of crop. The most affected traits due to salinity are grain yield, number of grains per head, shoot dry weight, harvest index, and leaf chloride, sodium, calcium, and potassium concentrations (Igartua et al. 1995). The Na^+ exclusion and grain K^+/Na^+ ratios are suggested as reliable traits for selecting salt tolerant plants. The carotenoid concentration, catalase enzyme activity, and K^+/Na^+ ratio was identified as biochemical trait associated with salinity tolerance of sorghum plants (Shakeri et al. 2020). The shoot and root dry weight, and root volume of sorghum seedlings, were recommended as important traits for screening for salinity tolerance (Hassanein and Azab 1993). A sorghum variety, Gadam, was found to be superior in seedling traits such as germination percentage, seedling chlorophyll and is suggested to be one of the best germplasm materials for future sorghum breeding programs (Mbinda and Kimtai 2019). There are numerous diseases that affect sorghum, but only a few are of economic importance globally. Globally important diseases affecting sorghum cultivation include grain mold, leaf blight, anthracnose, downy mildew, charcoal rot, gray leaf spot, rough spot, zonate leaf spot, and rust (Das and Rakshit 2016). Similarly, breeders continuously scout new germplasm materials for insects' resistance. The major insects are shootfly, aphids, stem borer, fall armyworm, and sorghum midge. Also, the parasitic weed striga is an important constraint for sorghum production especially in Africa.

Crop modelling and identifying suitable growing environments are vital for increasing the acreage of sorghums across the globe. Sorghum forage quality is comparable to corn, at the same time sorghums can be grown on marginal lands with less inputs (Elango et al. 2020a). So, identifying suitable growing regions and pushing its adaptation limits to temperate zones (high elevation areas, high latitudes) require adaptation of sorghum to low temperatures (Krishnamurthy et al. 2014; Elango et al. 2022). A current limiting factor in temperate sorghum production is cold tolerance among commercial hybrids and elite breeding lines compared to other cereals. Persistent temperature below 15 °C can reduce seed germination and plant emergence (Maulana and Tesso 2013). Modelling, prediction, and mapping of suitable growing regions should go hand in hand in order to reap the benefits of sorghums versatility. Understanding the genetic variation behind the adaptive features that allow sorghum to withstand challenging environments may help researchers make improvements to sorghum and other important crops.

15.9 Future Prospective in Sorghum Breeding

Considerable efforts have been made to develop climate smart sorghum cultivars for niche and wider adaptation to varied agroclimatic regions of the world. However, application of genomic selection and other approaches like genetic engineering through CRISPR-Cas9/12 and traditional transgenic approaches are lagging behind in sorghum crop improvement. There is a need to identify the constraints, and developments in gene editing technologies are the need of the hour. Breeder should focus on developing climate-smart sorghums; as global climate change intensifies, breeding sorghum varieties that are resistant to drought, heat, and salinity will become a matter of paramount importance along with nutritional enhancement for high protein content, increased micronutrients (like iron and zinc), and improved digestibility. Since most of the sorghum grains contain considerable amounts of health-promoting polyphenols, and moreover sorghums are gluten-free, making them an ideal crop for diversifying diets to reduce obesity and other diet-related illnesses. Recent study suggested that sorghum grain polyphenols have the potential to positively modulate the gut microbiomes of humans. Besides, traits like bioenergy, pest and disease resistance, and developing hybrids (CMS-based) and identification of ideal parental lines for rainy and postrainy season adaptations are crucial. In addition, developing varieties for specific uses, such as gluten-free products or brewing, and breeding for traits that reduce post-harvest losses, such as resistance to grain mold, and breeding sorghum that fits well with intercropping systems or rotations, benefiting soil health and pest management should go hand in hand with major breeding efforts.

References

Abberton M, Batley J, Bentley A et al (2016) Global agricultural intensification during climate change: a role for genomics. Plant Biotechnol J 14:1095–1098. https://doi.org/10.1111/pbi.12467

Adeyanju A, Little C, Yu J, Tesso T (2015) Genome-wide association study on resistance to stalk rot diseases in grain sorghum. G3 (Bethesda) 5:1165–1175. https://doi.org/10.1534/g3.114.016394

Altomare C, Norvell WA, Björkman T, Harman GE (1999) Solubilization of phosphates and micronutrients by the plant-growth-promoting and biocontrol fungus Trichoderma harzianum Rifai 1295-22. Appl Environ Microbiol 65:2926–2933. https://doi.org/10.1128/aem.65.7.2926-2933.1999

Andrade-Sanchez P, Gore MA, Heun JT et al (2014) Development and evaluation of a field-based high-throughput phenotyping platform. Funct Plant Biol 41:68. https://doi.org/10.1071/FP13126

Araus JL, Cairns JE (2014) Field high-throughput phenotyping: the new crop breeding frontier. Trends Plant Sci 19:52–61. https://doi.org/10.1016/j.tplants.2013.09.008

Awika JM (2008) Behavior of 3-deoxyanthocyanidins in the presence of phenolic copigments. Food Res Int 41:532–538. https://doi.org/10.1016/j.foodres.2008.03.002

Awika JM (2011) Sorghum flavonoids: unusual compounds with promising implications for health. Advances in cereal science: implications to food processing and health promotion. Am Chem Soc 9:171–200. https://doi.org/10.1021/bk-2011-1089.ch009

Awika JM, Rooney LW (2004) Sorghum phytochemicals and their potential impact on human health. Phytochemistry 65:1199–1221. https://doi.org/10.1016/j.phytochem.2004.04.001

Ayyangar GN, Ponnaiya BWX (1941) The occurrence and inheritance of a bloomless sorghum. Curr Sci 10:408–409

Bai Y, Findlay B, Sanchez Maldonado AF et al (2014) Novel pyrano and vinylphenol adducts of deoxyanthocyanidins in sorghum sourdough. J Agric Food Chem 62:11536–11546. https://doi.org/10.1021/jf503330b

Bao Y, Tang L, Breitzman MW et al (2019) Field-based robotic phenotyping of sorghum plant architecture using stereo vision. J Field Robot 36:397–415. https://doi.org/10.1002/rob.21830

Barker J, Zhang N, Sharon J et al (2016) Development of a field-based high-throughput mobile phenotyping platform. Comput Electron Agric 122:74–85. https://doi.org/10.1016/j.compag.2016.01.017

Basavaraju P, Shetty NP, Shetty SH et al (2009) Infection biology and defence responses in sorghum against lletotrichum sublineolum. J Appl Microbiol 107:404–415. https://doi.org/10.1111/j.1365-2672.2009.04234.x

Bashan Y, de Bashan LE (2010) How the plant growth-promoting bacterium azospirillum promotes plant growth—a critical assessment, 1st edn. Elsevier, Amsterdam

Bengston C, Larsson S, Liljenberg C (1978) Effects of water stress on cuticular transpiration rate and amount and composition of epicuticular wax in seedlings of six oat varieties. Physiol Plant 44:319–324

Bernardo R (2016) Bandwagons I, too, have known. Theor Appl Genet 129:2323–2332. https://doi.org/10.1007/s00122-016-2772-5

Bernardo R, Yu J (2007) Prospects for genomewide selection for quantitative traits in maize. Crop Sci 47:1082–1090. https://doi.org/10.2135/cropsci2006.11.0690

Blum A (1975) Effect of the Bm gene on epicuticular wax deposition and the spectral characteristics of sorghum leaves. SABRAO J 7:45–52

Boddu J, Svabek C, Ibraheem F et al (2005) Characterization of a deletion allele of a sorghum Myb gene yellow seed1 showing loss of 3-deoxyflavonoids. Plant Sci 169:542–552. https://doi.org/10.1016/j.plantsci.2005.05.007

Borrell AK, van Oosterom EJ, Mullet JE et al (2014) Stay-green alleles individually enhance grain yield in sorghum under drought by modifying canopy development and water uptake patterns. New Phytol 203:817–830. https://doi.org/10.1111/nph.12869

Bouchet S, Olatoye MO, Marla SR et al (2017) Increased power to dissect adaptive traits in global sorghum diversity using a nested association mapping population. Genetics 206:573–585. https://doi.org/10.1534/genetics.116.198499

Boyles RE, Brenton ZW, Kresovich S (2019) Genetic and genomic resources of sorghum to connect genotype with phenotype in contrasting environments. Plant J 97:19–39. https://doi.org/10.1111/tpj.14113

Brant EJ, Baloglu MC, Parikh A, Altpeter F (2021) CRISPR/Cas9 mediated targeted mutagenesis of LIGULELESS-1 in sorghum provides a rapidly scorable phenotype by altering leaf inclination angle. Biotechnol J 16:e2100237. https://doi.org/10.1002/biot.202100237

Brenton ZW, Cooper EA, Myers MT et al (2016) A genomic resource for the development, improvement, and exploitation of sorghum for bioenergy. Genetics 204:21–33. https://doi.org/10.1534/genetics.115.183947

Burow GB, Franks CD, Acosta-Martinez V, Xin Z (2009) Molecular mapping and characterization of BLMC, a locus for profuse wax (bloom) and enhanced cuticular features of Sorghum (Sorghum bicolor (L.) Moench.). Theor Appl Genet 118:423–431. https://doi.org/10.1007/s00122-008-0908-y

Caniato FF, Guimarães CT, Hamblin M et al (2011) The relationship between population structure and aluminum tolerance in cultivated sorghum. PLoS One 6:e20830. https://doi.org/10.1371/journal.pone.0020830

Casa AM, Pressoir G, Brown PJ et al (2008) Community resources and strategies for association mapping in sorghum. Crop Sci 48:30–40. https://doi.org/10.2135/cropsci2007.02.0080

Chatterjee D, Lesko T, Peiffer M et al (2023) Sorghum and maize flavonoids are detrimental to growth and survival of fall armyworm Spodoptera frugiperda. J Pest Sci 96:1551–1567. https://doi.org/10.1007/s10340-022-01535-y

Che P, Anand A, Wu E et al (2018) Developing a flexible, high-efficiency Agrobacterium-mediated sorghum transformation system with broad application. Plant Biotechnol J 16:1388–1395. https://doi.org/10.1111/pbi.12879

Che P, Wu E, Simon MK et al (2022) Wuschel2 enables highly efficient CRISPR/Cas-targeted genome editing during rapid de novo shoot regeneration in sorghum. Commun Biol 5:344. https://doi.org/10.1038/s42003-022-03308-w

Christopher JT, Christopher MJ, Borrell AK et al (2016) Stay-green traits to improve wheat adaptation in well-watered and water-limited environments. J Exp Bot 67:5159–5172. https://doi.org/10.1093/jxb/erw276

Cloutier M, Chatterjee D, Elango D et al (2020) Sorghum root flavonoid chemistry, cultivar, and frost stress effects on rhizosphere bacteria and fungi. Phytobiomes J 5:39. https://doi.org/10.1094/pbiomes-01-20-0013-fi

Cloutier M, Chatterjee D, Elango D et al (2021) Sorghum root flavonoid chemistry, cultivar, and frost stress effects on rhizosphere bacteria and fungi. Phytobiomes J 5:39–50. https://doi.org/10.1094/PBIOMES-01-20-0013-FI

Crossa J, Pérez-Rodríguez P, Cuevas J et al (2017) Genomic selection in plant breeding: methods, models, and perspectives. Trends Plant Sci 22:961–975. https://doi.org/10.1016/j.tplants.2017.08.011

Cuevas J, Crossa J, Soberanis V et al (2016) Genomic prediction of genotype × environment interaction kernel regression models. Plant Genome 9. https://doi.org/10.3835/plantgenome2016.03.0024

Das IK, Rakshit S (2016) Millets, their importance, and production constraints. In: Biotic stress resistance in millets. Elsevier, Amsterdam, pp 3–19

de Wet JMJ (1978) Systematics and evolution of sorghum sect sorghum (gramineae). Am J Bot 65:477. https://doi.org/10.2307/2442706

Deu M, Rattunde F, Chantereau J (2006) A global view of genetic diversity in cultivated sorghums using a core collection. Genome 49:168–180. https://doi.org/10.1139/G05-092

Djanaguiraman M, Gowsiga S, Govindaraj M, Habyarimana E, Senthil A, Thavaprakaash N, Jeyakumar P, Kokilavani J, Chellammal C (2023) Impact of root architecture and transpiration rate on drought tolerance in stay-green sorghum. Crop Sci:1–18. https://doi.org/10.1002/csc2.21108

Duca D, Lorv J, Patten CL et al (2014) Indole-3-acetic acid in plant-microbe interactions. Anton Leeuw Int J Gen Mol Microbiol 106:85–125. https://doi.org/10.1007/s10482-013-0095-y

Edwards PB (1982) Do waxes on juvenile eucalyptus leaves provide protection from grazing insects? Aust J Ecol 7:347–352

Eglinton G, Hamilton RJ (1967) Leaf epicuticular waxes. Science (1979) 156:1322–1335

Elango D (2020) Elite sorghum (Sorghum bicolor (L.) Moench) hybrids for postrainy season adaptation. Madras Agric J 107:1–6. https://doi.org/10.29321/MAJ

Elango D, Kawarazuka N (2019) Gender responsive participatory varietal selection for sustainable seed potato systems in Assam, India. International Potato Center, Lima

Elango D, Chopra S, Roth GW (2020a) Seeding and nitrogen fertilization effects on the yield and quality of brachytic dwarf brown midrib forage sorghum hybrids. Crop Forage Turfgrass Manag 6. https://doi.org/10.1002/cft2.20067

Elango D, Xue W, Chopra S (2020b) Genome wide association mapping of epi-cuticular wax genes in Sorghum bicolor. Physiol Mol Biol Plants 26:1727–1737. https://doi.org/10.1007/s12298-020-00848-5

Elango D, Wang X, Bhatnagar RS et al (2022) Association genetics of early season cold and late season frost tolerance in Sorghum bicolor. Crop Sci 62:1844–1865. https://doi.org/10.1002/csc2.20810

El-Otmani M, Arpaia ML, Coggins CWJ et al (1989) Developmental changes in 'Valencia' orange fruit epicuticular wax in relation to fruit position on the tree. Sci Hortic 41:69–81

Fahlgren N, Feldman M, Gehan MA et al (2015) A versatile phenotyping system and analytics platform reveals diverse temporal responses to water availability in Setaria. Mol Plant 8:1520–1535. https://doi.org/10.1016/j.molp.2015.06.005

Fiorani F, Schurr U (2013) Future scenarios for plant phenotyping. Annu Rev Plant Biol 64:267–291. https://doi.org/10.1146/annurev-arplant-050312-120137

Fitzpatrick CR, Copeland J, Wang PW et al (2018) Assembly and ecological function of the root microbiome across angiosperm plant species. Proc Natl Acad Sci U S A 115:E1157–E1165. https://doi.org/10.1073/pnas.1717617115

Garnica-Vergara A, Barrera-Ortiz S, Muñoz-Parra E et al (2016) The volatile 6-pentyl-2H-pyran-2-one from Trichoderma atroviride regulates Arabidopsis thaliana root morphogenesis via auxin signaling and ETHYLENE INSENSITIVE 2 functioning. New Phytol 209:1496–1512. https://doi.org/10.1111/nph.13725

Glick BR, Cheng Z, Czarny J, Duan J (2007) Promotion of plant growth by ACC deaminase-producing soil bacteria. Eur J Plant Pathol 119:329–339. https://doi.org/10.1007/s10658-007-9162-4

Grenier C, Bramel-Cox PJ, Hamon P (2001) Core collection of sorghum: I. Stratification based on eco-geographical data. Crop Sci 41:234–240. https://doi.org/10.2135/cropsci2001.411234x

Hardoim PR, van Overbeek LS, van Elsas JD (2008) Properties of bacterial endophytes and their proposed role in plant growth. Trends Microbiol 16:463–471. https://doi.org/10.1016/j.tim.2008.07.008

Harlan JR, de Wet JMJ (1971) Toward a rational classification of cultivated plants. Taxon 20:509–517. https://doi.org/10.2307/1218252

Hassan S, Mathesius U (2012) The role of flavonoids in root-rhizosphere signalling: opportunities and challenges for improving plant-microbe interactions. J Exp Bot 63:3429–3444. https://doi.org/10.1093/jxb/err430

Hassanein A-HM, Azab AM (1993) Salt tolerance of grain sorghum. In: Towards the rational use of high salinity tolerant plants. Springer, Berlin, pp 153–156

Heffner EL, Lorenz AJ, Jannink J, Sorrells ME (2010) Plant breeding with genomic selection: gain per unit time and cost. Crop Sci 50:1681–1690. https://doi.org/10.2135/cropsci2009.11.0662

Hickey JM, Chiurugwi T, Mackay I, Powell W (2017) Genomic prediction unifies animal and plant breeding programs to form platforms for biological discovery. Nat Genet 49:1297–1303. https://doi.org/10.1038/ng.3920

Hiruma K, Gerlach N, Sacristán S et al (2016) Root endophyte colletotrichum tofieldiae confers plant fitness benefits that are phosphate status dependent. Cell 165:464–474. https://doi.org/10.1016/j.cell.2016.02.028

House LR (1985) A guide to sorghum breeding, 2nd edn. International Crops Research Institute for the Semi-Arid Tropics (ICRISAT), Patancheru

Hufnagel B, Guimaraes CT, Craft EJ et al (2018) Exploiting sorghum genetic diversity for enhanced aluminum tolerance: allele mining based on the AltSB locus. Sci Rep 8:1–13. https://doi.org/10.1038/s41598-018-27817-z

Ibraheem F, Gaffoor I, Chopra S (2010) Flavonoid phytoalexin-dependent resistance to anthracnose leaf blight requires a functional yellow seed1 in Sorghum bicolor. Genetics 184:915–926. https://doi.org/10.1534/genetics.109.111831

Ibraheem F, Gaffoor I, Tan Q et al (2015) A sorghum MYB transcription factor induces 3-deoxyanthocyanidins and enhances resistance against leaf blights in maize. Molecules 20:2388–2404. https://doi.org/10.3390/molecules20022388

Igartua E, Gracia MP, Lasa JM (1995) Field responses of grain sorghum to a salinity gradient. Field Crop Res 42:15–25. https://doi.org/10.1016/0378-4290(95)00018-L

Jefferson PG, Johnson DA, Asay KH (1989) Epi cuticular wax production, water status and leaf temperature in Triticeae range grasses of contrasting visible glaucousness. Can J Plant Sci 69:513–519

Jiang W, Zhou H, Bi H et al (2013) Demonstration of CRISPR/Cas9/sgRNA-mediated targeted gene modification in Arabidopsis, tobacco, sorghum and rice. Nucleic Acids Res 41:e188. https://doi.org/10.1093/nar/gkt780

Jiao Y, Burke J, Chopra R et al (2016) A sorghum mutant resource as an efficient platform for gene discovery in grasses. Plant Cell 28:1551–1562. https://doi.org/10.1105/tpc.16.00373

Kariyat RR, Gaffoor I, Sattar S et al (2019) Sorghum 3-deoxyanthocyanidin flavonoids confer resistance against corn leaf aphid. J Chem Ecol 45:502–514. https://doi.org/10.1007/s10886-019-01062-8

Kayodé APP, Nout MJR, Linnemann AR et al (2011) Uncommonly high levels of 3-deoxyanthocyanidins and antioxidant capacity in the leaf sheaths of dye sorghum. J Agric Food Chem 59:1178–1184. https://doi.org/10.1021/jf103963t

Kayodé APP, Bara CA, Dalodé-Vieira G et al (2012) Extraction of antioxidant pigments from dye sorghum leaf sheaths. LWT Food Sci Technol 46:49–55. https://doi.org/10.1016/j.lwt.2011.11.003

Krishnamurthy L, Dinakaran E, Kumar AA, Reddy BVS (2014) Field technique and traits to assess reproductive stage cold tolerance in sorghum (Sorghum bicolor (L.) Moench). Plant Prod Sci 17:218–227. https://doi.org/10.1626/pps.17.218

Kwak MJ, Kong HG, Choi K et al (2018) Rhizosphere microbiome structure alters to enable wilt resistance in tomato. Nat Biotechnol 36:1100–1116. https://doi.org/10.1038/nbt.4232

Lasky JR, Upadhyaya HD, Ramu P et al (2015) Genome-environment associations in sorghum landraces predict adaptive traits. Sci Adv 1:1–14. https://doi.org/10.1126/sciadv.1400218

Liu G, Li J, Godwin ID (2019) Genome editing by CRISPR/Cas9 in sorghum through biolistic bombardment. Methods Mol Biol 1931:169–183

Lucy M, Reed E, Glick BR (2004) Applications of free living plant growth-promoting rhizobacteria. Antonie Van Leeuwenhoek 86:1–25. https://doi.org/10.1023/B:ANTO.0000024903.10757.6e

Lundberg DS, Lebeis SL, Paredes SH et al (2013) Defining the core Arabidopsis thaliana root microbiome. Nature 501:86–90

Mace ES, Jordan DR (2011) Integrating sorghum whole genome sequence information with a compendium of sorghum QTL studies reveals uneven distribution of QTL and of gene-rich regions with significant implications for crop improvement. Theor Appl Genet 123:169–191. https://doi.org/10.1007/s00122-011-1575-y

Mace E, Innes D, Hunt C et al (2019) The Sorghum QTL Atlas: a powerful tool for trait dissection, comparative genomics and crop improvement. Theor Appl Genet 132:751–766. https://doi.org/10.1007/s00122-018-3212-5

Madhusudhana R, Patil JV (2013) A major QTL for plant height is linked with bloom locus in sorghum [Sorghum bicolor (L.) Moench]. Euphytica 191:259–268. https://doi.org/10.1007/s10681-012-0812-z

Mahmud K, Makaju S, Ibrahim R, Missaoui A (2020) Current progress in nitrogen fixing plants and microbiome research. Plants 9:1–17. https://doi.org/10.3390/plants9010097

Maina F, Bouchet S, Marla SR et al (2018) Population genomics of sorghum (Sorghum bicolor) across diverse agroclimatic zones of Niger. Genome 61:223–232. https://doi.org/10.1139/gen-2017-0131

Mareque C, Taulé C, Beracochea M, Battistoni F (2015) Isolation, characterization and plant growth promotion effects of putative bacterial endophytes associated with sweet sorghum (Sorghum bicolor (L) Moench). Ann Microbiol 65:1057–1067. https://doi.org/10.1007/s13213-014-0951-7

Mareque C, Da Silva TF, Vollú RE et al (2018) The endophytic bacterial microbiota associated with sweet sorghum (Sorghum bicolor) is modulated by the application of chemical N fertilizer to the field. Int J Genomics 2018:7403670. https://doi.org/10.1155/2018/7403670

Maulana F, Tesso TT (2013) Cold temperature episode at seedling and flowering stages reduces growth and yield components in sorghum. Crop Sci 53:564–574. https://doi.org/10.2135/cropsci2011.12.0649

Mbinda W, Kimtai M (2019) Evaluation of morphological and biochemical characteristics of sorghum [Sorghum bicolor [L.] Moench] varieties in response salinity stress. Annu Res Rev Biol 33:1–9. https://doi.org/10.9734/arrb/2019/v33i130110

Meuwissen THE, Hayes BJ, Goddard ME (2001) Prediction of total genetic value using genome-wide dense marker maps. Genetics 157:1819–1829. https://doi.org/10.1093/genetics/157.4.1819

Mizuno H, Kawahigashi H, Ogata J et al (2013) Genomic inversion caused by gamma irradiation contributes to downregulation of a WBC11 homolog in bloomless sorghum. Theor Appl Genet 126:1513–1520. https://doi.org/10.1007/s00122-013-2069-x

Morris GP, Ramu P, Deshpande SP et al (2013) Population genomic and genome-wide association studies of agroclimatic traits in sorghum. Proc Natl Acad Sci U S A 110:453–458. https://doi.org/10.1073/pnas.1215985110

Müller K, Börger H (1940) Experimentelle untersuchungen über die Phytophthora-resistem der kartoffel, vol 23. Arb Biol Reichsanstalt Landw-Forstwirtsch, Berlin, pp 189–231

Naylor D, Coleman-Derr D (2018) Drought stress and root-associated bacterial communities. Front Plant Sci 8:2223. https://doi.org/10.3389/fpls.2017.02223

Neal AL, Ton J (2013) Systemic defense priming by Pseudomonas putida KT2440 in maize depends on benzoxazinoid exudation from the roots. Plant Signal Behav 8:120–124. https://doi.org/10.4161/psb.22655

Nicholson RL, Kollipara SS, Vincent JR et al (1987) Phytoalexin synthesis by the sorghum mesocotyl in response to infection by pathogenic and nonpathogenic fungi. Proc Natl Acad Sci USA 84:5520–5524. https://doi.org/10.1073/pnas.84.16.5520

Ortiz D, Litvin AG, Salas Fernandez MG (2018) A cost-effective and customizable automated irrigation system for precise high-throughput phenotyping in drought stress studies. PLoS One 13:e0198546. https://doi.org/10.1371/journal.pone.0198546

Paterson AH, Bowers JE, Bruggmann R et al (2009) The Sorghum bicolor genome and the diversification of grasses. Nature 457:551–556. https://doi.org/10.1038/nature07723

Pedersen WL, Chakrabarty K, Klucas RV, Vidaver AK (1978) Nitrogen fixation (acetylene reduction) associated with roots of winter wheat and sorghum in Nebraska. Appl Environ Microbiol 35:129–135. https://doi.org/10.1128/aem.35.1.129-135.1978

Percy KE, Baker EA (1990) Effects of simulated acid rain on epicuticular wax production, morphology, chemical composition and on cuticular membrane thick ness in two clones of Sitka spruce [Picea sitchensis (Bong.) Carr.]. New Phytol 116:79–87

Peterson GC, Krittika S, Weibel DE (1982) Inheritance of some bloomless and sparse-bloom mutants in sorghum. Crop Sci 22:63–67

Poland J, Rutkoski J (2016) Advances and challenges in genomic selection for disease resistance. Annu Rev Phytopathol 54:79–98. https://doi.org/10.1146/annurev-phyto-080615-100056

Reddy BVS, Kumar AA, Sharma HC et al (2012a) Sorghum improvement (1980–2010): status and way forward. J SAT Agric Res 10:1–14

Reddy BVS, Reddy PS, Sadananda AR et al (2012b) Postrainy season sorghum: constraints and breeding approaches. J SAT Agric Res 10:1–12

Reicosky DA, Hanover JW (1978) Physiological effects of surface waxes. I. Light reflectance for glaucous and non-glaucous Picea pungens. Plant Physiol 62:101–104

Remus-Emsermann MNP, Schlechter RO (2018) Phyllosphere microbiology: at the interface between microbial individuals and the plant host. New Phytol 218:1327–1333. https://doi.org/10.1111/nph.15054

Rhodes DH, Hoffmann L, Rooney William L et al (2014) Genome-wide association study of grain polyphenol concentrations in global sorghum [Sorghum bicolor (L.) Moench] germplasm. J Agric Food Chem 62:10916–10927. https://doi.org/10.1021/jf503651t

Rooney LW (1996) Sorghum and millets. In: Cereal grain quality. Springer, Dordrecht, pp 153–177

Salas Fernandez MG, Bao Y, Tang L, Schnable PS (2017) A high-throughput, field-based phenotyping technology for tall biomass crops. Plant Physiol 174:2008–2022. https://doi.org/10.1104/pp.17.00707

Santoyo G, Moreno-Hagelsieb G, del Carmen Orozco-Mosqueda M, Glick BR (2016) Plant growth-promoting bacterial endophytes. Microbiol Res 183:92–99. https://doi.org/10.1016/j.micres.2015.11.008

Shakeri E, Emam Y, Pessarakli M, Tabatabaei SA (2020) Biochemical traits associated with growing sorghum genotypes with saline water in the field. J Plant Nutr 43:1136–1153. https://doi.org/10.1080/01904167.2020.1724307

Shakoor N, Ziegler G, Dilkes BP et al (2016) Integration of experiments across diverse environments identifies the genetic determinants of variation in sorghum bicolor seed element composition. Plant Physiol 170:1989–1998. https://doi.org/10.1104/pp.15.01971

Singh SP (1985) Sources of cold tolerance in grain sorghum. Can J Plant Sci 65:251–257. https://doi.org/10.4141/cjps85-037

Snyder BA, Nicholson RL (1990) Synthesis of phytoalexins in sorghum as a site-specific response to fungal ingress. Science (1979) 248:1637–1639

Steenhoudt O, Vanderleyden J (2000) Azospirillum, a free-living nitrogen-fixing bacterium closely associated with grasses: genetic, biochemical and ecological aspects. FEMS Microbiol Rev 24:487–506. https://doi.org/10.1016/S0168-6445(00)00036-X

Stringlis IA, Proietti S, Hickman R et al (2018a) Root transcriptional dynamics induced by beneficial rhizobacteria and microbial immune elicitors reveal signatures of adaptation to mutualists. Plant J 93:166–180. https://doi.org/10.1111/tpj.13741

Stringlis IA, Yu K, Feussner K et al (2018b) MYB72-dependent coumarin exudation shapes root microbiome assembly to promote plant health. Proc Natl Acad Sci U S A 115:E5213–E5222. https://doi.org/10.1073/pnas.1722335115

Stringlis IA, Zamioudis C, Berendsen RL et al (2019) Type III secretion system of beneficial rhizobacteria pseudomonas simiae WCS417 and pseudomonas defensor WCS374. Front Microbiol 10:1–11. https://doi.org/10.3389/fmicb.2019.01631

Sukumaran S, Xiang W, Bean SR et al (2012) Association mapping for grain quality in a diverse sorghum collection. Plant Genome 5:126–135. https://doi.org/10.3835/plantgenome2012.07.0016

Tari I, Laskay G, Takács Z, Poór P (2013) Response of sorghum to abiotic stresses: a review. J Agron Crop Sci 199:264–274. https://doi.org/10.1111/jac.12017

Technow F, Riedelsheimer C, Schrag TA, Melchinger AE (2012) Genomic prediction of hybrid performance in maize with models incorporating dominance and population specific marker effects. Theor Appl Genet 125:1181–1194. https://doi.org/10.1007/s00122-012-1905-8

Thomas DA, Barber HN (1974) Studies on leaf characteristics of a cline of Eucalyptus urnigera from mount Wellington, Tasmania. II. Reflection, transmission and absorption of radiation. Aust J Bot 22:701–707

Too EJ, Onkware AO, Were BA et al (2018) Molecular markers associated with aluminium tolerance in Sorghum bicolor. Hereditas 155:20. https://doi.org/10.1186/s41065-018-0059-3

Upadhyaya HD, Pundir RPS, Dwivedi SL et al (2009) Developing a mini core collection of sorghum for diversified utilization of germplasm. Crop Sci 49:1769–1780. https://doi.org/10.2135/cropsci2009.01.0014

Vorholt JA (2012) Microbial life in the phyllosphere. Nat Rev Microbiol 10:828–840. https://doi.org/10.1038/nrmicro2910

Wang X, Singh D, Marla S et al (2018) Field-based high-throughput phenotyping of plant height in sorghum using different sensing technologies. Plant Methods 14:53. https://doi.org/10.1186/s13007-018-0324-5

Watanabe K, Guo W, Arai K et al (2017) High-throughput phenotyping of sorghum plant height using an unmanned aerial vehicle and its application to genomic prediction modeling. Front Plant Sci 8:421. https://doi.org/10.3389/fpls.2017.00421

White JW, Andrade-Sanchez P, Gore MA et al (2012) Field-based phenomics for plant genetics research. Field Crop Res 133:101–112. https://doi.org/10.1016/j.fcr.2012.04.003

Xiang L, Bao Y, Tang L et al (2019) Automated morphological traits extraction for sorghum plants via 3D point cloud data analysis. Comput Electron Agric 162:951–961. https://doi.org/10.1016/j.compag.2019.05.043

Xiong Y, Zhang P, Warner RD, Fang Z (2019) Sorghum grain: from genotype, nutrition, and phenolic profile to its health benefits and food applications. Compr Rev Food Sci Food Saf 18:2025–2046. https://doi.org/10.1111/1541-4337.12506

Xu Y, Li P, Zou C et al (2017) Enhancing genetic gain in the era of molecular breeding. J Exp Bot 68:2641–2666. https://doi.org/10.1093/jxb/erx135

Yang L, Browning JD, Awika JM (2009) Sorghum 3-deoxyanthocyanins possess strong phase II enzyme inducer activity and cancer cell growth inhibition properties. J Agric Food Chem 57:1797–1804. https://doi.org/10.1021/jf8035066

Yang L, Dykes L, Awika JM (2014) Thermal stability of 3-deoxyanthocyanidin pigments. Food Chem 160:246–254. https://doi.org/10.1016/j.foodchem.2014.03.105

Young SN, Kayacan E, Peschel JM (2019) Design and field evaluation of a ground robot for high-throughput phenotyping of energy sorghum. Precis Agric 20:697–722. https://doi.org/10.1007/s11119-018-9601-6

Zhang D, Li J, Compton RO et al (2015) Comparative genetics of seed size traits in divergent cereal lineages represented by sorghum (Panicoidae) and rice (Oryzoidae). G3 (Bethesda) 5:1117–1128. https://doi.org/10.1534/g3.115.017590

Zhang D, Tang S, Xie P et al (2022) Creation of fragrant sorghum by CRISPR/Cas9. J Integr Plant Biol 64:961–964. https://doi.org/10.1111/jipb.13232

Zhou Y, Srinivasan S, Mirnezami SV et al (2019) Semiautomated feature extraction from RGB images for sorghum panicle architecture GWAS. Plant Physiol 179:24–37. https://doi.org/10.1104/pp.18.00974

Zhu Y, Shi Z, Yao Y et al (2017) Antioxidant and anti-cancer activities of proanthocyanidins-rich extracts from three varieties of sorghum (Sorghum bicolor) bran. Food Agric Immunol 28:1530–1543. https://doi.org/10.1080/09540105.2017.1351526

Future Strategies for Sorghum Improvement Under Climate Change Scenario

16

Ali Devlet and Yunus Emre Arvas

Abstract

Since the importance of sorghum on a global scale is known, its sensitivity to various stress events caused by climate change is an important issue that should be considered in terms of developing different adaptation strategies and ensuring food and feed security in the future. Sorghum's ability to grow in different climatic conditions and its resistance to abiotic stresses makes it a strategic product in the face of climate change. In addition to threatening global agricultural systems, global climate change also has the disadvantage of reducing crop yields and disrupting the resilience of the ecosystem. Abiotic stress factors such as high temperature, drought, flood, and excessive rainfall negatively affect the production of basic food products. To overcome these challenges, it is crucial to focus on climate-adapted food crops that require lower inputs and can provide sustainable yields under changing climate conditions. Sorghum is resistant to abiotic stresses and can grow under difficult conditions, which is important for strategies that can provide sustainable yield in changing climate conditions. Sorghum is consumed as a staple food in many regions of Asia and Africa and provides necessary vitamins. Understanding the sensitivity of sorghum to climate change can help identify opportunities to improve its yield and quality, thereby reducing

A. Devlet
Faculty of Agriculture and Natural Science, Department of Field Crops, Bilecik Seyh Edebali University, Bilecik, Turkey

Y. E. Arvas (✉)
Faculty of Sciences, Department of Biology, Karadeniz Technical University, Trabzon, Turkey

Faculty of Science, Department of Molecular Biology and Genetics, Van Yüzüncü Yıl University, Tusba, Van, Turkey

E. Habyarimana et al. (eds.), *Omics and Biotechnological Approaches for Product Profile-Driven Sorghum Improvement*,
https://doi.org/10.1007/978-981-97-4347-6_16

risks associated with food and feed security in future climate change scenarios. Efforts should be made to develop different alternatives for different sorghum varieties against future climate change scenarios. It is thought that this research will help determine which varieties can offer better yields in the future and guide the development of appropriate management strategies. Additionally, locally simulating the potential impacts of climate change on cereal crops is crucial for developing adaptation and consumption strategies. In summary, raising awareness and sensitivity to stress events caused by climate change and prioritizing sorghum crops in crop improvement strategies are essential to pave the way toward climate-smart agriculture. Focusing on climate-compatible, high-yield crops such as sorghum can alleviate the negative effects of climate change, result in sustainable agricultural policies by reducing inputs, and ultimately contribute to food and feed security in the face of a changing climate.

Main Conclusions: In this study, global food insecurity caused by climate change, the sensitivity of sorghum to climate change, and efforts to promote climate-friendly agriculture were examined. Additionally, the development of potential crop improvement strategies available in sorghum is emphasized.

Keywords

Climate change · Crop improvement · Food security · Sorghum

16.1 Introduction

In the fifth assessment report (AR5) of the Intergovernmental Panel on Climate Change (IPCC) in 2014, they reemphasized that climate change poses a significant risk to global food security (FAO 1999; Schmidhuber and Tubiello 2007).

The prevalence of poverty in these countries dramatically limits their ability to combat climate change and natural disasters. Consequently, the imperative adaptation of agriculture to climate change is critical to progress in food security (Barnabás et al. 2008).

Sorghum (*Sorghum bicolor* L. Moench), the fifth most important cereal crop in the world, is grown in more than 30 countries and consumed as the staple food by more than 500 million people (ICRISAT 2009). Understanding the specific niche for enhancing strategic agricultural productivity in sorghum can improve food production sustainability, particularly in water-limited environments (Akinseye et al. 2017; Peng et al. 2019; Raza et al. 2019; United Nations 2019; Roy et al. 2018; Dilley et al. 2005; Chadalavada et al. 2021; Sultan et al. 2013).

It is essential to consider these climate change impacts in the context of other drivers of agricultural irrigation, such as the expansion of irrigation areas, crop choices, crop harvest intensity, and improvements in irrigation water efficiency. While these factors play a role in shaping agricultural water use and crop productivity, climate change can significantly exacerbate the challenges faced by agricultural systems.

Overall, understanding the impact of climate change on sorghum yield and integrating this knowledge with other drivers of agricultural water use is crucial for effectively addressing the challenges and ensuring the future sustainability of sorghum production in the face of climate change.

Climate change will affect crop production through changes in CO_2 concentration, temperature rise, changes in precipitation and evapotranspiration, and extreme precipitation events. The complex regional model with positive and negative impacts is the product of these factors, but these factors vary according to crops, regions, and water management methods (Nelson et al. 2009). Although the prediction results of crop yield will vary with different crops and scenarios, according to various climate change predictions from developed and developing countries, the yield of irrigated wheat and corn will decline by 2050, among which the yield of irrigated wheat in developing countries will decline by 20–28%.

Global climate change has a significant impact on agricultural water and crop water demand as well as agricultural production through water factors, thus affecting agricultural water management and its response strategies. Addressing climate change has become a significant challenge and an essential task for agricultural water management today, and it is urgent to increase attention and relevant research depth (Bruinsma 2003).

For many years, the genetic gain in breeding experiments and better crop management have affected yield growth in India. However, the planting area of sorghum has declined rapidly in recent years, especially in recent years, because many biological and abiotic stress factors have limited farmers' yield. The change in consumers' eating habits and the availability of high-quality grain also affected the demand for sorghum, decreasing the number of people engaged in cultivation (Rakshit and Patil 2014). With the latest technological developments, especially with the analysis of the whole genome sequence, the sorghum plant can be customized for specific genes. It will be possible to use genome analysis tools to meet various future needs (Hariprasanna and Patil 2015).

16.2 Crop Yield Simulations

Sorghum varieties, particularly in West Africa, play a significant role in agriculture, with approximately 23 million hectares of sorghum cultivation, making it the second most important crop after maize (corn). Over the years, the output of sorghum in West Africa has shown an increase, rising from 4.6 million tons in 1979 to 12.4 million tons in 2012, according to FAOSTAT data. The three best sorghum varieties from Mali were selected for field tests under drought and irrigation conditions by farmers in West Africa. These varieties were chosen to calibrate a crop model. The field tests were conducted in 2004 and 2005. The Rural Economic Research Institute (IER) research station in Sotoba near Bamako, Mali (at coordinates 12.17N; 7.57W) was used for these experiments.

As described in previous studies (such as Kourssey et al., 2008a, b), the three sorghum varieties selected for the study can be divided into two groups. The first group consists of GineLo and GuineaAm varieties. These two varieties exhibit photoperiod sensitivity and have the advantage of adjusting product cycle length. However, their relatively lower average efficiency is a disadvantage. The second group includes a "modern" variety called Caudat. This variety is characterized by early maturity, shorter growth period, and insensitivity to photoperiod (Traoré et al. 2011).

These details highlight the importance of selecting sorghum varieties that can synchronize their flowering period with the end of the rainy season, thereby avoiding potential issues related to drought, pests, and diseases. Photoperiod-sensitive varieties offer the advantage of allowing flexible seeding dates, which can be beneficial in managing these challenges (Sultan et al. 2013).

16.2.1 Climate Change Scenario

Figure 16.1 shows projected changes in annual precipitation and mean surface temperature from 2031 to 2060 under the RCP8.5 scenario. The data presented in the figures are based on the average of 9 General Circulation Model (GCM) simulations for each of the 13 stations. A clear west-east dipole pattern is observed in terms of precipitation changes. The central Sahel region, represented by eight stations, shows an increase in annual rainfall, while the western Sahel stations experience stagnant or decreasing rainfall.

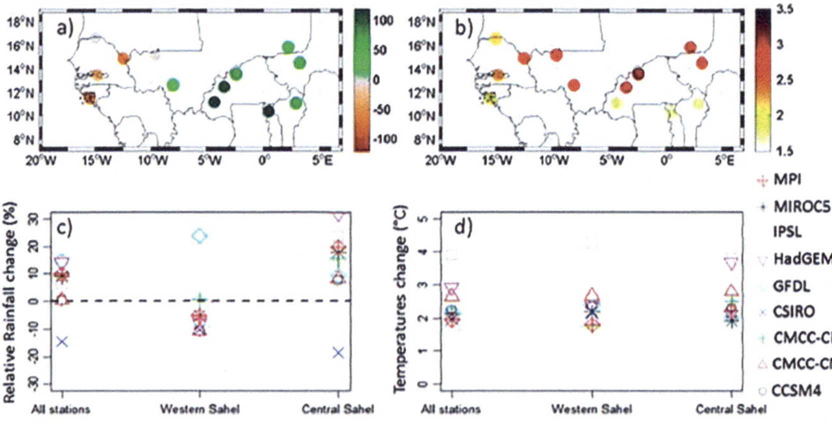

Fig. 16.1 The annual precipitation and average surface temperature vary depending on the RCP8.5 scenario. The top panel shows the relative change in precipitation (%) (**a**) between 2031–2060 and 1961–1990, and the absolute change in temperature (°C) (**b**) calculated as the average of 9 GCM simulations at 13 stations. The lower panels show the relative change in precipitation (%) (**c**) and the absolute change in temperature (°C) (**d**) calculated as the average of the individual meteorological observations, six from the West and six from the East

Figure 16.1c indicates regional mean precipitation variations, including meteorological stations and western and central Sahel subsets. Seven out of nine GCMs simulate future precipitation for the Western Sahel, and eight GCMs simulate precipitation increases for the Central Sahel. Although the exact location of the dividing line in the dipole pattern varies, different subsets of the CMIP5 and CMIP3 patterns consistently show this dipole pattern across precipitation changes across various emission scenarios and different future periods (Monerie et al. 2012, 2013; Biasutti 2013; Biasutti and Sobel 2009; Fontaine et al. 2011).

Figure 16.2 shows that the rainfall deficit becomes more pronounced during the first phase of the monsoon season in the western Sahel, particularly from June to July (Fig. 16.2a). However, from September to October (Fig. 16.2), there is a positive change in precipitation for the entire Sahelian monsoon season.

In contrast to the change in precipitation, the pattern of temperature change (as shown in Fig. 16.1b, d) is uniform along the longitude, with a latitudinal gradient.

The distribution of changes in precipitation and climate sensitivity varies between models, resulting in differences in soil moisture and cloud cover, with more precipitation being associated with colder surface temperatures. In all cases, however, and consistent with CMIP3 and Battisti and Naylor (2009), the warming climate is

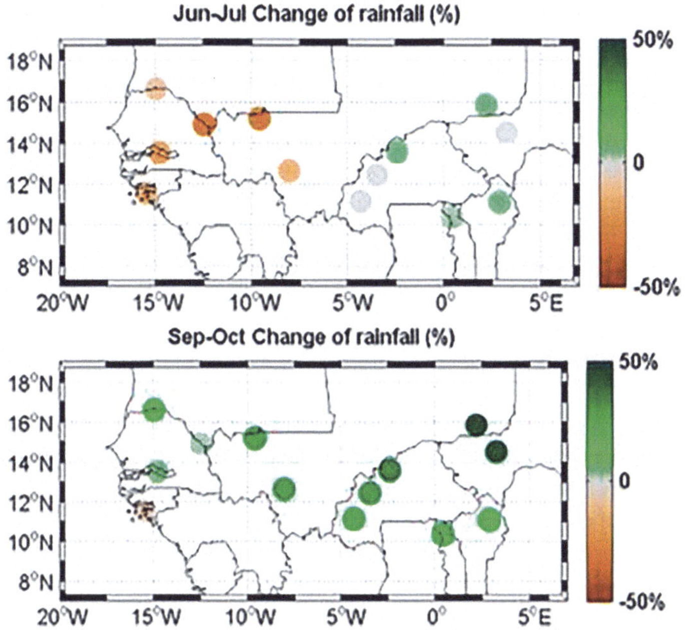

Fig. 16.2 Shows the relative change in precipitation between June and July and September and October in 2031–2060 and 1961–1990 under the RCP8.5 scenario. The calculation of the relative change in precipitation is based on the average of 9 GCM simulations for each of the 13 weather stations

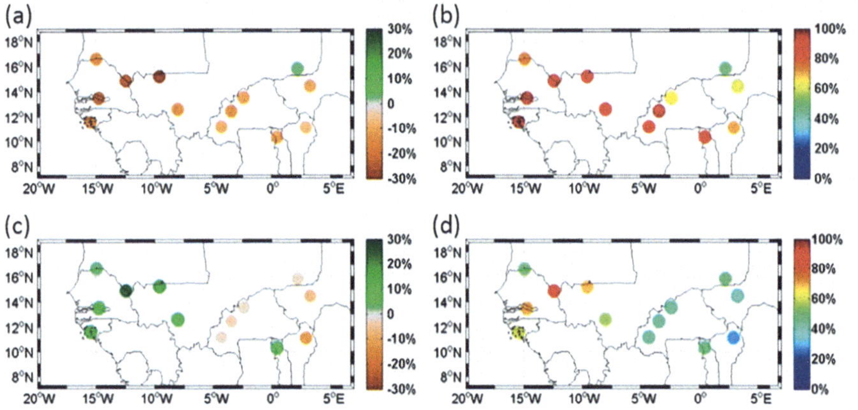

Fig. 16.3 Here we present simulations of mean yield and variability under RCP 8.5. The left panel shows the relative changes in average sorghum yield (**a**) and yield coefficient of variation (**c**) between 2031–2060 and 1961–1990 (%). The yield is the average of two crop models and three sorghum varieties. The right panel shows the consistency (%) of all models where the average yield (**b**) decreased and the coefficient of variation (**d**) increased between 2031 and 2060

predicted to make the Sahel hotter than any temperature experienced in the twentieth century.

The impact of climate change on sorghum yield was analyzed using 2 crop simulation models and 3 sorghum varieties, and the results showed a decrease in average sorghum yield in 12 out of 13 fields (Fig. 16.3a). The study also revealed that the west-east dipole change of precipitation contributes to this adverse effect. The average yield decline in 2031–2050 is a robust feature of our simulation because climate and crop models have good consistency regarding yield decline (Fig. 16.3b). More than 90% of the simulation results led to a decline in the output of four of the six stations in the western Sahel. The consistency between the predicted yield decline simulation results is low in the central Sahel.

16.3 Conclusion

Sorghum is a versatile crop with many potential uses and a wealth of untapped genetic resources, particularly in the wild genetic pool. The wild relatives of sorghum have many beneficial traits that can be used to improve the crop. However, the practical use of these wild sorghum species needs to be improved because of limited knowledge and genetic barriers to wild-cultivated gene transfer. However, recent advances in next-generation sequencing technologies have provided researchers with a wealth of genome data that could facilitate the development of sorghum improvement programs and harness the valuable genetic resources found in wild sorghum. These resources could assist in developing new crops within the Andrographaceae family by transferring functional genes to genera such as saccharin, or the domestication of additional sorghum species (Dillon et al. 2007).

For decision-makers, understanding whether the level of uncertainty in future forecasts will remain the same or decrease significantly in the coming decades is crucial. Global climate change, including increased rainfall, elevated carbon dioxide levels, and projected warming within the optimal range for crop growth and development, has enabled sorghum crops to maintain production during the post-rainy season. Sorghum crops exhibit relatively higher adaptability to harsh environmental conditions, so they are often cultivated in warm and dry regions. Despite the genetic gains achieved through breeding trials and improved crop management, sorghum cultivation area in India has been declining rapidly in recent years, mainly due to various biological and abiotic stress factors that limit farm-level yields. Changes in consumer dietary preferences and the availability of alternative grains have also impacted the demand for sorghum, leading to a decrease in the number of farmers engaged in sorghum cultivation. Therefore, deploying biotechnology tools is necessary to develop heterotic hybrids and open-pollinated varieties resistant to different biological stress factors, such as stem flies, Chilo suppressalis, and cereal molds (Rakshit and Patil 2014). In the 1930s, improvement efforts for sorghum in India primarily involved the selection of pure strains from local varieties. This approach continued until the 1960s and resulted in the development of improved varieties through pure strain selection. Sorghum improvement played a significant role in the dryland green revolution, showcasing its potential for agricultural development in arid regions. Local varieties in India were typically tall, photosensitive, late maturing, and flowering occurred after the cessation of rainfall. These varieties were widely adopted by local farmers but had low harvest indexes. Hybridization between the local varieties led to a modest yield increase of around 5% in experimental settings, but it substantially impacted production in the 1960s. Hybridization and selection among improved local varieties brought about some genetic improvement, although it did not reach a detectable level. However, these efforts helped retain the essential traits of local adaptation (Aruna 2014).

Significant genotypic changes occurred in the 1960s, marking a turning point in the improvement of sorghum. The interaction between variety selection and appropriate management practices resulted in substantial leaps in productivity and enhanced production stability. As a result, sorghum yields increased from 560 kg/ha in 1970 to 1000 kg/ha in 2000. Alongside the development of dwarf and high-yielding varieties, emphasis was placed on harnessing hybrid vigor, as demonstrated in the United States. The discovery of male genetic sterility (ms2) by Stephens in 1937 and, subsequently, cytoplasmic male sterility (msc1) by Stephens and Holland in 1954 opened up possibilities for developing sorghum hybrids. These findings paved the way for the successful development of sorghum hybrids by leveraging male sterility traits.

References

Akinseye FM, Adam M, Hoffmann MP, Traore PCS, Agele SO, Whitbread AM (2017) Assessing crop model improvements through comparison of sorghum (*Sorghum bicolor* L. moench) simulation models: a case study for West African cultivars. Field Crop Res 201:19–31. https://doi.org/10.1016/j.fcr.2016.10.015

Aruna C (2014) High-yielding sorghum cultivars for kharif. In: Chapke RR, Vinayagam SS, Patil JV (eds) Improved sorghum cultivation and value-addition perspectives. Directorate of Sorghum Research, Hyderabad, pp 89–93. ISBN 81-89335-50-2

Barnabás B, Jäger K, Fehér A (2008) The effect of drought and heat stress on reproductive processes in cereals. Plant Cell Environ 31:11–38

Battisti D, Naylor RL (2009) Historical warnings of future food insecurity with unprecedented seasonal heat. Science 323:240

Biasutti M (2013) Forced Sahel rainfall trends in the CMIP5 archive. J Geophys Res Atmos 118:1613–1623

Biasutti M, Sobel AH (2009) Delayed Sahel rainfall and global seasonal cycle in a warmer climate. Geophys Res Lett 36:L23707

Bruinsma J (2003) World agriculture: towards 2015/2030. An FAO perspective. Earthscan, London, 444 pp

Chadalavada K, Kumari BDR, Kumar TS (2021) Sorghum mitigates climate variability and change in crop yield and quality. Planta 253(5):113. https://doi.org/10.1007/s00425-021-03631-2

Dilley M, Chen RS, Deichmann U, Lerner-Lam AL, Arnold M (2005) Natural disaster hotspots: a global risk analysis. World Bank, Washington, DC

Dillon SL, Shapter FM, Henry RJ, Cordeiro G, Izquierdo L, Lee LS (2007) Domestication to crop improvement: sorghum and Saccharum (Andropogoneae) genetic resources. Ann Bot 100:975–989. https://doi.org/10.1093/aob/mcm192

FAO (1999) Food insecurity: when people must live with hunger and fear of starvation. The state of food insecurity in the world. FAO, Rome, p 32

Fontaine B, Roucou P, Monerie P-A (2011) Changes in the African monsoon region at medium-term time horizon using 12 AR4 coupled models under the A1b emissions scenario. Atmos Sci Lett 12:83–88

Hariprasanna K, Patil JV (2015) Sorghum: origin, classification, biology, and improvement. In: Sorghum molecular breeding. Springer, Berlin, pp 3–20. https://doi.org/10.1007/978-81-322-2422-8_1

ICRISAT (2009) Sorghum. International Crops Research Institute for the Semi-Arid Tropics, Patancheru. http://www.icrisat.org/sorghum/sorghum.htm

Kouressy M, Dingkuhn M, Vaksmann M, Heinemann AB (2008a) Adaptation to diverse semi-arid environments of sorghum genotypes having different plant type and sensitivity to photoperiod. Agric For Meteorol 148:357–371

Kouressy M, Dingkuhn M, Vaksmann M, Clément-Vidal A, Chantereau J (2008b) Potential contribution of dwarf and leaf longevity traits to improve photoperiod sensitive sorghum. Eur J Agron 28:195–209

Monerie P-A, Fontaine B, Roucou P (2012) Expected future changes in the African monsoon between 2030 and 2070 using some CMIP3 and CMIP5 models under a medium-low RCP scenario. J Geophys Res 117:D16111

Monerie P-A, Roucou P, Fontaine B (2013) Mid-century effects of climate change on African monsoon dynamics using the A1B emission scenario. Int J Climatol 33:881–896

Nelson GC, Rosegrant MW, Koo J et al (2009) Climate change impacts on agriculture and costs of adaptation. International Food Policy Research Institute, Washington, DC

Peng W, Berry EM, Ferranti P, Berry E, Anderson J (2019) The concept of food security. Encycl Food Secur Sustain 2:1–7

Rakshit S, Patil JV (2014) Sorghum. In: Chopra VL (ed) Breeding field crops II: advances. Studium Press (LLC), Houston, pp 79–96

Raza A, Razzaq A, Mehmood SS, Zou X, Zhang X, Lv Y, Xu J (2019) Impact of climate change on crops adaptation and strategies to tackle its outcome: a review. Plants 8(2):34

Roy JP, Tschakert H, Waisman S, Abdul Halim P, Antwi-Agyei P, Dasgupta B, Hayward M, Kanninen D, Liverman C, Okereke PF, Pinho K, Riahi AG, Suarez Rodriguez (2018) Sustainable development, poverty eradication and reducing inequalities. In: Masson-Delmotte V, Zhai P, Pörtner H-O, Roberts D, Skea J, Shukla PR, Pirani A, Moufouma-Okia W, Péan C, Pidcock R, Connors S, Matthews JBR, Chen Y, Zhou X, Gomis MI, Lonnoy E, Maycock T, Tignor M (eds) Waterfield T Global warming of 1.5°C. An IPCC special report on the impacts of global warming of 1.5°C above pre-industrial levels and related global greenhouse gas emission pathways, in the context of strengthening the global response to the threat of climate change, sustainable development, and efforts to eradicate poverty. IPCC, Geneva

Schmidhuber J, Tubiello FN (2007) Global food security under climate change. Proc Natl Acad Sci U S A 104:19703–19708

Stephens JC (1937) Male sterility in sorghum: its possible utilization in production of hybrid seed. J Am Soc Agron 29:690

Stephens JC, Holland RF (1954) Cytoplasmic malesterility for hybrid sorghum seed production. Agron J 46:20

Sultan B, Roudier P, Baron C, Quirion P, Muller B, Alhassane A, Ciais P, Guimberteau M, Traoré SB, Dingkuhn M (2013) Assessing climate change impacts on sorghum and millet yields in West Africa. Environ Res Lett 8:014040

Traoré SB et al (2011) Characterizing and modeling the diversity of cropping situations under climatic constraints in West Africa. Atmos Sci Lett 12:89–95

UN Report (2019) World population prospects—2019 highlights. https://www.un.org/development/desa/publications/world-population-prospects-2019-highlights.html. Accessed on 23 Jan 2021

Summary of the Potential and Exploitation of Omics and Biotechnological Breakthroughs in Sorghum

17

Sunita Gorthy, Pramod Sargar, Chaitra Hugar, Jieqin Li, and Ephrem Habyarimana

Abstract

Sorghum (*Sorghum bicolor* (L.) Moench) is a resilient cereal crop of substantial agricultural and industrial importance. Recent advancements in omics and biotechnology offer avenues for elevating sorghum productivity, nutritional content, and stress resilience. Omics technologies, including genomics, transcriptomics, proteomics, metabolomics, and genetic engineering, play a pivotal role in enhancing sorghum production. Multi-omics analyses offer a molecular understanding of sorghum by shedding light on the mechanisms controlling cell-specific regulation and stress tolerance. The potential and application of omics and biotechnological advances in sorghum are highlighted in this overview. It highlights how these technologies can be used to address global issues pertaining

S. Gorthy · E. Habyarimana (✉)
International Crops Research Institute for the Semi-Arid Tropics, Hyderabad, Telangana, India
e-mail: Ephrem.Habyarimana@icrisat.org; E.Habyarimana@cgiar.org

P. Sargar
International Crops Research Institute for the Semi-Arid Tropics (ICRISAT), Patancheru, India

Vasantrao Naik Marathwada Krishi Vidyapeeth, Parbhani, Maharashtra, India

C. Hugar
International Crops Research Institute for the Semi-Arid Tropics (ICRISAT), Patancheru, India

University of Agricultural Sciences, Raichur, Karnataka, India

J. Li
Anhui Science and Technology University (AHSTU), Fengyang, Anhui, China

E. Habyarimana et al. (eds.), *Omics and Biotechnological Approaches for Product Profile-Driven Sorghum Improvement*,
https://doi.org/10.1007/978-981-97-4347-6_17

to sustainability and food security. The importance of omics in advancing sorghum research is highlighted by important discoveries and applications like mapping studies, the discovery of differentially expressed genes, and the creation of improved sorghum varieties. Sorghum research is driven by the combination of biotechnological and omics technologies, which can be leveraged to address global food security and agriculture challenges. The integration of these technologies represents a major advancement in sorghum research and cultivation and holds great promise for sustainable agriculture and a secure food supply.

Keywords

Sorghum · Metabolomics · Transcriptomics · Genome editing · Genomics · Proteomics · Phenomics

17.1 Introduction

Sorghum, commonly denominated as the "camel of crops," occupies a prominent position in the realm of agriculture due to its notable adaptability across diverse environmental conditions, spanning from arid regions to temperate climates. Sorghum is an important crop for agriculture and the fifth most extensively cultivated crop in the world, after wheat, maize, rice, and barley. The U.S. Department of Agriculture (2020) reports that 4,8407,000 metric tons of sorghum were produced worldwide. Notably, Nigeria, Sudan, and Mexico are the world's top three producers of sorghum followed by the United States, Ethiopia, and India (Fig. 17.1). The integration of biotechnological methods and omics technology has enabled research and development efforts related to sorghum breeding and enhancement to accelerate in recent years. This chapter endeavors to synthesize and encapsulate noteworthy advancements in the field, elucidating key discoveries and applications of omics technologies in augmenting sorghum traits. Sorghum stands as a versatile crop with worldwide applications, serving as both biomass and grain for human consumption, as well as fodder and poultry feed. Notably, recent studies employing omics technologies have sought to unravel the potential roles of signaling molecules and metabolic pathways in plant stress responses, employing sorghum as an exemplary model crop (Ibraheem et al. 2018).

By utilizing whole-genome sequencing methods, the genetic potential of sorghum can be unlocked, and genes of agronomic importance can be identified. Furthermore, the application of multi-omics analysis contributes to the clarification of the molecular mechanisms behind sorghum's salt tolerance, providing essential insights into these mechanisms and enabling the improved use of saline lands (Ren et al. 2022). The International Crops Research Institute for the Semi-Arid Tropics (ICRISAT) and national programs are actively attempting to genetically enhance sorghum through the use of omics technologies, such as genotyping-by-sequencing, whole genome resequencing and marker-assisted breeding (see Fig. 17.2) (Ashok Kumar et al. 2011; Hu et al. 2019). Genotyping-by-sequencing has played a crucial

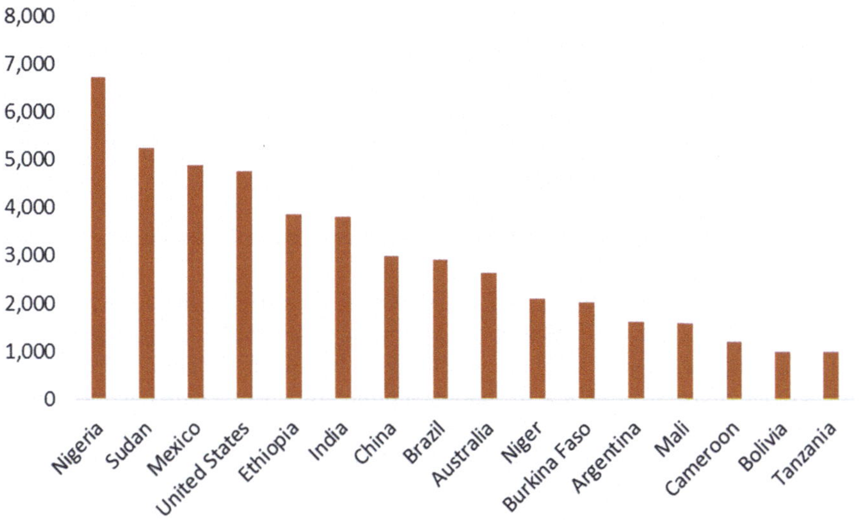

Fig. 17.1 Production (1000 MT) statistics and percentage share of the world's top 10 sorghum-producing countries (U.S. Department of Agriculture 2020)

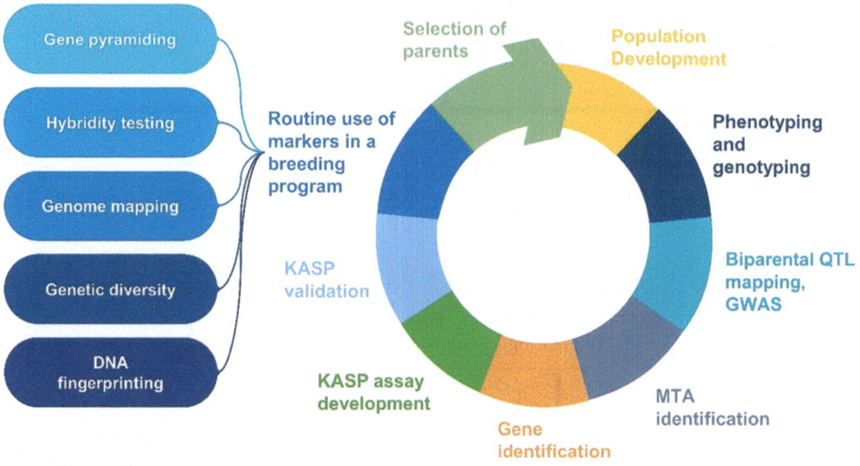

Fig. 17.2 Important stages of marker-assisted selection (MAS)

role in developing detailed polymorphism maps for sorghum, pinpointing the genomic regions linked to essential traits like grain iron (Fe) and zinc (Zn) concentrations.

In summary, the integration of omics technologies (Fig. 17.3) holds significant promise for substantially enhancing sorghum production and utilization. Recent biotechnological advancements utilizing these technologies have predominantly focused on elevating the crop's genetic potential, augmenting its nutritional value, and fortifying its resistance to both biotic and abiotic stresses. This chapter will elucidate select studies highlighting the improvements in the genetic potential, nutritional value, and stress resilience of sorghum achieved through the application of omics technology in sorghum research.

17.2 Genomics in Sorghum

17.2.1 Sorghum Genomics Resources: Structural Genomics

Genetic mapping initiatives in sorghum started in the early 1990s, as depicted in Table 17.1. The early utilization of the restriction fragment length polymorphism (RFLP) marker system in sorghum was reported by Tao et al. in 1998. Following this, a range of marker systems, including amplified fragment length polymorphism (AFLP), randomly amplified polymorphic DNA (RAPD), single sequence repeats (SSRs), and diversity array technology (DArT) markers, were introduced for sorghum (Haussmann et al. 2002; Bowers et al. 2003; Wu et al. 2006; Mace et al. 2008). These diverse marker systems were employed in quantitative trait loci (QTL) mapping, genetic diversity (GD) analysis, and genome fingerprinting within the sorghum species. The publication of the sorghum genome sequence, which contains

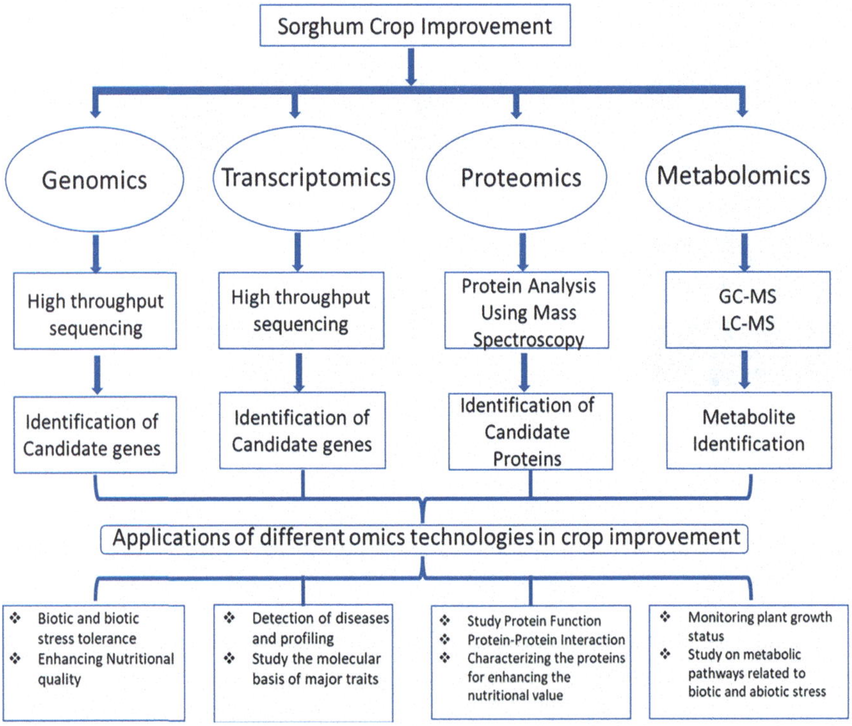

Fig. 17.3 Representation of different omics technologies in sorghum crop improvement

Table 17.1 Sorghum genetic linkage maps

Female	Male	Population size	Molecular markers	Map size[a] (cM)	Authors
	F2s				
M 91051	Shanqui Red	55	37 RFLPs	283 (8)	Hulbert et al. (1990)
IS 18809	IS 2482C	81	85 RFLPs, 7 isozymes	949 (13)	Whitkus et al. (1992)
IS 24756	IS 18729	149	35 RFLPs	440 (5)	Binelli et al. (1992)
IS 24756	IS 18729	55	96 RFLPs	709 (15)	Berhan et al. (1993)
CK 60	PI 229828	78	201 RFLPs	1530 (10)	Pereira et al. (1994)
IS 3620C	BTx 623	50	190 RFLPs	1789 (14)	Xu et al. (1994)
BTx 623	*Sorghum propinquum*	56	276 RFLPs	1445 (10)	Chittenden et al. (1994)
PI 229828	CK 60	152	111 RFLPs	1299 (10)	Pereira and Lee (1995)
BTx 623	*Sorghum propinquum*	370	202 RFLPs	935 (11)	Lin et al. (1995)
CK 60	PI 229828	68	7 SSRs	1575	Taramino et al. (1997)
Shihong 137	L-Tian	186	118 SSRs	1884.6 (15)	Guan et al. (2011)
B2 V4	1383-2	150	122 MSAP, 22 SSRs	483.6 (11)	Duan et al. (2009)
	F3				
BTx 623	BSC 35	93	71 RFLPs	633 (15)	Ragab et al. (1994)
	F5s				
379	IS 2807	110	145 RFLPs	977	Dufour et al. (1997)
IS 2807	249	91	133 RFLPs	877 (13)	Dufour et al. (1997)
IS 2807	379	91	133 RFLPs	877 (13)	Dufour et al. (1997)
QL 39	QL 41	128	155 RFLPs, 8 SSRs	1400 (21)	Tao et al. (1998)
IS 2807	379	110	298 RFLPs, 137 AFLPs	1899 (11)	Boivin et al. (1999)
B 35	Tx 430	96	142 RFLPs	1602 (14)	Crasta et al. (1999)
QL 41	QL 39	152	281 RFLPs, 5 morphological markers, 25 SSRs	>2750 (10)	Tao et al. (2000)

(continued)

Table 17.1 (continued)

Female	Male	Population size	Molecular markers	Map size[a] (cM)	Authors
	RILs (F6 onwards)				
Composite map	–	183 RFLPs	1096 (12)	Dufour et al. (1997)	
TX 7078	B 35	98	20 RFLPs, 150 RAPD	1580 (17)	Tuinstra et al. (1997)
IS 3620C	BTx 623	137	344 RFLPs	1364 (10)	Peng et al. (1999)
Tx 7000	B 35	98	3 SSRs, 214 RFLPs and 7 RAPD	1200 (10)	Subudhi et al. (2000)
BTx 623	IS 3620C	138	114 RFLPs, 31 SSRs	1287 (10)	Kong et al. (2000)
Tx 7000	B 35	98	162 RFLPs	837 (10)	Xu et al. (2000)
BTx 623	IS 3620C	139	323 RFLPs, 143 SSRs	1406 (10)	Bhattramakki et al. (2000)
SC 56	Tx 7000	125	144 RFLPs	1355 (10)	Kebede et al. (2001)
	Mixed map	–	154 RFLPs, 10 morphological markers, 34 SSRs	1450 (10)	Bennetzen et al. (2001)
BTx 623	IS 3620C	137	Morphological markers, 259 RFLPs, 226 SSRs, 303 DArTs	1528 (10)	Mace et al. (2008)
296B	IS 18551	168	38 EST-SSRs, 107 SSRs, 3 morphological markers, 10 Unigene SSRs	1143 (16)	Satish et al. (2009)
N 13	E 36-1	94	14 RFLPs, 45 genomic SSRs, 3 RAPD markers, 125 AFLPs, 55 EST-SSRs	2838 (10)	Ramu et al. (2009)
296B	IS 18551	168	100 SSRs, 28 EST-SSRs	1074.5 (15)	Srinivas et al. (2009a)
296B	IS 18551	168	100 SSRs, 38 EST-SSRs, 10 Unigene SSRs, 2 morphological markers	1098.5 (15)	Srinivas et al. (2009b)
R 890592	ICSV 745	119	234 DArTs, 244 RFLPs, 10 SSRs	1433 (10)	

(continued)

Table 17.1 (continued)

Female	Male	Population size	Molecular markers	Map size[a] (cM)	Authors
IS 8525	R 931945-2-2	146	47 SSRs, 357 DArTs, 188 AFLPs	1453 (10)	
SC 170-6-8	B 923296	88	13 SSRs and 170 DArTs	1138 (10)	
QL 12	BTx 642	94	117 DArTs	910 (10)	
SSM 249	SAR 10	183	131 SSRs, 627 DArTs, 47 RFLPs	1227 (10)	
	Consensus map	–	839, non-DArT markers and 1190 DArTs	1603.5 (10)	
M 71	SS 79	188	6 EST-SSR, 49 SSR, 102 AFLP	1029 (11)	Shiringani et al. (2010)
IS 2122	27B	210	127 SSRs, one morphological marker and 21 genic-SSRs	700 (10)	Aruna et al. (2011)
SS 79	M7 1	188	102 AFLP, 6 EST-SSR and 49 SSR	1029 (11)	Shiringani and Friedt (2011)
B 923296	SC 170-6-8	141	377 DArTs	2259 (10)	Mace and Jordan (2011)
M35-1	B 35	245	228 SSRs and 3 morphological markers	1235.5 (10)	Nagaraja Reddy et al. (2013)
BR 007	SC 283	90	255 DArTs, one RFLP marker 83 SSRs, and 5 sequence-tagged site (STS)	2034.9 (10)	Sabadin et al. (2012)
74LH 3213	MS 138B	133	247 SSRs	1697.2 (10)	Takai et al. (2012)
654	LTR 108	244	3418 bin markers, SNPs	1591.4 (10)	Zou et al. (2012)
BTx 623	*Sorghum propinquum*	163	141 SSRs	773.1 (10)	Kong et al. (2013)
ICSV 745	R 890562	119	234 DArTs, and 244 RFLPs	1487.96 (12)	Alam et al. (2014)
296B	PVK 801	342	10 SSRs, 1148 DArTs, 927 DArTSeqs and 13 SSRs	1355.52	Kotla et al. (2019)
Tx623A	Sudangrass	RILs	1065 SLAF markers	1191.7	Jin et al. (2021)

[a] In the genetic map length column, the numbers within parentheses represent the linkage groups

approximately 71,000 SSRs, provided opportunities for the expansion of SSR molecular markers. The combination of the complete sorghum genome sequence with next-generation sequencing (NGS) techniques like genotyping-by-sequencing (GBS) has streamlined the development of cost-effective marker systems. The release of initial reference genome accessions for sorghum in 2009 by Paterson et al., alongside advancements in NGS tools and rapid progress in computational tools (CTs), opened new avenues for applying genomic DNA markers for high-throughput genotyping (HTG) in sorghum. Recent advancements in sequencing and comprehensive-genome genotyping tools, such as NGS and whole-genome marker arrays (WGMA), have facilitated the accumulation of vast genomic resources for sorghum conservation (Morris et al. 2013a; McCormick et al. 2018). Notably, additional sorghum lines have undergone resequencing using NGS methods like Illumina Genome Analyzer sequencing technology (IGAST) and the whole-genome shotgun technique (WGST), combined with more efficient alignment software. This approach allows for the exploration of various structural variations and single nucleotide polymorphisms (SNPs) in sorghum genomes that have already been sequenced (Nelson et al. 2011; Morris et al. 2013b; Bekele et al. 2013). As outlined by Saxena et al. in 2014, these structural alterations encompass SNPs, copy number variations, insertions-deletions (InDels), duplications, translocations, inversions, and absence/presence variants.

Genotyping-by-sequencing (GBS) is a commonly used method in genome-wide association studies (GWAS). It uses restriction enzymes to simplify genome complexity and a multiplex approach to detect high-density SNPs (Elshire et al. 2011). The GBS tool has been extensively used to assess GD and structure in sorghum accessions (Morris et al. 2013a), facilitating the identification of novel alleles and genes associated with important traits. This enhances the quality of germplasm utilized in breeding programs. Sorghum research utilizes genomic resources like DNA markers, high-density genetic maps, DArT, SNPs, and fully sequenced genomes (Billot et al. 2012; Evans et al. 2013; Kong et al. 2013; Mace et al. 2013). RAD-seq enables high-resolution genotyping and SNP identification thus reducing the genomic complexity. RAD-seq approach was used to identify 283,000 SNPs from various sorghum accessions (Nelson et al. 2011). Understanding genome structure changes is crucial for biological and evolutionary studies, highlighting the need to explore genetic variations and plant performance.

17.2.2 Reference Genome of Sorghum

The initial reference genome of sorghum was obtained on BTx 623, by using Sanger sequencing method (Paterson et al. 2009) with subsequent modifications by McCormick et al. (2018). According to the reports, this diploid crop consists of 34,129 genes, retrotransposons, and DNA transposons along with many sequence variations, copy number variations (CNV), InDels, and the presence and absence variations (PAVs) (Zheng et al. 2011; Mace et al. 2013).

Numerous microarray datasets and SNP databases are available for sorghum, with technological advancements allowing for more genome sequences (Hua et al. 2019). The grain sorghum line Tx430, assembled using Oxford Nanopore sequencing and Bionano Genomics Direct Label and Stain optical mapping, has 39,510 predicted genes (Deschamps et al. 2018). Tx430 is different from BTx623 having 39,510 predicted genes, which is a greater number than the latter. Its median length is also shorter. Subsequently, the Rio genome of sweet sorghum was then created to improve our knowledge of its sugar metabolism. The genome size of Rio was 729.4 Mb, encompassing 35,476 predicted genes, including 54 specific to Rio. To gain insight into sugar metabolism, the genome of Rio's sweet sorghum was assembled at the chromosomal level. Each of these genome assemblies is specific to its corresponding accession and does not fully capture the GD within the species (Bayer et al. 2020).

Acknowledging the limitations of relying solely on a single reference genome to capture the GD within a species, researchers adopted a "pan-genome" strategy for sorghum. This approach involved compiling a comprehensive dataset using an ongoing mapping and assembly technique, incorporating 354 sequenced genomes, of which 176 had a coverage exceeding 10X. In contrast to the original reference assembly spanning 708 Mbp, the pan-genome has expanded by 20%, reaching 883 Mbp (Ruperao et al. 2021). The significant genetic variation observed across different cultivars likely contributes to this expanded sequence diversity (Cuevas et al. 2019). The newly established pan-genome encompasses an average of 32,795 genes across all cultivars, with a core set of 16,821 genes. Additional reference genomes are deemed crucial for gaining a comprehensive understanding of the genetic basis underlying the phenotypic diversity that emerged during the domestication and evolution of sorghum. Table 17.2 represents the reference genome released for sorghum so far, showcasing the improvements in genomic resources from different genotypes. The annotations are updated and refined. In Table 17.2 there are different sorghum genotypes with details on its annotations gene numbers and their corresponding protein transcripts. Version V5.1 is completely resequenced and updated using PacBio long read data. There are few additions in the version V3.1 resources, but with additional RNA-seq from JGI projects and full-length transcripts included to further improve the completeness of the gene set.

The fully sequenced reference genome (sorghum genome) has revolutionized genomic research in sorghum, serving as a foundation for diverse studies. The development of pan-genomes enables exploration of GD within sorghum populations, facilitating breeding efforts. Genomic data (Table 17.3) aids in understanding mechanisms governing complex traits related to yield, composition, and environmental adaptations. Recent studies, such as the revelation of the Rio sweet sorghum reference genome (Cooper et al. 2019) and the development of ten high-quality genome assemblies for biofuel improvement (Voelker et al. 2023), exemplify the utility of genomic data in advancing sorghum research. The rise of pan-genomic data for essential crop species offers new insights into plant diversity and structural alterations influencing agriculturally significant phenotypes (Zhou and Liu 2022). These studies collectively contribute to a profound understanding of genes, genomes,

Table 17.2 *Sorghum bicolor* reference genome released to date

	Btx 623	Wray	SC187	BTx642	RTx430	Rio
Assembly Source:	JGI	JGI	JGI	JGI	JGI	JGI
Assembly Version:	v5.0	v1.0	v1.0	v1.0	v2.0	v2.0
Annotation Version:	v5.1	v1.1	v1.1	v1.1	v2.1	v2.1
Total scaffold length (bp):	719,894,357	736,814,591	688,333,777	680,145,566	677,670,093	729,379,862
Number of scaffolds:	65	195	184	136	175	281
Min. number of scaffolds containing half of assembly (L50):	5	5	5	5	5	5
Shortest scaffold from L50 set (N50):	71,215,609	71,990,388	65,552,663	64,512,399	64,573,693	70,703,592
Total contig length (bp):	719,654,357	736,744,591	687,823,777	678,955,566	674,290,093	693,889,862
Number of contigs:	89	202	235	255	513	3830
Min. number of contigs containing half of assembly (L50):	6	6	11	18	51	423
Shortest contig from L50 set (N50):	50,741,355	63,876,658	22,956,415	12,093,548	4,014,886	396,665
Number of protein-coding transcripts:	44,210	55,249	48,344	49,348	46,881	41,048
Number of protein-coding genes:	32,160	30,610	32,101	35,115	34,601	35,490

Source: phytozome-next.jgi.doe.gov

Table 17.3 Key genomic resources for sorghum

Resource	Description	References
Sorghum genome	The fully sequenced sorghum reference genome	https://www.sorghumbase.org/; https://phytozome-next.jgi.doe.gov/
Pan-genome	Exploration of GD	https://ngdc.cncb.ac.cn/sorgsd/
QTL databases	Quantitative trait loci information	https://aussorgm.org.au/sorghum-qtl-atlas/; https://www.sorghumbase.org/guides?section=Genomic%20Resources/QTLs
SNP markers	Single nucleotide polymorphism markers	https://ngdc.cncb.ac.cn/sorgsd/

and complex pathways, paving the way for sustainable enhancements in sorghum crops.

At ICRISAT, a comprehensive sorghum pan-genome was constructed by integrating reference genomes with genetic information from 354 diverse sorghum accessions representing various races. This study systematically investigated the structural and functional attributes of the pan-genome, elucidating its practical applications in facilitating genetic advancements. The resultant pan-genome encompasses a total of 35,719 genes, comprising a core genome with 16,821 genes, and an average of 32,795 genes per cultivar. Notably, the variable genes within the pan-genome exhibit enrichment in environment-responsive genes, allowing for the classification of sorghum accessions based on their racial characteristics. Approximately 53% of the identified genes manifest presence-absence variation, and a subset of these variable genes is predicted to be functionally linked to traits related to drought adaptation.

Employing a dataset of over 2 million SNPs derived using pan-genome, 398 SNPs were significantly correlated with crucial agronomic traits, with 92 of these SNPs residing within gene loci. Further analysis focusing on drought-related gene expression revealed 1788 genes functionally associated with diverse conditions, with 79 genes notably absent from the reference genome assembly. This study thus contributes valuable genomic diversity resources for sorghum, offering a foundation for genome-assisted crop improvement. The integration of databases containing information on QTL and SNP markers enhances marker-assisted breeding strategies, facilitating the development of improved sorghum varieties endowed with enhanced traits.

Sorghum exhibits diverse applications encompassing food, feed, and bioenergy production, thus categorizing into four primary ideotypes: grain, sweet, cellulosic, and forage. Realizing the full potential of these ideotypes necessitates an in-depth comprehension of the genetic architecture influencing yield and local adaptability. The initial adoption of the restriction fragment length polymorphism (RFLP) marker system in sorghum marked a pioneering step, succeeded by the utilization of other marker systems such as AFLP, RAPD, SSRs, and DArT. These markers played pivotal roles in QTL mapping, GD analysis, and fingerprinting (Tao et al. 1998;

Table 17.4 Chronological order showing sequencing of sorghum genome

Year of publishing	Genotypes sequenced	Key findings	References
2009	1 (BTx623: reference genome)		Paterson et al. (2009)
2011	3	1,057,018 SNPS	Zheng et al. (2011)
		99,948 InDels	
		1500 genes	
		16,487 PAVs	
		17,111CNVs	
2013	44	4,946,038 SNPs	Mace et al. (2013)
		1,982,971 InDels	
		725 genes	
		120,929 CNVs	
2018	241	31,946,640 SNPs	Zhang et al. (2018)
		4,266,768 InDels	
2019	10,323 GBS sequencing		Hu et al. (2019)
2019	1 (Rio)	35,467 genes	Cooper et al. (2019)
		1,890,101 SNPs	
		391,466 InDels	
2021	13	44,079 gene families	Tao et al. (2019)
2022	*S. bicolor* × *S. halepense (SbxSh)* 172 genotypes	567,046,841 SNPs	Habyarimana et al. (2022)
		91,825,474 InDels	
		1,532,171 SVs	
		4,973,961 CNVs	
		5548 genes	
2023	10 (carbon partitioning-NAM population)	171,000 genes	Voelker et al. (2023)

Haussmann et al. 2002; Bowers et al. 2003; Wu et al. 2006; Mace et al. 2008). The accessibility of the sorghum genome sequence facilitated the generation of a substantial number of markers. Presently, advancements in sequencing technologies and whole-genome genotyping tools, including NGS and WGMA, have empowered the conservation of genomic resources in sorghum. These tools, coupled with computational methodologies, have been instrumental in resequencing diverse sorghum lines, investigating various structural variants, and SNPs within previously sequenced genomes (Table 17.4). Structural genomic mutations, encompassing SNPs, copy number variations, InDels, duplications, translocations, inversions, and absence/presence variants, have been elucidated through these approaches (Morris et al. 2013a; McCormick et al. 2018).

Studies outlined above underscore the significance of structural genomic mutations in shaping species diversity and evolution. In the context of sorghum, genomic investigations have linked structural mutations to key traits such as dwarfing, juicy stalks, chilling tolerance (CT), and flowering time. Comparative whole-genome analysis between the sweet sorghum genotype 'Rio' and 'BTx623' revealed numerous gene PAVs, including those associated with sucrose transporters (Cooper et al. 2019). Additionally, a GWAS exploring the genetic architecture of bioenergy-related traits identified a substantial deletion in a sorghum-specific iron transporter linked to stalk sugar accumulation (Brenton et al. 2016). A recent comprehensive survey of genome-wide deletions in a diverse panel of nearly 350 sorghum accessions unveiled large deletions in genes related to biotic and abiotic stress responses, demonstrating unique patterns corresponding to geographic origins and indicative of local adaptation (Songsomboon et al. 2021). Several well-known sorghum genes, including Dw1, Dw2, Dw3, and qPHT7.1 (Quinby and Karper 1954), exert significant influence on plant height, flowering duration, and seed coloration. Allelic variation in other genes, such as Ma1 and Tannin1, further suggests a role in stress adaptation (Sattler et al. 2014; Cuevas et al. 2016). These genes collectively represent the outcomes of diverse research endeavors aimed at enhancing the sorghum crop.

17.3 Transcriptomics

The investigation of molecular processes underlying the response of sorghum (*Sorghum bicolor*) to both biotic and abiotic stressors has been conducted employing advanced transcriptomics and proteomics technologies. These methodologies facilitate the identification of specific genes and proteins implicated in stress response mechanisms, thereby contributing to the development of stress-resistant sorghum cultivars. Transcriptome studies have been pivotal in elucidating sorghum's response to various abiotic stimuli, notably drought, and have enabled the identification of genes associated with traits like drought tolerance. In a particular study, microarrays were utilized to examine the transcriptional response of *Sorghum bicolor* to the individual and combined effects of heat and drought stress, revealing distinct transcriptional patterns for combined stress conditions (Johnson et al. 2014). Another investigation analyzed the transcriptome of field-drought-affected sorghum from seedling to maturity, identifying biotic and metabolic responses within a 1-week timeframe (Varoquaux et al. 2019).

A comparative study of transcriptome profiles between two sorghum lines, the drought-tolerant SC56 and drought-sensitive Tx7000, revealed differential gene expression associated with drought tolerance (Azzouz-Olden et al. 2020). Notably, SC56 exhibited superior performance in wet conditions by upregulating growth-related processes and ensuring homeostasis. Recent advancements in transcriptomic studies in 2023 expanded upon previous research by analyzing the expression of all identified sorghum genes, emphasizing differentially expressed genes linked to drought tolerance (Abdel-Ghany et al. 2020). Drought-resistant sorghum genotypes exhibited elevated expression of stress tolerance, detoxification, and damage control/repair genes. Another study in 2022 investigated transcriptome data of glutinous and nonglutinous sorghum varieties, revealing distinct patterns of starch accumulation, and identifying genes involved in starch biosynthesis and degradation (Ke et al. 2022). Furthermore, a 2023 study underscored the significance of drought-resistant sorghum genotypes by examining the expression of all annotated sorghum genes and identifying differentially expressed genes associated with drought resistance (Abdel-Ghany et al. 2020).

These recent advances in transcriptomic studies underscore the potential of sorghum in crop improvement. Transcriptomics enables the identification of genes responsive to diverse stressors and aids in the biosynthesis of essential nutrients, such as carotenoids, crucial for enhancing sorghum nutritional content. Additionally, insights into metabolic pathways related to sorghum growth and development gleaned from transcriptomic studies provide a foundation for strategies aimed at improving yield and quality. In summary, transcriptomic studies contribute significantly to the development of resilient sorghum varieties, with enhanced stress resistance, nutritional content, and increased yields.

17.4 Proteomics

Proteomics denotes the systematic investigation of proteins on a large scale, encompassing their structures and functional attributes. Sorghum, an integral cereal crop cultivated globally, has been subjected to extensive proteomic scrutiny, particularly concerning its responses to diverse environmental challenges such as abiotic stress, insect infestation, and drought. A study published in 2020 delved into the proteomic analysis of G-46 and CSV 44F sorghum genotypes, employing various protein extraction methods from sorghum leaves exposed to salt stress. Subsequently, the electrophoretic patterns of these proteins were scrutinized to comprehend the intricate dynamics of the proteome and transcriptome profiling in sorghum under salt-induced stress conditions (Punia et al. 2020). Another investigation in the same year utilized physiological and comparative proteomics approaches to elucidate the response of sorghum roots at the seedling stage to simulated drought stress. This study identified 49 distinct proteins exhibiting differential expression in response to the imposed stress conditions (Li et al. 2020).

In the context of insect pest resistance, an investigation employed label-free quantitative proteomics to discern the proteomic landscape of sorghum implicated in its defense against the spotted stem borer, a lepidopteran insect pest causing substantial economic ramifications. The study scrutinized and juxtaposed the proteomes of three sorghum genotypes—ICSV700, IS2205 (resistant), and Swarna (susceptible)—characterized by differential resistance/susceptibility to insect pests. Gene ontology analysis unveiled those proteins responsive to *Chilo partellus* in resistant *Sorghum bicolor* genotypes predominantly participated in stress and defense mechanisms, small molecule biosynthesis, amino acid metabolism, catalytic activity, and translation regulation (Tamhane et al. 2021). In the context of abiotic stress tolerance, particularly drought, a comprehensive examination of whole-plant transcriptome and proteome responses illuminated the involvement of cellular processes contributing to drought tolerance in sorghum (Ngara et al. 2021). A 2020 publication employed comparative physiological and root proteome analyses to discern responses to water limitation in two sorghum varieties, revealing intricate protein networks that potentially underlie sorghum's adaptive responses to water limitation (Goche et al. 2020). Collectively, proteomics has served as a valuable tool for investigating diverse facets of sorghum biology, encompassing its responses to stress, insect pests, and drought. These inquiries have facilitated the identification of pivotal proteins and pathways integral to sorghum's capacity to endure and flourish under challenging environmental conditions.

17.5 Metabolomics

Metabolomics, a scientific discipline concerned with the comprehensive analysis of small molecules participating in metabolic pathways, has played a pivotal role in elucidating the metabolic networks within the sorghum plant. Several important metabolites in sorghum such as tannins, starch, anthocyanins, lignin, etc., play the crucial role in various functions. The accrued insights have notably contributed to the advancement of sorghum varieties characterized by enhanced nutritional profiles and bioenergy-related attributes. This burgeoning field of metabolomics represents a novel avenue for refining our comprehension of the impact of GD on the phenotypic variation observed in plant species. Ongoing endeavors in metabolomics research have been specifically oriented toward augmenting the quality and yield-associated characteristics of various crops, with a particular focus on sorghum, as outlined by Kumar et al. in 2017. Given sorghum's paramount significance in global agriculture, metabolomics methodologies have been adeptly employed to scrutinize its performance under diverse field conditions, as evidenced by the work of Manuela Mandrone et al. in 2021. The application of metabolomics extends beyond mere analytical exploration; it plays a pivotal role in elucidating environment–gene interactions, mutant characterization, phenotyping, identification of biomarkers, and, ultimately, the enhancement of crop traits, as highlighted by Razzaq et al. in 2019.

Metabolomics has been used in sorghum to investigate the changes occurring in different metabolites among different cultivars under different stress conditions to uncover biomarkers indicative of stress responses and pinpoint metabolic deficiencies in salicylic acid biosynthesis under nitrogen stress (Sheflin et al. 2019; Sakurai 2022). Integrating metabolomics with quantitative genetics, transcriptomics, and genetic modification underscores its profound relevance in enhancing plant traits (Razzaq et al. 2019). This approach is very efficient in advancing sorghum breeding by elucidating stress response pathways and identifying biomarkers for selecting superior traits. In sorghum breeding lines, both primary and secondary metabolites have been identified through nontargeted metabolomics, particularly in leaves and roots. A study involving 11 sorghum breeding lines revealed significant variations in 21 morphophysiological traits across different sorghum types. Notably, metabolites such as chlorogenic and shikimic acid showed correlations with various traits, highlighting the potential of integrating metabolomics with morphophysiological datasets (Turner et al. 2016).

Understanding root exudation is pivotal for plants as it involves the release of small molecules into the rhizosphere. However, translating this process to crops in systems such as hydroponics or sterile media remains a challenge. A study focusing on sorghum (*Sorghum bicolor* L. Moench) utilized ultra-performance liquid chromatography and gas chromatography mass spectrometry-based nontargeted metabolomics to assess the variation in exudate composition among two genotypes across three substrates (sand, clay, and soil). The research revealed substantial variability in both plant growth and quantitative exudate composition depending on the substrate. Specifically, a greater abundance of rhizosphere-enhanced metabolites (REMs) and rhizosphere-abated metabolites (RAMs) was observed in sand and clay substrates compared to soil (Miller et al. 2019).

A study on ten sorghum entries under low nitrogen conditions revealed altered metabolic profiles, including metabolites important in nitrogen storage and amino acid synthesis. The rhizosphere microbiome, dominated by a single operational taxonomic unit (OTU) called pseudomonas, showed reduced defense response under low nitrogen conditions. The research highlights the importance of considering interactions between biotic and abiotic stresses, as they can impact plant defense mechanisms. These factors should be given precedence in breeding programs aimed at biofuel production on marginal soils (Sheflin et al. 2019).

17.6 Sorghum Phenomics

Sorghum phenomics is an emerging field that focuses on the study of the physical and biochemical traits of sorghum plants. It generally deals with a deeper understanding of different physical and biochemical traits. Early-season CT is one such example wherein phenomics and genomics approaches have been used (Marla et al. 2023), in addition to this there is a very well-established Consortium for Advanced Sorghum Phenomics (CASP) platform that utilizes advanced technologies for phenotyping in the field as well as in glass house. The CASP platform makes use of computational modeling techniques to predict the performance of sorghum genotypes and very early developmental states (https://doi.org/10.46936/staf.proj.2018.50615/60000073). With these advanced methodologies at hand, sorghum is increasingly recognized as a prime candidate for diverse genomics and functional genomics investigations (Baloch et al. 2023).

Comparative studies across various "omics" domains have expanded our understanding of cellular mechanisms for drought tolerance (Pacific Northwest National Laboratory (PNNL) | arpa-e.energy.gov). The project aims to measure drought tolerance using water stress mapping and sensors, understanding traits related to biomass yield and drought/salinity stress, and predicting them early in plant development. These advances in sorghum phenomics have contributed to our understanding of the physical and biochemical traits of sorghum plants, including their tolerance to abiotic stresses such as drought and chilling temperatures. They have also helped to speed up sorghum breeding and improve its performance. The CASP project is an example of how innovative technologies can be used to predict sorghum performance early on, which can help accelerate breeding efforts. This study explores the use of phenomics and genomics-enabled approaches for pre-breeding sorghum early-season CT in tropical-origin crops. The researchers used an uncrewed aircraft systems (UAS) high-throughput phenotyping platform to improve scalability. They found that first-generation markers (KASP) failed to function in independent breeding programs due to the CT allele being common in diverse breeding lines. Second-generation markers were successful in tracking the donor CT allele in diverse breeding lines. The findings demonstrate the effectiveness of these approaches in molecular breeding of complex adaptive traits. A detailed account on high-throughput phenotyping was presented in the Chap. 4 of this book.

17.7 Sorghum Genetic Engineering Advancements

Genetic engineering has enabled precise modifications to the sorghum genome, enhancing its agronomic traits. Five fundamental technologies have been developed for sorghum improvement, including in vitro protocols for plant regeneration, molecular markers, gene identification and cloning, genetic engineering for trait integration, and genomics databases. Notable successes include the creation of drought-tolerant and insect-resistant sorghum varieties (Maqbool et al. 2001). Additionally, biofortification efforts have led to sorghum cultivars with improved nutritional content. Genetic engineering has been used to improve the nutritional value, agronomic traits, and biofuel production of sorghum. Genetic engineering has been used to increase the protein digestibility and essential amino acid content of sorghum. This review discusses genetic engineering approaches to improve sorghum grain's protein digestibility and nutritional value. The focus is on the introduction of genetic constructs that cause the silencing of γ- and/or α-kafirins. Transgenic sorghum with these constructs exhibits reduced kafirin oligomers, increased flour protein digestibility, and increased essential amino acid proportion. The ultrastructure of endosperm protein bodies and floury texture are also observed. The study suggests that the environment influences the suppression of kafirin synthesis, suggesting the use of genome editing methods for improving kafirin digestibility and increasing sorghum grain nutritional value (Elkonin et al. 2018). This has been achieved by overexpressing genes involved in protein metabolism and amino acid biosynthesis.

Agrobacterium-mediated transformation is the most common and reliable technique till date as it produces single-copy inserts along with complete gene insertion (Girijashankar and Swathisree 2009). Selection of important elements such as promoters and genes are very crucial for sorghum transformation. Some of the essential promoters for sorghum transgenic expression includes ubi1 and mpiC1, while MARS elements help reduce positional silencing. Genes such as Cry, bar, HVA I, dhdps-raec 1, and rice chitinase1 are valuable for transferring to sorghum cultivars. However, creating transgenic sorghum is challenging due to transgene silencing and the crop's recalcitrant nature. Lu et al. (2009) suggests the possibility of marker-free transgenic sorghum plants, emphasizing the importance of efficient tissue culture protocols with MARS elements and *Agrobacterium*-mediated gene transfer for successful gene introduction and expression. A study found that introducing *Zea mays* gibberellin 20-oxidase cDNA into switchgrass increased biomass production, by doubling the yield. The research also optimized sorghum regeneration and transformation methods using binary vectors and bar gene as plant selectable markers, and achieving over 14% transformation frequency (Do 2016).

Purdue University researchers have pioneered the development of the first genetically modified sorghum plants using microprojectile bombardment. This breakthrough enables the successful introduction of a marker gene into two sorghum genotypes, offering new avenues for breeders to access a wider range of genes when creating new varieties. This advancement holds promise for transforming sorghum into a more sustainable alternative to corn in the United States, potentially reducing input requirements like pesticides and providing a cost-effective feed grain for farmers. Additionally, sorghum's significance as a vital food source for millions in Africa and Asia underscores the importance of this research. The use of a gene gun powered by pressurized helium facilitated this genetic modification without harming sorghum embryos, addressing a common challenge in genetic engineering.

17.8 Advances in Genome Editing in Sorghum

Genome editing technologies have the potential to improve agricultural productivity and quality by altering genes that influence traits such as stress tolerance and nutrient absorption (Kamthan et al. 2016; Anwar and Kim 2020). Jiang et al.'s (2013) work on sorghum genome editing targeted the DsRED2 coding region, integrating a 20 bp specific guide RNA and GFP-NptII gene for selectable markers. To enhance grain quality in sorghum, Li et al. (2018) presented a successful case of full gene family editing. They co-targeted the whole gene family using a binary vector (pBUN421) and a single sgRNA, reaching a high editing effectiveness of 92.4%. Furthermore, Char et al. (2020a) utilized two guide RNAs to target distinct endogenous genes, showcasing multiplex gene editing. In a separate study by Liu et al. (2019), a reliable knockout of the cinnamyl alcohol dehydrogenase (CAD) gene was achieved using a particle inflow gun (PIG), boasting a 25% editing efficiency. These technologies have enabled the efficient transfer of genes into sorghum and the generation of transgenic sorghum plants. In summary, genetic engineering has been used to enhance nutritional content, agronomic traits, and biofuel production of sorghum. This has been achieved by overexpressing genes involved in protein metabolism and amino acid biosynthesis, transferring traits such as drought tolerance and insect resistance, and enhancing the saccharification efficiency. Several transformation technologies have been developed for sorghum, enabling the efficient transfer of genes into sorghum and the generation of transgenic sorghum plants. Table 17.5 presents a comprehensive list of genome editing studies, encompassing diverse methodologies, and highlighting key findings that have significantly contributed to advancing our understanding of genetic modifications across various biological systems (Fig. 17.4).

Table 17.5 List of *Agrobacterium*-mediated genome editing studies

Promoter sgRNA/Cas	No of gRNA	Gene targeted	Selectable marker	Edit efficiency (%)	Phenotype	Authors
Maize ubi1	1	k1C5 and gKAF1	bar	–	Grain sorghum mutants with improved kafirin digestibility	Elkonin et al. (2023)
OsU6/ CaMv35S	2	SbLG1	nptII	33.3	Altered leaf inclination angle, ligule, and auricle size. Distinct phenotypes for WT, monoallelic and biallelic mutants	Brant et al. (2021)
OsU6/ ZmUbi1	2	SbFTSbGA2ox5	bar	33.3, 83.3	Delayed flowering. No phenotype, biallelic mutations potentially lethal	Char et al. (2020b)
TaU3/ ZmUbi	1	k1C gene family	nptII	92.4	Partial opacity in T1 seeds, reduced α-kafirin, improved grain protein digestibility and lysine content	Li et al. (2018)
ZmU6/ ZmUbi1	1	Sb-CENH3	nptII	37–40	NR. Biallelic frameshift mutations potentially lethal	Che et al. (2018)
OsU6/ OsAct1	1	mDsRED2	nptII	–	DsRED2 expression	Jiang et al. (2013)

17.9 Conclusion

The incorporation of biotechnological advancements alongside omics innovations has revolutionized both the breeding and research of sorghum. Omics technologies, such as transcriptomics, proteomics, metabolomics, genomics, and genetic engineering, have provided a deeper understanding of the molecular mechanisms governing sorghum's response to stress and developmental stages. This thorough investigation has raised stress tolerance, increased nutritional content, and increased sorghum productivity. Genome mapping was made easier by genomics research, which also revealed genetic variants that underlie important traits. The molecular

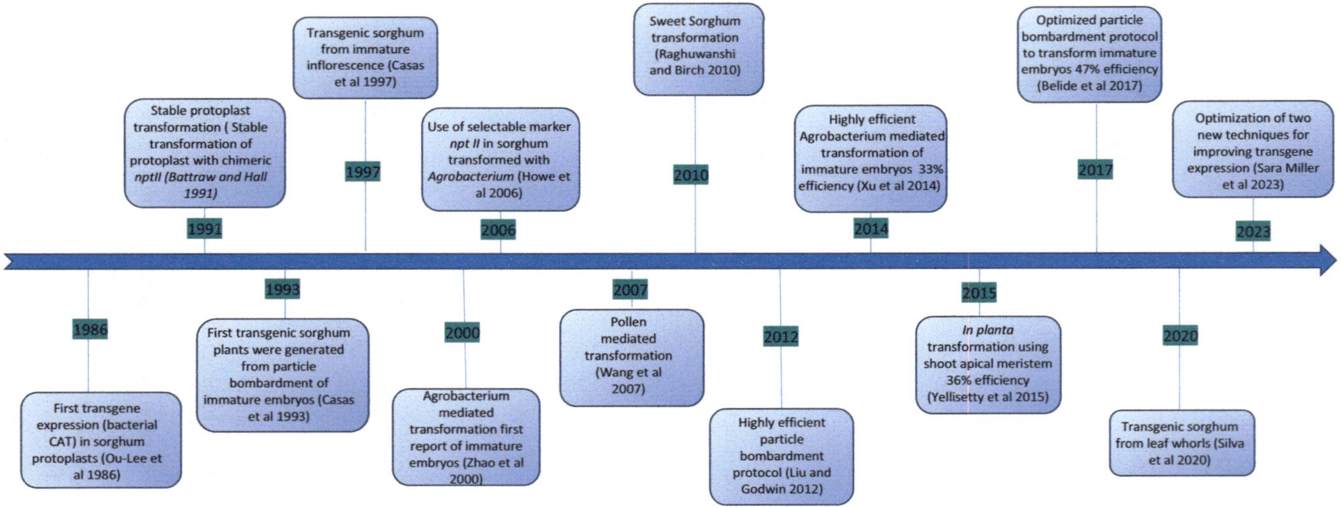

Fig. 17.4 Updated timeline for genetic transformation in sorghum

details of stress tolerance were revealed by transcriptomics, which also identified genes with differential expression. Proteins, metabolites, and pathways essential to the nutritional composition and stress response were revealed by proteomic and metabolomic analyses. Improved varieties were produced through targeted modifications made possible by precision genetic engineering. Sorghum research is being propelled by the convergence of biotechnological advancements and omics technologies, which present viable answers to worldwide agricultural problems. Advancements deepen our understanding of sorghum biology and expedite development of superior varieties with enhanced stress tolerance, nutrition, and biomass yield. Looking forward, opportunities for innovation abound, promising sustainable agriculture and a resilient food supply. In essence, omics and biotechnological tools represent a transformative leap in sorghum research, addressing urgent global challenges in food security and robust crop development.

References

Abdel-Ghany SE, Ullah F, Ben-Hur A, Reddy ASN (2020) Transcriptome analysis of drought-resistant and drought-sensitive sorghum (*Sorghum bicolor*) genotypes in response to peg-induced drought stress. Int J Mol Sci 21(3):772. https://doi.org/10.3390/ijms21030772

Alam MM, Mace ES, Van Oosterom EJ, Cruickshank A, Hunt CH, Hammer GL et al (2014) QTL analysis in multiple sorghum populations facilitates the dissection of the genetic and physiological control of tillering. Theor Appl Genet 127(10):2253–2266. https://doi.org/10.1007/s00122-014-2377-9

Anwar A, Kim JK (2020) Transgenic breeding approaches for improving abiotic stress tolerance: recent progress and future perspectives. Int J Mol Sci 21(8):2695. https://doi.org/10.3390/ijms21082695

Aruna C, Bhagwat VR, Madhusudhana R, Sharma V, Hussain T, Ghorade RB et al (2011) Identification and validation of genomic regions that affect shoot fly resistance in sorghum (*Sorghum bicolor* (L.) Moench). Theor Appl Genet 122(8):1617–1630. https://doi.org/10.1007/s00122-011-1559-y

Ashok Kumar A, Reddy BVS, Sharma HC, Hash CT, Rao PS, Ramaiah B, Reddy PS (2011) Recent advances in sorghum genetic enhancement research at ICRISAT. Am J Plant Sci 2(4):589–600. https://doi.org/10.4236/ajps.2011.24070

Azzouz-Olden F, Hunt AG, Dinkins R (2020) Transcriptome analysis of drought-tolerant sorghum genotype SC56 in response to water stress reveals an oxidative stress defense strategy. Mol Biol Rep 47:3291–3303. https://doi.org/10.1007/s11033-020-05396-5

Baloch FS, Altaf MT, Liaqat W, Bedir M, Nadeem MA, Cömertpay G, Çoban N, Habyarimana E, Barutçular C, Cerit I, Ludidi N, Karaköy T, Aasim M, Chung YS, Nawaz MA, Hatipoğlu R, Kökten K, Sun H-J (2023) Recent advancements in the breeding of sorghum crop: current status and future strategies for marker-assisted breeding. Front Genet 14:1150616. https://doi.org/10.3389/fgene.2023.1150616

Battraw M, Hall TC (1991) Stable transformation of Sorghum bicolor protoplasts with chimeric neomycin phosphotransferase II and β-glucuronidase genes. Theor Appl Genet 82:161–168

Bayer PE, Golicz AA, Scheben A, Batley J, Edwards D (2020) Plant pan-genomes are the new reference. Nat Plants 6(8):914–920. https://doi.org/10.1038/s41477-020-0733-0

Bekele WA, Wieckhorst S, Friedt W, Snowdon RJ (2013) High-throughput genomics in sorghum: from whole-genome resequencing to a SNP screening array. Plant Biotechnol J 11(9):1112–1125. https://doi.org/10.1111/pbi.12106

Belide S, Vanhercke T, Petrie JR, Singh SP (2017) Robust genetic transformation of sorghum (Sorghum bicolor L.) using differentiating embryogenic callus induced from immature embryos. Plant Methods 13:1–12

Bennetzen JL, Subramanian V, Xu J, Salimath SS, Subramanian S, Bhattramakki D et al (2001) A framework genetic map of sorghum containing RFLP, SSR and morphological markers. In: DNA-based markers plants. Springer, Berlin, pp 347–355. https://doi.org/10.1007/978-94-015-9815-6_20

Berhan AM, Hulbert SH, Butler LG, Bennetzen JL (1993) Structure and evolution of the genomes of *Sorghum bicolor* and Zea mays. Theor Appl Genet 86(5):598–604. https://doi.org/10.1007/BF00838715

Bhattramakki D, Dong J, Chhabra AK, Hart GE (2000) An integrated SSR and RFLP linkage map of *Sorghum bicolor* (L.) Moench. Genome 43(6):988–1002. https://doi.org/10.1139/g00-074

Billot C, Rivallan R, Sall MN, Fonceka D, Deu M, Glaszmann JC et al (2012) A reference microsatellite kit to assess the genetic diversity of Sorghum *bicolor* (Poaceae). Am J Bot 99(6):e245–e250. https://doi.org/10.3732/ajb.1100548

Binelli GLPE, Gianfranceschi L, Pe ME, Taramino G, Busso C, Stenhouse J et al (1992) Similarity of maize and sorghum genomes as revealed by maize RFLP probes. Theor Appl Genet 84(1):10–16. https://doi.org/10.1007/bf00223975

Boivin K, Deu M, Rami JF, Trouche G, Hamon P (1999) Towards a saturated sorghum map using RFLP and AFLP markers. Theor Appl Genet 98(2):320–328. https://doi.org/10.1007/s001220051076

Bowers JE, Abbey C, Anderson S, Chang C, Draye X, Hoppe AH et al (2003) A high-density genetic recombination map of sequence-tagged sites for sorghum, as a framework for comparative structural and evolutionary genomics of tropical grains and grasses. Genetics 165(1):367–386. https://doi.org/10.1093/genetics/165.1.367

Brant EJ, Baloglu MC, Parikh A, Altpeter F (2021) CRISPR/Cas9 mediated targeted mutagenesis of LIGULELESS-1 in sorghum provides a rapidly scorable phenotype by altering leaf inclination angle. Biotechnol J 16(11):2100237. https://doi.org/10.1002/biot.202100237

Brenton Z, Cooper EA, Myers MT, Boyles R, Shakoor N, Zielinski KJ, Rauh B, Bridges WC, Morris GP, Kresovich S (2016) A genomic resource for the development, improvement, and exploitation of sorghum for bioenergy. Genetics 204(1):21–33. https://doi.org/10.1534/genetics.115.183947

Casas AM, Kononowicz AK, Zehr UB, Tomes DT, Axtell JD, Butler LG, Bressan RA, Hasegawa PM (1993) Transgenic sorghum plants via microprojectile bombardment. Proc Natl Acad Sci 90(23):11212–11216

Casas AM, Kononowicz AK, Haan TG, Zhang L, Tomes DT, Bressan RA, Hasegawa PM (1997) Transgenic sorghum plants obtained after microprojectile bombardment of immature inflorescences. In Vitro Cell Dev Biol-Plant 33:92–100

Char SN, Lee H, Yang B (2020a) Use of CRISPR/Cas9 for targeted mutagenesis in sorghum. Curr Protoc plant Biol 5(2):e20112. https://doi.org/10.1002/cppb.20112

Char SN, Wei J, Mu Q, Li X, Zhang ZJ, Yu J et al (2020b) An Agrobacterium-delivered CRISPR/Cas9 system for targeted mutagenesis in sorghum. Plant Biotechnol J 18(2):319–321. https://doi.org/10.1111/pbi.13229

Che P, Anand A, Wu E, Sander JD, Simon MK, Zhu W et al (2018) Developing a flexible, high-efficiency Agrobacterium-mediated sorghum transformation system with broad application. Plant Biotechnol J 16(7):1388–1395. Portico. https://doi.org/10.1111/pbi.12879

Chittenden LM, Schertz KF, Lin YR, Wing RA, Paterson AH (1994) A detailed RFLP map of *Sorghum bicolor* x S. propinquum, suitable for high-density mapping, suggests ancestral duplication of Sorghum chromosomes or chromosomal segments. Theor Appl Genet 87(8):925–933. https://doi.org/10.1007/bf00225786

Cooper EA, Brenton ZW, Flinn BS et al (2019) A new reference genome for *Sorghum bicolor* reveals high levels of sequence similarity between sweet and grain genotypes: implications for the genetics of sugar metabolism. BMC Genomics 20:420. https://doi.org/10.1186/s12864-019-5734-x

Crasta O, Xu W, Rosenow DT, Mullet JE, Nguyen HT (1999) Mapping of post-flowering drought resistance traits in grain sorghum: association between QTLs influencing premature senescence and maturity. Mol Gen Genomics 262(3):579–588. https://doi.org/10.1007/s004380051120

Cuevas HE, Zhou C, Tang H, Khadke PP, Das S, Lin YR, Ge Z, Clemente TE, Upadhyaya HD, Hash CT, Paterson AH (2016) The evolution of photoperiod-insensitive flowering in sorghum, a genomic model for panicoid grasses. Mol Biol Evol 33(9):2417–2428. https://doi.org/10.1093/molbev/msw120

Cuevas HE, Fermin-Pérez RA, Prom LK, Cooper EA, Bean S, Rooney WL (2019) Genome-wide association mapping of grain mold resistance in the US sorghum association panel. Plant Genome 12(2):180070. https://doi.org/10.3835/plantgenome2018.09.0070

Deschamps S, Zhang Y, Llaca V, Ye L, Sanyal A, King M et al (2018) A chromosome-scale assembly of the sorghum genome using nanopore sequencing and optical mapping. Nat Commun 9(1):4844–4910. https://doi.org/10.1038/s41467-018-07271-1

Do PT (2016) Genetic engineering of sorghum and switchgrass for improved biofuel production. University of Missouri, Columbia

Duan Y, Qian J, Sun Y, Yi Z, Yan M (2009) Construction of methylation linkage map based on MSAP and SSR markers in *Sorghum bicolor* (L). IUBMB Life 61(6):663–669. https://doi.org/10.1002/iub.213

Dufour P, Deu M, Grivet L, D'Hont A, Paulet F, Bouet A et al (1997) Construction of a composite sorghum genome map and comparison with sugarcane, a related complex polyploid. Theor Appl Genet 94(3):409–418. https://doi.org/10.1007/s001220050430

Elkonin L, Italyanskaya J, Panin V (2018) Genetic modification of sorghum for improved nutritional value: state of the problem and current approaches. J Investig Genomics 5(1):39–48. https://doi.org/10.15406/jig.2018.05.00076

Elkonin LA, Gerashchenkov GA, Borisenko NV, Kenzhegulov OA, Sarsenova SK, Panin VM, Rozhnova NA (2023) SITE-directed mutagenesis for producing grain sorgum mutants with improved kafirine digestibility. Ecol Genet 21:48–49

Elshire RJ, Glaubitz JC, Sun Q, Poland JA, Kawamoto K, Buckler ES et al (2011) A robust, simple genotyping-by-sequencing (GBS) approach for high diversity species. PLoS One 6(5):e19379. https://doi.org/10.1371/journal.pone.0019379

Evans J, McCormick RF, Morishige D, Olson SN, Weers B, Hilley J et al (2013) Extensive variation in the density and distribution of DNA polymorphism in sorghum genomes. PLoS One 8(11):e79192. https://doi.org/10.1371/journal.pone.0079192

Girijashankar V, Swathisree V (2009) Genetic transformation of *Sorghum bicolor*. Physiol Mol Biol Plants 15(4):287–302. https://doi.org/10.1007/s12298-009-0033-7. Epub 2009 Dec 6. PMID: 23572939; PMCID: PMC3550352

Goche T, Shargie NG, Cummins I et al (2020) Comparative physiological and root proteome analyses of two sorghum varieties responding to water limitation. Sci Rep 10:11835. https://doi.org/10.1038/s41598-020-68735-3

Guan YA, Wang HL, Qin L, Zhang HW, Yang YB, Gao FJ et al (2011) QTL mapping of bio-energy related traits in sorghum. Euphytica 182(3):431–440. https://doi.org/10.1007/s10681-011-0528-5

Habyarimana E, Gorthy S, Baloch FS, Ercisli S, Chung G (2022) Whole-genome resequencing of *Sorghum bicolor* and *S. bicolor* × *S. halepense* lines provides new insights for improving plant agroecological characteristics. Sci Rep 12(1):5556. https://doi.org/10.1038/s41598-022-09433-0

Haussmann B, Hess D, Seetharama N, Welz H, Geiger H (2002) Construction of a combined sorghum linkage map from two recombinant inbred populations using AFLP, SSR, RFLP, and RAPD markers, and comparison with other sorghum maps. Theor Appl Genet 105(4):629–637. https://doi.org/10.1007/s00122-002-0900-x

Howe A, Sato S, Dweikat I, Fromm M, Clemente T (2006) Rapid and reproducible Agrobacterium-mediated transformation of sorghum. Plant Cell Rep 25:784–791

Hu Z, Olatoye MO, Marla S, Morris GP (2019) An integrated genotyping-by-sequencing polymorphism map for over 10,000 sorghum genotypes. Plant Genome 12(1). https://doi.org/10.3835/plantgenome2018.06.0044

Hua K, Zhang J, Botella JR, Ma C, Kong F, Liu B et al (2019) Perspectives on the application of genome-editing technologies in crop breeding. Mol Plant 12(8):1047–1059. https://doi.org/10.1016/j.molp.2019.06.009

Hulbert SH, Richter TE, Axtell JD, Bennetzen JL (1990) Genetic mapping and characterization of sorghum and related crops by means of maize DNA probes. Proc Natl Acad Sci 87(11):4251–4255. https://doi.org/10.1073/pnas.87.11.4251

Ibraheem O, Adigun RO, Olatunji IT (2018) Omics technologies in unraveling plant stress responses; using sorghum as a model crop, how far have we gone? Vegetos 31(2):1–18. https://doi.org/10.4172/2229-4473.1000405

Jiang W, Zhou H, Bi H, Fromm M, Yang B, Weeks DP (2013) Demonstration of CRISPR/Cas9/sgRNA-mediated targeted gene modification in Arabidopsis, tobacco, sorghum and rice. Nucleic Acids Res 41(20):e188. https://doi.org/10.1093/nar/gkt780

Jin P, Wang L, Zhao W, Zheng J, Wang YH, Liu Y et al (2021) Construction of high-density genetic map and QTL mapping in sorghum × sudangrass. Euphytica 217(8):162–211. https://doi.org/10.1007/s10681-021-02895-9

Johnson S, Lim FL, Finkler A, Fromm H, Slabas AR, Knight MR (2014) Transcriptomic analysis of Sorghum bicolor responding to combined heat and drought stress. BMC Genomics 15(1):456. https://doi.org/10.1186/1471-2164-15-456

Kamthan A, Chaudhuri A, Kamthan M, Datta A (2016) Genetically modified (GM) crops: milestones and new advances in crop improvement. Theor Appl Genet 129(9):1639–1655. https://doi.org/10.1007/s00122-016-2747-6

Ke F, Zhang K, Li Z et al (2022) Transcriptomic analysis of starch accumulation patterns in different glutinous sorghum seeds. Sci Rep 12:11133. https://doi.org/10.1038/s41598-022-15394-1

Kebede H, Subudhi PK, Rosenow DT, Nguyen HT (2001) Quantitative trait loci influencing drought tolerance in grain sorghum (*Sorghum bicolor* L. Moench). Theor Appl Genet 103(2):266–276. https://doi.org/10.1007/s001220100541

Kong L, Dong J, Hart GE (2000) Characteristics, linkage-map positions, and allelic differentiation of *Sorghum bicolor* (L.) Moench DNA simple-sequence repeats (SSRs). Theor Appl Genet 101(3):438–448. https://doi.org/10.1007/s001220051501

Kong W, Jin H, Franks CD, Kim C, Bandopadhyay R, Rana MK et al (2013) Genetic analysis of recombinant inbred lines for *Sorghum bicolor* × Sorghum propinquum. G3 (Bethesda) 3(1):101–108. https://doi.org/10.1534/g3.112.004499

Kotla A, Phuke R, Hariprasanna K, Mehtre SP, Rathore A, Gorthy S et al (2019) Identification of QTLs and candidate genes for high grain Fe and Zn concentration in sorghum (*Sorghum bicolor* (L.) Moench). J Cereal Sci 90(102):850. https://doi.org/10.1016/j.jcs.2019.102850

Kumar R, Bohra A, Pandey AK, Pandey MK, Kumar A (2017) Metabolomics for plant improvement: status and prospects. Front Plant Sci 8:1302. https://doi.org/10.3389/fpls.2017.01302

Li A, Jia S, Yobi A, Ge Z, Sato SJ, Zhang C et al (2018) Editing of an alpha-kafirin gene family increases digestibility and protein quality in sorghum. Plant Physiol 177(4):1425–1438. https://doi.org/10.1104/pp.18.00200

Li H, Li Y, Ke Q, Kwak SS, Zhang S, Deng X (2020) Physiological and differential proteomic analyses of imitation drought stress response in *Sorghum bicolor* root at the seedling stage. Int J Mol Sci 21(23):9174. https://doi.org/10.3390/ijms21239174. PMID: 33271965; PMCID: PMC7729455

Lin YR, Schertz KF, Paterson AH (1995) Comparative analysis of QTLs affecting plant height and maturity across the Poaceae, in reference to an interspecific sorghum population. Genetics 141(1):391–411. https://doi.org/10.1093/genetics/141.1.391

Liu G, Godwin ID (2012) Highly efficient sorghum transformation. Plant Cell Rep 31:999–1007

Liu G, Li J, Godwin ID (2019) Genome editing by CRISPR/Cas9 in sorghum through biolistic bombardment. In: Sorghum. Humana Press, New York, pp 169–183

Lu L, Xing-Rong W, Yin X, Morrand J, Chen X, Folk WR, Zhang ZJ (2009) Development of marker-free transgenic sorghum [Sorghum bicolor (L.) Moench] using standard binary vectors with bar as a selectable marker. Plant Cell Tissue Organ Cult 99(1):97–108. https://doi.org/10.1007/s11240-009-9580-4

Mace ES, Jordan DR (2011) Integrating sorghum whole genome sequence information with a compendium of sorghum QTL studies reveals uneven distribution of QTL and of gene-rich regions with significant implications for crop improvement. Theor Appl Genet 123(1):169–191. https://doi.org/10.1007/s00122-011-1575-y

Mace ES, Xia L, Jordan DR, Halloran K, Parh DK, Huttner E et al (2008) DArT markers: diversity analyses and mapping in *Sorghum bicolor*. BMC Genomics 9(1):26–11. https://doi.org/10.1186/1471-2164-9-26

Mace ES, Tai S, Gilding EK, Li Y, Prentis PJ, Bian L et al (2013) Whole-genome sequencing reveals untapped genetic potential in Africa's indigenous cereal crop sorghum. Nat Commun 4(1):2320–2329. https://doi.org/10.1038/ncomms3320

Mandrone M, Chiocchio I, Barbanti L, Tomasi P, Tacchini M, Poli F (2021) Metabolomic study of sorghum (*Sorghum bicolor*) to interpret plant behavior under variable field conditions in view of smart agriculture applications. J Agric Food Chem 69(3):1132–1145. https://doi.org/10.1021/acs.jafc.0c06533

Maqbool SB, Devi P, Sticklen MB (2001) Biotechnology: genetic improvement of sorghum (*Sorghum bicolor* (L.) Moench). In Vitro Cell Dev Biol Plant 37(5):504–515. http://www.jstor.org/stable/4293504

Marla S, Felderhoff T, Hayes C, Perumal R, Wang X, Poland J, Morris GP (2023) Genomics and phenomics enabled prebreeding improved early-season chilling tolerance in sorghum. G3 (Bethesda) 13(8):jkad116. https://doi.org/10.1093/g3journal/jkad116. PMID: 37232400; PMCID: PMC10411554

McCormick RF, Truong SK, Sreedasyam A, Jenkins J, Shu S, Sims D et al (2018) The *Sorghum bicolor* reference genome: improved assembly, gene annotations, a transcriptome atlas, and signatures of genome organization. Plant J 93(2):338–354. https://doi.org/10.1111/tpj.13781

Miller SB, Heuberger AL, Broeckling CD, Jahn CE (2019) Non-targeted metabolomics reveals sorghum rhizosphere-associated exudates are influenced by the belowground interaction of substrate and sorghum genotype. Int J Mol Sci 20:431. https://doi.org/10.3390/ijms20020431

Morris GP, Rhodes DH, Brenton Z, Ramu P, Thayil VM, Deshpande S et al (2013a) Dissecting genome-wide association signals for loss-of-function phenotypes in sorghum flavonoid pigmentation traits. G3 (Bethesda) 3(11):2085–2094. https://doi.org/10.1534/g3.113.008417

Morris GP, Ramu P, Deshpande SP, Hash CT, Shah T, Upadhyaya HD et al (2013b) Population genomic and genome-wide association studies of agroclimatic traits in sorghum. Proc Natl Acad Sci 110(2):453–458. https://doi.org/10.1073/pnas.1215985110

Nagaraja Reddy R, Madhusudhana R, Murali Mohan S, Chakravarthi DVN, Mehtre SP, Seetharama N et al (2013) Mapping QTL for grain yield and other agronomic traits in post-rainy sorghum (*Sorghum bicolor* (L.) Moench). Theor Appl Genet 126(8):1921–1939. https://doi.org/10.1007/s00122-013-2107-8

Nelson JC, Wang S, Wu Y, Li X, Antony G, White FF et al (2011) Single-nucleotide polymorphism discovery by high-throughput sequencing in sorghum. BMC Genomics 12(1):352–415. https://doi.org/10.1186/1471-2164-12-352

Ngara R, Goche T, Swanevelder DZH, Chivasa S (2021) Sorghum's whole-plant transcriptome and proteome responses to drought stress: a review. Life 11(7):704. https://doi.org/10.3390/life11070704

Ou-Lee TM, Turgeon R, Wu R (1986) Expression of a foreign gene linked to either a plant-virus or a Drosophila promoter, after electroporation of protoplasts of rice, wheat, and sorghum. Proc Natl Acad Sci 83(18):6815–6819

Paterson AH, Bowers JE, Bruggmann R, Dubchak I, Grimwood J, Gundlach H et al (2009) The *Sorghum bicolor* genome and the diversification of grasses. Nature 457(7229):551–556. https://doi.org/10.1038/nature07723

Peng Y, Schertz KF, Cartinhour S, Hart GE (1999) Comparative genome mapping of *Sorghum bicolor* (L.) Moench using an RFLP map constructed in a population of recombinant inbred lines. Plant Breed 118(3):225–235. https://doi.org/10.1046/j.1439-0523.1999.118003225.x

Pereira MG, Lee M (1995) Identification of genomic regions affecting plant height in sorghum and maize. Theor Appl Genet 90(3):380–388. https://doi.org/10.1007/bf00221980

Pereira MG, Lee M, Bramel-Cox P, Woodman W, Doebley J, Whitkus R (1994) Construction of an RFLP map in sorghum and comparative mapping in maize. Genome 37(2):236–243. https://doi.org/10.1139/g94-033

Punia H, Tokas J, Bhadu S, Mohanty AK, Rawat P, Malik A, Satpal (2020) Proteome dynamics and transcriptome profiling in sorghum [*Sorghum bicolor* (L.) Moench] under salt stress. 3 Biotech 10(9):412. https://doi.org/10.1007/s13205-020-02392-1. Epub 2020 Aug 29. PMID: 32904477; PMCID: PMC7456429

Quinby JR, Karper RE (1954) Inheritance of height in sorghum. Agron J 46:211–216

Ragab RA, Dronavalli S, Maroof MS, Yu YG (1994) Construction of a sorghum RFLP linkage map using sorghum and maize DNA probes. Genome 37(4):590–594. https://doi.org/10.1139/g94-084

Raghuwanshi A, Birch RG (2010) Genetic transformation of sweet sorghum. Plant Cell Rep 29:997–1005

Ramu P, Kassahun B, Senthilvel S, Ashok Kumar C, Jayashree B, Folkertsma RT et al (2009) Exploiting rice–sorghum synteny for targeted development of EST-SSRs to enrich the sorghum genetic linkage map. Theor Appl Genet 119(7):1193–1204. https://doi.org/10.1007/s00122-009-1120-4

Razzaq A, Sadia B, Raza A, Hameed MK, Saleem F (2019) Metabolomics: a way forward for crop improvement. Metabolites 9(12):303. https://doi.org/10.3390/metabo9120303

Ren G, Puyuan Y, Cui J, Gao Y, Yin C, Bai Y, Zhao D, Jin-Hua C (2022) Multiomics analyses of two sorghum cultivars reveal the molecular mechanism of salt tolerance. Front Plant Sci 13:886805. https://doi.org/10.3389/fpls.2022.886805

Ruperao P, Thirunavukkarasu N, Gandham P, Selvanayagam S, Govindaraj M, Nebie B, Manyasa E, Gupta R, Das RR, Odeny DA, Gandhi H, Edwards D, Deshpande SP, Rathore A (2021) Sorghum pan-genome explores the functional utility for genomic-assisted breeding to accelerate the genetic gain. Front Plant Sci 12:666342. https://doi.org/10.3389/fpls.2021.666342

Sabadin PK, Malosetti M, Boer MP, Tardin FD, Santos FG, Guimaraes CT et al (2012) Studying the genetic basis of drought tolerance in sorghum by managed stress trials and adjustments for phenological and plant height differences. Theor Appl Genet 124(8):1389–1402. https://doi.org/10.1007/s00122-012-1795-9

Sakurai N (2022) Recent applications of metabolomics in plant breeding. Breed Sci 72(1):56–65. https://doi.org/10.1270/jsbbs.21065. Released on J-STAGE March 08, 2022, Advance online publication February 03, 2022, Online ISSN 1347-3735, Print ISSN 1344-7610

Satish K, Srinivas G, Madhusudhana R, Padmaja PG, Nagaraja Reddy R, Murali Mohan S et al (2009) Identification of quantitative trait loci for resistance to shoot fly in sorghum (*Sorghum bicolor* (L.) Moench). Theor Appl Genet 119(8):1425–1439. https://doi.org/10.1007/s00122-009-1145-8

Sattler SE, Saballos A, Xin Z, Funnell-Harris DL, Vermerris W, Pedersen JF (2014) Characterization of novel sorghum *brown midrib* mutants from an EMS-mutagenized population. G3 (Bethesda) 4(11):2115–2124. https://doi.org/10.1534/g3.114.014001

Saxena RK, Edwards D, Varshney RK (2014) Structural variations in plant genomes. Brief Funct Genomics 13(4):296–307. https://doi.org/10.1093/bfgp/elu016

Sheflin AM, Chiniquy D, Yuan C, Goren E, Kumar I, Braud M, Brutnell T, Eveland AL, Tringe S, Liu P, Kresovich S, Marsh EL, Schachtman DP, Prenni JE (2019) Metabolomics of sorghum roots during nitrogen stress reveals compromised metabolic capacity for salicylic acid biosynthesis. Plant Direct 3(3):e00122. https://doi.org/10.1002/pld3.122. PMID: 31245765; PMCID: PMC6508800

Shiringani AL, Friedt W (2011) QTL for fibre-related traits in grain× sweet sorghum as a tool for the enhancement of sorghum as a biomass crop. Theor Appl Genet 123(6):999–1011. https://doi.org/10.1007/s00122-011-1642-4

Shiringani AL, Frisch M, Friedt W (2010) Genetic mapping of QTLs for sugar-related traits in a RIL population of *Sorghum bicolor* L. Moench. Theor Appl Genet 121(2):323–336. https://doi.org/10.1007/s00122-010-1312-y

Silva TN, Kelly ME, Vermerris W (2020) Use of Sorghum bicolor leaf whorl explants to expedite regeneration and increase transformation throughput. Plant Cell Tiss Org Cult (PCTOC) 141(2):243–255

Songsomboon K, Brenton Z, Heuser J, Kresovich S, Shakoor N, Mockler TC, Cooper EA (2021) Genomic patterns of structural variation among diverse genotypes of *Sorghum bicolor* and a potential role for deletions in local adaptation. G3 (Bethesda) 11(7):jkab154. https://doi.org/10.1093/g3journal/jkab154

Srinivas G, Satish K, Madhusudhana R, Seetharama N (2009a) Exploration and mapping of microsatellite markers from subtracted drought stress ESTs in *Sorghum bicolor* (L.) Moench. Theor Appl Genet 118(4):703–717. https://doi.org/10.1007/s00122-008-0931-z

Srinivas G, Satish K, Madhusudhana R, Reddy RN, Mohan S, Seetharama N (2009b) Identification of quantitative trait loci for agronomically important traits and their association with genic-microsatellite markers in sorghum. Theor Appl Genet 118(8):1439–1454. https://doi.org/10.1007/s00122-009-0993-6

Subudhi PK, Rosenow DT, Nguyen HT (2000) Quantitative trait loci for the stay green trait in sorghum (*Sorghum bicolor* L. Moench): consistency across genetic backgrounds and environments. Theor Appl Genet 101(5):733–741. https://doi.org/10.1007/s001220051538

Takai T, Yonemaru JI, Kaidai H, Kasuga S (2012) Quantitative trait locus analysis for days-to-heading and morphological traits in an RIL population derived from an extremely late flowering F1 hybrid of sorghum. Euphytica 187(3):411–420. https://doi.org/10.1007/s10681-012-0727-8

Tamhane VA, Sant SS, Jadhav AR et al (2021) Label-free quantitative proteomics of *Sorghum bicolor* reveals the proteins strengthening plant defense against insect pest *Chilo partellus*. Proteome Sci 19:6. https://doi.org/10.1186/s12953-021-00173-z

Tao YZ, Jordan DR, Henzell RG, McIntyre CL (1998) Construction of a genetic map in a sorghum recombinant inbred line using probes from different sources and its comparison with other sorghum maps. Aust J Agric Res 49(4):729–736. https://doi.org/10.1071/a97112

Tao YZ, Henzell RG, Jordan DR, Butler DG, Kelly AM, McIntyre CL (2000) Identification of genomic regions associated with staying green in sorghum by testing RILs in multiple environments. Theor Appl Genet 100(8):1225–1232. https://doi.org/10.1007/s001220051428

Tao Y, Zhao X, Wang X, Hathorn A, Hunt CH, Cruickshank A, Van Oosterom E, Godwin ID, Mace E, Jordan DR (2019) Large-scale GWAS in sorghum reveals common genetic control of grain size among cereals. Plant Biotechnol J 18(4):1093–1105. https://doi.org/10.1111/pbi.13284

Taramino G, Tarchini R, Ferrario S, Lee M, Pe ME (1997) Characterization and mapping of simple sequence repeats (SSRs) in Sorghum *bicolor*. Theor Appl Genet 95(1):66–72. https://doi.org/10.1007/s001220050533

Tuinstra MR, Grote EM, Goldsbrough PB, Ejeta G (1997) Genetic analysis of post-flowering drought tolerance and components of grain development in Sorghum bicolor (L.) Moench. Mol Breed 3(6):439–448. https://doi.org/10.1023/a:1009673126345

Turner MF, Heuberger AL, Kirkwood JS, Collins CC, Wolfrum EJ, Broeckling CD, Prenni JE, Jahn CE (2016) Non-targeted metabolomics in diverse sorghum breeding lines indicates primary and secondary metabolite profiles are associated with plant biomass accumulation and photosynthesis. Front Plant Sci 7:953. https://doi.org/10.3389/fpls.2016.00953

U.S. Department of Agriculture (2020). https://ipad.fas.usda.gov/ cropexplorer/cropview/commodityView.aspx?startrow=1andcropid=0459200andsel_ year=2020andrankby=Production. Accessed 26 Jun 2022

Varoquaux N, Cole B, Taylor JW, Pierroz G, Baker CM, Patel D, Madera M, Jeffers TL, Hollingsworth J, Sievert J, Yoshinaga Y, Owiti J, Singan V, DeGraaf S, Xu L, Blow MJ, Harrison MJ, Visel A, Jansson C et al (2019) Transcriptomic analysis of field-droughted sorghum from seedling to maturity reveals biotic and metabolic responses. Proc Natl Acad Sci U S A 116(52):27124–27132. https://doi.org/10.1073/pnas.1907500116

Voelker WG, Krishnan K, Chougule K, Alexander LC, Lu Z, Olson A, Ware D, Songsomboon K, Ponce CM, Brenton Z, Boatwright JL, Cooper EA (2023) Ten new high-quality genome assemblies for diverse bioenergy sorghum genotypes. Front Plant Sci 13:1040909. https://doi.org/10.3389/fpls.2022.1040909

Wang W, Wang J, Yang C, Li Y, Liu L, Xu J (2007) Pollen-mediated transformation of Sorghum bicolor plants. Biotechnol Appl Biochem 48(2):79–83

Whitkus R, Doebley J, Lee M (1992) Comparative genome mapping of sorghum and maize. Genetics 132(4):1119–1130. https://doi.org/10.1093/genetics/132.4.1119

Wu YQ, Huang Y, Tauer CG, Porter DR (2006) Genetic diversity of sorghum accessions resistant to greenbugs as assessed with AFLP markers. Genome 49(2):143–149. https://doi.org/10.1139/g05-095

Xu GW, Magill CW, Schertz KF, Hart GE (1994) A RFLP linkage map of *Sorghum bicolor* (L.) Moench. Theor Appl Genet 89(2):139–145. https://doi.org/10.1007/bf00225133

Xu K, Huang X, Wu M, Wang Y, Chang Y, Liu K, Zhang J, Zhang Y, Zhang F, Yi L, Li T (2014) A rapid, highly efficient and economical method of Agrobacterium-mediated in planta transient transformation in living onion epidermis. PLoS One 9(1):e83556

Xu W, Subudhi PK, Crasta OR, Rosenow DT, Mullet JE, Nguyen HT (2000) Molecular mapping of QTLs conferring stay-green in grain sorghum (*Sorghum bicolor* L. Moench). Genome 43(3):461–469. https://doi.org/10.1139/g00-003

Yellisetty V, Reddy LA, Mandapaka M (2015) In planta transformation of sorghum (Sorghum bicolor (L.) Moench) using TPS1 gene for enhancing tolerance to abiotic stresses. J Genet 94:425–434

Zhang HH, Xu N, Wu X, Wang J, Ma S, Li X et al (2018) Effects of four types of sodium salt stress on plant growth and photosynthetic apparatus in sorghum leaves. J Plant Interact 13(1):506–513. https://doi.org/10.1080/17429145.2018.1526978

Zheng LY, Guo XS, He B, Sun LJ, Peng Y, Dong SS et al (2011) Genome-wide patterns of genetic variation in sweet and grain sorghum (*Sorghum bicolor*). Genome Biol 12(11):R114–R115. https://doi.org/10.1186/gb-2011-12-11-r114

Zhao ZY, Cai T, Tagliani L, Miller M, Wang N, Pang H, Rudert M, Schroeder S, Hondred D, Seltzer J, Pierce D (2000) Agrobacterium-mediated sorghum transformation. Plant Mol Biol 44:789–798

Zhou X, Liu Z (2022) Unlocking plant metabolic diversity: a (pan)-genomic view. Plant Commun 3(2):100300. https://doi.org/10.1016/j.xplc.2022.100300

Zou G, Zhai G, Feng Q, Yan S, Wang A, Zhao Q et al (2012) Identification of QTLs for eight agronomically important traits using an ultra-high-density map based on SNPs generated from high-throughput sequencing in sorghum under contrasting photoperiods. J Exp Bot 63(15):5451–5462. https://doi.org/10.1093/jxb/ers205

Sorghum Environment Characterization and G × M Modeling Toolbox

18

Sunita Choudhary and Kaliamoorthy Sivasakthi

Abstract

Crop modeling plays a crucial role in modern agriculture, offering a strategic and scientific approach to crop improvement programs. By simulating and predicting the complex interactions between various factors influencing crop growth, including climate, soil, and genetic traits, crop models provide valuable insights. These insights enable researchers and breeders to optimize agricultural practices, predict crop performance under different conditions, and accelerate the development of improved varieties. Crop modeling enhances the precision of decision-making in areas such as planting schedules, resource management, and the selection of resilient genotypes. Furthermore, it contributes to the development of climate-resilient crops, helping agriculture adapt to changing environmental conditions. The efficiency, cost-effectiveness, and sustainability of crop improvement programs are significantly enhanced through the informed integration of crop modeling, ultimately contributing to global food security and agricultural sustainability.

Keywords

TPE · Sorghum · G × M Toolbox · Crop modeling

S. Choudhary (✉) · K. Sivasakthi
International Crops Research Institute for the Semi-Arid Tropics, Patancheru, Telangana, India
e-mail: Sunita.choudhary@icrisat.org

E. Habyarimana et al. (eds.), *Omics and Biotechnological Approaches for Product Profile-Driven Sorghum Improvement*,
https://doi.org/10.1007/978-981-97-4347-6_18

18.1 Introduction

Sorghum (*Sorghum bicolor* L. Moench) holds great agricultural significance in semi-arid and arid regions of South Asia and sub-Saharan Africa, where it frequently confronts challenges like drought, erratic rainfall, and high temperatures. Globally, sorghum is cultivated across 40.25 million hectares, producing around 58.70 million tons at an average productivity rate of 1458 kg/ha (Nagesh Kumar et al. 2022). The top 10 sorghum-producing countries, including the United States, Sudan, Mexico, Nigeria, India, Niger, Ethiopia, Australia, Brazil, and China, collectively contribute approximately 77% of the world's sorghum production (Aruna and Cheruku 2019).

In the 2020–2021 season, global sorghum production reached approximately 62.51 million tons, with the United States as the leading producer at 9.4 million tons, followed by Nigeria, Ethiopia, and Sudan. India ranked fifth, producing 4.78 million tons of sorghum on 4.39 million hectares. During the 2021–2022 kharif season, India produced an estimated 1.54 million tons of sorghum, with Andhra Pradesh making a significant contribution, yielding 4.10 lakh tons in 2020–2021 from 1.20 lakh hectares, with a productivity of 3428 kg/ha (https://angrau.ac.in/downloads/AMIC/OutlookReports/2021/9-SORGHUM_January%20to%20December%202021.pdf).

Sorghum is a dietary staple for over 200 million people in these regions, serving as a vital source of both food and fodder, especially in traditional smallholder farming systems (Visarada and Aruna 2019). Known as a "healthy cereal," sorghum is valued for its substantial carbohydrate content (68%), protein content (10%), and its richness in micronutrients and phytochemicals with potential nutraceutical benefits (Visarada and Aruna 2019). In major sorghum-growing regions of India, per capita consumption of sorghum is notably high at 75 kg per year, contributing to over 50% of the dietary iron and zinc requirements for low-income populations in India (Rao et al. 2006; Ashok Kumar et al. 2011).

18.2 Characterizing Drought Stress in Post-rainy Reason (Rabi) Sorghum in India

In the Deccan Plateau, India, post-rainy sorghum cultivation mainly occurs on shallow- and medium-deep soils, relying on retained soil moisture after the rainy season (Elangovan et al. 2010). However, reliance on residual moisture often leads to drought stress, resulting in reduced yields. Thus, understanding drought patterns is crucial for crop improvement efforts (Chenu et al. 2011; Balota et al. 2008; Blummel and Rao 2006). To achieve sustainable intensification of sorghum cultivation, a quantitative understanding of production environments is necessary. Crop simulation modeling (CSM) is increasingly used for environmental characterization and defining target population environments (TPEs) (Muchow et al. 1996; Chapman et al. 2000a, b; Chenu et al. 2013), which help reduce genotype-by-environment effects in breeding.

The classification of the sorghum cultivation regions in India into four production zones (Central, South, North, and Far South) was based on factors such as location, yield, and environmental conditions. Weather data from 19 stations, spanning 8–36 years, and soil information from databases such as the National Bureau of Soil Survey and Land Use Planning and the International Soil Reference and Information Centre were collected (Table 18.1). Kholova et al. (2013) employed crop simulations using the Agricultural Production Systems sIMulator (APSIM) sorghum model (version 7.3) to analyze primary locations in each zone. They assessed crop water status, identified common drought patterns using cluster analysis, quantified stress types, and evaluated their impact on grain and stover yields. Additionally, simulations assuming unlimited water and nitrogen supply were conducted to assess production potential.

1. *Information Regarding Simulated Site Yields, District Average Yield, and Simulated Potential Yield*: The model accurately depicted spatial variations in district yields across the rabi sorghum belt in India. Districts in the central zone consistently had the lowest average yield, while those in the North and South zones had higher averages. The far south zone had the highest predicted yield. These variations were linked to differences in soil quality and rainfall patterns. The strong correlation between predicted and observed yields validates the model's sensitivity to these factors, highlighting its utility in characterizing stress patterns.
2. *Drought Stress Pattern:* The study assessed water supply and demand ratios (S/D) across the crop cycle and identified five distinct environmental types within the rabi sorghum region. Despite initially high soil water content, most seasons faced water limitations, primarily terminal drought. Other stress types included post-flowering stress and consistently low-stress conditions. Surprisingly, seasons with low stress showed that standard nitrogen management practices led to growth restrictions due to nitrogen becoming insufficient before anthesis.
3. *Analysis of Environment Types and Their Impact on Sorghum Production:* The study investigated the frequencies and effects of drought stress patterns throughout the rabi belt (Table 18.2).
   (a) **Examination Within the Production Zones**: In the rabi sorghum belt, the central zone dominates grain production but exhibits the lowest long-term yield average, around 460 kg/ha, due to harsh environmental conditions. Simulated stress analysis revealed severe stress types, such as vegetative terminal drought (15%) and pre-flowering terminal drought (16%), causing significant yield reductions of 92% and 78%, respectively, compared to mild stress conditions. Other environment types showed yields similar to mild stress conditions (approximately 860 kg/ha). Total biomass production experienced similar yield reductions, with vegetative and pre-flowering stress causing 21% and 12% decreases, respectively.
   - *In the Southern production zone,* accounting for 30% of rabi sorghum production, average yields reached 580 kg/ha. Despite facing vegetative

(9%) and pre-flowering (15%) terminal droughts, which caused significant yield reductions of 83% and 62%, respectively, post-flowering stress types did not severely limit yields. Stover yield was only affected in the vegetative stress environment, with an 11% decline.

- *In the Northern zone*, contributing about 20% of rabi sorghum grain, with around 0.51 million tons from 0.90 million ha, average yields matched those of the Southern zone at 590 kg/ha. Vegetative stress occurred less often (5%), but pre-flowering terminal stress was more common (23%), resulting in a 57% reduction in grain yield. Biomass production also decreased, albeit to a lesser extent (24% and 6%, respectively).
- *In the Far South zone*, known for its highest average yields of around 760 kg/ha but the lowest production volume (about 10% of the rabi belt, equivalent to 0.24 million tons from 0.34 million ha), severe stress patterns were less prevalent. Vegetative and pre-flowering stress types occurred in only 4% and 13% of cases but significantly reduced average yields by 76% and 57%, respectively. Biomass production also saw a decline, albeit lesser, due to vegetative drought (12% reduction). Mild stress environments dominated this region (52%), with post-flowering drought minimally affecting yields.

(b) **Investigation Throughout the Entire Rabi Belt:** In our extensive analysis covering the entire rabi sorghum belt, we evaluated stress patterns and yields for each zone, considering their contribution to total grain production. Our weighted analysis revealed that 60% of the time, sorghum crops faced water limitations. Among these limitations, severe stress patterns such as vegetative and pre-flowering terminal droughts were prominent, resulting in notable reductions in grain yield. Interestingly, post-flowering stress had minimal impact on grain yields. Stover yields were less affected, showing only slight reductions in biomass under specific drought conditions. Overall, this analysis shed light on the challenges associated with rabi sorghum cultivation across different zones within the belt (Fig. 18.1).

4. *Sorghum Production Potential*: In summary, the analysis indicated significantly higher biomass and grain yield potential in the studied zones compared to typical dryland conditions with recommended nitrogen input. However, production potential varied among the northern, southern, central, and far south zones. The northern and southern zones demonstrated the highest average grain yields and biomass production, while the central and far south zones exhibited slightly lower yields under optimal water and nitrogen conditions. Despite similar average radiation levels, differences in in-crop radiation and temperature conditions were observed at specific sites (Tables 18.1 and 18.2, Fig. 18.1).

In conclusion, Kholova et al. (2013) provide a comprehensive insight into water limitation dynamics throughout the sorghum crop's life cycle in India's rabi region. They identify and categorize five distinct drought stress patterns, offering opportunities for crop improvement through management practices and genetic manipulation. However, trade-offs exist, such as those between stover and grain production,

Table 18.1 Overview of the main attributes concerning the four primary production zones within the rabi sorghum belt, encompassing weather data accessibility, soil type, soil depth, and plant-available water content (PAWC) for each location in these zones (data obtained from Kholova et al. 2013, Field Crops Research, 141, 38–46)

Zone	Average production (in thousand tons)	Average area (in thousands of ha)	Average yield (kg/ha)	Weather station (years)	Soil PAWC (mm)	Representative soil site/soil depth (cm)	Soil type
Central	826	1976	459	Jeur (25)	62.8	Solapur/50	Clayey entisol, isohyperthermic, lithic ustorthents montimorillontic
				Mohol (23)	62.8	Solapur/50	
				Pune (36)	68.9	Solapur/55	
				Rahuri (20)	62.8	Solapur/50	
				Solapur (23)	62.8	Solapur/50	
				Bheemarayanigudi (8)	68.9	Solapur/55	
				Bijapur (12)	68.9	Solapur/55	
South	659	1159	583	Gulbarga (24)	68.9	Solapur/55	Vertosol, mixed, typic rhodustalfs
				Patancheru (28)	98	Patancheru/70	
				Annigeri (24)	73.1	Bellary/65	
				Dharwad (21)	73.1	Bellary/65	
				Raichur (10)	73.2	Raichur/60	
				Akola (23)	85	Parbhani/75	
North	504	899	590	Aurangabad	73.5	Parbhani/65	The soil types include vertosol, sodic haplustert, nontmorillonitic, isohyperthermic, clayey skeletal entisol, typic haplustert, clay vertosol, montimorillonitic, isohyperthermic, calcareous, and typic chromusterts
				Parbhani (27)	85	Parbhani/75	
				Anantapur (33)	91.4	Raichur/75	
				Nandyal (8)	97.4	Raichur/80	
Far south	199	270	756	Hagari (23)	73.1	Bellary/65	The soil types consist of clayey skeletal entisol, typic haplustert, vertosol, sodic haplustert, nontmorillonitic, and isohyperthermic
				Bellary (23)	73.1	Bellary/65	

Table 18.2 Analyzing the environmental types in the rabi sorghum belt reveals distinct drought stress patterns in each production zone. It includes average simulated data for total biomass and grain yield within these zones, along with average productivity and potential productivity under optimal water and nitrogen conditions. Weighted means were calculated based on production proportions to derive overall rabi region values (data sourced from Kholova et al. 2013, Field Crops Research, 141, 38–46)

		TERMINAL DROUGHTS						
Production zone	Characteristic	Vegetative stress	Pre-flowering stress	Post-flowering Stress	Post-flowering relieved stress	Mild stress	Zone average	Zone potential
Central	Pattern frequency (%)	15	16	15	18	37	–	–
	Biomass (kg/ha)	2198	2444	3192	3115	2778	2756	8279
	Yield (kg/ha)	66	174	819	883	860	617	3014
South	Pattern frequency (%)	9	15	16	26	34		–
	Biomass (kg/ha)	3212	3620	4294	4006	3620	3792	8613
	Yield (kg/ha)	196	445	1110	1095	1173	945	3166
North	Pattern frequency (%)	5	23	21	11	41		–
	Biomass (kg/ha)	2594	3198	4269	3748	3408	3545	8714
	Yield (kg/ha)	0	456	1238	1205	1063	917	3213
Far south	Pattern frequency (%)	4	13	16	16	52		–
	Biomass (kg/ha)	3145	3445	4152	3892	3580	3671	6956
	Yield (kg/ha)	314	570	1405	1286	1324	1190	2704
Whole rabi-belt (weighted means)	Pattern frequency (%)	7	18	18	17	40		–
	Biomass (kg/ha)	2614	3005	3831	3561	3216	3354	8334
	Yield (kg/ha)	128	447	1190	1156	1128	847	3068

profitability, and risk. Future research will focus on conducting a detailed G × M × E simulation analysis to explore these trade-offs and enhance sorghum cultivation strategies in the region.

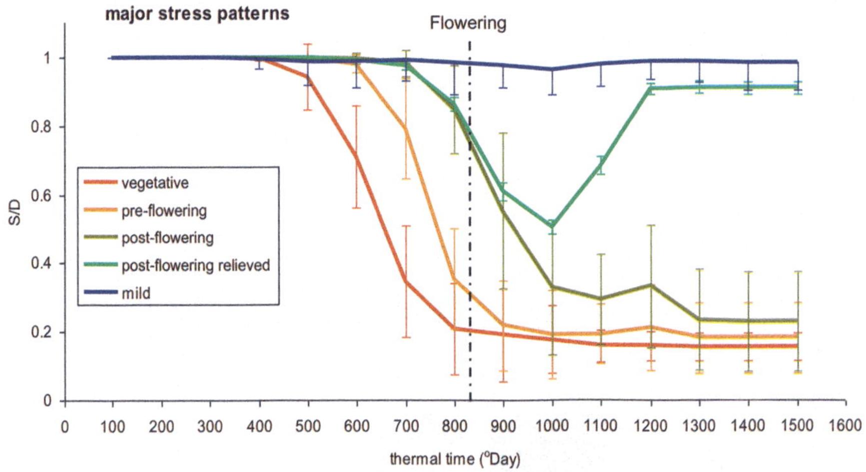

Fig. 18.1 The cluster analysis of simulated trajectories across the rabi sorghum belt revealed five distinct patterns in average crop water status (S/D index) during the crop cycle. Vertical bars denote standard errors within each cluster, with the vertical line marking the thermal time to flowering for cultivar M35-1 (830 °C d). This graph originates from the research conducted by Kholova et al., published in Field Crops Research in 2013 (Volume 141, pages 38–46)

Ronanki et al. (2022) extends our previous sorghum drought characterization research by aiming to enhance and stabilize post-rainy sorghum production through an expanded simulation setup. We introduce a digital approach to analyze the dynamics of post-rainy sorghum production systems in India. By examining spatio-temporal variations, we identify geography-specific and climate-responsive interventions using Genotype × Management (G × M) strategies. These evaluations encompass diverse agronomic practices across various soil types, considering the biological variability in sorghum cultivar characteristics. Leveraging crop growth simulations with APSIM, we pinpoint optimal G × M combinations to improve post-rainy sorghum production in India.

18.3 Identification of "Homogeneous System Units" Based on the Geographical Distribution of Optimal Genotype × Management

Our primary aim was to assist crop breeders, who often rely on crop modeling to pinpoint suitable regions for developing and testing crop products and management practices, in line with initiatives like the Breeding Program Assessment Tool (BPAT) review (Kholova et al. 2020).

Initially, we simplified data complexity using principal component analysis (PCA) and then applied a clustering method to establish distinct geographical classes termed "homogeneous system units" (HSUs). This innovative approach allowed us to categorize the tested regions (grids) into four HSUs based on shared

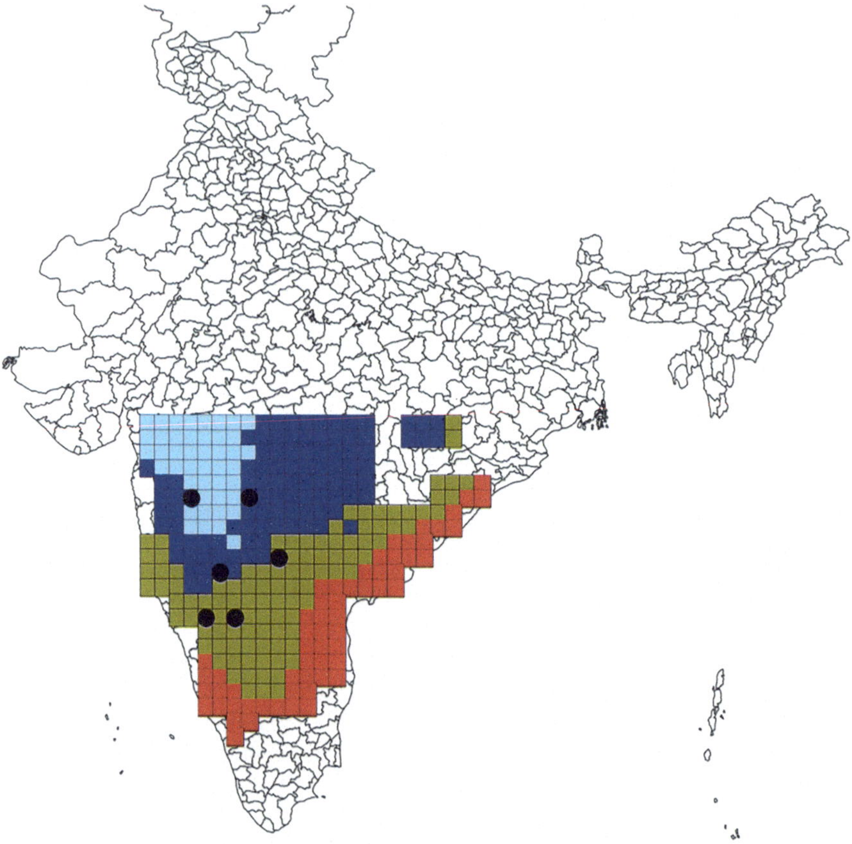

Fig. 18.2 The map of India displays four distinct "homogeneous system units" (HSU) in different colors, with summarized characteristics in Table 18.3. Grid segments within each HSU (0.5° × 0.5°) are expected to respond more uniformly to specific G × M interventions compared to segments in other HSUs. Black circles on the map mark sorghum testing sites for the post-monsoon season under the All India Coordinated Research Project (AICRP)

optimal production characteristics and responses to G × M interventions within the system (see Fig. 18.2). This marked a significant advancement over previous methods used in studies like Kholova et al. (2013), Hajjarpoor et al. (2018, 2022), Chauhan et al. (2013), Chenu et al. (2011), and Chapman et al. (2000a, b), which primarily focused on environmental characterization and targeting specific environments. The "HSU" analysis enabled us to differentiate geographies with high similarities within a geographic group and differences between groups, considering not only the modeled interactions of crop, environment, and management but also the system's responsiveness to G × M interventions.

The main system characteristics, along with the optimal G × M combinations within these geographical units, were summarized in Table 18.3. In essence, Table 18.3 illustrates that moving away from the central core (HSU_1) results in

Table 18.3 In Fig. 18.2, we analyze the summary statistics of production and stability indicators in the homogeneous system unit (HSU) clusters, achieved through optimized management (M) and genetic crop traits (G)

Cluster number	HSU_1 (Light Blue)	HSU_2 (Dark Blue)	HSU_3 (Brown)	HSU_4 (Red)
(M) Sowing window	16-23 sep	23-30 sep	16-23 sep	16-23
(M) Plant density (plant /m²)	12.54	12.58	13.18	13.66
(M) Nitrogen fertilization (kg /ha urea)	74-74	72.8-72.8	71-71	71.3-71.3
(G) Crop duration (tt_endjuv_to_ini)	181	178	197	220
(G) Rate of canopy growth, Vigor (power coefficient for TPLA max)	2.88	2.882	2.926	2.96
(G) Transpiration Responsiveness (Capacity of canopy to limit transpiration in High VPD)	0.9065	0.893	0.941	0.9485
Grain yield (kg/ha)	2231	2182	2675	3308
Stover yield (kg/ha)	4026	3982	4692	5986
Proportion of seasons with grain yield failure (%)	24	12	2	1
Deviation in total biomass production (kg/ha)	767	747	740	854

enhanced production potential and system stability, particularly toward HSU_4. This enhancement is evident in various aspects, including stover and grain yield production and system stability indicators. These trends are also mirrored in the optimized G × M parameters within each HSU, such as planting density, crop duration, and plant vigor. Additionally, a parameter indicating plant responsiveness to vapor pressure deficit (VPD) increases with greater distance from the production core, from HSU_1 to HSU_4. Across all HSUs, the majority of optimal G × M combinations favor early sowing windows and stable fertilizer doses, typically around 70–70 kg/ha (comprising both basal dose and top-dressing applications). See Table 18.3 for more details.

18.4 Development of Modeling Tools to Analyses Climate Risk for Cropping Systems

Crop yield results from intricate interactions among genotype, environment, and management (G × E × M), with acceptance influenced by farmer and consumer preferences. This complexity presents an opportunity when broken down into analyzable components, allowing for analysis, understanding, prediction, and prioritization of research investments to maximize returns. As part of this effort, crop modeling tools were developed to identify G × E × M options for enhancing the resilience and productivity of sorghum in India. This innovation involved implementing the APSIM modeling framework, enabling key stakeholders, particularly sorghum improvement teams, to grasp the spatial variability of post-rainy sorghum production systems in India and determine optimal G × M strategies to stabilize or enhance production under specific environmental conditions (Fig. 18.3).

The most successful crop improvement programs in complex agroecologies have pioneered the use of crop growth models to support production enhancement. This approach involves integrating traditional multi-location evaluations with models that allow for extrapolation of crop testing across spatiotemporal scales. By doing

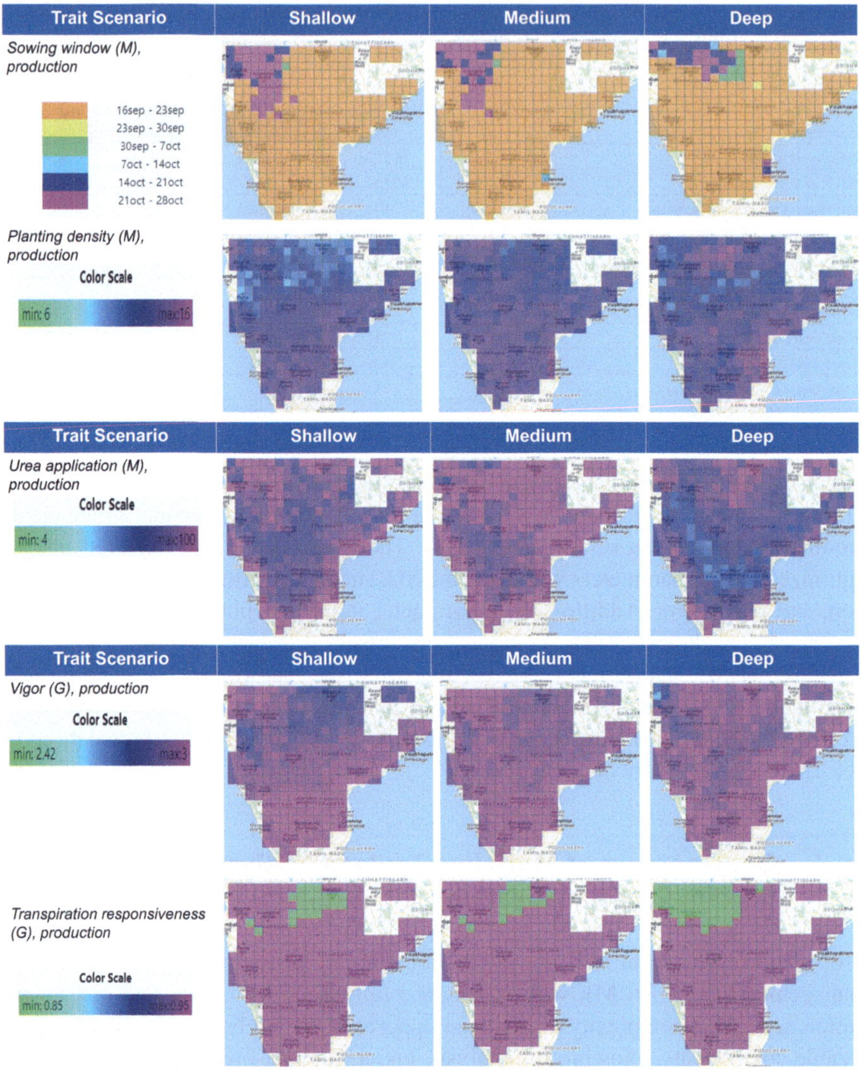

Fig. 18.3 The G × M Toolbox prototype (https://ls40.pef.czu.cz) demonstrating the impact of various genetic (G) option like plant vigor, transpiration responsiveness, maturity, and management (M) option like plant window, planting density, fertilizer application in Sorghum TPE. It is remained highly probabilistic; maximum resilience and economic benefits depended on drought frequency. It can help breeding programs to explore specific combinations of traits in sorghum growing regions to maximize the production options

so, the probability of selecting plant material and agricultural practices that better suit the geo-biophysical context of the production system increases significantly.

In essence, system modeling tools can achieve the necessary spatiotemporal scale required for evaluating cropping systems by:

- Understanding production potential, main limitations, and identifying TPEs.
- Identifying suitable G × E × M interventions with a high probability of maximizing/stabilizing production within a given region/TPE.

Although breeding programs typically have a basic understanding of the main constraints limiting crop production in target agroecosystems (biotic and abiotic stress), few programs rigorously quantify stress severity, frequency, and its effects on production across multiple locations and season—crucial for explaining yield losses. Crop simulation tools enable such analyses within relevant systems and facilitate the identification of bio-geo-physical units with higher degrees of similarity, i.e., TPEs. TPEs can then serve as geographically defined breeding units for developing specific G × M interventions to optimize crop production.

Moreover, crop models allow for the analysis of future environments, providing insights into strategies needed to adapt cropping systems to climatic changes. Identifying TPEs also helps optimize the distribution of multi-location testing sites and interpret the outcomes of trials within TPEs, thus avoiding testing crops in less relevant sites and selecting crops in abnormal years that do not adequately represent the most frequent environmental scenarios in a particular TPE.

The same simulation setup used to define TPEs can subsequently identify G × E × M interventions with a high probability of enhancing crop production/resilience within the system. The options identified in silico can be targeted for in vivo evaluation, allowing for testing specific plant types with specific management options in multi-location trials, as opposed to the conventional strategy of "testing everything everywhere." Therefore, models can serve as decision-making tools to quantitatively define breeding targets, evaluate and optimize the economics of breeding programs, and improve the efficiency of resource investments.

18.5 Modeling the Effect of Plant Water Use Traits on Yield and Stay-Green Expression in Sorghum

Sorghum production in the post-rainy season is vital for supporting livelihoods in semi-arid tropics, where drought poses a significant challenge. Expanding on previous simulation data by Kholova et al. (2013), the author introduced theoretical modifications to drought adaptation traits, including plant growth patterns, canopy and root water conductivity, and responses to drought stress. Subsequent simulations, outlined in Kholová et al. (2014), assessed the impact of these trait adjustments on growth, development, and yields of both grain and stover in a widely used post-rainy sorghum genotype across various scenarios (pre-flowering, flowering, post-flowering, post-flowering relieved, and no stress).

Simulation of Leaf Area Growth Dynamics

1. Simulating a smaller canopy size generally enhances yields in most water stress conditions in the rabi tract, except for low or no stress scenarios. Conversely, a larger canopy size has the opposite effect on grain and stover production forecasts. A smaller canopy delays stress onset in terminal drought stress scenarios and increases grain yield due to delayed water depletion, resulting in a stay-green phenotype.
2. Simulating high early vigor in plants reduced grain yield in most scenarios but increased stover production, except in unstressed conditions. Conversely, low early vigor led to a mild grain yield increase in stress scenarios, with a slight decrease in stover production across all conditions. These changes also contributed to water conservation. Modifying TPLA inflection and production ratio had minor effects on stress patterns and yield but improved stability.
3. Simulating longer phyllochron intervals extended crop duration but decreased grain yield in stressful scenarios, except pre-flowering stress, while stover yield increased. Shortening the phyllochron interval boosted grain yield significantly but reduced stover production. The decrease in stover yield was less than the gain in grain yield. Shorter duration crops depleted water resources faster in severe stress but improved water use after anthesis in milder stress scenarios. The difference lies in avoiding water limitations at critical stages rather than water conservation in this context.

Simulation of Limited Maximum Transpiration Rate (TR) and Soil Water Extraction Rate

- *Simulation of Maximum Transpiration Rate Limits:* Decreasing transpiration rates by about 10% in M35-1 crops notably boosted both grain and stover yields across all stress scenarios. Unlike other traits studied, limiting transpiration did not involve trade-offs between grain and stover production but instead enhanced overall water productivity (TE). The smallest yield improvements occurred in unstressed conditions. In severe stress scenarios, capping the maximum transpiration rate restored yields by improving water utilization after anthesis, preventing crop failure in many cases. However, the delay in stress onset due to limited transpiration was relatively minor compared to other traits, though sufficient to exhibit the stay-green phenotype.
- *Simulation of Soil Water Extraction Rate:* The virtual M35-1 crop, with reduced water-extracting capacity as soil moisture declines (lower kl), showed a moderate increase in stover yield in most stress scenarios while maintaining similar grain yields. Notably, the impact of kl varied significantly across specific stress scenarios. Plants with lower kl experienced slightly earlier water stress but could prolong the period of available water due to a simultaneous slowdown in biomass accumulation. This effect was particularly evident when stress occurred after flowering and was then relieved. Moreover, in severe stress conditions, the crop with low kl slightly enhanced TE by limiting water supply. Overall, while lower kl values led to earlier stress onset, they also extended the growth period, enhancing crop resilience and the potential for expressing the stay-green phenotype.

Simulation of Canopy Sensitivity to Declining Soil Water

- *Simulation of Leaf Expansion Sensitivity to Stress:* Slowing leaf expansion in drier soil conditions had minimal impact on yield but boosted stover productivity across all scenarios. Conversely, in wetter soil, reducing leaf expansion postponed stress onset and enhanced water utilization during grain filling, particularly in intermittent stress conditions. Crops that limited leaf expansion in wetter soil, especially under severe water stress, achieved higher grain yields compared to those with restricted leaf expansion in drier soil. Failure of crops with reduced leaf expansion in drier soil resulted from early vigorous growth followed by limited water utilization after flowering, potentially leading to premature senescence. However, variations in leaf expansion sensitivity to stress had negligible effects on grain and stover yields.
- In conclusion, the 2014 study by Kholova et al. showcased the positive effects of introducing variability in plant water-use traits, leading to improved grain yield and the promotion of the stay-green phenotype. Despite a negative impact on stover production, these changes ultimately enhanced overall crop production stability. By enhancing water productivity and limiting canopy transpiration under specific conditions, the study demonstrated the possibility of simultaneously improving grain and stover yields, along with crop resilience. This research offers valuable tools for developing customized crop ideotypes tailored to different zones and demands, providing crucial insights for breeding programs.

18.6 Modeling the Prospective Effects of Climate Change on Sorghum Yields in India's Post-rainy Season

In 2021, Chadalavada and her team conducted a study to assess the accuracy of the CMIP5 multi-climate model mean (MMM) in reproducing Indian monsoons and post-monsoon seasons. While the MMM successfully captured spatial patterns and annual temperature cycles, it displayed biases in precipitation and daily rainfall intensity, with less accurate simulations of precipitation and temperature trends compared to observed climate data. The study focused on evaluating the spatial distribution of southwest (SW) and northeast (NE) monsoon rainfall in India using the MMM, which closely aligned with data from the Indian Meteorological Department (IMD). However, it tended to underestimate rainfall totals in the western ghats. The NE monsoon, crucial for crop growth, was adequately simulated, along with temperature patterns during the SW and NE monsoon periods. The research also explored projected changes in precipitation and temperatures during India's monsoon seasons, indicating increased rainfall, particularly during the NE monsoon, and rising temperatures. These findings underscore the significant impact of climate change on India's monsoon seasons, highlighting the necessity for adaptation and mitigation measures.

Furthermore, the study utilized the CERES-sorghum model to simulate post-rainy season sorghum crop growth across India, with results generally surpassing observed yields due to favorable model assumptions. While performance varied slightly across states, the study supported the CERES-sorghum model's suitability for assessing potential climate change impacts on sorghum crops. Additionally, the study employed the CSM-CERES-sorghum model with DSSAT V4.7 to evaluate potential climate change impacts on sorghum growth and yields across various scenarios and regions in India. The results indicated substantial increases in sorghum yields, particularly under RCP 4.5, with variations among states. Southern states exhibited higher yields, suggesting that climate change, characterized by increased rainfall and suitable temperature ranges, could benefit smallholders by enhancing productivity. This resilience of sorghum in challenging conditions positions it as a valuable crop amidst climate change challenges during the post-rainy season (Figs. 18.4 and 18.5).

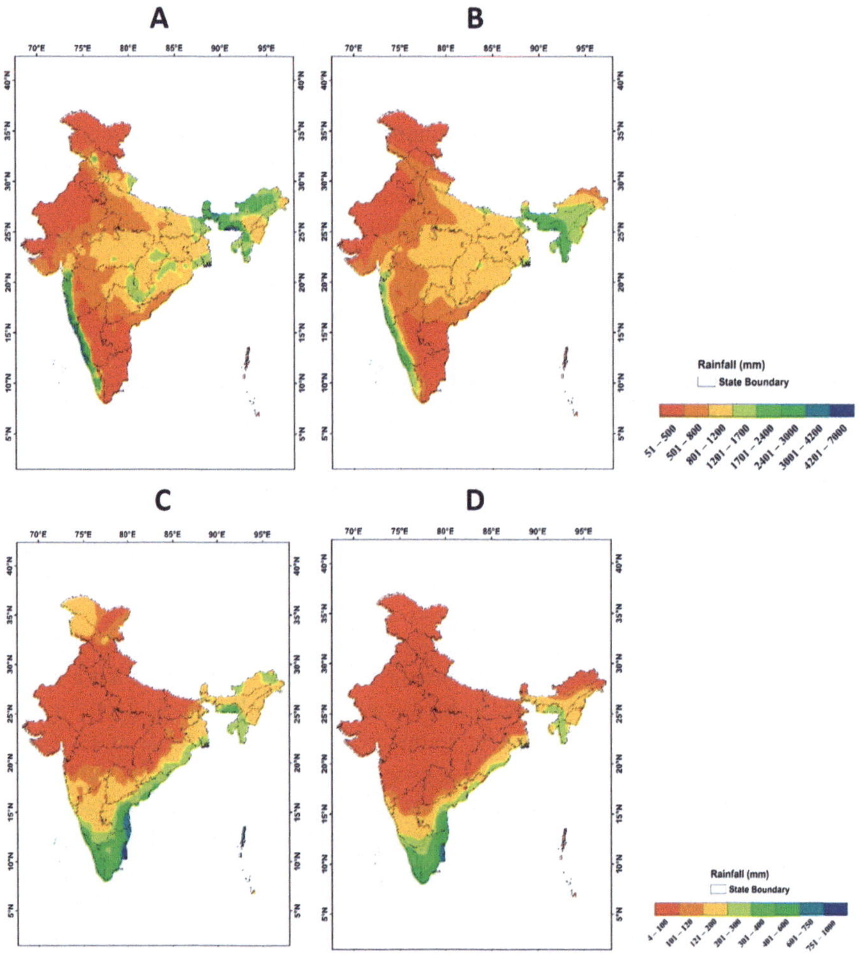

Fig. 18.4 Figure depicts a comparison between the spatial distribution of mean southwest (JJAS) and northwest (OND) rainfall totals over India from 1981 to 2005, as represented by IMD and NEX MMM. The figure includes the following subplots: (**a**) IMD for JJAS, (**b**) NEX_MMM for JJAS, (**c**) IMD for OND, and (**d**) NEX_MMM for OND

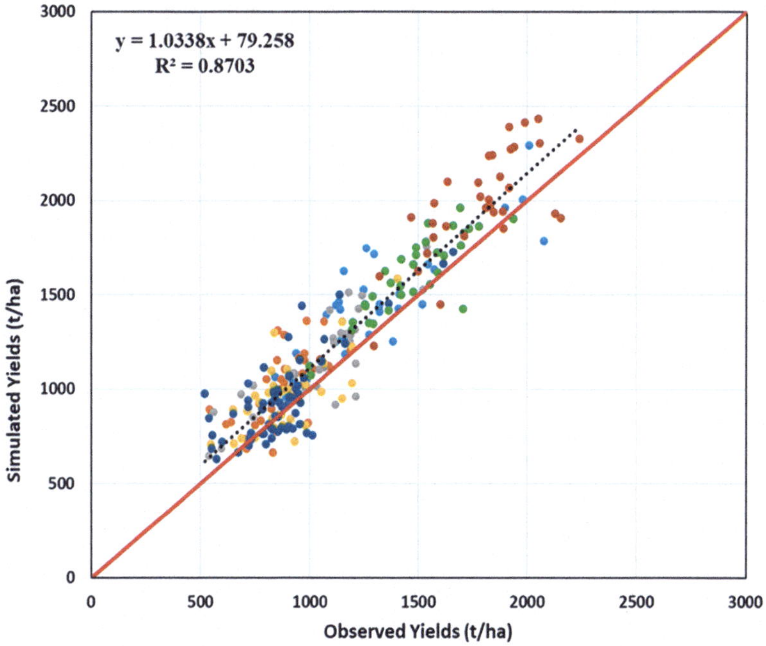

(a)

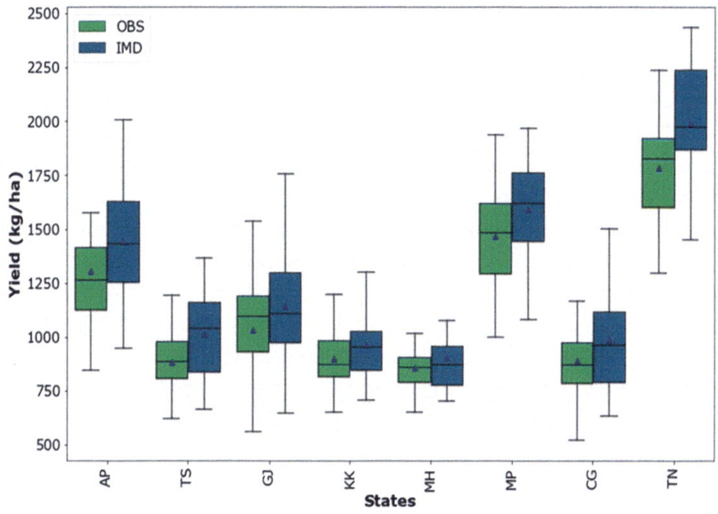

(b)

Fig. 18.5 Figure (**a**) depicts the relationship between observed and modeled sorghum yields following the rainy season from 1981 to 2015. Additionally, (**b**) provides a comparison of sorghum grain yields, both in terms of their average and variability, across the study region. The study encompassed eight states, with the following abbreviations: AP (Andhra Pradesh), TS (Telangana state), GJ (Gujarat), KK (Karnataka), MH (Maharashtra), MP (Madhya Pradesh), CG (Chhattisgarh), and TN (Tamil Nadu)

18.7 Modeling Adaptation of Sorghum in Ethiopia with APSIM: Opportunities with G × E × M

In 2023, Tirfessa and colleagues underscored sorghum's vital role as a staple food and feed crop in Ethiopia's arid lowland regions, where farmers cultivate both early-sown long-duration landraces and late-sown short-duration improved varieties. Given the varying timing and severity of drought stress across locations and seasons, understanding the interactions among genetic traits (G), environmental conditions (E), and management practices (M) is crucial for optimizing grain and forage yield with limited resources. The authors highlighted the value of crop simulation modeling in unraveling these complex G × E × M interactions and devising strategies to adapt to prevalent drought patterns in Ethiopia.

Their long-term simulation study focused on the impact of two sorghum production systems—early sowing with a late-maturing tall landrace type and late sowing with an early-maturing short variety—across various regions in Ethiopia, including Oromia, Amhara, Tigray, and Southern Ethiopia.

Key findings include:

- *Trade-off Between Biomass and Grain Yield:* The two production systems exhibited a trade-off between total biomass and grain yield, with late sowing and early-maturing varieties generally yielding more grain except in very favorable seasons, while producing less biomass except in poor seasons. Late sowing mitigated risks in poor seasons by yielding more biomass and grain.
- *Interaction with Growing Season Duration and Water Availability:* The choice of production system interacted with growing season length and water availability. Early planting of late-maturing landrace types led to longer growing seasons, resulting in higher biomass production but often reduced water availability during grain filling and consequently lower grain yield. Late planting with early-maturing varieties was more suitable in such scenarios.
- *Regional Differences:* Analysis of Oromia and Tigray, regions with distinct rainfall patterns, revealed that early sowing with late-maturing varieties performed better in regions with favorable rainfall like Tigray, while late sowing with early-maturing varieties had better grain yield outcomes in regions with reduced and bimodal rainfall like Oromia, albeit with a biomass trade-off.
- *Production Risk Trade-offs:* The study quantified production risk trade-offs linked to sorghum production systems, providing valuable insights for decision-making.
- *Model Application and Adaptation:* Utilizing a sorghum model tailored to Ethiopian conditions, the study showcased its potential for exploring G × E × M interactions to enhance sorghum production in dry lowland areas.
- *General Applicability of the Sorghum Model:* Similar studies conducted in Australia, the United States, India, and Mali underscored the model's robustness, underscoring the importance of adapting it to local conditions for detailed adaptation analysis.

In summary, the study evaluated trade-offs associated with different sorghum production systems across Ethiopian regions, shedding light on the influence of growing season duration, water availability, and regional disparities. These insights can inform decision-making in sorghum cultivation, aiming for improved yields and reduced production risks.

18.8 Modeling Future Climate Change Impacts on Sorghum Production with Best Management Options in Amhara Region, Ethiopia

Sorghum plays a critical role in Ethiopia's arid and semi-arid regions, but its yield is limited by climate change, poor soil fertility, insects, and weeds. The study by Mohammed and Misganaw in 2022 seeks to refine the CERES-sorghum model in DSSAT, forecast climate change effects on sorghum production in the 2030s and 2050s, and identify effective crop management strategies for sustaining sorghum production.

Projected Climate Changes in the Study Region: The study anticipates escalating mean annual temperatures by the 2030s and 2050s, under varying emission scenarios (RCP4.5 and RCP8.5), with mean annual maximum temperatures rising by 1.4–1.5 and 1.9–2.5 °C respectively. Mean annual minimum temperatures are also expected to increase by approximately 1.4–1.6 °C by the 2030s and 2–2.5 °C by the 2050s. Moreover, a projected rise in mean annual total rainfall by about 4–8% during the same periods is indicated. These findings, consistent with previous research, suggest potential adverse impacts on crop production in the semiarid environments of northeastern Ethiopia.

Projected Climate Change Impact on Phenology of Sorghum: The study investigated how projected climate change affects sorghum phenology at Harbu and Sirinka sites. Simulation results show significant decreases in anthesis and maturity dates of the Girana-1 sorghum cultivar by the 2030s and 2050s under both RCP4.5 and RCP8.5 scenarios. At Harbu, anthesis dates may decrease by 3.4–4.8% by the 2030s and 7.2% by the 2050s, while at Sirinka, the decrease could range from 10.5% to 20% by the 2050s. Reductions in maturity dates are more pronounced at Sirinka due to higher terminal water deficit and soil differences. Increased temperatures, particularly by the 2050s, may induce water deficit conditions, leading to earlier completion of sorghum's life cycle. These findings suggest potential impacts on crop productivity aligning with previous research on temperature-induced changes in sorghum phenology.

Projected Climate Change Impact on Sorghum Grain Yield: The study investigated the impact of projected climate change on sorghum grain yield at Harbu and Sirinka sites. Simulations revealed a decrease in grain yield by the 2030s and 2050s under both RCP4.5 and RCP8.5 scenarios compared to the baseline. At Harbu, yield may decrease by 8.8–19.6% by the 2030s and 13.2–13.2% by the 2050s, while at Sirinka, the decrease could range from 12.9% to 17.1% by the 2030s and 11.9% to 17.1% by the 2050s. The primary reason for this reduction was attributed to increased future temperatures, which could accelerate crop growth stages and lead to water shortages. Changes in soil properties may also contribute to the adverse effects of climate change on sorghum yield.

Crop Management Scenarios for Sorghum Production: The study evaluated various crop management strategies for sorghum production, focusing on sowing dates and supplemental irrigation at Harbu and Sirinka sites.

Key findings include:

- Significant effects of sowing dates and supplemental irrigation on sorghum grain yield under different climate scenarios.
- Early sowing, particularly under nonirrigated conditions, yielded the best results, aligning with peak water demand.
- Supplemental irrigation and early sowing can enhance sorghum production in semi-arid conditions.
- Delayed sowing, especially without irrigation, led to reduced grain yield.
- Supplemental irrigation was beneficial across different sowing dates.
- Early sowing and supplemental irrigation are recommended as effective adaptation strategies to increase sorghum yield in water-deficit environments.

Overall, the study emphasizes the importance of these management practices in mitigating climate change impacts and improving sorghum production in semi-arid regions, discouraging late sowing without irrigation due to its negative effects on yield.

Effect of nitrogen and supplemental irrigation on sorghum grain yield: The study investigated the impact of nitrogen application and supplemental irrigation on sorghum grain yield at Harbu and Sirinka sites.

Key findings include:

- Significant effects of nitrogen application and supplemental irrigation on grain yield under different climate scenarios.
- Highest simulated grain yield observed with 92 kg N/ha under irrigation, indicating a strong synergistic effect of nitrogen and irrigation.
- Simulated yields were consistently lower under nonfertilized and nonirrigated conditions.
- Soil moisture plays a crucial role in nutrient absorption, emphasizing the importance of optimal nitrogen application and supplemental irrigation for sustainable sorghum production.

In conclusion, the study highlights the critical role of nitrogen fertilizer and supplemental irrigation in enhancing sorghum yield, particularly in regions with limited soil moisture, essential for sustainable production amidst current and projected climate conditions in semi-arid areas.

18.9 Improving Sorghum Productivity in Sub-Saharan Africa Under Changing Climatic Conditions: A Modeling Approach

In 2020, Akinseye and colleagues underscored the profound effects of climate variability and change on smallholder farmers in sub-Saharan Africa, particularly those heavily dependent on agriculture. They highlighted the potential of crop modeling, specifically utilizing APSIM, to assess the repercussions of anticipated climate change on two sorghum varieties (early and medium maturing) under diverse agricultural management practices.

Analysis of historical and future climatic trends in the study locations revealed significant changes in rainfall patterns in Bamako, Mali, and projected rainfall increases in Kano, Nigeria, during the mid-century period (2040–2069). Both regions experienced warming trends, with greater temperature rises projected for the mid-century compared to the near future (2010–2039), notably in minimum temperatures.

Comparative analysis of simulated grain yield variability between the baseline period (1980–2009) and future climate scenarios (near-future, 2010–2039; and mid-century, 2040–2069) across 5 different climate models revealed varied impacts on sorghum yield. Early maturing varieties exhibited relatively stable, low yields, with declines projected in certain scenarios, while medium maturing varieties displayed higher yield potential with increased variability.

Furthermore, the study evaluated the sensitivity of a sorghum crop model to changes in cultivar parameters and sowing dates. It found that radiation use efficiency and light extinction coefficient significantly influenced grain yield for early maturing sorghum, while changes in these parameters and phyllochron traits played a major role for medium maturing sorghum. Sowing dates also had a substantial impact on yields, with the medium maturing variety showing yield increases in Bamako, Mali, and varying changes observed in Kano, Nigeria, depending on the sowing dates. These findings underscore the critical role of parameter selection and sowing date decisions in sorghum crop management.

18.10 Assessing Sorghum Yield Responses to Climate Change in Semi-arid Central Tanzania Using Crop Simulation Models

In 2014, Msongaleli and his team worked on calibrating and evaluating the DSSAT and APSIM models to simulate sorghum growth under current and future climate conditions in central Tanzania, especially for the Tegemeo sorghum variety. Using data from eight general circulation models (GCMs) and representative concentration pathway (RCP 4.5) scenarios for mid-century, the results reveal mixed sorghum yield responses, with both increases and decreases. The analysis shows rising maximum temperatures, varied rainfall predictions, and discrepancies with IPCC projections. It emphasizes the need for location-specific research to address climate change risks and vulnerabilities in crop production.

The study also explores the use of different cultivars, including short- and long-duration varieties, as potential adaptations to changing climate conditions. The simulations demonstrate that longer cultivar growth durations lead to higher yields, aligning with the resilience of local photo-period-sensitive cultivars. These findings suggest reintroducing photoperiod sensitivity into modern sorghum cultivars to provide farmers with more flexibility in sowing dates. Despite concerns about rising temperatures, the study's temperature projections for central Tanzania indicate that conditions for significant yield reductions may not be reached by the 2050s. By selecting appropriate management practices, including improved cultivars or local landraces, smallholder farmers can enhance their adaptive capacity, boost sorghum production, and improve food security and livelihoods in the region.

18.11 Conclusion

This chapter aims to provide readers with illustrative examples centered on sorghum crops, showcasing the current utilization of modeling approaches to enhance agricultural/cropping systems production and bolster resilience. We delve into specific instances of sorghum model development crucial for capturing the intricacies of agricultural systems in developing countries, addressing challenges unique to modeling sorghum crops in India and Africa. Additionally, we highlighted emerging directions in model development essential for tackling global developmental goals and challenges.

References

Aruna C, Cheruku D (2019) Genetic improvement of grain sorghum. In: Breeding sorghum for diverse end uses. Woodhead Publishing, Sawston, pp 157–173

Ashok Kumar A, Anuradha K, Ramaiah B (2011) Increasing grain Fe and Zn concentration in sorghum: progress and way forward. J SAT Agric Res 11(12):1–5

Balota M, Payne WA, Rooney W, Rosenow D (2008) Gas exchange and transpiration ratio in sorghum. Crop Sci 48(6):2361–2371

Blummel M, Rao PP (2006) Economic value of sorghum stover traded as fodder for urban and peri-urban dairy production in Hyderabad, India. Int Sorghum Millets Newslett 47:97–100

Chapman SC, Cooper M, Butler DG, Henzell RG (2000a) Genotype by environment interactions affecting grain sorghum. I. Characteristics that confound interpretation of hybrid yield. Aust J Agric Res 51(2):197–208

Chapman SC, Hammer GL, Butler DG, Cooper M (2000b) Genotype by environment interactions affecting grain sorghum. III. Temporal sequences and spatial patterns in the target population of environments. Aust J Agric Res 51(2):223–234

Chauhan YS, Solomon KF, Rodriguez D (2013) Characterization of north-eastern Australian environments using APSIM for increasing rainfed maize production. Field Crop Res 144:245–255

Chenu K, Cooper M, Hammer GL, Mathews KL, Dreccer MF, Chapman SC (2011) Environment characterization as an aid to wheat improvement: interpreting genotype–environment interactions by modelling water-deficit patterns in north-eastern Australia. J Exp Bot 62(6):1743–1755

Chenu K, Deihimfard R, Chapman SC (2013) Large-scale characterization of drought pattern: a continent-wide modelling approach applied to the Australian wheatbelt–spatial and temporal trends. New Phytol 198(3):801–820

Elangovan M, Prabhakar P, Chandra Sekara Reddy D (2010) Characterization and evaluation of sorghum [Sorghum bicolor (L.) Moench] germplasm from Karnataka, India. Karnataka J Agric Sci 20(4):840

Hajjarpoor A, Vadez V, Soltani A, Gaur P, Whitbread A, Babu DS, Kholová J (2018) Characterization of the main chickpea cropping systems in India using a yield gap analysis approach. Field Crop Res 223:93–104

Hajjarpoor A, Nelson WC, Vadez V (2022) How process-based modeling can help plant breeding deal with G x E x M interactions. Field Crop Res 283:108554

Kholova J, McLean G, Vadez V, Craufurd P, Hammer GL (2013) Drought stress characterization of post-rainy season (rabi) sorghum in India. Field Crop Res 141:38–46

Kholová J, Murugesan T, Kaliamoorthy S, Malayee S, Baddam R, Hammer GL, Vadez V (2014) Modelling the effect of plant water use traits on yield and stay-green expression in sorghum. Funct Plant Biol 41(11):1019–1034

Kholova J, Adam M, Diancoumba M, Hammer G, Hajjarpoor A, Chenu K, Jarolímek J (2020) Sorghum: general crop-modelling tools guiding principles and use of crop models in support of crop improvement programs in developing countries. Sorghum in the 21st century: food–fodder–feed–fuel for a rapidly changing world, pp 189–207

Muchow RC, Cooper M, Bidinger FR, Hammer GL, Borrell AK, Chapman SC (1996) Targeting sorghum improvement in drought-prone environments: approaches and progress

Nagesh Kumar MV, Ramya V, Govindaraj M, Dandapani A, Maheshwaramma S, Ganapathy KN, Kavitha K, Goverdhan M, Jagadeeshwar R (2022) India's rainfed sorghum improvement: three decades of genetic gain assessment for yield, grain quality, grain mold and shoot fly resistance. Front Plant Sci 13:1056040. https://doi.org/10.3389/fpls.2022.1056040

Rao PP, Birthal PS, Reddy BV, Rai KN, Ramesh S (2006) Diagnostics of sorghum and pearl millet grains-based nutrition in India. Int Sorghum Millets Newslett 47:93–96

Ronanki S, Pavlík J, Masner J, Jarolímek J, Stočes M, Subhash D, Kholová J (2022) An APSIM-powered framework for post-rainy sorghum-system design in India. Field Crop Res 277:108422

Visarada KBRS, Aruna C (2019) Sorghum: a bundle of opportunities in the 21st century. In: Breeding sorghum for diverse end uses. Woodhead Publishing, Sawston, pp 1–14